Chrysomeloidea II (Orsodacnidae, Megalopodidae, Chrysomelidae)

Volume 6/2/2

Catalogue of Palaearctic Coleoptera

VOLUME 6/2/2

The titles published in this series are listed at *brill.com/cpc*

Pachybrachis fimbriolatus Suffrian, 1844

Chrysomeloidea II (Orsodacnidae, Megalopodidae, Chrysomelidae)

Updated and Revised Second Edition

VOLUME 6/2/2

Edited by

J. Bezděk
L. Sekerka

BRILL

LEIDEN | BOSTON

Cover illustration: *Pachybrachis fimbriolatus* Suffrian, 1844 (© Jan Bezděk)

Date of publication: 23 May 2024.

The index of the species-group names is available on site of the Koninklijke Brill BV, Leiden (10.6084/m9.figshare.25540735).

The Library of Congress Cataloging-in-Publication Data is available online at https://catalog.loc.gov
LC record available at https://lccn.loc.gov/2015018977

Typeface for the Latin, Greek, and Cyrillic scripts: "Brill". See and download: brill.com/brill-typeface.

ISBN 978-90-04-70810-5 (hardback, set)
ISBN 978-90-04-70727-6 (hardback, vol. 2)
ISBN 978-90-04-70728-3 (e-book, vol. 2)
DOI 10.1163/9789004707283

This book is printed on acid-free paper and produced in a sustainable manner.

Contents

Volume 1

Volume 2

References

Abdullah, M. & Qureshi, S. S. 1968a: The Chrysomelidae, Coleoptera of Pakistan. Part III. – A key to the genera and species of the Galerucinae, with descriptions of new genera and species. *Pakistan Journal of Scientific and Industrial Research* **11** (4): 396–414. DP: [31] October 1968 (issue table of contents).

Abdullah, M. & Qureshi, S. S. 1968b: A new genus and species of the Galerucinae (Coleoptera: Chrysomelidae) from pear tree in West Pakistan. *Pakistan Journal of Scientific and Industrial Research* **11** (4): 423–424. DP: [31] October 1968 (issue table of contents).

Abdullah, M. & Qureshi, S. S. 1968c: Ashrafia anwarullahi, a new genus and species of the Galerucinae (Coleoptera: Chrysomelidae) from West Pakistan. *Pakistan Journal of Scientific and Industrial Research* **11** (4): 425–426. DP: [31] October 1968 (issue table of contents).

Abdullah, M. & Qureshi, S. S. 1969a: A key to the Pakistani genera and species of the Hispinae and Cassidinae (Coleoptera: Chrysomelidae), with description of a new species from West Pakistan including the economic importance. *Pakistan Journal of Scientific and Industrial Research* **12** (1/2): 95–104. DP: [30] April 1969 (issue table of contents).

Abdullah, M. & Qureshi, S. S. 1969b: A key to the Pakistani genera and species of the Chrysomelinae and Halticinae (Coleoptera: Chrysomelidae), with description of new genera and species including the economic importance. *Pakistan Journal of Scientific and Industrial Research* **12** (1/2): 105–120. DP: [30] April 1969 (issue table of contents).

Abeille de Perrin, E. 1877: [Note sur trois espèces de Stylosomus européens]. *Bulletin de la Société Entomologique de France* **1877** (5): 58–60. DP: 22 March 1877 (Evenhuis 2002).

Abeille de Perrin, E. 1895a: Cinq nouvelles altises du Nord de l'Afrique (Col.). *Bulletin de la Société Entomologique de France* **1895** (20): cdi–cdiv. DP: [31] December 1895 (footer on p. cccxcvii).

Abeille de Perrin, E. 1895b: Notes sur quelques chrysomélines de Syrie (Col.). *Bulletin de la Société Entomologique de France* **1895** (20): cdiv–cdvi. DP: [31] December 1895 (footer on p. cccxcvii).

Abeille de Perrin, E. 1896: Notes sur certains Longitarsus, suivies de la description de deux espèces nouvelles d'halticides (Col.). *Bulletin de la Société Entomologique de France* **1896** (3): 52–54. DP: 18 May 1896 (*Bibl. Zool.* 1896, 1: 271).

Abeille de Perrin, E. 1897: Description de deux Entomoscelis paléarctiques (Col.). *Bulletin de la Société Entomologique de France* **1897** (3): 42–44. DP: 14 June 1897 (*Bibl. Zool.* 1897, 2: 353).

Abeille de Perrin, E. 1901: Nouvelles espèces de coléoptères français. *L'Échange, Revue Linnéenne* **17** (201): 68–70. DP: [30] September 1901 (issue cover).

Abeille de Perrin, E. 1905: Diagnoses de trois coléoptères français nouveaux. *Bulletin de la Société Entomologique de France* [**1904**] (19): 280–282. DP: 7 January 1905 (*dates d'apparition* at the end of volume).

Abeille de Perrin, E. 1907: Description de sept halticides réputées nouvelles. *Compte-Rendu Sommaire des Séances de la Société des Sciences Naturelles de Provence* **1907**: lxxix–lxxxii. DP: [31 December] 1907 (volume title page).

 | DOI:10.1163/9789004707283_002

Abeille de Perrin, E. 1909: Coléoptères d'Algérie jugés nouveaux. *Bulletin de la Société Entomologique de France* **1909** (10): 179–181. DP: 19 June 1909 (*dates d'apparition* at the end of volume).

Achard, J. 1913: Description de trois espèces nouvelles de Chlamys (Col. Phytophaga) de la faune africaine. *Bulletin de la Société Entomologique de France* **1913** (8): 183–189. DP: 14 May 1913 (*dates de publication* at the end of volume).

Achard, J. 1914a: Rectifications à la nomenclature des Chrysomelidae (Col.). *Bulletin de la Société Entomologique de France* **1914** (2): 71–72. DP: 11 February 1914 (*dates de publication* at the end of volume).

Achard, J. 1914b: Un genre nouveau de coléoptères phytophages. *Bulletin de la Société Entomologique de France* **1914** (9): 288–290. DP: 27 May 1914 (*dates de publication* at the end of volume).

Achard, J. 1916a: Description d'un nouveau genre de coléoptères phytophages de Madagascar. *Bulletin de la Société Entomologique de France* **1916** (1): 47–48. DP: 23 February 1916 (*dates de publication* at the end of volume).

Achard, J. 1916b: Révision des coléoptères phytophages de Madagascar. *Annales de la Société Entomologique de France* **85** [1916–1917] (1): 1–18. DP: 9 May 1916 (*dates d'apparition* on p. 631).

Achard, J. 1919: Description de deux espèces nouvelles du genre Chlamys. *Annales de la Société Entomologique de Belgique* **59** [1919–1920] (3): 36–40. DP: 6 June 1919 (issue cover).

Achard, J. 1920: Description d'un Colobaspis nouveau du Yunnan (Col. Chrysomelidae). *Bulletin de la Société Entomologique de France* **1920** (3): 48–49. DP: 16 March 1920 (*dates de publication* at the end of volume).

Achard, J. 1921a: Synonymie de quelques Chrysomelidae (Col.). *Bulletin de la Société Entomologique de France* **1921** (5): 61–62. DP: 13 April 1921 (*dates de publication* at the end of volume).

Achard, J. 1921b: Descriptions d'espèces nouvelles de Trichochrysea (Col. Chrysomelidae). *Bulletin de la Société Entomologique de France* **1921** (11): 171–172. DP: 20 July 1921 (*dates de publication* at the end of volume).

Achard, J. 1922a: Description d'un galérucide nouveau (Col. Chrysomelidae). *Bulletin de la Société Entomologique de France* **1922** (3): 37. DP: 10 March 1922 (*dates de publication* on p. 348).

Achard, J. 1922b: Descriptions de nouveaux Chrysomelini (Col. Chrysomelidae). *Fragments Entomologiques* (Prague) **1**: 13–16 & **2**: 17–28. DP (vol. 1): [30] September 1922 (issue cover), DP (vol. 2): 30 October 1922 (footer on p. 17) [see also the comments in the Dating of publications chapter].

Achard, J. 1923a: Nouveaux Chrysomelini d'Amérique (Col. Chrysomelidae). *Fragments Entomologiques* (Prague) **5**: 65–80 & **6**: 81–93. DP (vol. 5): 30 April 1923 (footer on p. 65), DP (vol. 6): 31 May 1923 (footer on p. 81) [see also the comments in the Dating of publications chapter].

Achard, J. 1923b: Un nouveau clytride du Caucase (Col. Chrysomelidae). *Fragments Entomologiques* (Prague) **6**: 93. DP: 31 May 1923 (footer on p. 81) [see also the comments in the Dating of publications chapter].

Achard, J. 1923c: Variations de quelques chrysomélides de l'Amérique tropicale. *Fragments Entomologiques* (Prague) 8: 120–126. DP: 15 August 1923 (Achard's inscription on the p. 113 in NMPC copy) [see also the comments in the Dating of publications chapter].

Achard, J. 1923d: Notes sur les Colaspidema et genres voisins (Col. Chrysomelidae). *Fragments Entomologiques* (Prague) 8: 126–128 & 9: 129–137. DP (vol. 8): 15 August 1923, DP (vol. 9): 30 October 1923 (Achard's inscription on the pp. 113 and 129 in NMPC copy) [see also the comments in the Dating of publications chapter].

Achard, J. 1923e: Descriptions de Phaedon nouveaux (Col. Chrysomelidae). *Fragments Entomologiques* (Prague) 9: 137–140. DP: 30 October 1923 (Achard's inscription on the p. 129 in NMPC copy) [see also the comments in the Dating of publications chapter].

Achard, J. 1924: Les Phytodecta et leurs variations. Espèces et variétés nouvelles. *Časopis Československé Společnosti Entomologické* 21 (1/2): 31–37. DP: 25 March 1924 (issue cover).

Adams, M. F. 1817: Descriptio insectorum novorum Imperii Russici, inprimis Caucasi et Sibiriae. *Mémoires de la Société Impériale des Naturalistes de Moscou* 5: 278–314. DP: [31 December] 1817 (Bousquet & Davies 2021).

Agassiz, J. L. R. 1846a: *Nomenclator zoologicus, continens nomina systematica generum animalium, tam viventium quam fossilium, secundum ordinem alphabeticum disposita, adjectis auctoribus, libris, in quibus reperiuntur, anno editionis, etymologia et familiis, ad quas pertinen, in singulis classibus. Fasciculus XI. Continens Coleoptera.* Soloduri: Jent & Gassmann, xi + [1] + 170 pp. DP: 9 October 1846 (Bousquet 2016).

Agassiz, J. L. R. 1846b: *Nomenclatoris zoologici index universalis, continens nomina systematica classium, ordinum, familiarum et generum animalium omnium, tam viventium quam fossilium, secundum ordinem alphabeticum unicum disposita, adjectis homonymiis plantarum, nec non variis adnotationibus et emendationibus.* Soloduri: Jent & Gassmann, viii + 383 pp. DP: 29 December 1846 (Bousquet 2016).

Ahrens, A. 1810: Beiträge zu einer Monographie der Rohrkäfer. *Neue Schriften der Naturforschenden Gesellschaft zu Halle* 1 (3): 9–48. DP: 13 May 1810 (*Allg. Verz. Bücher, Ostermesse* 1810: 154 and Evenhuis 2014).

Ahrens, A. 1812a: *Fauna Insectorum Europae. Fasciculus primus.* Halae: Car. Aug. Kümmel, [2] + [25] pp. + 25 pls. DP: 27 March 1812 (Bousquet 2016).

Ahrens, A. 1812b: Beiträge zur Kenntniss deutscher Käfer. *Neue Schriften der Naturforschenden Gesellschaft zu Halle* 2 (2): 3–40. DP: 4 October 1812 (*Allg. Verz. Bücher, Michaelismesse* 1812: 333 and Evenhuis 2014).

Aldridge, R. J. W. & Pope, R. D. 1986: The British species of Bruchidius Schilsky (Coleoptera: Bruchidae). *Entomologist's Gazette* 37 (3): 181–193. DP: 28 August 1986 (issue colophon).

Ali, A. H. & Ali, A. Z. M. 1987: [New species of Bruchidius from Iraq]. *Journal of Biological Scientific Research* (Baghdad) 18 (2): 1–8 (in Arab). DP: [31] May 1987 (issue cover).

Ali, A. H. & Ali, A. Z. M. 1988: [New species of legum beetle of genus Caryedon (Bruchidae: Coleoptera) from Iraq]. *Journal of Biological Scientific Research* (Baghdad) 19 (1): 1–6 (in Arab). DP: [31] January 1988 (issue cover).

Ali, A. H. & Kadhim, A. A. 1986: [Description of some new species of the genus Spermophagus in Iraq]. *Journal of Biological Scientific Research* (Baghdad) **17** (2): 43–57 (in Arab). DP: [31] May 1986 (issue cover).

Allard, E. 1859a: Diagnoses latines qui suivent de quelques haltises nouvelles. *Bulletin de la Société Entomologique de France* **1859** (2): c–cii. DP: 24 August 1859 (Evenhuis 2022).

Allard, E. 1859b: Diagnoses qui suivent de quelques haltises de France. *Bulletin de la Société Entomologique de France* **1859** (2): cv–cvi. DP: 24 August 1859 (Evenhuis 2022).

Allard, E. 1859c: Diagnoses suivantes des espèces d'halticites d'Europe et d'Algérie du genre Graptodera. *Bulletin de la Société Entomologique de France* **1859** (3): clxv–clxviii. DP: 9 November 1859 (Evenhuis 2022).

Allard, E. 1860a: Quelques détails relatifs au genre Orestia. *Bulletin de la Société Entomologique de France* [**1859**] (4): ccxli–ccxlii. DP: 1 March 1860 (Evenhuis 2022).

Allard, E. 1860b: Diagnoses latines suivantes de quatre espèces nouvelles d'altisites. *Bulletin de la Société Entomologique de France* [**1859**] (4): cclx–cclxi. DP: 1 March 1860 (Evenhuis 2022).

Allard, E. 1860c: Essai monographique sur les galerucites anisopodes Latr. ou description des altises d'Europe et des bords de la Mer Méditerranée. *Annales de la Société Entomologique de France, Troisème Série* **8** (1): 39–144. DP: 13 June 1860 (Evenhuis 2022).

Allard, E. 1860d: Essai monographique sur les galerucites anisopodes Latr. ou description des altises d'Europe et des bords de la Mer Méditerranée. *Annales de la Société Entomologique de France, Troisème Série* **8** (2): 369–418. DP: 12 September 1860 (issue cover).

Allard, E. 1860e: Essai monographique sur les galerucites anisopodes Latr. ou description des altises d'Europe et des bords de la Mer Méditerranée. *Annales de la Société Entomologique de France, Troisème Série* **8** (3): 539–578. DP: 26 December 1860 (issue cover).

Allard, E. 1861: Essai monographique sur les galerucites anisopodes Latr. ou description des altises d'Europe et des bords de la Mer Méditerranée. *Annales de la Société Entomologique de France, Troisème Série* **8** [1860] (4): 785–834. DP: 15 May 1861 (issue cover).

Allard, E. 1862: Catalogue complémentaire des diverses espèces d'altises qui ont été décrites tant dans cet ouvrage, par. E. Allard, que par MM. Foudras, Wollaston, Kutschera, etc., et qui proviennent d'Europe et du nord de l'Afrique. *Annales de la Société Entomologique de France, Quatrième Série* **1** [1861] (3): 307–348. DP: 22 January 1862 (Lefèvre 1885).

Allard, E. 1863: [new taxon]. In: Grenier, A.: *Catalogue des coléoptères de France et matériaux pour servir à la faune des coléoptères français*. Paris: A. Grenier, iv + 3–79 + 135 pp. DP: 6 August 1863 (book cover, Grenier 1864, see also the comments in the Dating of publications chapter).

Allard, E. 1864: Description de deux nouvelles espèces d'altises. *Annales de la Société Entomologique de France, Quatrième Série* **3** [1863] (4): 659–660. DP: 23 March 1864 (Lefèvre 1885).

Allard, E. 1865: Description d'une nouvelle espèce de Clythra et tableau synoptique du sous-genre Lacnaea. *Annales de la Société Entomologique de France, Quatrième Série* **4** [1864] (3): 383–386. DP: 25 January 1865 (Lefèvre 1885).

Allard, E. 1866: Monographie des galérucides d'Europe, du Nord de l'Afrique et de l'Asie. Monographie des alticides tribu de la famille des phytophages. *L'Abeille, Mémoires d'Entomologie* 3 [1865–1867]: 169–508. DP (pp. 169–180): 27 December 1865 (*Bull. Soc. Ent. Fr.* 1865: lxvii), (pp. 181–428): [31 December] 1866 (verso of the volume title page), (pp. 429–508): [28] February 1867 (verso of the volume title page).

Allard, E. 1867: Monographie des galérucides d'Europe, du Nord de l'Afrique et de l'Asie. Monographie des alticides tribu de la famille des phytophages. *L'Abeille, Mémoires d'Entomologie* 3 [1865–1867]: 169–508. DP (pp. 169–180): 27 December 1865 (*Bull. Soc. Ent. Fr.* 1865: lxvii), (pp. 181–428): [31 December] 1866 (verso of the volume title page), (pp. 429–508): [28] February 1867 (verso of the volume title page).

Allard, E. 1868: Mélanges entomologiques. Étude sur le groupe des Bruchites d'Europe et du bassin de la Méditerrannée. *Annales de la Société Entomologique de Belgique* 11 [1867–1868]: 83–124. DP: 27 September 1868 (*Ann. Soc. Ent. Belg.* 1868, 12 (Comptes rendus): viii).

Allard, E. 1869: Description de quelques coléoptères nouveaux et notes diverses. *L'Abeille, Journal d'Entomologie* 5 [1868–1869]: 465–478. DP: [31 December] 1869 (volume cover and footer on p. 469).

Allard, E. 1870a: Petites nouvelles. *Petites Nouvelles Entomologiques* 1 [1869–1875] (2) (13): 49–50. DP: 1 January 1870 (issue title page).

Allard, E. 1870b: [new taxon]. In: Heyden, L. F. J. D. von: Entomologische Reise nach dem südlichen Spanien, der Sierra Guadarrama und Sierra Morena, Portugal und den Cantabrischen Gebirge. *Berliner Entomologische Zeitschrift* 14 **(Beiheft)**: [1] + 218 pp., 2 pls. DP: [31 December] 1870 (title page, Bousquet 2016).

Allard, E. 1875: [new taxa]. In: Marseul, M.: Répertoire des coléoptères d'Europe décrits isolément depuis 1863. Troisième partie. *L'Abeille, Journal d'Entomologie* 12 [1874–1875]: 1–456 pp. DP (pp. 1–288): [31 December] 1874, (pp. 289–456): [31 December] 1875 (footers throughout the volume).

Allard, E. 1876: Descriptions d'espèces nouvelles. *L'Abeille, Journal d'Entomologie* 14 [1874–1876]: 21–25. DP: [31 December] 1876 (volume cover).

Allard, E. 1878: [new taxon]: In: Schneider, O. & Leder, H.: *Beiträge zur Kenntniss der kaukasischen Käferfauna*. Brünn: W. Burkart, 359 + [1] pp., 6 pls. DP: [31 December] 1878 (Döberl 2009, see also the comments in the Dating of publications chapter).

Allard, E. 1884a: Mélanges entomologiques. *Annales de la Société Entomologique de Belgique* 27 [1883] (1): 5–52. DP: 2 February 1884 (*Bull. Compt.-Rend. Séances Soc. Ent. Belg., Sér. III* 1884 (42): lvii).

Allard, E. 1884b: Diagnoses de quatre nouveaux coléoptères. *Wiener Entomologische Zeitung* 3 (8): 248. DP: 30 September 1884 (footers throughout the issue).

Allard, E. 1887: [Diagnoses de quelques espèces nouvelles de coléoptères phytophages]. *Bulletin de la Société Entomologique de France* 1887 (23): cc–cci. DP: 24 December 1887 (Evenhuis 2002).

Allard, E. 1889a: Synopsis des galerucines à corselet sillonné transversalement, 1re partie. *Annales de la Société Entomologique de France, 6e Série* 8 [1888] (3): 305–332. DP: 9 January 1889 (Lefèvre 1895).

Allard, E. 1889b: Diagnoses de coléoptères nouveaux. *Le Naturaliste, 2ᵉ Série* 11 (46): 33. DP: 1 February 1889 (issue title page).

Allard, E. 1889c: Diagnoses de coléoptères nouveaux. *Le Naturaliste, 2ᵉ Série* 11 (47): 43. DP: 15 February 1889 (issue title page).

Allard, E. 1889d: Note sur les galérucides, coléoptères phytophages. *Bulletin ou Comptes-Rendus des Séances de la Société Entomologique de Belgique, Série III* 1889 (112): lxvi–lxxxiii. DP: 12 June 1889 (*Bull. Soc. Ent. Fr.* 1889: cxxiii).

Allard, E. 1889e: Nouvelle note sur les phytophages à la suite d'un examen des galérucides appartenant au Musée royal de Belgique. *Bulletin ou Comptes-Rendus des Séances de la Société Entomologique de Belgique, Série III* 1889 (114): cii–cxvii. DP: 14 August 1889 (*Bull. Soc. Ent. Fr.* 1889: clxvi).

Allard, E. 1889f: Contributions à la faune indo-chinoise, 5ᵉ mémoire, galérucides et alticides. *Annales de la Société Entomologique de France, 6ᵉ Série* 9 (2): 303–312. DP: 31 October 1889 (Lefèvre 1895).

Allard, E. 1890: Troisième note sur les galérucides. *Bulletin ou Comptes-Rendus des Séances de la Société Entomologique de Belgique, Série IV* 1890 (6): lxxx–xciv. DP: 11 June 1890 (*Bull. Soc. Ent. Fr.* 1890: cxiv).

Allard, E. 1891: Coléoptères. Phytophages. Pp. 229–234. In: Collection d'insectes formée dans l'Indo-Chine par M. Pavie Consul de France au Cambodge. *Nouvelles Archives du Muséum d'Histoire Naturelle, Troisième Série* 3 (2): 205–276. DP: [31 December] 1891 (issue cover).

Allard, E. 1895: Notes sur les bruchides recueillis dans l'Inde anglaise, par M. Andrewes, de Londres. *Annales de la Société Entomologique de Belgique* 39 (4): 225–228. DP: 30 April 1895 (issue cover).

Allen, A. A. 1967: Two new species of Longitarsus Latr. (Col., Chrysomelidae) in Britain. *Entomologist's Monthly Magazine* 103 (1235/1237): 75–82. DP: 20 October 1967 (issue cover).

Allen, A. A. 1976: Notes on some British Chrysomelidae (Col.) including amendments and additions to the list. *Entomologist's Records and Journal of Variation* 88 (9): 220–225. DP: 15 September 1976 (heading on p. 220).

Allenspach, V. 1945: Zur Kenntnis von Chrysomela asclepiadis Villa. *Mitteilungen der Schweizerischen Entomologischen Gesellschaft* 19 [1943–1946] (9): 338–341. DP: 1 July 1945 (issue title page).

Allibert, A. 1847: Notes sur divers insectes coléoptères trouves dans des graines de légumineuses rapportées de Canton par M. Yvan, médecin de l'ambassade française en Chine, et sur quelques autres espèces qui ont vécu dans des haricots venant du Brésil. *Revue Zoologique, par la Société Cuvierienne* 10 (1): 11–19. DP: 15 February 1847 (*Compt. Rend. Hebdom. Séances Acad. Sci.* 1847, 24: 251).

Al-Nadawi, F. A. M. 2019: First record of mint leaf beetle, Chrysolina herbacea (Duftschmid, 1825), (Coleoptera: Chrysomelidae) in Iraq. *Baghdad Science Journal* 16 (4): 831–842. DP: 1 December 2019 (paper title page).

Alonso-Zarazaga, M. A. (ed.) 2017: *Cooperative catalogue of Palaearctic Coleoptera Curculionoidea. Monografías electrónicas S.E.A., volume 8*. Zaragoza: Sociedad Entomológica Aragonesa, 729 pp. DP: 2 November 2017 (colophon).

An, S. L. 1988: Two new species of Chrysomelidae (Coleoptera) from Korea. *Korean Journal of Systematic Zoology* 4 (1): 43–46. DP: 1 June 1988 (paper title page).

An, S. L. & Kwon, Y. J. 2002: Classification of the leafbeetles from Korea. Part IV. Subfamily Zeugophorinae (Coleoptera: Chrysomelidae). *Insecta Koreana* **19** (3/4): 271–276. DP: 31 December 2002 (paper title page).

An, S. L., Kwon, Y. J. & Lee, S. M. 1985: Classification of the leafbeetles from Korea. Part I. Subfamily Hispinae (Coleoptera: Chrysomelidae). *Insecta Koreana* **5**: 1–9. DP: 29 May 1985 (volume cover).

Anand, R. K. & Cox, M. L. 1986: Taxonomic revision of the genus Aulacophora Chevr. I. Species with yellow elytra (Coleoptera, Chrysomelidae, Galerucinae). *Entomologische Abhandlungen* 50 (1): 81–91. DP: 15 December 1986 (paper title page).

Ancey, F. 1876: Description d'une espèce nouvelle de Chrysomela. *Petites Nouvelles Entomologiques* 2 [1876–1879] (8) (162): 94. DP: 15 December 1876 (issue title page).

Anonymous 1940: [new taxon]. In: Smith, M. A. (ed.): *The zoological record. Volume the seventy-sixth being the records of zoological literature relating chiefly to the year 1939. Insecta.* London: Zoological Society, 420 pp. DP: [31] July 1940 (Evenhuis et al. 2008).

Antoine, M. 1937: Notes d'entomologie marocaine. XXIV. Deuxième contribution à la faune du Djebel Rhat. *Bulletin de la Société des Sciences Naturelles du Maroc* **16** [1936] (3): 239–245. DP: 15 April 1937 (noted on the p. 272).

Antoine, M. 1939: Notes d'entomologie marocaine, XXX. – Coléoptères nouveaux du Moyen Atlas. *Bulletin de la Société Entomologique de France* **44** (9/10): 145–152. DP: 12 July 1939 (*dates de parution* on p. 240).

Anton, K.-W. 1991: Neuzumeldende Samenkäfer-Arten für Mitteleuropa (Coleoptera: Bruchidae). *Mitteilungen der Entomologischen Gesellschaft Basel* **41** (2/3): 97–100. DP: 30 September 1991 (table of contents).

Anton, K.-W. 1993: Description of two new species of the genus Spermophagus Schoenherr (Coleoptera, Bruchidae, Amblycerinae) from the Oriental Region. *Entomologica Basiliensia* **16**: 377–383. DP: 29 December 1993 (table of contents).

Anton, K.-W. 1994a: The Bruchidae (Coleoptera) of Saudi Arabia, with descriptions of two new species. *Fauna of Saudi Arabia* **14**: 97–104. DP: 30 September 1994 (table of contents).

Anton, K.-W. 1994b: The Bruchidae (Coleoptera) of Oman, with descriptions of a new genus and two new species. *Fauna of Saudi Arabia* **14**: 105–112. DP: 30 September 1994 (table of contents).

Anton, K.-W. 1996a: Two new species of the Bruchidius halodendri group from Turkmenia (Coleoptera: Bruchidae: Bruchinae). *Genus* 7 (3): 475–480. DP: 30 October 1996 (paper title page).

Anton, K.-W. 1996b: Spermophagus borowieci n. sp. from Pakistan (Coleoptera: Bruchidae: Amblycerinae). *Genus* 7 (3): 481–483. DP: 30 October 1996 (paper title page).

Anton, K.-W. 1998: Revision of the genus Bruchidius. Part I: The B. seminarius group (Coleoptera: Bruchidae: Bruchinae). *Stuttgarter Beiträge zur Naturkunde, Serie A (Biologie)* **573**: 1–13. DP: 31 August 1998 (paper title page).

Anton, K.-W. 1999a: Results of the Czechoslovak-Iranian entomological expeditions to Iran 1970, 1973 and 1977. Coleoptera: Bruchidae. *Časopis Národního Muzea, Řada přírodovědná* (Prague) **167** [1998] (1/4): 73–90. DP: [31] January 1999 (colophon).

Anton, K.-W. 1999b: Notes on Borowiecius Anton, with the description of a new species (Coleoptera: Bruchidae: Bruchinae). *Genus* 10 (1): 65–72. DP: 31 March 1999 (paper title page).

Anton, K.-W. 1999c: Two new species of the Bruchidius halodendri group from Central Asia (Coleoptera: Bruchidae: Bruchinae). *Linzer Biologische Beiträge* **31** (1): 443–448. DP: 30 July 1999 (paper title page).

Anton, K.-W. 1999d: Bruchidius tenerifensis, a new species of the B. cisti group from the Canary Islands (Coleoptera: Bruchidae: Bruchinae). *Genus* 10 (3): 399–402. DP: 31 October 1998 (paper title page).

Anton, K.-W. 1999e: Revision of the genus Sulcobruchus Chujo 1937, and description of Parasulcobruchus nov. gen. (Coleoptera, Bruchidae, Bruchinae). *Linzer Biologische Beiträge* **31** (2): 629–650. DP: 31 December 1999 (paper title page).

Anton, K.-W. 1999f: Two new species of the Bruchus brachialis group from the Mediterranean region (Coleoptera: Bruchidae: Bruchinae). *Linzer Biologische Beiträge* **31** (2): 655–660. DP: 31 December 1999 (paper title page).

Anton, K.-W. 2000a: Five new species of the Callosobruchus chinensis group from the Oriental region and Australia (Coleoptera: Bruchidae: Bruchinae). *Genus* 11 (1): 13–28. DP: 11 April 2000 (paper title page).

Anton, K.-W. 2000b: Descriptions of four new species of the genus Bruchidius Schilsky, 1905 (Coleoptera: Bruchidae) from the Arabian Peninsula. *Fauna of Arabia* **18**: 245–252. DP: 30 November 2000 (paper title page).

Anton, K.-W. 2004: Bruchidius niger, a new species from southwestern Mediterranean region (Coleoptera: Bruchidae: Bruchinae). *Genus* **15** (3): 381–385. DP: 10 October 2004 (paper title page).

Anton, K.-W. 2010a: New acts and comments. Chrysomelidae: Bruchinae. Pp. 62–64. In: Löbl, I. & Smetana, A. (eds.): *Catalogue of Palaearctic Coleoptera. Volume 6. Chrysomeloidea.* Stenstrup: Apollo Books, 924 pp. DP: 22 February 2010 (colophon).

Anton, K.-W. 2010b: Chrysomelidae: Bruchinae. Pp. 339–353. In: Löbl, I. & Smetana, A. (eds.): *Catalogue of Palaearctic Coleoptera. Volume 6. Chrysomeloidea.* Stenstrup: Apollo Books, 924 pp. DP: 22 February 2010 (colophon).

Anton, K.-W. & Delobel, A. 2003: African species of the Bruchidius centromaculatus group with "eyed" female pygidium (Coleoptera: Bruchidae: Bruchinae). *Genus* **14** (2): 159–190. DP: 30 June 2003 (paper title page).

Anton, K.-W. & Delobel, A. 2004: Description of five new species in the genus Caryedon Schoenherr, with a taxonomical note on C. angeri (Semenov) (Coleoptera: Bruchidae: Pachymerinae). *Genus* 15 (1): 65–90. DP: 30 March 2004 (paper title page).

Anton, K.-W. & Delobel, A. 2017: Three new Asian species of Bruchidius (Coleoptera: Chrysomelidae: Bruchinae). *Acta Entomologica Musei Nationalis Pragae* 57 (1): 161–172. DP: 30 June 2017 (paper title page).

Apfelbeck, V. 1906: Nove vrste koleoptera s Balkanskog poluostrva. (Paeninsulae balcanicae coleopterorum species novae). *Glasnik Zemaljskog Muzeja u Bosni i Hercegovini* 17 [1905]: 239–251. DP: [31 December] 1906 (volume title page).

Apfelbeck, V. 1912: Komponente balkanske faune iz roda Chrysomelidae (Col.). Zoogeografski prinos fauni balkanskog Poluostrova. *Glasnik Zemaljskog Muzeja u Bosni i Hercegovini* 24 (1): 235–263. DP: [31 December] 1912 (volume title page).

Apfelbeck, V. 1914: Komponente balkanske faune iz roda Chrysomelidae (Col.). *Glasnik Zemaljskog Muzeja u Bosni i Hercegovini* 26 (3): 435–450. DP: [31 December] 1914 (issue title page).

Apfelbeck, V. 1916: Fauna insectorum balcanica. VI. 2. Die Komponenten der Balkanfauna aus der Familie der Chrysomelidae (Col.). (Ein zoogeographischer Beitrag zur Fauna der Balkanhalbinsel). *Wissenschaftliche Mitteilungen aus Bosnien und der Herzegowina* 13: 354–396. DP: [31] October 1916 (*Natur. Novit.* 1916, 38: 238).

Aragona, L. A. 1830: *De quibusdam coleopteris Italiae novis aut rarioribus tentamen inaugurale quod annuentibus magnifico domino rectore illustrissimo facultatis directore spectabili d. decano ac clarissimis D. D. professoribus auspice J. M. Zendrini historiae naturalis spec. prof. ord. pro doctoris medicinae laurea rite capessenda Aloysius Aragona Ticinensis in celebratiss. I. R. Archigymnasio Ticinensi publicae disquisitioni submittit. Mense Aprilis MDCCCXXX in thesis adnexas disputabitur in Universitatis Aedibus*. Ticini Regii: Bizzoni, 31 pp. DP: [30] April 1830 (title page and Bousquet 2016).

Arnold, W. 1938: Beitrag zur Kenntnis der Gattung Cryptocephalus. *Entomologische Blätter* 33 [1937] (6): 474–476. DP: 20 January 1938 (issue cover).

Arora, G. L. 1977: Taxonomy of the Bruchidae (Coleoptera) of Northwest India. Part I. Adults. *Oriental Insects* **Supplement** 7: 1–132. DP: [31] July 1977 (supplement title page).

Arora, G. L. 1980: *A study of the biology and taxonomy of the genus Bruchidius (Coleoptera: Bruchidae) from India. Final Technical Report (1974–1979). U.S.PL-480 Reseach Project A7-ENT-103*. Chandigarh: Department of Zoology, Panjab University, 96 pp. DP: 17 April 1980 (last page of the book).

Arora, G. L. & Singal, S. K. 1979: A new species of Conicobruchus from Kashmir (Coleoptera: Bruchidae). *Oriental Insects* 13 (1/2): 55–59. DP: [31] August 1979 (footer on the paper title page).

Askevold, I. S. 1991: Classification, reconstructed phylogeny, and geographic history of the New World members of Plateumaris Thomson, 1859 (Coleoptera: Chrysomelidae: Donaciinae).

Memoirs of the Entomological Society of Canada 157: 1–175. DP: 22 November 1991 (journal website).

Aslam, N. A. 1967: Nomenclatorial notes on Chrysomeloidea (Coleoptera). *Journal of Natural History* 2 [1968] (1): 127–129. DP: 29 December 1967 (Evenhuis 2003).

Aslan, E. G., Legalov, A. A., Gahari, H., Warchałowski, A., Colonnelli E. & Jedryczkowski, W. B. 2018: New records of Chrysomelidae and Curculionidae (Coleoptera) from Iran. *Scientific Bulletin of the Uzhhorod University, Series Biology* 44: 58–63. DP: [31 December] 2018 (paper title page).

Aslan, I., Gruev, B. & Özbek, H. 1999: A preliminary review of the subfamily Alticinae (Coleoptera, Chrysomelidae) in Turkey. *Turkish Journal of Zoology* 23 (4): 373–414. DP: [31 December] 1999 (paper title page).

Aslan, I. & Warchałowski, A. 1998: Eine neue kleinasiatische Art der Untergattung Burlinius Lopatin, 1965 (Coleoptera: Chrysomelidae: Cryptocephalinae). *Genus* 9 (3): 343–350. DP: 30 September 1998 (paper title page).

Aslan, I. A. & Warchałowski, A. 2005: Longitarsus ozbeki sp. n., a new species from Asia Minor (Coleoptera: Chrysomelidae: Alticinae). *Entomologica Fennica* 16: 221–224. DP: 16 December 2005 (paper title page).

Aubé, C. 1843: Sur quelques Altica confondues sous le nom d'Altica oleracea. *Annales de la Société Entomologique de France, Deuxième Série* 1 (1): 5–10. DP: 3 May 1843 (Evenhuis 2022).

Aubé, C. 1862: Coléoptères nouveaux d'Europe et observations entomologiques. *Annales de la Société Entomologique de France, Quatrième Série* 2 (1): 71–74. DP: 5 June 1862 (Lefèvre 1885).

Aurivillius, C. 1917: *Skalbaggar. Coleoptera. Första familjegruppen: Växtbaggar. Phytophaga. Svensk Insektfauna* 9. Uppsala: Almquist & Wiksells, 119 pp. DP: [31 December] 1917 (book cover).

Auel, H. 1909: Eine Varietät von Melasoma 20-punctata Scop. (Col.). *Zeitschrift für Wissenschaftliche Insektenbiologie* 5 (10): 317–318. DP: 16 October 1909 (issue title page).

Bacal, S., Munteanu, N. & Toderaş, I. 2013: Checklist of beetles (Insecta: Coleoptera) of the Republic of Moldova. *Brukenthal Acta Musei* 8 (3): 415–450. DP: [31 December] 2013 (issue title page).

Bach, M. 1855: Lieferung 5. Pp. 1–142. In: *Käferfauna für Nord- und Mitteldeutschland mit besonderer Rücksicht auf die preussischen Rheinlande. Dritter Band.* [1856], Coblenz: J. Hölscher, 317 pp. DP: 6 December 1855 (Bousquet 2016).

Bach, M. 1856: Nachträge und Verbesserungen zur Käferfauna von Nord- und Mitteldeutschland. *Entomologische Zeitung* (Stettin) 17 (7/8): 241–247. DP: [31] August 1856 (issue title page).

Bach, M. 1858: Lieferung 6. Pp. 143–317. In: *Käferfauna für Nord- und Mitteldeutschland mit besonderer Rücksicht auf die preussischen Rheinlande. Dritter Band.* [1856], Coblenz: J. Hölscher, 317 pp. DP: 21 October 1858 (Bousquet 2016).

Bach, M. 1867: Nachträge, Zusätze und Verbesserungen zum 3. Bande der Käferfauna. Pp. 319–364. In: *Käferfauna für Nord- und Mitteldeutschland mit besonderer Rücksicht auf die preussischen Rheinlande. Schlusslieferung. Enthält Nachträge zu Band 1–4 nebst System. Verzeichniss*. Coblenz: J. Hölscher, 415–523 + 393–493 + 319–364 + 293–308 pp. DP: 24 April 1867 (Bousquet 2016).

Báguena, L. 1958: Observaciones sobre algunos Chrysomelidae. *Eos, Revista Española de Entomología* 34 (1): 9–24. DP: 31 March 1958 (issue cover).

Báguena, L. 1959: Cuatro novedades y un comentario sobre coleópteros de España. *Eos, Revista Española de Entomología* 35 (3): 209–214. DP: 30 September 1959 (volume contents).

Báguena, L. 1960a: Notas críticas y descriptivas sobre algunos Clytrini (Col. Chrysomelidae). *Graellsia* 18 (1/3): 15–28. DP: [31 December] 1960 (issue cover).

Báguena, L. 1960b: Clave practica para la determinacion de los Clytrini ibericos (Col. Chrysomelidae). *Graellsia* 18 (4/6): 123–145. DP: [31 December] 1960 (issue cover).

Balbi, E. 1892a: Diagnosi e descrizione di due specie dei generi Leptura (Linneo) e Timarcha (Latreille). *Rivista Italiana di Scienze Naturali* 12 (4): 49–50. DP: 1 April 1892 (paper title page).

Balbi, E. 1892b: Descrizione d'una nuova specie italiana del genere Hispa (Linneo). *Bollettino della Società Romana per gli Studi Zoologici* (Roma) 1 (3/5): 195–196. DP: [31] August 1892 (*Natur. Novit.* 1892, 14: 287).

Ballion, E. 1878: Verzeichniss der im Kreise vom Kuldsha gesammelten Käfer. *Bulletin de la Société Impériale des Naturalistes de Moscou* 53 (2): 253–389. DP: 30 November 1878 (Bousquet & Davies 2021).

Ballion, E. 1890: Einiges aus meinen Notizblättern. *Societas Entomologica* 5 [1890–1891] (5): 33. DP: 1 June 1890 (paper title page).

Balthasar, V. 1934: Ein Beitrag zur Kenntnis der palaearktischen Donaciini. *Entomologisches Nachrichtenblatt* (Troppau) 8 (4): 128–130. DP: [31] December 1934 (footer on each page).

Baly, J. S. 1859a: *Catalogue of Hispidae in the collection of the British Museum. Part I.* [1858], London: Trustees of the British Museum, x + 172 pp., 9 pl. DP: [31] January 1859 (Bousquet 2016).

Baly, J. S. 1859b: Descriptions of new species of phytophagous beetles. *Annals and Magazine of Natural History, Third Series* 3 (15): 195–209, 1 pl. DP: 1 March 1859 (Evenhuis 2003).

Baly, J. S. 1859c: Description of new genera and species of phytophagous insects. *Annals and Magazine of Natural History, Third Series* 4 (19): 55–61. DP: 1 July 1859 (Evenhuis 2003).

Baly, J. S. 1859d: Description of new genera and species of phytophagous insects. *Annals and Magazine of Natural History, Third Series* 4 (20): 124–128. DP: 1 August 1859 (Evenhuis 2003).

Baly, J. S. 1860a: Descriptions of new species of phytophagous insects. *Transactions of the Entomological Society of London, New Series* 5 [1858–1861] (4): 146–161. DP: 4 January 1860 (Wheeler 1912).

Baly, J. S. 1860b: Descriptions of new genera and species of Eumolpidae. *Journal of Entomology* 1 [1860–1862] (1): 23–36, pl. I. DP: [30] April 1860 (issue title page).

Baly, J. S. 1860c: Descriptions of six new species of Chrysomela from the East. *Journal of Entomology* 1 [1860–1862] (2): 93–97. DP: [31] October 1860 (issue title page).

Baly, J. S. 1860d: Descriptions of some new species of Sagra; remarks on that genus; and the characters of Cheiloxena, a new genus belonging to the same family. *Transactions of the*

Entomological Society of London, New Series 5 [1858–1861] (6 & 7): 236–260, pl. XIV. DP (pp. 236–248): 7 August 1860, (pp. 249–260): 3 November 1860 (Wheeler 1912).

Baly, J. S. 1861a: Descriptions of new genera and species of Phytophaga. *Journal of Entomology* 1 [1860–1862] (3): 193–206, pl. XI. DP: 3 June 1861 (*Proc. Ent. Soc. London* 1861: 11; "February" erroneously mentioned on issue first page is corrected to "June" on p. 491).

Baly, J. S. 1861b: Descriptions of new genera and species of Phytophaga. *Journal of Entomology* 1 [1860–1862] (4): 275–302, pls. XII–XIII. DP: [31] December 1861 (issue title page).

Baly, J. S. 1862a: Descriptions of new species of phytophagous beetles. *Annals and Magazine of Natural History, Third Series* 10 (55): 17–29. DP: 1 July 1862 (Evenhuis 2003).

Baly, J. S. 1862b: Descriptions of new genera and species of Phytophaga. *Journal of Entomology* 1 [1860–1862] (6): 450–459, pl. XXI. DP: [30] September 1862 (issue title page).

Baly, J. S. 1863a: Descriptions of new species of Cassididae, together with a list of all the species belonging to the same family collected by the late M. Mouhot in Siam and Cambodia. *Journal of Entomology* 2 [1863–1866] (7): 6–14. DP: [31] January 1863 (issue title page).

Baly, J. S. 1863b: An attempt at a classification of the Eumolpidae. *Journal of Entomology* 2 [1863–1866] (9): 143–163. DP: [30] November 1863 (issue title page).

Baly, J. S. 1864a: Descriptions of new Phytophaga. *Transactions of the Entomological Society of London, Third Series* 1 [1862–1864] (8): 611–624. DP: 25 January 1864 (Wheeler 1912).

Baly, J. S. 1864b: *Descriptions of new genera and species of Phytophaga*. Entered at Stationer's Hall: J. S. Baly, 16 pp. DP: 19 April 1864 (Bousquet 2016).

Baly, J. S. 1864c: Descriptions of new genera and species of Eumolpidae. *Journal of Entomology* 2 [1863–1866] (10): 219–223. DP: [30] June 1864 (issue title page).

Baly, J. S. 1864d: Descriptions of uncharacterised genera and species of Phytophaga. *Transactions of the Entomological Society of London, Third Series* 2 [1864–1866] (3): 223–243. DP: 28 November 1864 (Wheeler 1912).

Baly, J. S. 1864e: Descriptions of new genera and species of Phytophaga. *Annals and Magazine of Natural History, Third Series* 14 (84): 433–442. DP: 1 December 1864 (Evenhuis 2003).

Baly, J. S. 1865a: Descriptions of new genera and species of Phytophaga. *Annals and Magazine of Natural History, Third Series* 15 (85): 33–38. DP: 1 January 1865 (Evenhuis 2003).

Baly, J. S. 1865b: Descriptions of a new species of Zeugophora, and a new species of Sphaeroderma. *Entomologist's Monthly Magazine* 1 [1864–1865] (8): 183–184. DP: [31] January 1865 (no direct evidence, derived from the fact that the journal was published once a month).

Baly, J. S. 1865c: Phytophaga Malayana. A revision of the phytophagous beetles of the Malay Archipelago, with descriptions of the new species collected by Mr. A. R. Wallace. *Transactions of the Entomological Society of London, Third Series* 4 [1865–1868] (1): 1–76 & (2): 77–300, 6 pls. DP (pp. 1–76): 3 July 1865, (pp. 77–300): 2 September 1867 (Wheeler 1912).

Baly, J. S. 1865d: Descriptions of new species of Crioceridae. *Annals and Magazine of Natural History, Third Series* 16 (93): 153–160. DP: 1 September 1865 (Evenhuis 2003).

Baly, J. S. 1865e: Descriptions of new genera and species of Phytophaga. *Transactions of the Entomological Society of London, Third Series* 2 [1864–1866] (5): 427–440. DP: 27 September 1865 (Wheeler 1912).

Baly, J. S. 1865f: Descriptions of new genera and species of Gallerucidae. *Annals and Magazine of Natural History, Third Series* **16** (94): 247–255. DP: 1 October 1865 (Evenhuis 2003).

Baly, J. S. 1865g: Descriptions of new genera and species of Gallerucidae. *Entomologist's Monthly Magazine* **2** [1865–1866] (17): 97–101. DP: 2 October 1865 (*Trans. Ent. Soc. London* (3) **2**: 115 (*Proc.*)).

Baly, J. S. 1865h: Attempt at a classification of the Eumolpidae. *Journal of Entomology* **2** [1863–1866] (13): 433–442. DP: [30] November 1865 (issue title page).

Baly, J. S. 1865i: Descriptions of new genera and species of Gallerucidae. *Entomologist's Monthly Magazine* **2** [1865–1866] (18): 127–128. DP: [30] November 1865 (heading on p. 128).

Baly, J. S. 1865j: Descriptions of new genera and species of Gallerucidae. *Annals and Magazine of Natural History, Third Series* **16** (96): 402–410. DP: 1 December 1865 (Evenhuis 2003).

Baly, J. S. 1865k: Descriptions of new genera and species of Gallerucidae. *Entomologist's Monthly Magazine* **2** [1865–1866] (19): 147–148. DP: 4 December 1865 (*Trans. Ent. Soc. London* (3) **2**: 129 (*Proc.*)).

Baly, J. S. 1867: Phytophaga Malayana. A revision of the phytophagous beetles of the Malay Archipelago, with descriptions of the new species collected by Mr. A. R. Wallace. *Transactions of the Entomological Society of London, Third Series* **4** [1865–1868] (1): 1–76 & (2): 77–300, 6 pls. DP (pp. 1–76): 3 July 1865, (pp. 77–300): 2 September 1867 (Wheeler 1912).

Baly, J. S. 1869: Descriptions of new genera and species of Hispidae; with notes on some previously described species. *Transactions of the Entomological Society of London* **1869** (5): 363–382. DP: 14 December 1869 (Wheeler 1912).

Baly, J. S. 1873: Catalogue of the phytophagous Coleoptera of Japan, with descriptions of the species new to science. *Transactions of the Entomological Society of London* **1873** (1): 69–99. DP: 31 March 1873 (Wheeler 1912).

Baly, J. S. 1874: Catalogue of the phytophagous Coleoptera of Japan, with descriptions of the species new to science. *Transactions of the Entomological Society of London* **1874** (2): 161–217. DP: 27 April 1874 (Wheeler 1912).

Baly, J. S. 1875a: Descriptions of new genera and species of Phytophaga. *Transactions of the Entomological Society of London* **1875** (1): 23–31. DP: 29 May 1875 (Wheeler 1912).

Baly, J. S. 1875b: Geänderte Namen. *Coleopterologische Hefte* **14**: 213. DP: [31 December] 1875 (volume title page).

Baly, J. S. 1876a: Diagnoses of undescribed species of Phytophaga. *Entomologist's Monthly Magazine* **13** [1876–1877] (148): 79–82. DP: 6 September 1876 (*Trans. Ent. Soc. London* 1876: xxv (*Proc.*)).

Baly, J. S. 1876b: Diagnoses of hitherto uncharacterized species of Phytophaga. *Entomologist's Monthly Magazine* **13** [1876–1877] (150): 126–129. DP: 1 November 1876 (*Trans. Ent. Soc. London* 1876: xxxi (*Proc.*)).

Baly, J. S. 1876c: Descriptions of new genera and species of Halticinae. *Transactions of the Entomological Society of London* **1876** (3): 433–449. DP: 2 November 1876 (Wheeler 1912).

Baly, J. S. 1876d: Descriptions of a new genus and of new species of Halticinae. *Transactions of the Entomological Society of London* **1876** (4): 581–602. DP: 23 December 1876 (Wheeler 1912).

Baly, J. S. 1877a: Descriptions of new genera and species of Galerucidae. *Entomologist's Monthly Magazine* **13** [1876–1877] (154): 224–227. DP: 14 March 1877 (*Bull. Soc. Ent. Fr.* 1877: 69).

Baly, J. S. 1877b: Descriptions of new genera and species of phytophagous beetles belonging to the family Cryptocephalidae, together with diagnoses and remarks on previously described genera. *Transactions of the Entomological Society of London* **1877** (1): 23–36. DP: 17 April 1877 (Wheeler 1912).

Baly, J. S. 1877c: Descriptions of new species of phytophagous beetles belonging to the family Eumolpidae; and a monograph of the genus Eumolpus. *Transactions of the Entomological Society of London* **1877** (1): 37–56. DP: 17 April 1877 (Wheeler 1912).

Baly, J. S. 1877d: Descriptions of new genera and of uncharacterized species of Halticinae. *Transactions of the Entomological Society of London* **1877** (2): 157–184. DP: 2 July 1877 (Wheeler 1912).

Baly, J. S. 1877e: Descriptions of new genera and species of Cryptocephalidae. *Transactions of the Entomological Society of London* **1877** (3): 211–232. DP: 16 October 1877 (Wheeler 1912).

Baly, J. S. 1877f: Characters of new genera and of some undescribed species of phytophagous beetles. *Annals and Magazine of Natural History, Fourth Series* **20** (119): 377–386. DP: 1 November 1877 (Evenhuis 2003).

Baly, J. S. 1877g: Descriptions of new genera and of uncharacterized species of Halticinae. *Transactions of the Entomological Society of London* **1877** (4): 283–323. DP: 31 December 1877 (Wheeler 1912).

Baly, J. S. 1878a: Characters of new genera and some undescribed species of phytophagous beetles. *Annals and Magazine of Natural History, Fifth Series* **1** (1): 38–44. DP: 1 January 1878 (Evenhuis 2003).

Baly, J. S. 1878b: Diagnoses of a new genus and some undescribed species of African Phytophaga. *Entomologist's Monthly Magazine* **14** [1877–1878] (165): 204–206. DP: 13 February 1878 (*Bull. Soc. Ent. Fr.* 1878: 31).

Baly, J. S. 1878c: Descriptions of genera and species of Australian phytophagous beetles. *Journal of the Linnean Society of London, Zoology* **13** [1876–1878] (72): 458–479. DP: 27 February 1878 (verso of the volume title page).

Baly, J. S. 1878d: Descriptions of a new genus and of new species of Halticinae. *Annals and Magazine of Natural History, Fifth Series* **1** (4): 312–322. DP: 1 April 1878 (Evenhuis 2003).

Baly, J. S. 1878e: Descriptions of some uncharacterized species of Crioceridae. *Cistula Entomologica* **2** [1875–1882] (19): 305–316. DP: 30 June 1878 (footer on p. 269).

Baly, J. S. 1878f: Description of new species and genera of Eumolpidae. *Journal of the Linnean Society, Zoology* **14** [1877–1879] (75): 246–264. DP: 31 August 1878 (verso of the volume title page).

Baly, J. S. 1878g: Characters of undescribed species of Halticinae. *Annals and Magazine of Natural History, Fifth Series* **2** (9): 223–233. DP: 1 September 1878 (Evenhuis 2003).

Baly, J. S. 1878h: Descriptions of the phytophagous Coleoptera collected by the late Dr. F. Stoliczka during Forsyth's expedition to Kashgar in 1873–74. *Cistula Entomologica* **2** [1875–1882] (20): 369–383. DP: 25 September 1878 (footer on p. 317).

Baly, J. S. 1878i: Descriptions of new species of phytophagous Coleoptera. *Journal of the Linnean Society, Zoology* 14 [1877–1879] (76): 336–356. DP: 31 October 1878 (verso of the volume title page).

Baly, J. S. 1878j: Descriptions of new genera and species of Gallerucinae. *Annals and Magazine of Natural History, Fifth Series* 2 (11): 411–422. DP: 1 November 1878 (Evenhuis 2003).

Baly, J. S. 1879a: List of the phytophagous Coleoptera collected in Assam by A. W. Chennell, Esq., with notes and descriptions of the uncharacterized genera and species. *Cistula Entomologica* 2 [1875–1882] (21): 435–465. DP: 15 February 1879 (footer on p. 389).

Baly, J. S. 1879b: Descriptions of new genera and species of Gallerucinae. *Annals and Magazine of Natural History, Fifth Series* 4 (20): 108–120. DP: 1 August 1879 (Evenhuis 2003).

Baly, J. S. 1879c: An attempt to point out the differential characters of some closely-allied species of Chrysomela, principally those contained in Suffrian's 11th group; also descriptions of some hitherto uncharacterized forms belonging to the same and other genera of the family. *Transactions of the Entomological Society of London* 1879 (2): 171–197, pl. II. DP: 7 August 1879 (Wheeler 1912).

Baly, J. S. 1881: Descriptions of new species of Galerucidae. *Transactions of the Entomological Society of London* 1881 (1): 51–59. DP: 14 April 1881 (Wheeler 1912).

Baly, J. S. 1886a: Descriptions of new genera and species of Galerucidae. *Transactions of the Entomological Society of London* 1886 (1): 27–39. DP: 8 April 1886 (Wheeler 1912).

Baly, J. S. 1886b: Descriptions of a new genus and of some new species of Galerucinae, also diagnostic notes on some of the older described species of Aulacophora. *Journal of the Linnean Society, Zoology* 20 [1886–1890] (116): 1–27. DP: 27 October 1886 (verso of the volume title page).

Baly, J. S. 1887: Description of a new species of phytophagous Coleoptera, alleged to be destructive to the dhan crops in the Chittagong District. *Journal of the Asiatic Society of Bengal, Part II, Natural history &c.* 55 [1886–1887] (4): 412. DP: 7 March 1887 (last page of the contents).

Baly, J. S. 1888a: Descriptions of three new Phytophaga from the East. *Entomologist's Monthly Magazine* 25 [1888–1889] (292): 85–86. DP: 1 September 1888 (registered at NHMUK library).

Baly, J. S. 1888b: Descriptions of some genera and species of Galerucinae. *Journal of the Linnean Society of London, Zoology* 20 [1886–1890] (119): 156–188. DP: 1 November 1888 (verso of the volume title page).

Baly, J. S. 1888c: Viaggio di L. Fea in Birmania e regioni vicine. XIII. List of the Hispidae collected in Burmah and Tenasserim by Mr. L. Fea, together with descriptions of some of the new species. *Annali del Museo Civico di Storia Naturale di Genova* 26: 653–666. DP: 12 December 1888 (Poggi 2010).

Baly, J. S. 1890a: Descriptions of two new genera, and of some uncharacterised species of Galerucinae. *Entomologist's Monthly Magazine, Second Series* 1 (1): 12–14. DP: 2 January 1890 (registered at NHMUK library).

Baly, J. S. 1890b: Contributions à la faune indo-chinoise. 6e mémoire. Sagridae, Crioceridae, Chrysomelidae, Hispidae. *Annales de la Société Entomologique de France, 6e Série* 9 [1889] (4): 485–492. DP: 28 May 1890 (Lefèvre 1895).

Bardin, O. D. & Timraleev, Z. A. 2007: A contribution to the fauna and ecology of leaf beetles (Coleoptera, Chrysomelidae) in the Republic of Mordovia. *Entomological Review* **87** (6): 670–676. DP: 1 October 2007 (journal website).

Baselga, A. 2006: The northeastern Palaearctic light coloured Neocrepidodera Heikertinger, 1911 (Coleoptera: Chrysomelidae), with description of a new species. *Zootaxa* **1246**: 55–68. DP: 29 June 2006 (footer on the paper title page).

Baselga, A. & Novoa, F. 2002: New species of Aphthona (Coleoptera: Chrysomelidae: Alticinae) and key to Iberian species of Aphthona hammarstroemi group. *Canadian Entomologist* **134** (1): 1–7. DP: [28] February 2002 (issue cover).

Baselga, A. & Novoa, F. 2003: A new species of Psylliodes (Coleoptera: Chrysomelidae) and key to the wingless species from the Iberian Peninsula. *Annals of the Entomological Society of America* **96** (6): 689–692. DP: 1 November 2003 (journal website).

Baselga, A. & Novoa, F. 2005: The western Palaearctic Neocrepidodera (Coleoptera: Chrysomelidae) of the N. impressa and N. ferruginea species groups. *Annals of the Entomological Society of America* **98** (6): 896–907. DP: 1 November 2005 (journal website).

Baselga, A. & Novoa, F. 2006: Diversity of Chrysomelidae (Coleoptera) in Galicia, Northwest Spain: estimating the completeness of the regional inventory. *Biodiversity and Conservation* **15**: 205–230. DP: [31] January 2006 (journal website).

Baselga, A. & Ruiz-García, J. 2007: Revision of the Lachnaia tristigma (Lacordaire, 1848) species-group (Coleoptera: Chrysomelidae) and description of a new species. *Zootaxa* **1630**: 39–46. DP: 5 November 2007 (footer on the paper title page).

Bassi, C. 1834: Description de quelques nouvelles espèces de coléoptères de l'Italie. *Annales de la Société Entomologique de France* **3** (3): 463–472, pl. XI. DP: [30] September 1834 (Evenhuis 2022).

Bastazo, G. 1998: Two new species of Longitarsus Berthold, 1827 from southern Spain (Coleoptera: Chrysomelidae: Alticinae). *Elytron* **11** [1997]: 31–38. DP: [31] March 1998 (colophon).

Bastazo, G. & Garcia Raso, J. E. 1985: Longitarsus caroli n. sp., from the south of the Iberian Peninsula (Coleoptera Chrysomelidae Alticinae). *Atti della Società Italiana di Scienze Naturali e del Museo Civile di Storia Naturale di Milano* **126** (1/2): 94–100. DP: 15 June 1985 (issue cover).

Basu, C. R. 1985: [Fauna of Namdapha] Insecta: Coleoptera: Chrysomelidae. *Records of the Zoological Survey of India* **82** (1/4): 201–214. DP: [30] September 1985 (colophon).

Basu, C. R. 1991: The Indian species of the genus Asiorestia Jacobson (Coleoptera: Chrysomelidae: Alticinae) with descriptions of three new species. *Records of the Zoological Survey of India* **89** (1/4): 123–129. DP: [31] December 1991 (colophon).

Basu, C. R. 1996: Insecta: Coleoptera: Chrysomelidae. Pp. 559–773. In: Ghosh, A. K. (ed.): *Fauna of West Bengal. Part 6B (Insecta: Coleoptera). State Fauna Series 3*. Calcutta: Zoological Survey of India, 449–866 pp. DP: [29] February 1996 (colophon).

Basu, C. R. 1999: On the collection of Hispinae (Coleoptera: Chrysomelidae) preserved in the Zoological Survey of India, Calcutta. *Records of the Zoological Survey of India* **97** (3): 149–190. DP: [30] September 1999 (colophon).

Basu, C. R. & Saha, G. N. 1977: Seven new species of Dactylispa (Coleoptera: Chrysomelidae: Hispinae) from Darjeeling District, West Bengal, India. *Oriental Insects* **11** (3): 379–388. DP: [30] November 1977 (footer on the paper title page).

Basu, C. R. & Sengupta, T. 1980: On a new genus and a new species of Alticinae (Coleoptera: Chrysomelidae) from Arunachal Pradesh. *Records of the Zoological Survey of India* **76** (1/4): 143–146. DP: [31] August 1980 (colophon).

Basu, C. R. & Sengupta, T. 1982: Little known Alticinae (Coleoptera: Chrysomelidae) with descriptions of five new species from India. *Oriental Insects* **15** [1981] (3): 263–276. DP: [31] January 1982 (footer on the paper title page).

Basu, C. R. & Sengupta, T. 1985: A new species of Orthaltica (Coleoptera: Chrysomelidae: Alticinae) from India. *Bulletin of the Zoological Survey of India* **7** (1): 45–48. DP: [31] August 1985 (colophon).

Bates, H. W. 1866: On a collection of Coleoptera from Formosa, sent home by R. Swinhoe, Esq., H.B.M. Consul, Formosa. *Proceedings of the Zoological Society of London* **1866** (2): 339–355. DP: [30] September 1866 (Duncan 1937).

Batiashvili, I. D. & Elerdashvili, N. L. 1969: K izucheniyu fauny zernovok (Coleoptera, Bruchidae) Gruzii. (On the knowledge of the fauna of bean weevils (Coleoptera, Bruchidae) from Georgia). *Entomologicheskoe Obozrenie* **48** (3): 524–531. SFP: 20 October 1969, DP: 17 November 1969 (registered at ZIN library).

Baudi di Selve, F. 1886: Rassegna delle specie della famiglia die Mylabridi (Bruchidi degli autori) viventi in Europa e regioni finitime. *Il Naturalista Siciliano* **5** [1885–1886] (10): 1–?12, (11): ?13–?32, (12): ?33–52 & **6** [1886–1887]: (1): 53–72, (2): 73–92, (3): 93–112, (4–5): 113–136 + [2]. DP: (pp. 1–?12): 1 July 1886, (pp. ?13–?32): 1 August 1886, (pp. ?33–52): 1 September 1886, (pp. 53–72): 1 October 1886, (pp. 73–92): 1 November 1886, (pp. 93–112): 1 December 1886, (pp. 113–136 + [2]): 1 February 1887 (Bousquet 2016).

Baudi di Selve, F. 1887: Mylabridum seu Bruchidum (Lin. Schön. All.) europeae et finitimarum regionum Faunae recensitio. *Deutsche Entomologische Zeitschrift* **31** (1): 33–80. DP: [30] June 1887 (table of contents).

Baudi di Selve, F. 1889: *Catalogo dei coleotteri del Piemonte*. Torino: Camilla e Bertolero, 226 pp. DP: [31 December] 1889 (Bousquet, 2016) [preprint from *Annali della Reale Accademia d'Agricoltura di Torino* **32** [1889]: 51–274. DP: 15 May 1890 (Bousquet 2016)].

Baudi di Selve, F. 1890: Mylabridum seu Bruchidum recensioni a Flaminio Baudi conscriptae addenda. *Deutsche Entomologische Zeitschrift* **1890** (2): 337–347. DP: [31] October 1890 (table of contents).

Bauduer, P. 1874a: [Descriptions suivantes de coléoptères nouveaux]. *Bulletin de la Société Entomologique de France* **1874** (17): 197–198. DP: 18 September 1874 (Evenhuis 2002).

Bauduer, P. 1874b: [Descriptions d'une nouvelle espèce de coléoptère]. *Bulletin de la Société Entomologique de France* **1874** (18): 211. DP: 30 September 1874 (Evenhuis 2002).

Baviera, C. 2007: First record of Gonioctena (Coleoptera, Chrysomelidae, Chrysomelinae) in Sicily, with the description of Gonioctena theae n. sp. *Italian Journal of Zoology* **74** (4): 389–393. DP: [31] December 2007 (issue table of contents).

Bechyné, J. 1942: O variabilitě Plateumaris consimilis Schrank. (Col. Donaciidae). De variatione Plateumaris consimilis Schrank. (Col. Donaciidae). *Acta Entomologica Musei Nationalis Pragae* 20: 232–237. DP: 1 December 1942 (volume title page).

Bechyné, J. 1944a: Cassididae a Hispidae v Čechách a na Moravě. Bohemiae et Moraviae Cassididarum et Hispidarum conspectus. *Entomologické Listy* 7 (3): 74–86. DP: 30 June 1944 (issue colophon).

Bechyné, J. 1944b: Revise druhů skupiny Timarcha sardea Villa. Specierum sectionis Timarcha sardea Villa revisio (Col. Phytoph. Chrysomelidae). *Acta Entomologica Musei Nationalis Pragae* 21–22 [1943–1944]: 139–146. DP: 1 November 1944 (volume title page).

Bechyné, J. 1944c: Příspěvek k poznání severoafrických druhů rodu Timarcha Latr. (Col. Chrysomelidae). Notulae ad cognitionem specierum Africae septentrionalis generis Timarcha Latr. (Col. Chrysomelidae). *Acta Entomologica Musei Nationalis Pragae* 21–22 [1943–1944]: 211–221. DP: 1 November 1944 (volume title page).

Bechyné, J. 1944d: Příspěvek k poznání severoafrických druhů rodu Timarcha Latr. (Col. Phytoph. Chrysomelidae). Addimenta ad cognitionem specierum Africae borealis generis Timarcha Latr. (Col. Phytoph. Chrysomelidae). *Acta Entomologica Musei Nationalis Pragae* 21–22 [1943–1944]: 284–293. DP: 1 November 1944 (volume title page).

Bechyné, J. 1944e: O druzích rodu Timarcha Latr. ze skupiny T. laevigata Linn. De speciebus generis Timarcha Latr. sectionis T. laevigata Linn. (Col. Phytophaga, Chrysomelidae). *Acta Entomologica Musei Nationalis Pragae* 21–22 [1943–1944]: 360–372. DP: 1 November 1944 (volume title page).

Bechyné, J. 1945a: O druzích rodu Timarcha skupiny T. tenebricosa F. De specibus generis Timarcha Latr. sectionis T. tenebricosa F. (Col. Phytoph. Chrysomelidae). *Entomologické Listy* 8 (1): 5–10. DP: 1 January 1945 (issue title page).

Bechyné, J. 1945b: Příspěvek k poznání našich mandelinek. Additamenta ad cognitionem Phytophagorum centralieuropaeorum. *Entomologické Listy* 8 (3): 87–91. DP: 15 March 1945 (issue title page).

Bechyné, J. 1945c: Timarcha pfefferi n. sp. (Col. Phytoph. Chrysomelidae). *Časopis Československé Společnosti Entomologické* 42 (1/4): 64–65. DP: 1 September 1945 (issue cover).

Bechyné, J. 1945d: De sectionis Timarchae latipes Linn. speciebus (Col. Phytoph. Chrysomelidae). O druzích sekce Timarcha latipes L. *Časopis Československé Společnosti Entomologické* 42 (1/4): 101–103. DP: 1 September 1945 (issue cover).

Bechyné, J. 1945e: O odchylkách druhu Phytodecta fornicatus B. De Phytodectae fornicati Brüggm. variationibus (Col. Phytophaga, Chrysomelidae). *Entomologické Listy* 8 (4): 111–112. DP: 15 October 1945 (issue title page).

Bechyné, J. 1946a: Nomenklatorické a faunistické poznámky k některým druhům rodu Chrysolina Motsch. Notulae synonymicae et faunisticae specierum nonnullarum generis Chrysolina Motsch. (Col. Phytoph. Chrysomelidae). *Entomologické Listy* 9 [1945–1946] (3): 105–112. DP: 30 September 1946 (issue cover).

Bechyné, J. 1946b: Nový druh a několik nových variet rodu Timarcha Latr. Descriptions d'une espèce et des variétés nouvelles du genre Timarcha Latr. (Col. Phytophaga, Chrysomelidae). *Časopis Československé Společnosti Entomologické* **43** (1/4): 27–30. DP: [31 December] 1946 (issue cover).

Bechyné, J. 1946c: Příspěvek k poznání evropských druhů rodu Timarcha Latr. Additamenta ad cognitionem europae specierum generis Timarcha Latr. (Col. Phytophaga, Chrysomelidae). *Acta Entomologica Musei Nationalis Pragae* **24**: 129–139. DP: [31 December] 1946 (volume title page).

Bechyné, J. 1947a: Druhý příspěvek k poznání rodu Chrysolina Motsch. Notulae ad cognitionem generis Chrysolina Motsch. II. (Col. Phytophaga, Chrysomelidae). *Entomologické Listy* **10** (2): 58–64. DP: 30 May 1947 (issue cover).

Bechyné, J. 1947b: Poznámky ke studii o rodu Timarcha Latr. (11.). Notulae ad cognitionem specierum generis Timarcha Latr. XI. (Col. Phytophaga, Chrysomelidae). *Časopis Československé Společnosti Entomologické* **44** (1/2): 3–9. DP: 1 June 1947 (issue cover).

Bechyné, J. 1947c: Notulae ad cognitionem generis Timarcha Latr. VIII. 8. Příspěvek k poznání rodu Timarcha Latr. (Col. Phytoph. Chrysomelidae). *Acta Entomologica Musei Nationalis Pragae* **25**: 55–61. DP: [31 December] 1947 (issue cover).

Bechyné, J. 1948a: Příspěvek k poznání rodu Phytodecta Kirby. Additamenta ad cognitionem specierum generis Phytodecta Kirby. (Col. Phytoph. Chrysomelidae). *Sborník Národního Musea v Praze* **3B** [1947] (3): 89–158, 5 pls. DP: [31] January 1948 (colophon).

Bechyné, J. 1948b: Třetí příspěvek k poznání rodu Chrysolina Motsch. Notulae ad cognitionem generis Chrysolina Motsch. III. *Entomologické Listy* **11** (1): 10–15. DP: 30 April 1948 (issue cover).

Bechyné, J. 1948c: Noví halticidi z Madagascaru. De novis speciebus halticidarum madagascariensium. (Col. Phytophaga). *Časopis Československé Společnosti Entomologické* **45** (1/2): 3–11. DP: 1 May 1948 (issue cover).

Bechyné, J. 1948d: Příspěvek k poznání phyllogenese a zoogeografie rodu Timarcha. Contribution à la connaissance du genre Timarcha Latr. 12.: Etudes phylogénétiques et zoogéographiques. (Col. Phytophaga, Chrysomelidae). *Sborník Národního Musea v Praze* **4B** (2): 1–62. DP: [30] November 1948 (colophon at the end of the issue).

Bechyné, J. 1948e: Les Timarcha des Pyrénées françaises (Coléoptères Chrysomelidae). *Revue Française d'Entomologie* **15** (4): 197–208. DP: 15 December 1948 (footer on p. 197).

Bechyné, J. 1949a: Čtvrtý příspěvek k poznání rodu Chrysolina Motsch. Notulae ad cognitionem generis Chrysolina Motsch. IV. *Entomologické listy* **12** (2): 48–55. DP: 15 May 1949 (issue cover).

Bechyné, J. 1949b: Contribution à la connaissance du groupe de Chrysolina carnifex (Col.). *Bulletin Mensuel de la Société Linnéenne de Lyon* **18** (5): 84–86. DP: [31] May 1949 (issue title page).

Bechyné, J. 1950a: Les Chrysomelidae vrais du Congo belge (Col. Phytophaga). *Revue de Zoologie et de Botanique Africaines* **44** (1): 119–131. DP: 30 December 1950 (paper title page).

Bechyné, J. 1950b: 7e Contribution à la connaissance du genre Chrysolina Motsch. (Col. Phytophaga Chrysomelidae). *Entomologische Arbeiten aus dem Museum G. Frey* 1: 47–185. DP: 31 December 1950 (table of contents).

Bechyné, J. 1952a: Notes sur quelques Chrysolina des Pyrénées (Col. Chrysomelidae). *Revue Française d'Entomologie* **19** (3): 171–173. DP: 10 September 1952 (last page of the issue).

Bechyné, J. 1952b: Achter Beitrag zur Kenntnis der Gattung Chrysolina Motsch. (Col. Phytoph. Chrysomelidae). *Entomologische Arbeiten aus dem Museum G. Frey* **3** (2): 351–385. DP: 31 December 1952 (table of contents).

Bechyné, J. 1953: Notes sur les Timarcha (Col. Phytoph. Chrysomelidae). *Eos, Revista Española de Entomología* **29** (2/4): 85–99. DP: [31 December] 1953 (volume title page).

Bechyné, J. 1954a: Note sui Chrysomelidae veri d'Europa (Coleopt.). *Bollettino de la Società Entomologica Italina* **84** (5/6): 90–91. DP: 20 July 1954 (issue cover, Poggi 2008).

Bechyné, J. 1954b: Beiträge zur Kenntnis der echten Chrysomeliden (Col. Phytophaga). *Entomologische Arbeiten aus dem Museum G. Frey* **5** (2): 581–674, pls. XXIV–XXX. DP: 5 November 1954 (table of contents).

Bechyné, J. 1955a: Notes sur les Chrysolinini (Coleopt., Chrysom.) de la Mediterranee orientale. *Bulletin of the Research Council of Israel* **4** (4): 347–350. DP: [31] March 1955 (reprint title page).

Bechyné, J. 1955b: Über die westafrikanischen Alticiden (Col. Phytophaga). *Entomologische Arbeiten aus dem Museum G. Frey* **6** (2): 486–568. DP: 1 December 1955 (table of contents).

Bechyné, J. 1956a: Note su Chrysomelidae veri (Coleopt.). *Bollettino de la Società Entomologica Italiana* **86** (1/2): 16–17. DP: 17 March 1956 (issue cover, Poggi 2008).

Bechyné, J. 1956b: Über die Alticiden-Sammlung Heikertinger (Col. Phytophaga). *Entomologischen Arbeiten aus dem Museum G. Frey* 7 (2): 577–598. DP: 1 November 1956 (issue table of contents).

Bechyné, J. 1957: Notes sur quelques Chrysomeloidea palearctiques recueillis par M. G. Fagel (Coleoptera Phytophaga). *Bulletin d'Institut Royal des Sciences Naturelles de Belgique* **33** (31): 1–4. DP: [31] July 1957 (paper title page).

Bechyné, J. 1958a: Contribution a l'étude des Chrysomeloidea des Iles Mascareignes. II. Alticidae. *Mauritius Institute Bulletin* **5** (3): 83–93. DP: 15 May 1958 (reprint title page).

Bechyné, J. 1958b: Über die taxonomische Valenz der Namen von Oreina s. str. (Col. Phytophaga). *Mitteilungen der Schweizerischen Entomologischen Gesellschaft* **31** (1): 79–95. DP: 30 May 1958 (paper title page).

Bechyné, J. 1958c: Notizen zu den neotropischen Chrysomeloidea (Col. Phytophaga). *Entomologische Arbeiten aus dem Museum G. Frey* **9** (2): 478–706. DP: 31 May 1958 (issue table of contents).

Bechyné, J. 1959: Coleoptera Chrysomelidae II. Results of the Lund Expedition in 1950–1951. *South African Animal Life* **5** [1958]: 227–238. DP: 14 March 1959 (Evenhuis 2022).

Bechyné, J. 1960a: Notes sur les alticides africains des collections de l'Institut royal des sciences naturelles de Belgique (Coleoptera, Phytophaga). *Bulletin d'Institut Royal des Sciences Naturelles de Belgique* **36** (8): 1–32. DP: [31] January 1960 (paper title page).

Bechyné, J. 1960b: Alticidae (Coleoptera Phytophagoidea). *Exploration du Parc National de l'Upemba, Mission G. F. de Witte en collaboration avec W. Adam, A. Janssens, L. van Meel et R. Verheyen (1946–1948)* **59** [1959] (3): 39–114. DP: 30 June 1960 (last page of the paper).

Bechyné, J. 1962: Notes sur quelques Chrysomelides vrais (Coléoptères). *Bulletin Mensuel de la Société Linnéenne de Lyon* **31** (6): 182–185. DP: [30] June 1962 (issue cover).

Bechyné, J. 1964: Notizen zu den madagassischen Chrysomeloidea (Col. Phytophaga). *Mitteilungen der Münchner Entomologischen Gesellschaft* **54**: 68–161. DP: 1 December 1964 (volume cover).

Bechyné, J. 1969: Contribution à la faune du Congo (Brazzaville). Mission A. Villiers et A. Descarpentries LXXXI. Coléoptères Alticidae. *Bulletin de l'Institut Français d'Afrique Noire, Série A* **30** [1968] (4): 1687–1728. DP: [31] January 1969 (last page of the issue).

Beck, J. M. 1817: *Beiträge zur baierischen Insektenfaune, oder Beschreibung und Abbildung neuentdeckter Käfer, mit angehängtem Namensverzeichnisse der Eleuteraten des Landgerichtbezirks Zusmeshausen.* Augsburg: J. Wolff, 45 + 4 pp., 7 pls. DP: 25 March 1817 (Bousquet 2016).

Becker, A. 1880: Beiträge zu meinen Verzeichnissen der um Sarepta und am Bogdo vorkommenden Pflanzen und Insekten, und Beschreibung einer Mylabris-Larve. *Bulletin de la Société Impériale de Naturalistes de Moscou* **55** (1): 145–156. DP: 30 September 1880 (Bousquet & Davies 2021).

Becker, A. 1892: Neue Pflanzen- und Insektenentdeckungen in der Umgegend von Sarepta und Zusammenstellung der Raupen und Käfer, die nur von einer Pflanzenart, und zwei, drei Pflanzenarten leben, die aber zu einer Familie gehören. *Bulletin de la Société Impériale des Naturalistes de Moscou, Nouvelle Série* **6** (1): 62–70. DP: 13 August 1892 (stamp on the BHL copy).

Bedel, L. 1891: *Faune des coléoptères du bassin de la Seine. Tome v. Phytophaga.* [1889–1901], Paris: Annales Société Entomologique de France, Publication hors série: [3] + 423 pp. DP (pp. 105–120): 22 September 1891, (pp. 121–136): 23 December 1891 (Bousquet 2016).

Bedel, L. 1892a: Supplément au catalogue des coléoptères de l'Yonne. Excursions dans l'arrondissement d'Avallon. *Annales de la Société Entomologique de France* **60** [1891] (4): 575–592. DP: 13 April 1892 (issue cover).

Bedel, L. 1892b: *Faune des coléoptères du bassin de la Seine. Tome v. Phytophaga.* [1889–1901], Paris: Annales Société Entomologique de France, Publication hors série: [3] + 423 pp. DP (pp. 137–152): 12 October 1892, (pp. 153–160): 28 December 1892 (Bousquet 2016).

Bedel, L. 1897: *Faune des coléoptères du bassin de la Seine. Tome v. Phytophaga.* [1889–1901], Paris: Annales Société Entomologique de France, Publication hors série: [3] + 423 pp. DP (pp. 161–180): 28 April 1897 (Bousquet 2016).

Bedel, L. 1898: *Faune des coléoptères du bassin de la Seine. Tome v. Phytophaga.* [1889–1901], Paris: Annales Société Entomologique de France, Publication hors série: [3] + 423 pp. DP (pp. 181–212): 26 October 1898 (2016).

Bedel, L. 1899: *Faune des coléoptères du bassin de la Seine. Tome v. Phytophaga.* [1889–1901], Paris: Annales Société Entomologique de France, Publication hors série: [3] + 423 pp. DP (pp. 229–260): 10 May 1899 (Bousquet 2016).

Bedel, L. 1900: *Faune des coléoptères du bassin de la Seine. Tome v. Phytophaga.* [1889–1901], Paris: Annales Société Entomologique de France, Publication hors série: [3] + 423 pp. DP (pp. 277–308): 13 June 1900 (Bousquet 2016).

Bedel, L. 1901: *Faune des coléoptères du bassin de la Seine. Tome v. Phytophaga.* [1889–1901], Paris: Annales Société Entomologique de France, Publication hors série: [3] + 423 pp. DP (pp. 341–356): 25 June 1901, (pp. 357–423): 23 October 1901 (Bousquet 2016).

Bedel, L. 1905: Synonymies de coléoptères paléarctiques. *L'Abeille, Journal d'Entomologie* **30** [1900–1906]: 235–236. DP: [31] January 1905 (*Bull. Soc. Ent. Fr.* 1905: 30).

Bedel, L. 1915: Notes sur les Entomoscelis méditerranéens (Col. Chrysomelidae). *Bulletin de la Société Entomologique de France* **1915** (13): 206–207. DP: 7 August 1915 (*dates de publication* at the end of volume).

Bedel, L. 1921: Coléoptères nouveaux du Maroc. Récoltes du Dr. M. Bedel dans le Moyen Atlas, 1916–1917. *Bulletin de la Société Entomologique de France* **1921** (5): 59–61. DP: 13 April 1921 (*dates de publication* at the end of volume).

Beenen, R. 2002: Revisional notes on Galeruca 2 (Coleoptera, Chrysomelidae). *Entomologische Blätter* **98** (1): 21–28. DP: 31 July 2002 (paper title page).

Beenen, R. 2003: Galeruca (Galemira n. subgen.) subcostata n. sp. from Pakistan (Coleoptera: Chrysomelidae). *Stuttgarter Beiträge zur Naturkunde, Serie A (Biologie)* **648**: 1–9. DP: 4 June 2003 (paper title page).

Beenen, R. 2005a: The identity of Galeruca circassica Reitter and G. bicolor Kollar & Redtenbacher (Coleoptera, Chrysomelidae). (Revisional notes on Galeruca 4). *Entomologische Blätter für Biologie und Systematik der Käfer* **101** (1): 13–20. DP: 30 May 2005 (paper title page).

Beenen, R. 2005b: Cerophysa erberi n. sp. from Swāt in Pakistan (Coleoptera: Chrysomelidae: Galerucinae). *Genus* **16** (3): 383–388. DP: 15 September 2005 (paper title page).

Beenen, R. 2008: Taxonomical and nomenclatural changes in Palaearctic Galerucinae and description of new species (Chrysomelidae). *Entomologische Blätter für Biologie und Systematik der Käfer* **103/104**: 63–80. DP: 16 October 2008 (paper title page).

Beenen R. 2010a: [new taxa]. In: New acts and comments. Chrysomelidae: Galerucinae. Pp. 74–75. In: Löbl, I. & Smetana, A. (eds.): *Catalogue of Palaearctic Coleoptera. Volume 6. Chrysomeloidea.* Stenstrup: Apollo Books, 924 pp. DP: 22 February 2010 (colophon).

Beenen, R. 2010b: Chrysomelidae: Galerucinae. Pp. 443–491. In: Löbl, I. & Smetana, A. (eds.): *Catalogue of Palaearctic Coleoptera. Volume 6. Chrysomeloidea.* Stenstrup: Apollo Books, 924 pp. DP: 22 February 2010 (colophon).

Beenen, R. 2010c: [new taxon]. In: Beenen, R. & Lee, C.-F. 2010: Two new Erganoides species from P.R. China and Taiwan (Coleoptera: Chrysomelidae: Galerucinae). *Genus* **21** (2): 257–264. DP: 30 July 2010 (paper title page).

Beenen, R. 2014: Key to the species of Radymna Reitter, 1913 with taxonomic and faunistic comments and description of two new species (Coleoptera, Chrysomelidae, Galerucinae). *Entomologische Blätter und Coleoptera* **110**: 87–100. DP: [31] December 2014 (volume cover).

Beenen, R. 2019: Faunistic, taxonomic and nomenclatural notes on Palaearctic Galerucinae (Coleoptera, Chrysomelidae) with description of six new species and a new genus. *Entomologische Blätter und Coleoptera* 115: 1–20. DP: [31] December 2019 (volume cover).

Beenen, R. & Bezděk, J. 2007: A new species of Scelolyperus from Central Asia and a key to the Palaearctic species (Coleoptera: Chrysomelidae). *Entomologische Blätter* 102 [2006] (1/3): 87–93. DP: 27 March 2007 (paper title page).

Beenen, R. & Warchałowski, A. 2010: Charaea pseudominutum n. sp., an undescribed but not unknown galerucine beetle (Coleoptera, Chrysomelidae, Galerucinae). *Entomologische Blätter für Biologie und Systematik der Käfer* 106: 57–62. DP: 30 December 2010 (paper title page).

Beenen, R. & Yang, X.-K. 2007: A new species of the genus Galeruca Müller from China (Coleoptera: Chrysomelidae, Galerucinae). *Acta Zootaxonomica Sinica* 32 (1): 67–69. DP: [31] January 2007 (paper title page).

Bekman, Y. I. 1929: Nablyudeniya nad zavozom zernovykh vrediteley. (Über Verschleppung der Samenschädlinge). *Izvestiya po Prikladnoy Entomologii* 4 (1): 151–166. DP: 1 October 1929 (registered at ZIN library).

Bellevoye, A. 1870: Observations sur les moeurs de plusieurs espèces de coléoptères qui vivent sur les plantes aquatiques et qui n'avaient été trouvées que très-rarement dans la département de la Moselle. *Bulletin de la Société d'Histoire Naturelle du Département de la Moselle* 12: 91–104, 1 pl. DP: 27 April 1870 (*Bull. Soc. Ent. Fr.* 1870: xxxix).

Bellier, J.-B. E. 1870: [Descriptions qui suivent de coléoptères nouveaux propres aux Basses-Alpes]. *Bulletin de la Société Entomologique de France* 1870 (1): xxvi–xxviii. DP: 13 July 1870 (Lefèvre 1885).

Benedikt, S. & Mantič, M. 2006: Nové druhy brouků (Coleoptera) pro Slovensko z území Tematínských vrchů. (First records of beetles (Coleoptera) in Slovakia from the territory of Tematínské vrchy hills). *Entomofauna Carpathica* 18 (1/2): 26–28. DP: 10 May 2006 (colophon).

Bergeal, M. 2001: Description de Neophaedon poneli, espèce nouvelle de Crète (Coleoptera, Chrysomelidae). *Nouvelle Revue d'Entomologie, Nouvelle Série* 18 (2): 153–156. DP: 15 December 2001 (last page of the cover).

Bergeal, M. & Doguet, S. 1989: Description d'une espèce nouvelle d'Aphthona d'Afrique du nord (Col. Chrysomelidae). *Nouvelle Revue d'Entomologie, Nouvelle Série* 6 (1): 50. DP: 22 June 1989 (last page of the cover).

Bergeal, M. & Doguet, S. 1991: Description d'une espèce nouvelle de Longitarsus du groupe de L. candidulus, changement de statut et désignation de lectotype pour L. candidulus thymelaearum Peyerimhoff (Col. Chrysomelidae). *Nouvelle Revue d'Entomologie, Nouvelle Série* 8 (2): 221–222. DP: 10 December 1991 (last page of the cover).

Bergeal, M. & Doguet, S. 2005: Contribution à l'étude des altises d'Espagne avec description d'une nouvelle espèce de Chaetocnema (Coleoptera, Chrysomelidae, Alticinae). *Nouvelle Revue d'Entomologie, Nouvelle Série* 22 (1): 29–32. DP: 30 June 2005 (last page of the cover).

Bergsträsser, J. A. B. 1778: *Nomenclatur und Beschreibung der Insecten in der Grafschaft Hanau-Münzenberg wie auch der Wetterau und der angränzenden Nachbarschaft dies und jenseits des Mains mit erleuchteten Kupfern. Erster Jahrgang.* [1777–1778], Hanau: [6] + 88 pp, 14 pls. DP (pp. 49–72 + pls. 13–14): 5 January 1778 (Bousquet 2016).

Berkenhout, J. 1769: *Outlines of the natural history of Great Britain and Ireland. Containing a systematic arrangement and concise description of all the animals, vegetables, and fossiles which have hitherto been discovered in these kingdoms. In three volumes. Vol. I. Comprehending the animal kingdom.* London: P. Elmsly, xiii + 233 pp. DP: [31] March 1769 (Bousquet 2016).

Berlioz, J. 1918: Coléoptères Eumolpides recueillis par M. R. Vitalis de Salvaza au Laos et dans le Haut-Tonkin. *Annales de la Société Entomologique de France* **86** [1917] (2/3): 307–316. DP: 13 February 1918 (*dates d'apparition* on p. 583).

Berlioz, J. 1921: Note sur une espèce nouvelle d'eumolpide (Col. Chrysomélides) de la Chine méridionale. *Bulletin du Muséum National d'Histoire Naturelle* (Paris) **27** (5): 332–333. DP: [31 December] 1921 (issue cover).

Berthold, A. A. 1827: *Latreille's Natürliche Familien des Thierreichs. Aus dem Französischen. Mit Anmerkungen und Zusätzen.* Weimar: Landes-Industrie-Comptoirs, x + 606 pp. DP: 22 September 1827 (Bousquet 2016).

Berti, N. & Daccordi, M. 1974a: Les Cyrtonastes (Col. Chrysomelidae Chrysomelinae) (1re partie). *Annales de la Société Entomologique de France France, Nouvelle Série* 10 (3): 593–610. DP: 17 December 1974 (stamp on the Gallica copy).

Berti, N. & Daccordi, M. 1974b: Les Cyrtonastes (Col. Chrysomelidae Chrysomelinae) (2e partie). *Annales de la Société Entomologique de France France, Nouvelle Série* 10 (4): 853–864. DP: [31 December] 1974 (issue title page).

Berti, N. & Doguet, S. 1994: Un Thelyterotarsus (Anodontelytrus) nouveau du Maroc (Coleoptera, Chrysomelidae). *Nouvelle Revue d'Entomologie, Nouvelle Série* **11** (2): 151–162. DP: 10 November 1994 (last page of the cover).

Berti, N. & Rapilly, M. 1973: Contribution à la faune de l'Iran. Voyages de MM. R. Naviaux et M. Rapilly (Col. Chrysomelidae). *Annales de la Société Entomologique de France, Nouvelle Série* **9** (4): 861–894. DP: [31 December] 1973 (issue title page).

Berti, N. & Rapilly, M. 1976: Faune d'Iran. Liste d'espèces et révision du genre Lilioceris Reitter (Col. Chrysomelidae). *Annales de la Société Entomologique de France, Nouvelle Série* **12** (1): 31–73. DP: 30 June 1976 (stamp on the Gallica copy).

Berti, N. & Rapilly, M. 1979: Contribution à la faune d'Iran. 4. – Cryptocephalinae (Coléoptères, Chrysomelidae). *Nouvelle Revue d'Entomologie* **9** (3): 239–269. DP: 15 November 1979 (last page of the cover).

Berti, N. & Rapilly, M. 1983: Une nouvelle espèce de Galerima pour la faune espagnole: G. villiersi (Coléoptères, Chrysomelidae, Galerucinae). *L'Entomologiste* **39** (6): 273–279. DP: [31] December 1983 (issue cover).

Bezděk, J. 1998: Monolepta anatolica sp. n. – a new species of leaf beetle from Turkey (Coleoptera: Chrysomelidae: Galerucinae). *Klapalekiana* **34** (3/4): 149–152. DP: 31 December 1998 (paper title page).

Bezděk, J. 2002: A review of the Gonioctena tredecimmaculata (Jacoby, 1888) group (Coleoptera, Chrysomelidae, Chrysomelinae). *Entomologica Basiliensia* **24**: 7–22. DP: 24 September 2002 (table of contents).

Bezděk, J. 2003a: Studies on Asiatic Apophylia. Part 1: new synonyms, lectotype designations, redescriptions, descriptions of new species and notes (Coleoptera, Chrysomelidae, Galerucinae). *Genus* **14** (1): 69–102. DP: 15 April 2003 (paper title page).

Bezděk, J. 2003b: Studies on Asiatic Apophylia (Insecta: Coleoptera: Chrysomelidae). Part 2: Revision of the aeruginosa and lebongana species groups. *Stuttgarter Beiträge zur Naturkunde, Serie A (Biologie)* **649**: 1–11. DP: 4 June 2003 (paper title page).

Bezděk, J. 2003c: Studies an Asiatic Apophylia. Part 3: Revisional study of type materials and descriptions of eight species new to science (Coleoptera: Chrysomelidae: Galerucinae). *Genus* **14** (2): 191–230. DP: 30 June 2003 (paper title page).

Bezděk, J. 2003d: Studies on Asiatic Apophylia. Part 5. Revisional study of type materials and descriptions of five species new to science (Chrysomelidae: Galerucinae). *Genus* **14** (4): 479–510. DP: 28 December 2003 (paper title page).

Bezděk, J. 2005a: New and interesting Apophylia species from South-East Asia (Coleoptera: Chrysomelidae: Galerucinae). *Raffles Bulletin of Zoology* **53** (1): 35–45. DP: 29 June 2005 (issue cover).

Bezděk, J. 2005b: Revisional study on African Apophylia (Coleoptera: Chrysomelidae: Galerucinae). Part 4. *Acta Entomologica Musei Nationalis Pragae* **45**: 165–182. DP: 31 December 2005 (paper title page).

Bezděk, J. 2006a: Revisional study on African Apophylia. Part 5 (Coleoptera: Chrysomelidae: Galerucinae). *Genus* **17** (2): 211–244. DP: 30 June 2006 (paper title page).

Bezděk, J. 2006b: Three new species and a new record of Apophylia Thomson, 1858 from eastern India (Coleoptera: Chrysomelidae: Galerucinae). *Annales Zoologici* (Warszawa) **56** (2): 259–264. DP: 30 June 2006 (verso of the issue cover).

Bezděk, J. 2008: New species and subspecies of Nymphius (Coleoptera: Chrysomelidae: Galerucinae) from Iran and Turkey. *Acta Entomologica Musei Nationalis Pragae* **48** (1): 79–93. DP: 15 August 2008 (paper title page).

Bezděk, J. 2009: Revisional study on the genus Mimastra (Coleoptera: Chrysomelidae: Galerucinae): species with unmodified protarsomeres in male. Part 1. *Acta Entomologica Musei Nationalis Pragae* **49** (2): 819–840. DP: 15 December 2009 (paper title page).

Bezděk, J. 2010a: Revisional study on the genus Mimastra (Coleoptera: Chrysomelidae: Galerucinae). Part 2. *Annales Zoologici* (Warszawa) **60** (1): 35–46. DP: 30 March 2010 (verso of the issue cover).

Bezděk, J. 2010b: Phyllobrotica malinka sp. nov. from Turkey and Iran and a review of allied species (Coleoptera: Chrysomelidae: Galerucinae). *Acta Entomologica Musei Nationalis Pragae* **50** (2): 563–575. DP: 15 December 2010 (paper title page).

Bezděk, J. 2011: [new taxon]. In: Bezděk, J. & Batelka, J. 2011: Order Coleoptera, family Chrysomelidae. Additions and description of a new species. Pp. 250–273. In: Harten, A. van (ed.): *Arthropod fauna of the UAE. Volume 4*. Abu Dhabi: Dar Al Ummah Printing, Publishing, Distribution & Advertising, 816 pp. DP: 31 May 2011 (paper title page).

Bezděk, J. 2012a: Revision of Hoplasoma (Coleoptera: Chrysomelidae: Galerucinae) of the Philippines, with descriptions of five new species. *Zootaxa* **3382**: 1–19. DP: 9 July 2012 (paper title page).

Bezděk, J. 2012b: Taxonomic and faunistic notes on Oriental and Palaearctic Galerucinae and Cryptocephalinae (Coleoptera: Chrysomelidae). *Genus* **23** (3): 375–418. DP: 15 October 2012 (paper title page).

Bezděk, J. 2012c: Tituboea purcharti sp. nov., the first representative of Clytrini from Socotra Island (Coleoptera: Chrysomelidae: Cryptocephalinae). Pp. 395–401. In: Hájek, J. & Bezděk, J. (eds.): Insect biodiversity of the Socotra Archipelago. *Acta Entomologica Musei Nationalis Pragae* **52** (supplementum 2): i–vi + 1–557. DP: 17 December 2012 (colophon and paper title page).

Bezděk, J. 2012d: Galerucinae (Coleoptera: Chrysomelidae) of Socotra Island, with a review of taxa recorded from Yemen. Pp. 403–428. In: Hájek, J. & Bezděk, J. (eds.): Insect biodiversity of the Socotra Archipelago. *Acta Entomologica Musei Nationalis Pragae* **52** (supplementum 2): i–vi + 1–557. DP: 17 December 2012 (colophon and paper title page).

Bezděk, J. 2013a: [new taxon]. In: Bezděk, J., Şen, I. & Gök, I. 2013: Calomicrus velai sp. nov. from Iran and redescription of C. koenigi (Jacobson, 1897) (Chrysomelidae, Galerucinae, Luperini). *Annales Zoologici* (Warszawa) **63** (2): 365–370. DP: 30 June 2013 (verso of the issue cover).

Bezděk, J. 2013b: Revision of the genus Hesperopenna (Coleoptera: Chrysomelidae: Galerucinae). I. Generic redescription, definition of species groups and taxonomy of H. medvedevi species group. *Acta Entomologica Musei Nationalis Pragae* **53** (2): 715–746. DP: 15 November 2013 (paper title page).

Bezděk, J. 2015: A review of Palaearctic Scelolyperus (Coleoptera: Chrysomelidae: Galerucinae), with description of S. perreus sp. nov. from Turkey. *Annales Zoologici* (Warszawa) **65** (1): 21–39. DP: 30 March 2015 (verso of the issue cover).

Bezděk, J. 2016: Revision of the Clytra subfasciata species group (Coleoptera: Chrysomelidae: Cryptocephalinae: Clytrini). *Zoology in the Middle East* **62** (2): 148–157. DP: 4 May 2016 (journal website).

Bezděk, J. 2017: Charaea Baly, 1878 (Coleoptera: Chrysomelidae: Galerucinae) of Vietnam, Laos, Thailand, Myanmar and Peninsular Malaysia. *Journal of Asia-Pacific Entomology* **20** (2): 477–489. DP: [30] June 2017 (journal website).

Bezděk, J. 2018a: Revision of Labidostomis (Coleoptera: Chrysomelidae: Cryptocephalinae: Clytrini) with large united elytral spot from Iran and adjacent countries. *Zootaxa* **4457** (4): 501–519. DP: 10 August 2018 (footer on the paper title page).

Bezděk, J. 2018b: Contribution to the knowledge of the Clytrini of the Eastern Mediterranean, the Near East and the Arabian Peninsula, with descriptions of four new species (Coleoptera: Chrysomelidae: Cryptocephalinae). *European Journal of Taxonomy* **481**: 1–37. DP: 4 December 2018 (paper last page).

Bezděk, J. 2020a: Revision of Agelopsis (Coleoptera: Chrysomelidae: Galerucinae). *Zootaxa* **4731** (2): 223–248. DP: 5 February 2020 (footer on the paper title page).

Bezděk, J. 2020b: Chikatunolepta n. gen. – a new genus of Saharo-Arabian Monoleptites (Coleoptera: Chrysomelidae: Galerucinae). *Israel Journal of Entomology* 50 (2): 11–29 DP: 14 July 2020 (paper title page).

Bezděk, J. 2021a: Updated checklist of Slovakian leaf-beetles (Coleoptera: Orsodacnidae, Megalopodidae, Chrysomelidae), with comments on the occurrence of some species. *Klapalekiana* 57 (1/2): 1–79. DP: 28 May 2021 (paper title page).

Bezděk, J. 2021b: Redefinition of Liroetis, with descriptions of two new species and an annotated list of species (Coleoptera: Chrysomelidae: Galerucinae). *Acta Entomologica Musei Nationalis Pragae* **61** (2): 529–614. DP: 31 December 2021 (paper title page).

Bezděk, J. & Baselga, A. 2015: Revision of western Palaearctic species of the Oulema melanopus group, with description of two new species from Europe (Coleoptera: Chrysomelidae: Criocerinae). *Acta Entomologica Musei Nationalis Pragae* **55** (1): 273–304. DP: 1 June 2015 (paper title page).

Bezděk, J. & Beenen, R. 2010: [new taxa]. In: Beenen, R.: New acts and comments. Chrysomelidae: Galerucinae. Pp. 74–75. In: Löbl, I. & Smetana, A. (eds.): *Catalogue of Palaearctic Coleoptera. Volume 6. Chrysomeloidea*. Stenstrup: Apollo Books, 924 pp. DP: 22 February 2010 (colophon).

Bezděk, J. & Beenen, R. 2020: Taxonomical changes and comments on Palaearctic and Oriental Chrysomelidae (Coleoptera). *Caucasian Entomological Bulletin* **16** (2): 319–327. DP: 17 December 2020 (last page of the paper).

Bezděk, J., Daccordi, M. & Kantner, F. 2012: Centroscelis kadleci sp. nov. from Yemen and a new synonymy in the genus Centroscelis (Coleoptera: Chrysomelidae: Chrysomelinae). *Acta Entomologica Musei Nationalis Pragae* **52** (1): 237–244. DP: 30 June 2012 (paper title page).

Bezděk, J., El-Torkey, A. M. & Al Dhafer, H. M. 2017: A new species of Cryptocephalus from Saudi Arabia (Coleoptera: Chrysomelidae: Cryptocephalinae). *Annales Zoologici* (Warszawa) **67** (1): 69–77. DP: 30 March 2017 (verso of the issue cover).

Bezděk, J. & Kantner, F. 2010: Revision of the genus Paraclytra (Coleoptera: Chrysomelidae: Cryptocephalinae: Clytrini), with description of P. cervenkai sp. nov. from Sultanate of Oman. *Zootaxa* **2353**: 1–33. DP: 8 February 2010 (footer on the paper title page).

Bezděk, J. & Lee, C.-F. 2009: Apophylia kaoi sp. nov. from Taiwan (Coleoptera: Chrysomelidae: Galerucinae). *Genus* 20 (3): 429–434. DP: 15 October 2009 (paper title page).

Bezděk, J. & Lee, C.-F. 2011: Revisional study on the genus Mimastra (Coleoptera: Chrysomelidae: Galerucinae). Part 4. *Annales Zoologici* (Warszawa) **61** (4): 709–729. DP: 30 December 2011 (verso of the issue cover).

Bezděk, J. & Lee, C.-F. 2014: Revision of Charaea (Coleoptera: Chrysomelidae: Galerucinae) from Taiwan. *Zootaxa* **3861** (1): 1–39. DP: 11 September 2014 (footer on the paper title page).

Bezděk, J. & Lee, C.-F. 2019: Revision of Pyrrhalta (Coleoptera: Chrysomelidae: Galerucinae) species with maculate elytra. *Zootaxa* **4664** (1): 518–534. DP: 5 September 2019 (footer on the paper title page).

Bezděk, J., Löbl, I. & Konstantinov, A. S. 2010: New acts and comments. Chrysomelidae: Lamprosomatinae. P. 80. In: Löbl, I. & Smetana, A. (eds.): *Catalogue of Palaearctic Coleoptera. Volume 6. Chrysomeloidea*. Stenstrup: Apollo Books, 924 pp. DP: 22 February 2010 (colophon).

Bezděk, J. & Mlejnek, R. 2016: Icones Insectorum Europae centralis, Megalopodidae, Orsodacnidae, Chrysomelidae: Donaciinae, Criocerinae. *Folia Heyrovskyana, Series B* 27: 1–63. DP: 30 September 2016 (paper title page).

Bezděk, J. & Regalin, R. 2015: Identity of species-group taxa of the Western Palaearctic Clytrini (Coleoptera: Chrysomelidae) described by Maurice Pic and Louis Kocher. *Acta Entomologica Musei Nationalis Pragae* 55 (supplementum): 1–114. DP: 15 November 2015 (colophon) [note: due to typographic error 15 October 2015 is given on the title page, while the correct date 15 November 2015 is mentioned on the wrappers and reflects when the paper was registered at ZooBank].

Bezděk, J. & Regalin, R. 2017: A review of Labidostomis species similar to L. longimana from south-eastern Europe with descriptions of two new species from Greece and Turkey (Coleoptera: Chrysomelidae: Cryptocephalinae: Clytrini). *Zootaxa* 4317 (2): 321–337. DP: 4 September 2017 (footer on the paper title page).

Bezděk, J. & Silfverberg, H. 2010: New acts and comments. Megalopodidae. P. 61. In: Löbl, I. & Smetana, A. (eds.): *Catalogue of Palaearctic Coleoptera. Volume 6. Chrysomeloidea*. Stenstrup: Apollo Books, 924 pp. DP: 22 February 2010 (colophon).

Bezděk, J. & Zhang, L.-J. 2006: Two new species of Apophylia from China (Coleoptera: Chrysomelidae: Galerucinae). *Acta Entomologica Musei Nationalis Pragae* 46: 145–150. DP: 6 November 2006 (paper title page).

Bezděk, J. & Zhang, L.-J. 2007: Taxonomical changes in the genera Hoplasoma and Haplosomoides (Coleoptera: Chrysomelidae: Galerucinae). *Acta Entomologica Musei Nationalis Pragae* 47: 189–193. DP: 4 December 2007 (paper title page).

Bhasin, G. D. 1943: A new species of Chrysomelidae from Burma. *Indian Journal of Entomology* 4 [1942] (1): 9–10. DP: 7 January 1943 (issue table of contents).

Bielz, E. A. 1852: Entomologische Beiträge. *Verhandlungen und Mittheilungen des Siebenbürgischen Vereins für Naturwissenschaften zu Hermannstadt* 3 (1): 13–16. DP: [31] January 1852 (issue title page).

Bieńkowski, A. O. 1997: Some surprising discoveries of Chrysolina relucens (Coleoptera, Chrysomelidae) on the White Sea shore, in Siberia and in the Far East. *Entomologica Fennica* 7: 195–199. DP: 14 February 1997 (paper title page).

Bieńkowski, A. O. 1998: Revision of the subgenus Anopachys Motschulsky, 1860 of the genus Chrysolina Motschulsky, 1860 (Coleoptera: Chrysomelidae: Chrysomelinae). *Genus* 9 (2): 95–153. DP: 30 June 1998 (paper title page).

Bieńkowski, A. O. 2001: A study on the genus Chrysolina Motschulsky, 1860, with a checklist of all the described subgenera, species, subspecies, and synonyms (Coleoptera: Chrysomelidae: Chrysomelinae). *Genus* 12 (2): 105–235. DP: 30 June 2001 (paper title page).

Bieńkowski, A. O. 2004: *Leaf-beetles (Coleoptera: Chrysomelidae) of the eastern Europe. New key to subfamilies, genera and species*. Moscow: Mikron-print, 278 pp. DP: [31 December] 2004 (book cover).

Bieńkowski, A. O. 2005: A review of the subgenus Sibiriella L. Medvedev, 1999 of the genus Chrysolina Motschulsky, 1860 (Coleoptera: Chrysomelidae: Chrysomelinae). *Russian Entomological Journal* **14** (1): 3–13. DP: 22 September 2005 (journal website).

Bieńkowski, A. O. 2007: *A monograph of the genus Chrysolina Motschulsky, 1860 (Coleoptera: Chrysomelidae) of the world. Part 1*. Moscow: Techpolygraphcentre, 417 pp. DP: 2 May 2007 (colophon).

Bieńkowski, A. O. 2008: Obzor gruppy vidov Chrysolina lineella podroda Anopachys Motschulsky, 1860 roda Chrysolina Motschulsky, 1860 (Coleoptera: Chrysomelidae: Chrysomelinae). (A review of the Chrysolina lineella species-group of the subgenus Anopachys Motschulsky, 1860, genus Chrysolina Motschulsky, 1860 (Coleoptera: Chrysomelidae: Chrysomelinae)). *Entomologicheskoe Obozrenie* 87 (2): 425–437. SFP: 26 May 2008, DP: 30 June 2008 (based on invoice deposited in Russian Entomological Society, A. G. Moseyko 2022, pers. comm.).

Bieńkowski, A. O. 2011: *Zhuki-listoedy (Coleoptera: Chrysomelidae) evropeyskoy chasti Rossii. [Leaf-beetles (Coleoptera: Chrysomelidae) of European part of Russia]*. Saarbrucken: Lambert Academic Publishing, 544 pp. DP: 22 July 2011 (publisher website).

Bieńkowski, A. O. 2012: Obzor zhukov-listoedov roda Chrysolina Motschulsky (Coleoptera, Chrysomelidae) Rossii i evropeyskikh stran blizhnego zarubezhia. III. Zamechania po sistematike i rasprostraneniyu vidov. (Review of the leaf-beetle genus Chrysolina Motschulsky (Coleoptera, Chrysomelidae) from Russia and European countries of the former USSR. III. Remarks on the systematics and distribution of the species). *Entomologicheskoe Obozrenie* **91** (3): 648–657. SFP: 3 September 2012, DP: 29 October 2012 (registered at ZIN library).

Bieńkowski, A. O. 2014: *Listoedy-raduzhnitsy (Coleoptera: Chrysomelidae: Donaciinae)*. Livny: G. V. Mukhametov, 380 pp. DP: [31 December] 2014 (colophon).

Bieńkowski, A. O. 2019: *Chrysolina of the world – 2019 (Coleoptera: Chrysomelidae). Taxonomic review*. Livny: G. V. Mukhametov, 920 pp. DP: 14 October 2019 (colophon).

Bieńkowski, A. O. 2022: A new unusual subgenus of the genus Chrysolina (Coleoptera: Chrysomelidae: Chrysomelinae) from the highland forests of China, Yunnan province. *Forests* **14** (66): 1–22. DP: 29 December 2022 (paper title page).

Bieńkowski, A. O. 2023a: A new species of the subgenus Timarchoptera Motschulsky, 1860 of the genus Chrysolina Motschulsky, 1860 (Coleoptera, Chrysomelidae). *Acta Zoologica Academiae Scientiarum Hungaricae* **69** (1): 25–37. DP: 28 February 2023 (last page of the paper).

Bieńkowski, A. O. 2023b: Two new species of the subgenus Pleurosticha of the genus Chrysolina (Coleoptera: Chrysomelidae) from the mountains of southern Siberia (Altai and Tyva). *Zoosystematica Rossica* **32** (1): 75–84. DP: 6 June 2023 (paper title page).

Bieńkowski, A. O. 2023c: Three amazing new species of the genus Chrysolina (Coleoptera: Chrysomelidae: Chrysomelinae) from China. *Acta Zoologica Academiae Scientiarum Hungaricae* **69** (3): 199–211. DP: 25 August 2023 (last page of the paper).

Bieńkowski, A. O. & Orlova-Bienkowskaya, M. Ya. 2011: Podvidy Chrysolina limbata (Coleoptera, Chrysomelidae). (Subspecies of Chrysolina limbata (Coleoptera, Chrysomelidae)). *Zoologicheskiy Zhurnal* 90 (8): 942–958. SFP: 5 July 2011, DP: 11 October 2011 (registered at ZIN library).

Bieńkowski, A. O. & Orlova-Bienkowskaya, M. Ya. 2023: On the quantitative criteria of subspecies in insects. Case study of Entomoscelis adonidis (Coleoptera: Chrysomelidae) in European Russia and the Caucasus. *Zootaxa* Zootaxa **5293** (2): 201–229. DP: 22 May 2023 (footer on the paper title page).

Billberg, G. J. 1808: [new taxon]. In: Schoenherr, C. J.: *Synonymia insectorum, oder: Versuch einer Synonymie aller bisher bekannten Insecten; nach Fabricii Systema Eleutheratorum &c. geordnet. Erster Band. Eleutherata oder Käfer. Zweiter Theil. Spercheus – Cryptocephalus*. Stockholm: C. F. Marquard, ix + 424 pp., 1 pl. DP: [31 December] 1808 (Bousquet 2016).

Billberg, G. J. 1820: *Enumeratio insectorum in museo Gust. Joh. Billberg*. Stockholm: Gadelianis, [2] + 138 pp. DP: [31 December] 1820 (Bousquet 2016).

Binaghi, G. 1938: Il sottogenere Protorina Weise ed alcune considerazioni sulle Chrysochloa dell'Appenino (Col. Chrys.). *Bollettino de la Società Entomologica Italiana* 70 (3): 34–41. DP: 21 March 1938 (Poggi 2008).

Binaghi, G. 1947: Una nuova Crepidodera delle Alpi Graie e Pennine (Col. Chrysom.). *Bollettino della Società Entomologica Italiana* 77 (5/6): 47–48. DP: 30 September 1947 (issue title page, Poggi 2008).

Binaghi, G. 1973: Le Chrysochloa del Monte Gottero nell'Appenino Ligure Orientale (Col. Chrysomelidae). *Doriana* 5 (208): 1–5. DP: 23 December 1973 (issue title page).

Biondi, M. 1982a: Contributo alla conoscenza delle Alticinae della fauna del Marocco, con descrizione di due nuove specie. 1ª nota. (Coleoptera, Chrysomelidae). *Fragmenta Entomologica* 16 (2): 229–237. DP: 15 October 1982 (table of contents in the cover).

Biondi, M. 1982b: Le specie appenniniche del genere Crepidodera Chevrolat sensu Foudras (Coleoptera, Chrysomelidae). *Bollettino del Museo Civico di Storia Naturale di Verona* 8 [1981]: 45–66. DP: 30 November 1982 (table of contents).

Biondi, M. 1984: Longitarsus idilphilus n. sp. del Marocco e considerazione sulle specie del gruppo suturalis (Coleoptera, Chrysomelidae, Alticinae). *Fragmenta Entomologica* 17 (2): 379–394. DP: 15 October 1984 (table of contents in the cover).

Biondi, M. 1986a: Lo status sistematico del complesso Minota obesa: un approccio morfologico, statistico e biogeografico (Coleoptera, Chrysomelidae, Alticinae). *Annali del Museo Civico di Storia Naturale di Genova* 86 [1986–1987]: 45–64. DP: 25 June 1986 (Poggi 2010).

Biondi, M. 1986b: Contributo alla conoscenza della fauna delle Isole Canarie. Revisione delle specie del gruppo del Longitarsus isoplexidis con descrizione di due nuovi taxa (Coleoptera, Chrysomelidae, Alticinae). *Vieraea* 16 (1/2): 351–360. DP: [30] June 1986 (paper title page).

Biondi, M. 1987: Contributo alla conoscenza dei Chrysomelidae Alticinae delle isole Canarie, con descrizione di una nuova specie di Psyllliodes (Coleoptera). *Vieraea* 17 (1/2): 93–97. DP: [30] June 1987 (paper title page).

Biondi, M. 1989: Contributo alla conoscenza dei crisomelidi alticini dei Monti Nebrodi e Peloritani (Sicilia Nord Orientale) e descrizione di una nuova specie di Longitarsus (Coleoptera, Chrysomelidae, Alticinae). *Animalia* **15** [1988] (1/3): 129–149. DP: [30] September 1989 (last page of the issue).

Biondi, M. 1991: Contributo alla conoscenza dei Crisomelidi Alticini della Macaronesia con descrizione di una nuova specie delle Canarie (Coleoptera, Chrysomelidae, Alticinae). *Vieraea* **20**: 33–38. DP: [31] December 1991 (paper title page).

Biondi, M. 1992: Note sui Crisomelidi Alticini della fauna di Turchia, con descrizione di tre nuove specie (Coleoptera, Chrysomelidae, Alticinae). *Fragmenta Entomologica* **23** [1991] (2): 341–354. DP: 30 April 1992 (table of contents in the cover).

Biondi, M. 1994: Phyllotreta gloriae a new species from southern Spain (Coleoptera, Chrysomelidae, Alticinae). *Eos, Revista Española de Entomología* **69** [1993]: 111–114. DP: 25 May 1994 (volume cover).

Biondi, M. 1995: Descrizione di una nuova specie di Cryptocephalus del sottogenere Burlinius Lopatin dell'Italia meridionale (Coleoptera Chrysomelidae). *Bollettino della Società Entomologica Italiana* **127** (1): 39–44. DP: 30 June 1995 (issue cover, Poggi 2008).

Biondi, M. 1996: The Longitarsus anchusae complex in Near Eastern and description of a new species (Coleoptera, Alticinae). *Nouvelle Revue d'Entomologie, Nouvelle Série* **12** [1995] (4): 259–271. DP: 12 March 1996 (last page of the cover).

Biondi, M. 1997: Longitarsus aubozaorum and Psylliodes littoralis, two new flea beetle species from Turkey (Coleoptera, Chrysomelidae). *Fragmenta Entomologica* **29** (2): 383–390. DP: 30 December 1997 (table of contents in the cover).

Biondi, M. & D'Alessandro, P. 2010: Genus-group names of Afrotropical flea beetles (Coleoptera: Chrysomelidae: Alticinae): annotated catalogue and biogeographical notes. *European Journal of Entomology* **107** (3): 401–424. DP: 20 July 2010 (verso of the issue cover).

Biondi, M. & D'Alessandro, P. 2017a: [new taxon]. In: Biondi, M., Frasca, R., Grobbelaar, E. & D'Alessandro, P.: Supraspecific taxonomy of the flea beetle genus Blepharida Chevrolat, 1836 (Coleoptera: Chrysomelidae) in the Afrotropical Region and description of Afroblepharida subgen. nov. *Insect Systematics & Evolution* **48** (2): 97–155. DP: 14 April 2017 (paper title page).

Biondi, M. & D'Alessandro, P. 2017b: Psylliodes urbaniae: a new species from Central Apennines (Coleoptera: Chrysomelidae: Galerucinae: Alticini). *Onychium* **13**: 121–130. DP: 20 April 2017 (paper title page).

Biondi, M. & D'Alessandro, P. 2017c: Longitarsus doeberli, a wingless new species from Socotra Island (Coleoptera: Chrysomelidae). Pp. 165–172. In: Hájek, J. & Bezděk, J. (eds.): Insect biodiversity of the Socotra Archipelago III. *Acta Entomologica Musei Nationalis Pragae* **57** (supplementum): i–vi + 1–225. DP: 31 December 2017 (paper title page).

Biondi, M., Daccordi, M., Regalin, R. & Zampetti, M. 1995: Coleoptera Polyphaga XV (Chrysomelidae, Bruchidae). In: Minelli, A., Ruffo, S. & La Posta, S. (eds.): *Checklist delle specie della fauna italiana. Volume 60.* [1994], Bologna: Calderini, 36 pp. DP: [31] May 1995 (colophon).

Biondi, M., Ianella, M. & D'Alessandro, P. 2019: Unravelling the taxonomic assessment of an interesting new species from Socotra Island: Blepharidina socotrana sp. nov. (Coleoptera: Chrysomelidae). *Acta Entomologica Musei Nationalis Pragae* **59** (2): 499–505. DP: 31 October 2019 (paper title page).

Blackburn, T. 1889: Further notes on Australian Coleoptera, with descriptions of new genera and species. *Proceedings of the Linnean Society of New South Wales, Second Series* **3** [1888]: 1387–1506. DP: 22 March 1889 (table of contents).

Blackburn, T. 1890a: Notes on Australian Coleoptera, with descriptions of new species. *Proceedings of the Linnean Society of New South Wales, Second Series* **5** (2): 303–366. DP: 29 September 1890 (table of contents).

Blackburn, T. 1890b: Notes on Australian Coleoptera, with descriptions of new genera and species. *Transactions of the Royal Society of South Australia* **13** [1889–1890] (2): 121–160. DP: [31] December 1890 (issue cover and table of contents).

Blackburn, T. 1896: Further notes on Australian Coleoptera, with descriptions of new genera and species. XIX. *Transactions of the Royal Society of South Australia* **20** [1895–1896] (1): 35–109. DP: [30] September 1896 (issue cover).

Blackburn, T. 1897: Revision of the genus Paropsis. Part I. *Proceedings of the Linnean Society of New South Wales* **21** [1896] (4): 637–693. DP: 31 May 1897 (table of contents).

Blackburn, T. 1899: Further notes on Australian Coleoptera, with descriptions of new genera and species. XXV. *Transactions of the Royal Society of South Australia* **23** [1898–1899] (1): 22–101. DP: [31] August 1899 (issue cover and table of contents).

Blake, D. H. 1939: A study of LeConte's types of the beetles in the genus Monoxia, with descriptions of new species. *Proceedings of the United States National Museum* **87** (3072): 145–171, pls. 18–19. DP: 5 October 1939 (table of contents).

Blanc, É. 1889: [Description d'une nouvelle espèce de bruchide de Tunisie]. *Bulletin de la Société Entomologique de France* **1889** (4): xlii–xliv. DP: 9 March 1889 (Evenhuis 2002).

Blanchard, C. E. 1845a: [Recherches entomologiques faites en Sicile et en Calabre]. *Bulletin de la Société Entomologique de France* **[1844]** (4): lxxix–lxxxvi. DP: 23 April 1845 (Evenhuis 2022).

Blanchard, C. E. 1845b: *Histoire des insectes, traitant de leurs moeurs et de leurs métamorphoses en general et comprenant une nouvelle classification fondée sur leurs rapports naturels. Coléoptères, Orthoptères, Thysanoptères, Nevroptères, Lépidoptères, Hémiptères, Aphaniptères, Strepsiptères, Diptères, Anoplures et Thysanures. Tome II*. Paris: Firmin Didot frères, 524 pp., pls. 1–20. DP: 11 June 1845 (Bousquet 2016).

Blanchard, C. E. 1845c: [Communication sur quelques Eumolpites recueillis en Sicile]. *Bulletin de la Société Entomologique de France* **1845** (1): iv. DP: [31] August 1845 (Evenhuis 2022).

Blanchard, C. E. 1846: [pls. 18–19]. In: *Voyage au Pole Sud et dans l'Océanie sur les corvettes l'Astrolabe et la Zélée; exécuté par ordre du roi pendant les années 1837–1838–1839–1840, sous le commandement de M. J. Dumont-d'Urville, Capitaine de vaisseau; publié par ordre du gouvernement, sous la direction supérieure de M. Jacquinot, Capitaine de vaisseau, commandant de la

Zélée. Atlas zoologique. [1842–1854], Paris: Gide et Cie, 140 plates. DP (pls. 18–19): 6 July 1846 (Bousquet 2016).

Blanchard, C. E. 1851: Insectos. Coleopteros. Pp. 285–564. In: Gay, C.: *Historia fisica y politica de Chile segun documentos adquiridos en esta republica durante doce años de residencia en ella. Zoologia. Tomo quinto.* Paris: C. Gay & Santiago: Museo de Historia Natural de Santiago, 564 + 3 pp. DP: [31 December] 1851 (Bousquet 2016).

Blanchard, C. E. 1853: Description des insectes. In: Hombron, J. B. & Jacquinot, H.: *Zoologie, Tome quatrième.* In: *Voyage au Pole Sud et dans l'Océanie sur les corvettes l'Astrolabe et la Zélée; exécuté par ordre du Roi pendant les années 1837–1838–1839–1840, sous le commandement de M. J. Dumont-d'Urville, Capitaine de vaisseau; publié par ordre du gouvernement, sous la direction supérieure de M. Jacquinot, Capitaine de vaisseau, commandant de la Zélée.* Paris: Gide et J. Baudry, [5] + 422 pp. DP: [31 December] 1853 (Bousquet 2016).

Blatchley, W. S. 1918: On some new or noteworthy Coleoptera from the West coast of Florida. – IV. *Canadian Entomologist* 50 (2): 52–59. DP: 8 February 1918 (last page of the issue).

Blatchley, W. S. 1921: Notes on Indiana Halticini with characterization of a new genus and descriptions of new species. *Journal of the New York Entomological Society* **29** (1): 16–27. DP: [31] March 1921 (issue cover).

Böber, R. 1795: Ueber einige entomologische Merkwürdigkeiten von Taurien. *Magazin des Thierreichs* **1** (3): 135–140. DP: 26 April 1795 (*Allg. Verz. Bücher, Ostermesse* 1795: 81 and Evenhuis 2014).

Bodemeyer, B. von 1916: Neuheiten der paläarktischen Coleopterenfauna. *Deutsche Entomologische Zeitschrift* **1916** (2): 111–112. DP: 1 July 1916 (table of contents).

Bodor, J. 1965: Négy uj Crioceris Fourc. változat (Coleoptera, Chrysomelidae). Vier neue Aberrationen der Chrysomeliden-Gattung Crioceris Fourcr. (Coleoptera). *Folia Entomologica Hungarica* **18** (1): 300–303. DP: 30 July 1965 (colophon).

Bogachev, A. V. 1947: Quelques espèces nouvelles des coleoptères de l'Iran. *Entomologie et Phytopathologie Appliquées* (Teheran) **4**: 15–18. DP: [30] June 1947 (volume cover).

Bogachev, A. V. 1948: Novyy vid listoedov roda Luperus Geoffr. (Chrysomelidae, Galerucinae). [A new species of leaf-beetles of the genus Luperus Geoffr. (Chrysomelidae, Galerucinae)]. *Doklady Akademii Nauk Armyanskoy SSR* **9** (3): 131–133. DP: [31 December] 1948 (reprint cover).

Bogachev, A. V. 1961: Novye vidy zhukov-chernotelok (Tenebrionidae) iz Tadzhikskoy SSR, drugikh respublik Sredney Azii i sopredelnykh stran. [New species of darkling-beetles (Tenebrionidae) from Tajik SSR, other Central Asian republics and surrounding countries]. *Trudy Instituta Zoologii i Parazitologii Akademii Nauk Tadzhikskoy SSR* **20**: 107–121. SFP: 27 March 1961, DP: 29 June 1961 (registered at VKP).

Bogachev, A. V. & Antsupova, T. E. 1980: Listoedy-pyavitsy, lukovye i sparzhevye treshchalki Krasnodarskogo kraya i ikh kormovye svyazi. [Criocerine leaf beetles of the genera Lema, Oulema, Lilioceris and Crioceris of Krasnodar Territory and their trophic relations]. *Trudy*

Kubanskogo Selskokhozyastvennogo Instituta **194**: 55–63. SFP: 27 December 1980, DP: [31 December] 1980 (issue cover).

Boheman, C. H. 1828: Novae Coleopterorum species. *Nouveaux Mémoires de la Société Impériale des Naturalistes de Moscou* **1** [1829]: 101–133. DP: [31] October 1828 (Bousquet & Davies 2021).

Boheman, C. H. 1833: [new taxa]. In: Schoenherr, C. J.: *Genera et species curculionidum, cum synonymia hujus familiae. Tomus primus. Pars prima.* Paris: A. Roret, xv + 381 pp. DP: [30] June 1833 (Bousquet 2016).

Boheman, C. H. 1839: [new taxa]. In: Schoenherr, C. J.: *Genera et species curculionidum, cum synonymia hujus familiae; species novae aut hactenus minus cognitae, descriptionibus Dom. L. Gyllenhal, C. H. Boheman, O. J. Fahraeus et entomologis aliis illustratae. Tomus quintus. Pars prima. Supplementum continens.* Paris: A. Roret, Lipsiae: A. F. Fleischer, viii + 456 pp. DP: 15 June 1839 (Bousquet 2016).

Boheman, C. H. 1847a: [new taxa]. In: Lucas, P. H.: Pls. 44, 45. In: *Exploration scientifique de l'Algérie pendant les années 1840, 1841, 1842 publiée par ordre du Gouvernement et avec le concours d'une Commission Académique. Sciences Physiques. Zoologie. II. Histoire naturelle des animaux articulés. Deuxième partie. Insectes.* [1846–1849], Paris: A. Bertrand, 590 pp. + 47 pls. DP (pls. 44, 45): 15 May 1847 (Evenhuis 2012, Bousquet 2016).

Boheman, C. H. 1847b: [new taxa]. In: Lucas, P. H.: *Exploration scientifique de l'Algérie pendant les années 1840, 1841, 1842 publiée par ordre du Gouvernement et avec le concours d'une Commission Académique. Sciences Physiques. Zoologie. II. Histoire naturelle des animaux articulés. Deuxième partie. Insectes.* [1846–1849], Paris: A. Bertrand, 590 pp. + 47 pls. DP (pp. 361–536): 20 November 1847 (Evenhuis 2012, Bousquet 2016).

Boheman, C. H. 1850: *Monographia Cassididarum. Tomus primus.* Holmiae: Officina Norstedtiana, [6] + 452 pp., 4 pls. DP: 11 September 1850 (Bousquet 2016).

Boheman, C. H. 1851: Bidrag till Gottlands Insekt-Fauna. *Kongliga Vetenskaps-Akademiens Handlingar* [**1849**]: 193–267. DP: 12 March 1851 (*Öfv. Kongl. Vetens.-Akad. Forhand.* 1851: 120).

Boheman, C. H. 1853: Entomologiska anteckningar under en resa i Södra Sverige 1851. *Kongliga Vetenskaps-Academiens Handlingar* [**1851**]: 53–210. DP: 13 April 1853 (*Öfv. Kongl. Vetens.-Akad. Forhand.* 1853: 76).

Boheman, C. H. 1854: *Monographia Cassididarum. Tomus secundus.* Holmiae: Officina Norstedtiana, 506 pp., pls. 5, 6. DP: 10 May 1854 (Bousquet 2016).

Boheman, C. H. 1855: *Monographia Cassididarum. Tomus tertius.* Holmiae: Officina Norstedtiana, 543 pp., pl. 7. DP: 10 October 1855 (Bousquet 2016).

Boheman, C. H. 1857: *Catalogue of Coleopterous Insects in the collection of the British Museum. Part IX. Cassididae.* [1856], London: Trustees of the British Museum, 225 pp. DP: 14 February 1857 (Sherborn 1934).

Boheman, C. H. 1858: Coleoptera. Species novas descripsit. Pp. 1–217, pls. I–II. In: Virgin, C. (ed.): *Kongliga Svenska Fregatten Eugenies Resa omkring jorden under befäl af C. A. Virgin, ären 1851–1853. Vetenskapliga Iakttagelser på H. M. Konung Oscar den Förstes befallning utgifna af*

K. Svenska Vetenskaps Akademien. Andra delen. Zoologi. 1. Insecta. Stockholm: P. A. Norstedt & Söner, 617 pp., 9 pls. DP (pp. 1–112): [31 December] 1858 (Bousquet 2016).

Boheman, C. H. 1859: Coleoptera. Species novas descripsit. Pp. 1–217, pls. I–II. In: Virgin, C. (ed.): *Kongliga Svenska Fregatten Eugenies Resa omkring jorden under befäl af C. A. Virgin, ären 1851–1853. Vetenskapliga Iakttagelser på H. M. Konung Oscar den Förstes befallning utgifna af K. Svenska Vetenskaps Akademien. Andra delen. Zoologi. 1. Insecta.* Stockholm: P. A. Norstedt & Söner, 617 pp., 9 pls. DP (pp. 113–217): [31] August 1859 (Bousquet 2016).

Boheman, C. H. 1862: *Monographia Cassididarum. Tomus quartus (Supplementum)*. Holmiae: Officina Norstedtiana, [4] + 504 pp. DP: 14 May 1862 (Bousquet 2016).

Boieldieu, A. A. 1852: [new taxon]. *Bulletin de la Société Entomologique de France* **1852** (3): lii, pl. 11. DP: 13 December 1852 (Evenhuis 2022).

Boieldieu, A. A. 1859: Descriptions d'espèces nouvelles de coléoptères. *Annales de la Société Entomologique de France, Troisième Série* 7 (3): 461–482, pl. 8. DP: 9 November 1859 (Evenhuis 2022).

Boisduval, J. B. A. 1835: *Voyage de découvertes de l'Astrolabe exécuté par ordre du Roi, pendant les années 1826–1827–1828–1829, sous le commandement de M. J. Dumont d'Urville. Faune entomologique de l'Océan Pacifique, avec l'illustration des insectes nouveaux recueillis pendant le voyage. Deuxième partie. Coléoptères et autres ordres.* Paris: J. Tastu, viii + 716 pp. DP: 27 March 1835 (Bousquet 2016).

Bolz, H. & Wagner, T. 2012: Neobarombiella, a diverse, newly described genus of Afrotropical Galerucinae (Coleoptera, Chrysomelidae). *Zootaxa* **3463**: 1–112. DP: 7 September 2012 (paper title page).

Bolz, H. & Wagner, T. 2014: A new Afrotropical Neobarombiella species from Socotra Island (Coleoptera: Chrysomelidae: Galerucinae). Pp. 277–281. In: Hájek, J. & Bezděk, J. (eds.): Insect biodiversity of the Socotra Archipelago II. *Acta Entomologica Musei Nationalis Pragae* 54 (supplementum): i–vi + 1–440. DP: 30 December 2014 (colophon and paper title page).

Bonelli, F. A. 1812: Specimen faunae subalpinae sistens Insecta pedemontii huc usque inedita, aut rariora, aut ea quae commodi damnive gratia quod inferunt, prudentis agricolae magis interest cognoscere. Fasciculus I. Coleoptera plerumque inedita comprehendens. *Memorie della Società di Agricoltura di Torino* 9: 149–183, pls. I–VI. DP: [31 December] 1812 (paper title page).

Bontems, C. 1978: Nomenclature du genre Oreina Chevrolat (Col., Chrysomelidae, Chrysomelinae). *Nouvelle Revue d'Entomologie* 8 (1): 69–80. DP: 31 March 1978 (issue cover).

Bontems, C. 1981: Les espèces de Linné et Fabricius du genre Oreina Chevrolat, 1837 (Col., Chrysomelidae, Chrysomelinae). *Nouvelle Revue d'Entomologie* **11** (1): 93–109. DP: 15 April 1981 (issue cover).

Bontems, C. 1984: Les Allorina de France et des régions limitrophes (Coleoptera, Chrysomelidae). *Nouvelle Revue d'Entomologie, Nouvelle Série* 1 (2): 179–201. DP: 30 September 1984 (issue cover).

Bontems, C. 2001: Les Oreina de Sibérie (Coleóptera, Chrysomelidae, Chrysomelinae). *Bulletin de la Société Entomologique de France* 106 (1): 65–77. DP: 22 February 2001 (*Bull. Soc. Ent. Fr.* 2001, 106 (5): 531).

Bontems, C. 2005: Les Oreina cacaliae de France (Coleoptera, Chrysomelidae). *Bulletin de la Société Entomologique de France* 110 (2): 141–151. DP: 31 May 2005 (*Bull. Soc. Ent. Fr.* 2005, 110 (4/5): 526).

Bontems, C. 2009: Les Oreina de Suffrian (I) (Coleoptera, Chrysomelidae). *Nouvelle Revue d'Entomologie, Nouvelle Série* 25 [2008] (4): 329–351. DP: 26 November 2009 (last page of the cover).

Bontems, C. 2010: Les Oreina de Suffrian (II) (Coleoptera, Chrysomelidae). *Nouvelle Revue d'Entomologie, Nouvelle Série* 26 [2009] (2): 133–154. DP: 7 June 2010 (last page of the cover).

Bontems, C. 2021: Sur la validité d'Oreina (Chrysochloa) cacaliae tristis (Coleoptera, Chrysomelidae). *Bulletin de la Société Entomologique de France* 126 (3): 329–335. DP: 17 September 2021 (paper title page).

Bordy, B. 1995: Cassida leucanthemi n. sp., espèce jumelle de C. sanguinosa Suffrian, 1844, (Coleoptera, Chrysomelidae). *Bulletin de la Société Entomologique de France* 100 (4): 377–383. DP: 19 December 1995 (*Bull. Soc. Ent. Fr.* 1995, 100 (5): 548).

Bordy, B. 1996: Cassida bergeali, espèce nouvelle du groupe de Cassida vibex L. 1767 (Col. Chrysomelidae). *Nouvelle Revue d'Entomologie, Nouvelle Série* 12 [1995] (4): 243–248. DP: 12 March 1996 (last page of the cover).

Bordy, B. 1997: Etude de l'apodème du sac éjaculateur des Cassidinae de France (Coleoptera, Chrysomelidae). *Nouvelle Revue d'Entomologie, Nouvelle Série* 14 (1): 53–67. DP: 30 August 1997 (last page of the cover).

Bordy, B. 2000: *Coléoptères Chrysomelidae, Volume 3. Hispinae et Cassidinae. Faune de France 85.* Paris: Fédération française des sociétés de sciences naturelles, xii + 250 pp. + 26 pl. DP: [30] April 2000 (DL at the end of book).

Bordy, B. 2008: Identification des espèces du complexe Hypocassida subferuginea (Schrank, 1776) (Coleoptera Chrysomelidae). *Nouvelle Revue d'Entomologie, Nouvelle Série* 25 (1): 45–59. DP: 31 December 2008 (last page of the cover).

Bordy, B. & Doguet, S. 1987: Contribution à la connaisance des Cassidinae de France. Étude de leur spermathèque (Coleoptera, Chrysomelidae). *Nouvelle Revue d'Entomologie, Nouvelle Série* 4 (2): 161–176. DP: 30 June 1987 (last page of the cover).

Borowiec, L. 1980a: Eine neue paläarktische Gattung der Samenkäfer (Coleoptera, Bruchidae). *Polskie Pismo Entomologiczne* 50 (1): 127–131. DP: 30 March 1980 (paper title page).

Borowiec, L. 1980b: A new species of Acanthoscelides Schilsky from Bulgaria (Coleoptera, Bruchidae). *Polskie Pismo Entomologiczne* 50 (2): 167–170. DP: 30 May 1980 (paper title page).

Borowiec, L. 1984a: A new species of Caryedon Schoenherr (Coleoptera, Bruchidae, Pachymerinae) from Hongkong. *Polskie Pismo Entomologiczne* 53 [1983] (4): 519–522. DP: 29 February 1984 (paper title page).

Borowiec, L. 1984b: Two new genera and species of seed-beetles from the Oriental region (Coleoptera, Bruchidae, Bruchinae). *Polskie Pismo Entomologiczne* **54** (1): 115–129. DP: 30 April 1984 (paper title page).

Borowiec, L. 1985a: Notes on the Palaearctic Spermophagus Schoenherr (Coleoptera, Bruchidae, Amblycerinae), with description of two new species. *Polskie Pismo Entomologiczne* **55** (1): 3–24. DP: 30 March 1985 (paper title page).

Borowiec, L. 1985b: Aspidomorpha gruevi n. sp. (Coleoptera, Chrysomelidae, Cassidinae) from Yemen. *Polskie Pismo Entomologiczne* **55** (3): 451–456. DP: 30 October 1985 (paper title page).

Borowiec, L. 1985c: A new genus of Palaearctic seed-beetles (Coleoptera, Bruchidae, Bruchinae). *Polskie Pismo Entomologiczne* **55** (3): 457–462. DP: 30 October 1985 (paper title page).

Borowiec, L. 1985d: Two new species of Spermophagus Schoenherr (Coleoptera, Bruchidae, Amblycerinae) from Israel and Iran. *Polskie Pismo Entomologiczne* **55** (3): 463–468. DP: 30 October 1985 (paper title page).

Borowiec, L. 1985e: The seminarius-group of Bruchidius Schilsky (Coleoptera, Bruchidae), with description of three new species. *Polskie Pismo Entomologiczne* **55** (4): 767–779. DP: 30 December 1985 (paper title page).

Borowiec, L. 1985f: On the Oriental Spermophagus Schoenherr (Coleoptera, Bruchidae, Amblycerinae), with description of four new species. *Polskie Pismo Entomologiczne* **55** (4): 781–790. DP: 30 December 1985 (paper title page).

Borowiec, L. 1986a: The specific status of Spermophagus variolosopunctatus Gyllenhal, 1833 (Coleoptera, Bruchidae, Amblycerinae), with description of a new species. *Polskie Pismo Entomologiczne* **56** (1): 161–164. DP: 30 April 1986 (paper title page).

Borowiec, L. 1986b: A new species and a new subspecies of tortoise-beetles from the Middle East (Coleoptera, Chrysomelidae, Cassidinae). *Polskie Pismo Entomologiczne* **56** (3): 573–576. DP: 30 September 1986 (paper title page).

Borowiec, L. 1986c: The seed-beetles from Yugoslavia, Albania and Greece (Coleoptera, Bruchidae). *Polskie Pismo Entomologiczne* **56** (3): 577–592. DP: 30 September 1986 (paper title page).

Borowiec, L. 1987: The genera of seed-beetles (Coleoptera, Bruchidae). *Polskie Pismo Entomologiczne* **57** (1): 3–207. DP: 30 March 1987 (paper title page).

Borowiec, L. 1990a: New records and new synonyms of Asiatic Cassidinae (Coleoptera, Chrysomelidae). *Polskie Pismo Entomologiczne* **59** [1989] (4): 677–711. DP: 30 March 1990 (paper title page).

Borowiec, L. 1990b: New species, new synonyms and new records in the genus Caryedon Schönh. (Coleoptera, Bruchidae, Pachymerinae). *Annales Zoologici* (Warszawa) **43** (19): 373–393. DP: 31 December 1990 (paper title page).

Borowiec, L. 1991: Revision of the genus Spermophagus Schoenherr (Coleoptera: Bruchidae: Amblycerinae). *Genus* **Supplement** [1]: 3–198. DP: [31] December 1991 (page of the respective publication at www.researchgate.net; accessed on: 6 April 2023).

Borowiec, L. 1993: Three new species of Basiprionotini from Oriental region (Coleoptera: Chrysomelidae: Cassidinae). *Genus* **4** (2): 79–85. DP: 30 September 1993 (paper title page).

Borowiec, L. 1994a: A monograph of the Afrotropical Cassidinae (Coleoptera: Chrysomelidae). Part I. Introduction, morphology, key to the genera, and reviews of the tribes Epistictinini, Basiprionotini and Aspidimorphini (except the genus Aspidimorpha). *Genus* **Supplement 2:** 3–276. DP: [31] January 1994 (page of the respective publication at www.researchgate.net; accessed on: 6 April 2023).

Borowiec, L. 1994b: New synonyms in the Cassidinae (Coleoptera: Chrysomelidae). *Genus* **5** (3): 153–159. DP: 30 October 1994 (paper title page).

Borowiec, L. 1995: Spermophagus atrispinus n. sp. from India and Thailand (Coleoptera: Bruchidae: Amblycerinae). *Genus* **6** (2): 99–102. DP: 30 June 1995 (paper title page).

Borowiec, L. 1999: *A world catalogue of the Cassidinae (Coleoptera: Chrysomelidae)*. Wroclaw: Biologica Silesiae, 476 pp. DP: 31 July 1999 (verso of book title page).

Borowiec, L. 2000: Hypocassida convexipennis, a new species from Iran (Coleoptera: Chrysomelidae: Cassidinae). *Genus* **11** (4): 601–605. DP: 26 December 2000 (paper title page).

Borowiec, L. 2001a: Oxylepus boroveci, a new species from Tunisia (Coleoptera: Chrysomelidae: Cassidinae). *Genus* **12** (3): 349–352. DP: 30 September 2001 (paper title page).

Borowiec, L. 2001b: New records of Asian and Australopapuan Cassidinae, with a description of five new species of Cassida L. from Thailand (Coleoptera: Chrysomelidae: Cassidinae). *Genus* **12** (4): 493–562. DP: 28 December 2001 (paper title page).

Borowiec, L. 2005a: Macrocoma doboszi, a new species from Turkey (Coleoptera: Chrysomelidae: Eumolpinae). *Genus* **16** (3): 373–377. DP: 15 September 2005 (paper title page).

Borowiec, L. 2005b: A new species of Scelolyperus Crotch, 1874 from Kirgizstan (Coleoptera: Chrysomelidae: Galerucinae). *Genus* **16** (3): 379–382. DP: 15 September 2005 (paper title page).

Borowiec, L. 2007: Two new species of Cassida Linnaeus, 1758 (Coleoptera: Chrysomelidae: Cassidinae) from Madagascar and notes on subgenera of the genus Cassida. *Zootaxa* **1586:** 47–58. DP: 17 September 2007 (paper title page).

Borowiec, L. 2009: New records of Asian and Australopapuan tortoise beetles (Coleoptera: Chrysomelidae: Cassidinae). *Genus* **20** (3): 435–484. DP: 15 October 2009 (paper title page).

Borowiec, L. 2012: On the synonymy of two neglected species of Cassida L. described by Ritter Böber, 1793 (Coleoptera: Chrysomelidae: Cassidinae). *Genus* **23** (1): 115–117. DP: 30 April 2012 (paper title page).

Borowiec, L. & Cho, H.-W. 2011: On the subgenus Lasiocassis Gressitt (Coleoptera: Chrysomelidae: Cassidinae), with description of a new species from South Korea. *Annales Zoologici* (Warszawa) **61** (3): 445–451. DP: 30 September 2011 (verso of the issue cover).

Borowiec, L. & Ghahari, H. 2021: An annotated checklist of the Iranian Cassidinae (Coleoptera Chrysomelidae). *Redia* **104:** 9–19. DP: [31 December] 2021 (paper title page).

Borowiec, L. & Lee, C.-F. 2009: A new species of Thlaspida Weise from Taiwan, and notes on distribution and host plants of Cassida insulana Gressitt (Coleoptera: Chrysomelidae: Cassidinae). *Genus* **20** (2): 349–353. DP: 15 July 2009 (paper title page).

Borowiec, L. & Sassi, D. 2002: A new species of Cassida L. from Palaearctic China (Coleoptera: Chrysomelidae: Cassidinae). *Genus* **13** (1): 143–147. DP: 10 April 2002 (paper title page).

Borowiec, L., Ścibior, R. & Kubisz, D. 2011: Critical check-list of the Polish Chrysomeloidea, excluding Cerambycidae (Coleoptera: Phytophaga). *Genus* **22** (4): 579–608. DP: 27 December 2011 (paper title page).

Borowiec, L. & Sekerka, L. 2010a: New acts and comments. Chrysomelidae: Cassidinae. Pp. 64–65. In: Löbl, I. & Smetana, A. (eds.): *Catalogue of Palaearctic Coleoptera. Volume 6. Chrysomeloidea.* Stenstrup: Apollo Books, 924 pp. DP: 22 February 2010 (colophon).

Borowiec, L. & Sekerka, L. 2010b: Chrysomelidae: Cassidinae. Pp. 368–390. In: Löbl, I. & Smetana, A. (eds.): *Catalogue of Palaearctic Coleoptera. Volume 6. Chrysomeloidea.* Stenstrup: Apollo Books, 924 pp. DP: 22 February 2010 (colophon).

Borowiec, L., Sultan, A. & Rafi, M. A. 2008: [new taxon]. In: Borowiec, L., Sultan, A., Rafi, M. A. & Naz, F.: Notosacantha pakistanica, a new species from Pakistan (Coleoptera: Chrysomelidae: Cassidinae: Notosacanthini). *Genus* **19** (2): 285–290. DP: 31 July 2008 (paper title page).

Borowiec, L. & Świętojańska, J. 2001a: Cassida undecimnotata – a species complex (Coleoptera: Chrysomelidae: Cassidinae). *Genus* **12** (1): 63–79. DP: 10 April 2001 (paper title page).

Borowiec, L. & Świętojańska, J. 2001b: The palaearctic species of the genus Rhytidocassis Spaeth, 1941 (Coleoptera: Chrysomelidae: Cassidinae). *Annales Zoologici* (Warszawa) **51** (3): 325–329. DP: 28 September 2001 (verso of the issue cover).

Borowiec, L. & Świętojańska, J. 2022: A monograph of the Afrotropical Cassidinae (Coleoptera: Chrysomelidae). Part 6. Revision of the tribe Cassidini 3, the genus Cassida L. *Zootaxa* **5171** (1): 1–250. DP: 4 August 2022 (paper title page).

Borowiec, L. & Takizawa, H. 1991: Notes on chrysomelid beetles (Coleoptera) of India and its neighboring areas. Part 10. *Japanese Journal of Entomology* **59** (3): 637–654. DP: 25 September 1991 (paper title page).

Bouchard, P. & Bousquet, Y. 2020: Additions and corrections to "Family-group names in Coleoptera (Insecta)". *ZooKeys* **922**: 65–139. DP: 25 March 2020 (paper title page).

Bouchard, P., Bousquet, Y., Aalbu, R. L., Alonso-Zarazaga, M. A., Merkl, O. & Davies, A. E. 2021: Review of genus-group names in the family Tenebrionidae (Insecta, Coleoptera). *ZooKeys* **1050**: 1–633. DP: 26 July 2021 (paper title page).

Bouchard, P., Bousquet, Y., Davies, A. E., Alonso-Zarazaga, M. A., Lawrence, J. F., Lyal, C. H. C., Newton, A. F., Reid, C. A. M., Schmitt, M., Ślipiński, S. A. & Smith, A. B. T. 2011: Family-group names in Coleoptera (Insecta). *ZooKeys* **88**: 1–972. DP: 4 April 2011 (paper title page).

Bouchard, P., Bousquet, Y., Davies, A. E. & Cai, C.-Y. 2024: On the nomenclatural status of type genera in Coleoptera (Insecta). *ZooKeys* **1194**: 1–981. DP: 13 March 2024 (paper title page).

Bourdonné, J.-C. 1977: Note sur Chrysolina (Bechynia) platypoda Bechyné (Coleoptera Chrysomelidae Chrysomelinae). *Nouvelle Revue d'Entomologie* **7** (3): 329–334. DP: 15 October 1977 (issue cover).

Bourdonné, J.-C. 1986: Chrysolina (Bechynia) substrangulata, nouvelle espèce de Hongrie (Coleoptera, Chrysomelidae). *Nouvelle Revue d'Entomologie, Nouvelle Série* **3** (2): 235–241. DP: 20 October 1986 (last page of the cover).

Bourdonné, J.-C. 1994: Les Cryptocephalus rouges ouest méditerranéens du groupe de C. bimaculatus Fabricius (Coleoptera, Chrysomelidae, Cryptocephalinae). *Nouvelle Revue d'Entomologie, Nouvelle Série* 11 (2): 185–209. DP: 10 November 1994 (last page of the cover).

Bourdonné, J.-C. 1997: Chrysolina (Mimophaedon) pourtoyi nouveau sous-genre et nouvelle espèce, endémique pyrénéen (Coleoptera, Chrysomelidae). *Nouvelle Revue d'Entomologie, Nouvelle Série* 13 [1996] (4): 341–351. DP: 7 May 1997 (last page of the cover).

Bourdonné, J.-C. 1999: Les types de Chrysomela de Suffrian appartenant au sous-genre Maenadochrysa Bechyné 1950 (Coleoptera, Chrysomelidae). *Nouvelle Revue d'Entomologie, Nouvelle Série* 16 (1): 27–58. DP: 17 September 1999 (last page of the cover).

Bourdonné, J.-C. 2005: Révision du sous-genre Taeniosticha Motschulsky, 1860 du genre Craspeda Motschulsky, 1860 (Coleoptera Chrysomelidae) – 1ère partie. *Nouvelle Revue d'Entomologie, Nouvelle Série* 21 [2004] (4): 297–363. DP: 31 March 2005 (last page of the cover).

Bourdonné, J.-C. 2008: Mise au point et compléments sur la révision des Taeniosticha du genre Craspeda (Coleoptera, Chrysomelidae, Chrysomelinae). *Nouvelle Revue d'Entomologie, Nouvelle Série* 25 (1): 17–26. DP: 31 December 2008 (last page of the cover).

Bourdonné, J.-C. 2012: Révision du genre Chalcoidea Motchulsky, 1860 avec description d'une espèce nouvelle, Chalcoidea (s. str.) hollandei (1ère partie) (Coleoptera, Chrysomelidae, Chrysomelinae). *Nouvelle Revue d'Entomologie, Nouvelle Série* 28 (2): 99–181. DP: 28 June 2012 (last page of the cover).

Bourdonné, J.-C. & Doguet, S. 1979: Contribution a l'étude des Gonioctena Chevr. (Phytodecta Kirby) d'Afrique du nord. *Nouvelle Revue d'Entomologie* 9 (1): 49–58. DP: 15 April 1979 (issue cover).

Bourdonné, J.-C. & Doguet, S. 1986: Matériaux pour un catalogue des coléoptères des Pyrénées. 2e Note – Chrysomelidae. *L'Entomologiste* 42 (1): 13–22. DP: [28] February 1986 (issue cover).

Bourdonné, J.-C. & Doguet, S. 1991: Données sur la biosystématique des Chrysolina l. s. (Coleoptera: Chrysomelidae: Chrysomelinae). *Annales de la Société Entomologique de France, Nouvelle Série* 27 (1): 29–64. DP: 30 March 1991 (table of contents).

Bourdonné, J.-C. & Doguet, S. 2017: Les Cryptocephalus du groupe de tristigma Charpentier, 1825 (Coleoptera, Chrysomelidae). *Le Coléoptériste* 20 (2): 115–134. DP: 31 August 2017 (table of contents).

Bourdonné, J.-C., Doguet, S. & Petitpierre, E. 2013: Chrysolina (Stichoptera) oceanoripensis nova species, endémique des dunes françaises des Landes de Gascogne et considérations sur le sous-genre Stichoptera Motschulsky, 1860 (Coleoptera, Chrysomelidae). *Nouvelle Revue d'Entomologie, Nouvelle Série* 29 (1): 33–52. DP: 20 August 2013 (last page of the cover).

Bousquet, Y. & Bouchard, P. 2013: The genera in the second catalogue (1833–1836) of Dejean's Coleoptera collection. *ZooKeys* 282: 1–219. DP: 2 April 2013 (paper title page).

Bouyon, H., Doguet, S. & Debreuil, M. 2010: Description d'une sous-espèce de Corse de Labidostomis (Chlorostola) centromaculata Gené, 1839 (Coleoptera, Chrysomelidae, Clytrinae). *Le Coléoptériste* 13 (2): 102–110. DP: 31 August 2010 (table of contents).

Bowditch, F. C. 1913: Notes on Chlamydae with descriptions of a few new forms. *Transactions of the American Entomological Society* 39 (1): 1–21. DP: 10 March 1913 (table of contents).

Bowditch, F. C. 1925: Notes on Galerucinae in my collection. *Psyche* 32 (4/5): 244–264. DP: [31] October 1925 (issue cover).

Brahm, N. J. 1790: *Insektenkalender für Sammler und Oekonomen. Erster Theil.* Mainz: Kurfürstl. Privil. Universitätsbuchhandlung, xcii + 248 pp. DP: 12 May 1790 (Bousquet 2016).

Brancsik, K. 1899: Aliquot Coleoptera nova Russiae asiaticae. *Jahresheft des Naturwissenschaftlichen Vereines des Trencsiner Comitates* 21–22 [1898–1899]: 97–105 + 1, pl. 4. DP: [31] December 1899 (footer on the paper title page).

Brancsik, K. 1910: Species et varietates novae Coleopterorum exoticorum ac palaearcticorum. *Jahresheft des Naturwissenschaftlichen Vereines des Trencsiner Comitates* 31–33 [1908–1910]: 176–190, pl. 1. DP: [31 December] 1910 (volume cover).

Brancsik, K. 1914: Coleoptera nova. *Bericht des Museumvereines für das Comitat Trencsén* 1914: 58–69. DP: [31 December] 1914 (volume cover).

Breit, J. 1908: Eine koleopterologische Sammelreise auf Mallorka (Balearen). *Verhandlungen der Kaiserlich-Königlichen Zoologisch-Botanischen Gesellschaft in Wien* 58 (2/3): (52)–(67). DP: 22 April 1908 (volume title page).

Breit, J. 1913: Beiträge zur Kenntnis der paläarktischen Coleopterenfauna. *Entomologische Blätter* 9 (11/12): 292–299. DP: 10 November 1913 (issue title page).

Breit, J. 1917: Zwei neue zentralasiatische Käferarten. *Coleopterologische Rundschau* 6 (10/12): 65–67. DP: 28 December 1917 (issue cover).

Breit, J. 1918a: Beitrag zur Kenntnis der Arten des Genus Cryptocephalus Geoffr. *Wiener Entomologische Zeitung* 37 (1/3): 35–52. DP: 25 March 1918 (footers throughout the paper).

Breit, J. 1918b: Eine neue Stylosomus-Art nebst Übersicht der bekannten mittelasiatischen Arten. *Koleopterologische Rundschau* 7 (3/8): 118–119. DP: 30 October 1918 (issue cover).

Breit, J. 1919: Chrysomela-Studien. *Koleopterologische Rundschau* 8 [1919–1920] (1/3): 14–19. DP: 20 September 1919 (issue cover).

Breit, J. 1920: Chrysomela-Studien II. *Koleopterologische Rundschau* 8 [1919–1920] (7/12): 81–88. DP: 20 December 1920 (issue cover).

Breit, J. 1921: Die mittelasiatischen Pachybrachys-Arten aus der Verwandtschaft des P. glycyrrhizae Ol. und vermicularis Suffr. *Entomologische Blätter* 17 (4/6): 64–74. DP: 20 June 1921 (issue cover).

Brelih, S., Döberl, M., Drovenik, B. & Pirnat, A. 2003: Materialien zur Käferfauna (Coleoptera) Slowenien. 1. Beitrag: Polyphaga: Chrysomeloidea (= Phytophaga): Chrysomelidae: Alticinae. *Scopolia* 50: 1–279. DP: [31] October 2003 (colophon).

Bremi-Wolf, J. J. 1855: Beschreibungen einiger, muthmasslich neuer, oder doch noch nicht beschreibener schweizerischer Käfer. *Entomologische Zeitung* (Stettin) 16 (7): 196–200. DP: [31] July 1855 (issue title page).

Brenske, E. 1890: Eine neue Chrysomela aus Italien. *Entomologische Nachrichten* 16 (1): 14–15. DP: [31] January 1890 (issue title page).

Bridwell, J. C. 1929: A preliminary generic arrangement of the palm bruchids and allies (Coleoptera) with descriptions of new species. *Proceedings of the Entomological Society of Washington* **31** (8): 141–160. DP: 26 December 1929 (verso of the volume title page).

Bridwell, J. C. 1932: The subfamilies of the Bruchidae. *Proceedings of the Entomological Society of Washington* **34** (6): 100–106. DP: 15 July 1932 (verso of the volume title page).

Bridwell, J. C. 1938: Specularius erythrinae, a new bruchid affecting seeds of Erythrina (Coleoptera). *Journal of the Washington Academy of Sciences* **28** (2): 69–76. DP: 15 February 1938 (issue cover).

Bridwell, J. C. 1946: The genera of beetles of the family Bruchidae in America north of Mexico. *Journal of the Washington Academy of Sciences* **36** (2): 52–57. DP: 15 February 1946 (issue cover).

Brisout de Barneville, C. N. F. 1860: Descriptions de quelques coléoptères nouveaux propres à la faune française et spécialement aux environs de Paris. *Annales de la Société Entomologique de France, Troisième Série* **8** (2): 339–350. DP: 12 September 1860 (issue cover).

Brisout de Barneville, C. N. F. 1863: [new taxa]. Grenier A.: *Catalogue des coléoptères de France et matériaux pour servir à la faune des coléoptères français*. Paris: A. Grenier, iv + 3–79 + 135 pp. DP: 6 August 1863 (book cover, Grenier 1864, see also the comments in the Dating of publications chapter).

Brisout de Barneville, C. N. F. 1866: Coléoptères nouveaux trouvés en Espagne pendant l'excursion de la Société en 1865. *Annales de la Société Entomologique de France, Quatrième Série* **6** (3): 355–426. DP: 31 December 1866 (Lefèvre 1885).

Brisout de Barneville, C. N. F. 1873a: [Description d'une nouvelle espèce de coléoptère phytophage]. *Bulletin de la Société Entomologique de France* **1873** (2): 5. DP: 10 May 1873 (Evenhuis 2002).

Brisout de Barneville, C. N. F. 1873b: [Note]. *Bulletin de la Société Entomologique de France* **1873** (16): 5–6. DP: 5 December 1873 (Evenhuis 2002).

Brisout de Barneville, C. N. F. 1880a: [Descriptions de cinq nouvelles espèces de coléoptères d'Europe, dont nous croyons devoir donner les diagnoses]. *Bulletin de la Société Entomologique de France* **1880** (3): 34–35. DP: 20 February 1880 (Evenhuis 2002).

Brisout de Barneville, C. N. F. 1880b: Descriptions de coléoptères nouveaux d'Europe. *Annales de la Société Entomologique de France, 5e Série* **10** (3): 231–235. DP: 8 December 1880 (Lefèvre 1885).

Brisout de Barneville, C. N. F. 1882: [Descriptions suivantes de deux nouveuux coléoptères français]. *Bulletin de la Société Entomologique de France* **1882** (23): 245–246. DP: 21 December 1882 (Evenhuis 2002).

Brisout de Barneville, C. N. F. 1884: Description de trois coléoptères nouveaux d'Algérie. *Revue d'Entomologie* **3** (4): 88–90. DP: [30] April 1884 (issue title page).

Brisout de Barneville, C. N. F. 1891: [Description d'un nouveau coléoptère de France]. *Bulletin de la Société Entomologique de France* **1891** (20): clxxxv. DP: 31 December 1891 (Evenhuis 2002).

Broun, T. 1893: *Manual of the New Zealand Coleoptera. Parts* V, VI and VII. Wellington: New Zealand Institute, xvii + 975–1504 pp. DP: 28 September 1893 (Bousquet 2016).

Broun, T. 1910: On the Coleoptera of the Kermadec Islands. *Transactions of the New Zealand Institute* 42 [1909]: 291–306. DP: 1 June 1910 (volume title page).

Brovdij, V. M. 1983: *Fauna Ukraini. Tom 19. Zhuki. Vypusk 20. Zhuki-listoedy. Shitonoski i shiponoski. [Fauna of Ukraine. Volume 19. Beetles. Release 20. Leaf-beetles. Tortoise beetles and porcupine beetles]*. Kiev: Naukova Dumka, 187 pp. SFP: 28 October 1983, DP: [31 December] 1983 (book title page).

Brown, W. J. 1942: The American species of Entomoscelis and Hippuriphila (Coleoptera, Chrysomelidae). *Canadian Entomologist* 74 (9): 172–176. DP: 21 October 1942 (footer on p. 176).

Brown, W. J. 1952: Some species of Phytophaga (Coleoptera). *Canadian Entomologist* 84 (11): 335–342. DP: 28 November 1952 (footer on p. 356).

Brown, W. J. 1962: The American species of Chrysolina Mots. (Coleoptera: Chrysomelidae). *Canadian Entomologist* 94 (1): 58–74. DP: 5 January 1962 (footer on p. 112).

Brüggemann, F. 1873: Systematisches Verzeichniss der bisher in der Gegend von Bremen gefundenen Käferarten. *Abhandlungen Herausgegeben vom Naturwissenschaftlichen Vereine zu Bremen* 3 (4): 441–524. DP: [31] December 1873 (table of contents).

Brullé, G. A. 1833: [Livraison 5–6, pp. 193–288]. In: IVe classe. Insectes. Pp. 64–395. In: Bory de Saint-Vincent, J. B. G. M.: *Expédition scientifique de Morée. Section des sciences physiques. Tome III.–I. re Partie. Zoologie. Deuxième Section. – Des animaux articulés.* [1832–1833], Paris, Strasburg: F. L. Levrault, [1] + 400 + [2 (Errata)] pp., pls. 27–53. DP: 9 January 1833 (Bousquet 2016).

Brullé, G. A. 1839: Insectes. Pp. 53–95. In: Barker-Webb, P. & Berthelot, S. (eds.): *Histoire naturelle des Iles Canaries. Tome deuxième. Deuxième partie. Contenant la zoologie.* Paris: Béthune, 119 pp., 7 pls. DP (pp. 53–56): 12 August 1839, (pp. 57–72): 16 September 1839, (pp. 73–88): 30 September 1839, (pp. 89–95): 28 October 1839 (Evenhuis 2019a).

Bryant, G. E. 1923: New injurious Phytophaga from India and Brazil. *Bulletin of Entomological Research* 13 [1922–1923] (3): 261–265. DP: [31] January 1923 (issue cover).

Bryant, G. E. 1924a: New species of Phytophaga. *Annals and Magazine of Natural History, Ninth Series* 13 (75): 299–304. DP: 1 March 1924 (Evenhuis 2003).

Bryant, G. E. 1924b: New species of Phytophaga (Coleopt.) *Annals and Magazine of Natural History, Ninth Series* 14 (82): 247–252. DP: 1 August 1924 (Evenhuis 2003).

Bryant, G. E. 1925: New species of Phytophaga from the Fiji Islands. *Annals and Magazine of Natural History, Ninth Series* 15 (89): 590–600. DP: 1 May 1925 (Evenhuis 2003).

Bryant, G. E. 1926: New species of Phytophaga (Col.). *Annals and Magazine of Natural History, Ninth Series* 17 (102): 403–410. DP: 1 April 1926 (Evenhuis 2003).

Bryant, G. E. 1930: Some new injurious Phytophaga from Somaliland and Uganda. *Bulletin of Entomological Research* 21 (3): 361–363. DP: 9 October 1930 (dates of publication on p. 599).

Bryant, G. E. 1934: New species of Indian Chrysomelidae (Col.). *Stylops* 3 (10): 225–228. DP: 15 October 1934 (issue cover).

Bryant, G. E. 1937: Notes on synonymy in the Phytophaga (Coleoptera). *Annals and Magazine of Natural History, Tenth Series* 20 (115): 97–101. DP: 1 July 1937 (Evenhuis 2003).

Bryant, G. E. 1939: Entomological results from the Swedish expedition 1934 to Burma and British India. Coleoptera: Chrysomelidae, collected by René Malaise. *Arkiv för Zoologi* **31A** (21): 1–20. DP: 14 July 1939 (last page of the paper).

Bryant, G. E. 1940: New species of African Phytophaga (Coleopt.). *Proceedings of the Royal Entomological Society of London, Series B Taxonomy* **9** (3): 41–48. DP: 15 March 1940 (issue cover).

Bryant, G. E. 1941a: New species of Chrysomelidae (Coleoptera) from Malaya and Java. *Annals and Magazine of Natural History, Eleventh Series* **8** (46): 283–288. DP: 1 October 1941 (Evenhuis 2003).

Bryant, G. E. 1941b: New species of Indian Chrysomelidae (Coleopt.). *Annals and Magazine of Natural History, Eleventh Series* **8** (47): 403–414. DP: 1 November 1941 (Evenhuis 2003).

Bryant, G. E. 1942a: Notes on the genus Pagria (Eumolpinae, Coleopt.). *Bulletin of Entomological Research* **33** (1): 31–34. DP: 22 April 1942 (table of contents).

Bryant, G. E. 1942b: New species of African Chrysomelidae (Coleoptera). *Annals and Magazine of Natural History, Eleventh Series* **9** (55): 513–516. DP: 15 July 1942 (Evenhuis 2003).

Bryant, G. E. 1946: New species of African Cryptocephalus (Chrysomelidae, Col.). *Annals and Magazine of Natural History, Eleventh Series* **12** [1945] (93): 609–615. DP: 10 July 1946 (Evenhuis 2003).

Bryant, G. E. 1948: New species of Chrysomelidae from Senegal and the Ivory Coast. *Annals and Magazine of Natural History, Twelfth Series* **1** (1): 58–65. DP: 27 August 1948 (Evenhuis 2003).

Bryant, G. E. 1952: Nepal Expedition, British Museum 1949–637, Chrysomelidae collected by Mr. O. Pulunin. Presented by Mr. R. W. Lloyd. *Annals and Magazine of Natural History, Twelfth Series* **5** (54): 601–608. DP: 1 June 1952 (Evenhuis 2003).

Bryant, G. E. 1954a: Entomological results from the Swedish expedition 1934 to Burma and British India. Coleoptera. Chrysomelidae, Galerucinae collected by René Malaise. *Arkiv för Zoologi, Andra Serien* **6** (5) (18): 413–424. DP: 14 January 1954 (last page of the paper).

Bryant, G. E. 1954b: Entomological results from the Swedish expedition 1934 to Burma and British India. Coleoptera: Chrysomelidae, Eumolpinae, Halticinae, Galerucinae collected by René Malaise. *Arkiv för Zoologi, Andra Serien* **6** (6) (29): 541–553. DP: 22 April 1954 (last page of the paper).

Bryant, G. E. 1954c: New species of Oriental Cryptocephalus (Col. Chrysomelidae). *Annales and Magazine of Natural History, Twelfth Series* **7** (78): 448–454. DP: 18 June 1954 (Evenhuis 2003).

Bryant, G. E. 1957a: New species of Chlamisus (Chlamisinae, Col. Chrysomelidae) from India, E. Indies and Queensland. *Annals and Magazine of Natural History, Twelfth Series* **10** (111): 215–220. DP: 27 June 1957 (Evenhuis 2003).

Bryant, G. E. 1957b: Coleoptera: Chrysomelidae of South-West Arabia. *Annals and Magazine of Natural History, Twelfth Series* **10** (113): 353–363. DP: 20 September 1957 (Evenhuis 2003).

Bua, G. 1953: I° contributo alla conoscenza dei crisomelidi del genere Arima Chap. Cenni morfologici e biologici sull'Arima marginata Fabr. e descrizione dell'Arima maritima sp. n.

Bollettino del Laboratorio di Entomologia Agraria "Filippo Silvestri" Portici **12**: 154–185. DP: 24 September 1953 (table of contents).

Bukejs, A., Telnov, D. & Barševskis, A. 2009: Review of Cassidinae (Coleoptera: Chrysomelidae) of the Fauna of Latvia. *Latvijas Entomologs* **47**: 27–57. DP: [31 December] 2009 (paper title page).

Bullock, E. 1928: Coleoptera from the Killarney district of County Kerry, Ireland. *Entomologist's Monthly Magazine, Third Series* **64** (768): 102–104. DP: 9 May 1928 (registered at NHMUK library).

Buquet, J. B. L. 1840: Note sur quelques coléoptères nouveaux d'Algérie et particulièrement de Constantine. *Revue Zoologique, par la Société Cuvierienne* **3** (8): 240–243. DP: 7 September 1840 (*Compt. Rend. Hebdom. Séances Acad. Sci.* 1840, 11: 450).

Buquet, J. B. L. 1869: [new taxon]. *Bulletin de la Société Entomologique de France* [**1868**] (4): cv. DP: 14 April 1869 (Lefèvre 1885).

Burgeon, L. 1942: Clytrides nouveaux du Congo belge (Col. Chrysom.). *Revue de Zoologie et de Botanique Africaines* **36** (1): 16–55. DP: 14 August 1942 (paper title page).

Burlini, M. 1942a: Forme nuove di coleotteri. *Bollettino della Società Entomologica Italiana* **74** (1): 13. DP: 31 January 1942 (issue cover, Poggi 2008).

Burlini, M. 1942b: Nuove forme di coleotteri italiani. *Bollettino della Società Entomologica Italiana* **74** (9): 126–128. DP: 10 November 1942 (issue cover, Poggi 2008).

Burlini, M. 1946: Nuove forme di coleotteri italiani. *Bollettino della Società Entomologica Italiana* **76** (9/10): 78–79. DP: 31 December 1946 (issue cover, Poggi 2008).

Burlini, M. 1949: Nuove specie di Cryptocephalus italiani (Coleoptera, Chrysomelidae). *Memorie del Museo Civico di Storia Naturale di Verona* **1** [1947–1948]: 221–229. DP: [31] March 1949 (last page of the volume).

Burlini, M. 1950: Una nuova specie di Cryptocephalus italiano (Coleoptera: Chrysomelidae). *Bollettino della Società Entomologica Italiana* **80** (7/8): 54–57. DP: 20 September 1950 (issue cover, Poggi 2008).

Burlini, M. 1951: Una nuova specie di Cryptocephalus italiano. Cryptocephalus lostiai (Coleoptera: Chrysomelidae). *Bollettino della Società Entomologica Italiana* **81** (3/4): 45–47. DP: 7 July 1951 (issue cover, Poggi 2008).

Burlini, M. 1953: Une nouvelle espèce de Cryptocephalus du Maroc: Cryptoc. Kocheri Burlini (Coleoptera Chrysomelidae). *Comptes Rendus des Séances Mensuelles de la Société des Sciences Naturelles du Maroc* **1953** (5): 73–75. DP: 22 July 1953 (last page of the issue).

Burlini, M. 1956a: Ricerche zoologiche sul Massiccio del Pollino (Lucania-Calabria). XIX. Coleoptera. 9. Due nuovi Pachybrachys italiani (Chrysomelidae). *Annuario dell'Istituto e Museo di Zoologia della Università di Napoli* **8** (3): 1–7. DP: [31] March 1956 (issue title page).

Burlini, M. 1956b: Revisione dei Cryptocephalus italiani e della maggior parte delle specie di Europe (Col. Chrysomelidae). *Memorie della Società Entomologica Italiana* **34** [1955]: 1–287. DP: 10 December 1956 (Poggi 2008).

Burlini, M. 1957a: Un nuovo Pachybrachys dell'isola di Cefalonia (Col. Chrysomelidae). *Bollettino della Società Entomologica Italiana* 87 (1/2): 25–26. DP: 6 February 1957 (issue cover, Poggi 2008).

Burlini, M. 1957b: Un nuovo Stylosomus asiatico (Col. Chrysomelidae). *Bollettino della Società Entomologica Italiana* 87 (3/4): 61–63. DP: 20 April 1957 (issue cover, Poggi 2008).

Burlini, M. 1958: Aggiunte e correzioni alla "Revisione dei Cryptocephalus italiani" (Col. Chrysomelidae). *Bollettino della Società Entomologica Italiana* 88 (7/8): 122–123. DP: 25 October 1958 (issue cover, Poggi 2008).

Burlini, M. 1959a: Deux nouvelles variétés de Cryptocephalus du Maroc (Coléoptères Chrysomélides). *Comptes Rendus de la Société des Sciences Naturelles du Maroc* 24 [1958] (8): 196. DP: 20 February 1959 (footer on p. 196).

Burlini, M. 1959b: Note su alcuni Pachybrachys appenninici e descrizione di una nuova razza (Coleoptera Chrysomelidae). *Memorie del Museo Civico di Storia Naturale di Verona* 7: 199–201. DP: 31 December 1959 (table of contents).

Burlini, M. 1960: Ricerche sui Cryptocephalus dell'Africa settentrionale (Col. Chrysomelidae). *Eos, Revista Española de Entomología* 36 (2): 137–139. DP: 30 June 1960 (issue cover).

Burlini, M. 1961: Contributo alla conoscenza dei Cryptocephalus (Col. Chrysomelidae). *Bollettino della Società Entomologica Italiana* 91 (5/6): 94–96. DP: 25 June 1961 (issue cover, Poggi 2008).

Burlini, M. 1962: Nouve forme di Cryptocephalus paleartici (Coleoptera, Chrysomelidae) appartenenti al Museo Nazionale di Storia Naturale di Budapest. *Annales Historico-Naturales Musei Nationalis Hungarici* 54: 327–328. DP: 31 December 1962 (Merkl et al. 2008).

Burlini, M. 1963a: Due note sulle Cryptocephalus (Col., Chrysomelidae). *Folia Entomologica Hungarica, Series Nova* 16 (1): 157–158. DP: 20 June 1963 (table of contents).

Burlini, M. 1963b: Contributo alla conoscenza dei Cryptocephalini paleartici. Pachybrachys lindbergi n. sp., della Penisola Iberica (Col. Chrysomelidae). *Bollettino della Società Entomologica Italiana* 93 (9/10): 148–150. DP: 20 December 1963 (issue cover, Poggi 2008).

Burlini, M. 1963c: Nota sul Pachybrachys siculus Weise (Coleoptera Chrysomelidae). *Memorie del Museo Civico di Storia Naturale di Verona* 9: 133–134. DP: 31 December 1963 (table of contents).

Burlini, M. 1965: Il contributo alla conoscenza dei Cryptocephalini paleartici. Cryptocephalus españoli, n. sp., della Penisola Iberica (Coleoptera, Chrysomelidae). *Bollettino della Società Entomologica Italiana* 95 (1/2): 17–19. DP: 20 February 1965 (issue cover, Poggi 2008).

Burlini, M. 1967: Addenda alla revisione dei Cryptocephalus. XXI contributo alla conoscenza dei Cryptocephalini (Coleopt.). *Bollettino della Società Entomologica Italiana* 97 (7/8): 129–132. DP: 20 October 1967 (issue cover, Poggi 2008).

Burlini, M. 1968: Revisione delle specie Italiane e della maggior parte delle specie Europee del genere Pachybrachis Chevr. (Coleoptera Chrysomelidae). XXII contributo alla conoscenza dei Cryptocephalini. *Memorie della Società Entomologica Italiana* 47: 11–116. DP: 31 December 1968 (Poggi 2008).

Burlini, M. 1969a: Cryptocephalus fulvus Goeze subsp. schatzmayri nova. XXIII contributo alla conoscenza dei Cryptocephalini (Coleoptera Chrysomelidae). *Bollettino della Società Entomologica Italiana* 99–101 (5/6): 114–115. DP: 20 June 1969 (issue cover, Poggi 2008).

Burlini, M. 1969b: Cryptocephalus subgen. Cryptodontus n., Cryptoc. subgen. Cerodens n. nov. e tabella dei sottogeneri di Cryptocephalus. XXIV contributo alla conoscenza dei Cryptocephalini (Coleoptera Chrysomelidae). *Memorie della Società Entomologica Italiana* **48** (3): 535–540. DP: 31 December 1969 (Poggi 2008).

Burlini, M. 1970: Una nuova sottospecie nord Africana del Cryptocephalus pygmaeus F. e note sul Cryptocephalus (Cerodens Burl.) emiliae Burl. XXV contributo alla conoscenza dei Cryptocephalini (Coleoptera Chrysomelidae). *Bollettino della Società Entomologica Italiana* **102** (1/2): 38–40. DP: 20 February 1970 (issue cover, Poggi 2008).

Buysson, H. du 1908a: Une espèce nouvelle du genre Thyamis Steph. *Miscellanea Entomologica* **15** [1907] (5/8): 1–3. DP: 16 July 1908 (stamp on the ZIN copy, see also the comments in the Dating of publications chapter).

Buysson, H. 1908b: Recherches sur l'epoque d'éclosion des altises, et sur la biologie et les caractères de quelques Thyamis Steph. de la faune de France (Col.). *Miscellanea Entomologica* **15** [1907] (10/11, supplement): 1–39. DP: 23 September 1908 (see the comments in Dating of publications).

Buysson, H. 1924: Chrysomela menthastri Suffr. varietates novae. *Miscellanea Entomologica* **27** (12): 89. DP: 10 July 1924 (footer on p. 89).

Caillol, H. 1894: [Diagnoses de trois espèces nouvelles de Mylabris (Bruches)]. *Bulletin de la Société Entomologique de France* **1894** (19): cclxix–cclxxi. DP: 22 December 1894 (Evenhuis 2002).

Caillol, H. 1908: [new taxon]. *Compte-Rendu Sommaire des Séances de la Société des Sciences Naturelles de Provence* **1908**: viii–ix. DP: [31 December] 1908 (title page of volume).

Caillol, H. 1919: Description d'un Acanthoscelides nouveau, de Timbouctou (Col. Bruchidae). *Bulletin de la Société Entomologique de France* **1919** (2): 53–54. DP: 7 April 1919 (*dates de publication* at the end of volume).

Caillol, H. 1921: Catalogue des coléoptères de Provence. 3me Partie. *Mémoires de la Société Linnéenne de Provence* **1925** [1914–1925]: 1–594 pp. DP (pp. 509–524): [31] December 1921 (footer on p. 509).

Caillol, H. 1925: Catalogue des coléoptères de Provence. 3me Partie. *Mémoires de la Société Linnéenne de Provence* **1925** [1914–1925]: 1–594 pp. DP (pp. 525–540): [31] March 1925 (footer on p. 525).

Calwer, C. G. 1857: *Käferbuch. Allgemeine und specielle Naturgeschichte der Käfer Europa's. Nebst der Anweisung sie zu sammeln und aufzubewahren.* [1858], Stuttgart: Krais & Hoffmann, xxviii + 788 pp. + 49 pls. DP: 5 November 1857 (Bousquet 2016).

Cantonnet, F. 1968: Révision des espèces françaises du genre Phytodecta et description d'une espèce nouvelle (Col. Chrysomelidae). *L'Entomologiste* **24** (2): 38–49. DP: [30] April 1968 (issue cover).

Capra, F. 1933: Note sugli Sphaeroderma Steph. (Col. Chrysom.). *Bollettino della Società Entomologica Italiana* **65** (7): 164–166. DP: 26 July 1933 (Poggi 2008).

Casey, T. L. 1924: *Memoirs on the Coleoptera XI*. Lancaster: Lancaster Press, 347 pp. DP: 20 May 1924 (noticed on p. 347).

Čepelák, R. 1928: Nové aberrace Osphya bipunctata Fabr. [New aberrations of Osphya bipunctata Fabr.]. *Časopis Československé Společnosti Entomologické* **24** [1927] 5/6: 123. DP: 15 February 1928 (footer on the paper title page).

Champion, G. C. 1919: Some Indian Coleoptera. *Entomologist's Monthly Magazine, Third Series* **55** (665): 236–240 & **55** (666): 241–246. DP (pp. 236–240): 11 October 1919, (pp. 241–246): 7 November 1919 (registered at NHMUK library).

Champion, G. C. 1925: Some Indian (and Tibetan) Coleoptera. *Entomologist's Monthly Magazine, Third Series* **61** (738): 260–264 & **61** (739): 265–273. DP (pp. 260–264): 16 November 1925, (pp. 265–273): 9 December 1925 (registered at NHMUK library).

Chapuis, F. 1874: Famille des Phytophages. In: Lacordaire, T. & Chapuis, F.: *Histoire naturelle des insectes. Genera des coléoptères ou exposé méthodique et critique de tous les genres proposés jusqu'ici dans cet ordre d'insectes. Tome dixième.* Paris: Roret, iv + 455 pp. DP: 1 July 1874 (Bousquet 2016).

Chapuis, F. 1875: Famille des Phytophages. In: Lacordaire, T. & Chapuis, F.: *Histoire naturelle des insectes. Genera des coléoptères ou exposé méthodique et critique de tous les genres proposés jusqu'ici dans cet ordre d'insectes. Tome onzième.* Paris: Roret, 420 pp. DP: 15 September 1875 (Bousquet 2016).

Chapuis, F. 1876: Relevé des Hispides des Isles Philippines avec les descriptions de quelques espéces inédites. *Comptes-Rendus des Séances de la Société Entomologique de Belgique* **19** [1876–1877] (1): xvii–xxviii. DP: 16 September 1876 (issue cover).

Chapuis, F. 1877a: Diagnoses des espèces du genre Aulacophora recueillies aux îles Philippines par le Dr Semper. *Comptes-Rendus des Séances de la Société Entomologique de Belgique* **19** [1876–1877] (3): xcix–ci. DP: 10 January 1877 (*Bull. Soc. Ent. Fr.* 1877: 24).

Chapuis, F. 1877b: Cryptocéphalides inédits du Musée Civique de Gênes. *Annali del Museo Civico di Storia Naturale di Genova* **9** [1876–1877]: 335–350. DP: 26 March 1877 (Poggi 2010).

Chapuis, F. 1877c: Espèces inédites de la tribu des hispides. Deuxième partie. *Annales de la Société Entomologique de Belgique* **20** (1): 47–57. DP: 22 August 1877 (*Bull. Soc. Ent. Fr.* 1877: 192).

Chapuis, F. 1879: Phytophages abyssiniens du Musée Civique d'Histoire Naturelle de Gênes. *Annali del Museo Civico di Storia Naturale di Genova* **15** [1879–1880]: 5–31. DP: 18 November 1879 (Poggi 2010).

Charpentier, T. de 1825: *Horae entomologicae, adjectis tabulis novem coloratis.* Wratislaviae: A. Gosohorsky, xvi + 255 + [5] pp., 9 pls. DP: 20 September 1825 (Bousquet 2016).

Chashchina, O. E. 2006: Novyy vid listoedov-bloshek roda Altica Geoffr. (Coleoptera, Chrysomelidae, Alticinae) s yuzhnogo Urala. (A new species of the flea-beetle genus Altica Geoffr. (Coleoptera, Chrysomelidae, Alticinae) from southern Urals). *Entomologicheskoe Obozrenie* **85** (4): 858–861. SFP: 24 September 2006, DP: 28 December 2006 (registered at ZIN library).

Chen, S.-H. 1931a: Descriptions de trois espèces nouvelles de Chrysomelini de l'Asie orientale. *Bulletin du Muséum National d'Histoire Naturelle, 2e Série* (Paris) **3** (1): 110–112. DP: 19 March 1931 (last page of the issue).

Chen, S.-H. 1931b: Description de trois Chrysomelinae nouveaux de l'Asie orientale. *Bulletin de la Société Entomologique de France* **1931** (9): 127–131. DP: 12 June 1931 (*dates d'apparition* on p. 299).

Chen, S.-H. 1932a: Description d'un Lamprosoma nouveaux de la Chine occidentale (Col. Chrysomelidae). *Bulletin de la Société Entomologique de France* 37 (7): 108. DP: 19 May 1932 (*dates d'apparition* on p. 320).

Chen, S.-H. 1932b: Nouvelles espèces d'Halticini (Col. Chrysomelidae) appartenant aux genres Hespera Weise et Parhespera, n. g. *Bulletin de la Société Entomologique de France* 37 (13/14): 193–197. DP: 10 September 1932 (*dates d'apparition* on p. 320).

Chen, S.-H. 1933a: Study of Chinese Halticinae beetles with descriptions of some exotic new species. *Sinensia* 3 (9): 211–254. DP: [31] March 1933 (issue cover).

Chen, S.-H. 1933b: Tableau synoptique des espèces du genre Neorthaea Maulik (Orthaea Jac.) appartenant à la famille des Chrysomelidae (Col.), avec descriptions d'espèces nouvelles. *Bulletin de la Société Entomologique de France* 38 (6): 88–96. DP: 20 April 1933 (*dates d'apparition* on p. 344).

Chen, S.-H. 1933c: Descriptions de deux Halticinae nouveaux de la Chine et du Japon (Col. Chrysomelidae). *Bulletin de la Société Entomologique de France* 38 (9): 143–144. DP: 16 June 1933 (*dates d'apparition* on p. 344).

Chen, S.-H. 1933d: Deux halticides nouveaux du Japon. *Bulletin de la Société Entomologique de France* 38 (12): 187–189. DP: 27 July 1933 (*dates d'apparition* on p. 344).

Chen, S.-H. 1933e: Chrysomelidae (Coleoptera) nouveaux de l'Asie tropicale. 1re note. *Bulletin du Muséum National d'Histoire Naturelle, 2e Série* (Paris) 5 (5): 381–388. DP: 3 August 1933 (last page of the issue).

Chen, S.-H. 1933f: Some species of Halticinae from Canton. *Peking Natural History Bulletin* 8 [1933–1934] (1): 43–58. DP: [30] September 1933 (*Ent. Current Lit.* 1933, 2 (6): 11).

Chen, S.-H. 1933g: Tableaux des espèces tonkinoises appartenant aux genres Hyphasoma Jacoby et Sebaethe Baly (Col. Chrysomelidae). *Bulletin de la Société Entomologique de France* 38 (17): 273–278. DP: 10 December 1933 (*dates d'apparition* on p. 344).

Chen, S.-H. 1933h: Chrysomelidae (Coleoptera) nouveaux de l'Asie tropicale. 2e note. *Bulletin du Muséum National d'Histoire Naturelle, 2e Série* (Paris) 5 (6): 449–456. DP: 26 December 1933 (last page of the issue).

Chen, S.-H. 1934a: Schwedisch-chinesische wissenschaftliche Expedition nach den nordwestlichen Provinzen Chinas unter Leitung von Dr. Sven Hedin und Prof. Sü Ping-chang. Insekten gesammelt vom schwedischen Arzt der Expedition Dr. David Hummel 1927–1930. 19. Coleoptera. 5. Chrysomelinae & Halticinae. *Arkiv för Zoologi* 27A [1936] (5): 1–13. DP: 24 January 1934 (last page of the paper).

Chen, S.-H. 1934b: On some species of Chrysomelidae (Col.) in the British Museum. *Stylops* 3 (3): 66–72. DP: 15 March 1934 (issue cover).

Chen, S.-H. 1934c: On some species of Chrysomelidae (Col.) in the British Museum. *Stylops* 3 (4): 73–78. DP: 14 April 1934 (issue cover).

Chen, S.-H. 1934d: *Recherches sur les Chrysomelinae de la Chine et du Tonkin. Thèses présentées a la Faculté des Sciences de l'Université de Paris pour obtenir le titre de docteur de l'Université de Paris. Série A N°. 291*. Paris: Société Entomologique de France, 105 pp. DP: 23 June 1934 (*Ann. Soc. Ent. Fr.* 1935, 104 (2): 127).

Chen, S.-H. 1934e: Coléoptères Halticinae recueillis par M. H. Sauter a Formose. *Annales de la Société Entomologique de France* 103 (2): 175–185. DP: 30 June 1934 (issue title page).

Chen, S.-H. 1934f: Deux Chrysomelidae nouveaux de l'Asie Orientale. *Annales de la Société Entomologique de France* 103 (2): 186. DP: 30 June 1934 (issue title page).

Chen, S.-H. 1934g: Description of some new Asiatic Chrysomelidae. *Transactions of the Science Society of China* 8 (1): 61–65. DP: [31] October 1934 (mentioned as published in Chen 1934h: 411).

Chen, S.-H. 1934h: Revision of the Halticinae (Col. Chrysomelidae) of Yunnan and Tonkin. *Sinensia* 5 (3/4): 225–416. DP: [31] October 1934 (verso of the volume title page).

Chen, S.-H. 1934i: On the Chinese species of the genus Corynodes (Col. Eumolpinae). *Sinensia* 5 (5/6): 484–504. DP: [31] December 1934 (verso of the volume title page).

Chen, S.-H. 1935a: Coleoptera Halticinae de la collection du Muséum recueillis par le Dr J. Harmand au Sikkim. *Bulletin de la Société Entomologique de France* 40 (5): 75–80. DP: 12 April 1935 (*dates de parution* on p. 300).

Chen, S.-H. 1935b: Classification of Asiatic Phytodecta (Col. Chrysomelinae). *Chinese Journal of Zoology* 1: 125–133. DP: [31] May 1935 (volume title page).

Chen, S.-H. 1935c: Study on Chinese eumolpid beetles. *Sinensia* 6 (3): 221–387. DP: [30] June 1935 (issue cover).

Chen, S.-H. 1935d: New and rare Chinese Coleoptera. *Sinensia* 6 (6): 768–781. DP: [31] December 1935 (issue cover).

Chen, S.-H. 1936a: Notes on some flea-beetles from tropical Asia (II). *Sinensia* 7 (1): 80–88. DP: [29] February 1936 (issue cover).

Chen, S.-H. 1936b: Recherches sur les Chrysomelinae de la Chine et du Tonkin. *Annales de la Société Entomologique de France* 105 (1/2): 145–176. DP: 30 May 1936 (issue title page).

Chen, S.-H. 1936c: Catalogue des Chrysomelinae de la Chine, de l'Indochine et du Japon. *Notes d'Entomologie Chinoise* 3 (5): 63–102. DP: 18 August 1936 (table of contents).

Chen, S.-H. 1936d: Notes on some flea-beetles from tropical Asia (III). *Sinensia* 7 (5): 594–600. DP: [31] October 1936 (issue cover).

Chen, S.-H. 1936e: The Chrysomelid genus Ambrostoma Motsch. *Sinensia* 7 (6): 713–729. DP: [31] December 1936 (issue cover).

Chen, S.-H. 1938a: Chrysomelidae et Halticinae récoltés par Monsieur Kjell Kolthoff dans le Province du Kiangsu (Chine). *Arkiv för Zoologi* 30**B** [1939] (4): 1–4. DP: 4 February 1938 (last page of the paper).

Chen, S.-H. 1938b: Recherches sur les Chrysomelinae de la Chine et du Tonkin. *Annales de la Société Entomologique de France* 106 [1937] (3/4): 282–323. DP: 10 February 1938 (issue title page).

Chen, S.-H. 1939a: Flea beetles collected at Kwangsi. *Sinensia* 10 (1/6): 20–55. DP: [31] December 1939 (issue cover).

Chen, S.-H. 1939b: New genera and species of Chinese Halticinae. *Sinensia* 10 (1/6): 56–91. DP: [31] December 1939 (issue cover).

Chen, S.-H. 1940a: On the Coleoptera Chlamydinae of China. *Sinensia* 11 (3/4): 189–206. DP: [30] April 1940 (issue cover).

Chen, S.-H. 1940b: New genera of Eumolpidae from Oriental Asia. *Sinensia* 11 (3/4): 207–212. DP: [30] April 1940 (issue cover).

Chen, S.-H. 1940c: Notes on Chinese Eumolpidae. *Sinensia* 11 (5/6): 483–528. DP: [30] September 1940 (issue cover).

Chen, S.-H. 1941a: Notes on donaciine beetles. *Sinensia* 12 (1/6): 1–17. DP: [31] December 1941 (issue cover).

Chen, S.-H. 1941b: Two new Stylosomus from North China (Col. Cryptocephalidae). *Sinensia* 12 (1/6): 19–21. DP: [31] December 1941 (issue cover).

Chen, S.-H. 1941c: New leaf beetles from China. *Sinensia* 12 (1/6): 189–198. DP: [31] December 1941 (issue cover).

Chen, S.-H. 1942a: Galerucinae nouveaux de la faune chinoise. *Notes d'Entomologie Chinoise* 9 (3): 9–67. DP: 18 March 1942 (table of contents).

Chen, S.-H. 1942b: Synopsis of the Coleoptera Sagrinae of China and Indochina. *Sinensia* 13 (1/6): 105–107. DP: [31] December 1942 (verso of the issue cover).

Chen, S.-H. 1942c: Synopsis of the coleopterous genus Cryptocephalus of China. *Sinensia* 13 (1/6): 109–124. DP: [31] December 1942 (verso of the issue cover).

Chen, S.-H. 1961: New species of Chinese Chrysomelidae. *Acta Entomologica Sinica* 10 (4/6): 429–435. DP: [31] December 1961 (paper title page).

Chen, S.-H. 1963: Results of the entomological expedition to Tibet in 1960–61. Coleoptera Chrysomelidae. *Acta Entomologica Sinica* 12 (4): 447–457. DP: [30] September 1963 (paper title page).

Chen, S.-H. 1964: New genera and species of Galerucinae from China. *Acta Entomologica Sinica* 13 (2): 201–211. DP: [31] March 1964 (paper title page).

Chen, S.-H. 1965: On the Chinese species of the chrysomeline genus Phratora. *Acta Zootaxonomica Sinica* 2 (3): 218–224. DP: [31] July 1965 (paper title page).

Chen, S.-H. 1966: Notes on Chinese Donaciinae. *Acta Entomologica Sinica* 15 (2): 137–147. DP: [30] April 1966 (paper title page).

Chen, S.-H. 1974a: New chrysomelid beetles from west China. *Acta Entomologica Sinica* 17 (1): 43–48. DP: [28] February 1974 (paper title page).

Chen, S.-H. 1974b: Key to the species and subspecies of the chrysomelid genus Gastrolina. *Acta Entomologica Sinica* 17 (2): 195–197. DP: [31] May 1974 (paper title page).

Chen, S.-H. 1976: [new taxa]. In: Chen, S.-H., Yu, P.-Y., Wang, S.-Y. & Jiang, S.-Q.: New leaf-beetles from West China. *Acta Entomologica Sinica* 19 (2): 205–224. DP: [31] May 1976 (paper title page).

Chen, S.-H. 1978: Atysa cinnamomi – a new galerucine beetle injurious to camphor trees in Fukien. *Acta Entomologica Sinica* **21** (1): 55–56. DP: [28] February 1978 (paper title page).

Chen, S.-H. 1992: [new taxa]. In: Chen, S.-H. & Yang, X.-K.: Descriptions of new and first recorded species of the genus Gallerucida (Coleoptera: Chrysomelidae: Galerucinae). *Zoological Research* **13** (2): 133–137. DP: [31] March 1992 (journal website).

Chen, S.-H. 1997: [new taxon]. In: Yang, X.-K., Li, W.-Z. & Yao, J.: The galerucine beetles of Xishuangbanna, Yunnan province (Coleoptera: Chrysomelidae). *Acta Zootaxonomica Sinica* **22** (4): 384–391. DP: [31] October 1997 (paper title page).

Chen, S.-H. 1998: [new taxon]. In: Wang, J.-S. & Yang, X.-K. 1998 (eds.): *Fauna of Chrysomelidae of Wuyishan Nature Reserve in China*. Beijing: China Forestry Publishing House, 213 pp., 8 pls. DP: [31] March 1999 (colophon).

Chen, S.-H. 2002: [new taxon]. In: Yang, X.-K.: Chrysomelidae: Galerucinae. Pp. 621–663. In: Huang, P.-K. (ed.): *Fauna of insects in Fujian province of China. Vol. 6*. Fuzhou: Fujian Science & Technology Press, 894 pp. DP: [31] December 2002 (colophon).

Chen, S.-H. 2008: [new taxon]. In: Zhang, L.-J., Li, W.-Z., Zhang, Y.-L. & Yang, X.-K.: A new species of genus Aplosonyx Chevrolat (Coleoptera: Chrysomelidae: Galerucinae) from China, with a key to the Chinese known species. *Zootaxa* **1898**: 63–68. DP: 13 October 2008 (footer on the paper title page).

Chen, S.-H. & & Jiang, S.-Q. 1976: [new taxa]. In: Chen, S.-H., Yu P.-Y., Wang S.-Y. & Jiang, S.-Q.: New leaf-beetles from West China. *Acta Entomologica Sinica* **19** (2): 205–224. DP: [31] May 1976 (paper title page).

Chen, S.-H. & Jiang, S.-Q. 1981: Coleoptera: Chrysomelidae – Galerucinae. Pp. 457–489. In: Qinghai-Xizang Plateau Expedition Group, Chinese Academy of Sciences (eds.): *The series of the scientific expedition to the Qinghai-Xizang Plateau. Insects of Xizang. Vol. 1*. Peking: Science Press, xi + 600 pp. DP: [31] October 1981 (colophon).

Chen, S.-H. & Jiang, S.-Q. 1984: New species of Chinese Galerucinae (Coleoptera: Chrysomelidae). *Entomotaxonomia* **6** (2/3): 83–92. DP: [30] September 1984 (paper title page).

Chen, S.-H. & Jiang, S.-Q. 1985: [new taxa]. In: Huang, D.-S., Han, Y. & Zhang, X.: The insect fauna of the Mt. Tuomuer area in Tianshan. Pp. 53–165. In: *Dengshan Kexue Kaochadui. [Biota of Tuomuer Region, Tianshan]*. Beijing: Xinjiang People's Press, 353 pp. DP: [31] March 1985 (colophon).

Chen, S.-H. & Jiang, S.-Q. 1986a: On the Chinese species of the galerucine genus Japonitata (Coleoptera: Chrysomelidae). *Acta Zootaxonomica Sinica* **11** (1): 72–79. DP: [31] January 1986 (paper title page).

Chen, S.-H. & Jiang, S.-Q. 1986b: Two new species of Galerucinae from Sichuan and Yunnan provinces (Coleoptera: Chrysomelidae). *Acta Zootaxonomica Sinica* **11** (2): 198–200. DP: [30] April 1986 (paper title page).

Chen, S.-H. & Jiang, S.-Q. 1986c: [new taxa]. In: Chen, S.-H., Wang, S.-Y. & Jiang, S.-Q.: New leaf beetles from Mount Tuomuer, Xinjiang. *Entomotaxonomia* 8 (1/2): 55–58. DP: [30] June 1986 (paper title page).

Chen, S.-H. & Jiang, S.-Q. 1987: New galerucine beetles from Xizang (Coleoptera: Chrysomelidae). *Sinozoologia* 5: 53–59. DP: [31] May 1987 (paper title page).

Chen, S.-H. & Jiang, S.-Q. 1988: Coleoptera: Chrysomelidae – Galerucinae. Pp. 337–345. In: Huang, F.-S., Wang, P.-Y., Yin, W.-Y., Yu, P.-Y., Lee, T.-S., Yang, C.-K. & Wang, X.-J. (eds.): *Insects of Mt. Namjagbarwa region of Xizang*. Beijing: Science Press, xii + 621 pp. DP: [30] June 1988 (colophon).

Chen, S.-H., Jiang, S.-Q. & Wang, S.-Y. 1987: New alpine Galerucinae from Hengduan mountains of Yunnan and Sichuan (Coleoptera: Chrysomelidae). *Sinozoologia* 5: 61–71. DP: [31] May 1987 (paper title page).

Chen, S.-H. & Kung, Y.-T. 1955: Notes on the Chinese species of the halticine genus Phyllotreta Steph. *Acta Entomologica Sinica* 5 (3): 317–326. DP: [30] September 1955 (paper title page).

Chen, S.-H. & Kung, Y. T. 1959: Notes on Chinese cucumber beetles of the genus Aulacophora Chevrolat. *Acta Entomologica Sinica* 9 (4): 373–387. DP: [31] August 1959 (paper title page).

Chen, S.-H. & Pu, F.-J. 1962a: Revision of Chinese Sagrinae. *Acta Entomologica Sinica* 11 (supplement): 103–113. DP: [31] December 1962 (paper title page).

Chen, S.-H. & Pu, F.-J. 1962b: Notes on Chinese Megalopodinae. *Acta Entomologica Sinica* 11 (supplement): 114–119. DP: [31] December 1962 (paper title page).

Chen, S.-H. & Pu, F.-J. 1980: New species of cryptocephaline beetles from Yunnan and Fujian (Coleoptera: Eumolpidae). *Entomotaxonomia* 2 (2): 109–115. DP: [30] June 1980 (paper title page).

Chen, S.-H. & Sun, T.-H. 1962: [new taxa]. In: Chen, S.-H., Tan, C.-C. [= J.-J.], Yu, P.-Y. & Sun, T.-H.: Results of the zoologico-botanical expedition to southwest China 1955–57 (Coleoptera, Hispinae II). *Acta Entomologica Sinica* 11 (supplement): 120–138. DP: [31] December 1962 (paper title page).

Chen, S.-H. & Sun, T.-H. 1964: [new taxa]. In: Chen, S.-H., Sun, T.-H. & Yu, P.-Y.: New hispine beetles from China and Vietnam. *Acta Zootaxonomica Sinica* 1 (1): 106–121. DP: [31] July 1964 (paper title page).

Chen, S.-H. & Tan, C.-C. [= J.-J.] 1961: [new taxa]. In: Chen, S.-H., Tan, C.-C. [= J.-J.] & Yu, P.-Y.: Results of the zoologico-botanical expedition to Southwest China 1955–57 (Coleoptera, Hispinae I). *Acta Entomologica Sinica* 10 (4/6): 457–481. DP: [31] December 1961 (paper title page).

Chen, S.-H. & Tan, C.-C. [= J.-J.] 1962: [new taxa]. In: Chen, S.-H., Tan, C.-C. [= J.-J.], Yu, P.-Y. & Sun, T.-H.: Results of the zoologico-botanical expedition to southwest China 1955–57 (Coleoptera, Hispinae II). *Acta Entomologica Sinica* 11 (supplement): 120–138. DP: [31] December 1962 (paper title page).

Chen, S.-H. & Tan, C.-C. [= J.-J.] 1964: New species of Dactylispa from China (Coleoptera, Hispinae). *Acta Entomologica Sinica* 13 (3): 414–427. DP: [31] May 1964 (paper title page).

Chen, S.-H. & Tan, C.-C. [= J.-J.] 1986: [new taxon]. In: Chen, S.-H., Yu, P.-Y., Sun, C.-H., Tan, C.-C. [= J.-J.] & Zia, Y.-Z.: *Coleoptera Hispidae. Fauna Sinica. Insecta Vol. 2*. Beijing: Science Press, xv + 653 pp., 15 pls. DP: [31] December 1986 (colophon).

Chen, S.-H. & Tan, J.-J. 1985: A new species of Dactylispa from Yunnan, China (Coleoptera: Hispidae). *Acta Zoologica Sinica* **31** (3): 269–271 DP: [30] September 1985 (paper title page).

Chen, S.-H. & Wang, S.-Y. 1962: On the distribution and desert adaptations of the chrysomelid beetles of Sinkiang. *Acta Zoologica Sinica* **14** (3): 337–354 DP: [30] September 1962 (paper title page).

Chen, S.-H. & Wang, S.-Y. 1976: [new taxa]. In: Chen, S.-H., Yu, P.-Y., Wang, S.-Y. & Jiang, S.-Q.: New leaf-beetles from West China. *Acta Entomologica Sinica* **19** (2): 205–224. DP: [31] May 1976 (paper title page).

Chen, S.-H. & Wang, S.-Y. 1980: New genera and species of Chinese Alticinae. *Entomotaxonomia* **2** (1): 1–25. DP: [31] March 1980 (paper title page).

Chen, S.-H. & Wang, S.-Y. 1981a: Coleoptera: Chrysomelidae – Alticinae. Pp. 491–508. In: Qinghai-Xizang Plateau Expedition Group, Chinese Academy of Sciences (eds.): *The series of the scientific expedition to the Qinghai-Xizang Plateau. Insects of Xizang. Vol. 1.* Peking: Science Press, xi + 600 pp. DP: [31] October 1981 (colophon).

Chen, S.-H. & Wang, S.-Y. 1981b: [new taxa]. In: Wang, S.-Y. & Chen, S.-H.: Coleoptera: Chrysomelidae – Chrysomelinae. Pp. 509–516. In: Qinghai-Xizang Plateau Expedition Group, Chinese Academy of Sciences (eds.): *The series of the scientific expedition to the Qinghai-Xizang Plateau. Insects of Xizang. Vol. 1.* Peking: Science Press, xi + 600 pp. DP: [31] October 1981 (colophon).

Chen, S.-H. & Wang, S.-Y. 1984a: New chrysomeline beetles from Hengduan Mountains, Yunnan (Coleoptera: Chrysomelidae). *Acta Zootaxonomica Sinica* **9** (2): 170–175. DP: [30] April 1984 (paper title page).

Chen, S.-H. & Wang, S.-Y. 1984b: Flea-beetles from Hengduan Mountains, Yunnan. Genera Hespera and Yunohespera (Coleoptera: Chrysomelidae). *Acta Entomologica Sinica* **27** (3): 308–322. DP: [31] August 1984 (paper title page).

Chen, S.-H. & Wang, S.-Y. 1985: [new taxon]. In: Huang, D.-S., Han, Y. & Zhang, X.: The insect fauna of the Mt. Tuomuer area in Tianshan. Pp. 53–165. In: *Dengshan Kexue Kaochadui.* [*Biota of Tuomuer Region, Tianshan*]. Beijing: Xinjiang People's Press, 353 pp. DP: [31] March 1985 (colophon).

Chen, S.-H. & Wang, S.-Y. 1986a: [new taxon]. In: Chen, S.-H., Wang, S.-Y. & Jiang, S.-Q.: New leaf beetles from Mount Tuomuer, Xinjiang. *Entomotaxonomia* 8 (1/2): 55–58. DP: [30] June 1986 (paper title page).

Chen, S.-H. & Wang, S.-Y. 1986b: Notes on Chinese flea-beetles of the genus Hespera Weise (Coleoptera: Chrysomelidae). *Acta Zootaxonomica Sinica* **11** (3): 283–306. DP: [31] July 1986 (paper title page).

Chen, S.-H. & Wang, S.-Y. 1987a: Two alpine genera of flea-beetles from Yunnan (Coleoptera: Chrysomelidae). *Acta Entomologica Sinica* **30** (2): 196–200. DP: [31] May 1987 (paper title page).

Chen, S.-H. & Wang, S.-Y. 1987b: Coleoptera: Chrysomelidae – Alticinae. Pp. 57–63. In: *Agricultural insects, spiders, plant diseases and weeds of Xizang. Vol. II.* Lasa: Xizang People Publishing House, 416 pp. DP: [31] August 1987 (paper title page).

Chen, S.-H. & Wang, S.-Y. 1988: Coleoptera: Chrysomelidae – Alticinae. Pp. 347–353. In: Huang, F.-S., Wang, P.-Y., Yin, W.-Y., Yu, P.-Y., Lee, T.-S., Yang, C.-K. & Wang, X.-J. (eds.): *Insects of Mt. Namjagbarwa region of Xizang*. Beijing: Science Press, xii + 621 pp. DP: [30] June 1988 (colophon).

Chen, S.-H., Wang, S.-Y. & Jiang, S.-Q. 1985: A new genus of Galerucinae from West China (Coleoptera: Chrysomelidae). *Acta Zoologica Sinica* **31** (4): 372–376. DP: [31] December 1985 (paper title page).

Chen, S.-H., Wang, S.-Y. & Jiang, S.-Q. 1986: The galerucine genus Capula from West China (Coleoptera: Chrysomelidae). *Acta Zootaxonomica Sinica* **11** (4): 398–400. DP: [31] October 1986 (paper title page).

Chen, S.-H. & Young, B. 1941: The coleopterous genus Asiphytodecta Chen. *Sinensia* **12** (1/6): 199–210. DP: [31] December 1941 (issue cover).

Chen, S.-H. & Yu, P.-Y. 1961: [new taxa]. In: Chen, S.-H., Tan, C.-C. [= J.-J.] & Yu, P.-Y.: Results of the zoologico-botanical expedition to Southwest China 1955–57 (Coleoptera, Hispinae I). *Acta Entomologica Sinica* **10** (4/6): 457–481. DP: [31] December 1961 (paper title page).

Chen, S.-H. & Yu, P.-Y. 1962: [new taxa]. In: Chen, S.-H., Tan, C.-C. [= J.-J.], Yu, P.-Y. & Sun, T.-H.: Results of the zoologico-botanical expedition to southwest China 1955–57 (Coleoptera, Hispinae II). *Acta Entomologica Sinica* **11** (supplement): 120–138. DP: [31] December 1962 (paper title page).

Chen, S.-H. & Yu, P.-Y. 1964: [new taxa]. In: Chen, S.-H., Sun, T.-H. & Yu, P.-Y.: New hispine beetles from China and Vietnam. *Acta Zootaxonomica Sinica* **1** (1): 106–121. DP: [31] July 1964 (paper title page).

Chen, S.-H. & Yu, P.-Y. 1976: [new taxa]. In: Chen, S.-H., Yu, P.-Y., Wang, S.-Y. & Jiang, S.-Q.: New leaf-beetles from West China. *Acta Entomologica Sinica* **19** (2): 205–224. DP: [31] May 1976 (paper title page).

Chen, S.-H., Yu, P.-Y., Sun, C.-H., Tan, C.-C. [= J.-J.] & Zia, Y.-Z. 1986: *Coleoptera Hispidae. Fauna Sinica. Insecta Vol. 2*. Beijing: Science Press, xv + 653 pp., 15 pls. DP: [31] December 1986 (colophon).

Chen, S.-H. & Zia, Y. 1961: Results of the zoologico-botanical expedition to Southwest China 1955–1957 (Coleoptera, Cassidinae). *Acta Entomologica Sinica* **10** (4/6): 439–451. DP: [31] December 1961 (paper title page).

Chen, S.-H. & Zia, Y. 1964: New cassidine beetles from China. *Acta Zootaxonomica Sinica* **1** (1): 122–138. DP: [31] July 1964 (paper title page).

Chen, S.-H. & Zia, Y. 1966a: On the Chinese citrus flea-beetles. *Acta Zootaxonomica Sinica* **3** (1): 67–75. DP: [31] January 1966 (paper title page).

Chen, S.-H. & Zia, Y. 1966b: Two new species of flea-beetles from Tsinghai. *Acta Zootaxonomica Sinica* **3** (2): 148–150. DP: [30] April 1966 (paper title page).

Chen, S.-H. & Zia, Y. 1974: A new genus and species of flea-beetles from Szechwan. *Acta Entomologica Sinica* **17** (4): 479–481. DP: [30] November 1974 (paper title page).

Chen, S.-H. & Zia, Y.-Z. 1984a: A new genus and species of Cassidinae from Yunnan (Coleoptera: Hispidae). *Entomotaxonomia* **6** (2/3): 79–82. DP: [30] September 1984 (paper title page).

Chen, S.-H. & Zia, Y.-Z. 1984b: A new species of Cassidinae from Sichuan (Coleoptera: Hispidae). *Acta Zoologica Sinica* 30 (4): 383–384. DP: [31] December 1984 (paper title page).

Chevin, H. 1985: Contribution à la biologie des Timarcha (Col. Chrysomelidae). II – Timarcha tenebricosa F. *Cahiers Liaison OPIE* 19 (2): 7–14. DP: [31 December] 1985 (paper title page).

Chevin, H. 1986: Contribution à la biologie des Timarcha (Col. Chrysomelidae): Timarcha nicaeensis Villa. *Cahiers Liaison OPIE* 20 (1): 17–21. DP: [31 December] 1986 (paper title page).

Chevin, H. 1994a: Contribution à la biologie des Timarcha (Col. Chrysomelidae). *L'Entomologiste* 50 (3): 161–168. DP: [30] June 1994 (issue cover).

Chevin, H. 1994b: Food selection and life-cycle of the old world Timarcha Latreille, 1829 (Col. Chrysomelinae). Pp. 533–539. In: Jolivet, P. H., Cox, M. L. & Petitpierre, E. (eds.): *Novel aspects of the biology of Chrysomelidae*. Dordrecht: Kluwer, 582 pp. DP: [31 December] 1994 (colophon).

Chevrolat, L. A. A. 1833: [new taxon, pl. 47, livraison 32]. In: *Iconographie du règne animal de G. Cuvier, ou représentation d'après nature de l'une des espèces les plus remarquables et souvent non encore figurées, de chaque genre d'animaux. Avec un texte descriptif mis au courant de la science. Ouvrage pouvant servir d'atlas à tous les traités de zoologie. Tome II. Planches des animaux invertébrés*. [1829–1844]. Paris & Londres: J. B. Bailière, 450 pls. DP (pls. 47–49): [30] September 1833 (Bousquet 2016).

Chevrolat, L. A. A. 1835a: [new taxon]. In: Dejean, P. F. M. A.: *Catalogue des coléoptères de la collection de M. le Comte Dejean*. [1833–1836], Paris: Méquignon-Marvis Père et Fils, 443 pp. DP (pp. 257–360): 22 August 1835 (Bousquet 2016).

Chevrolat, L. A. A. 1835b: *Coléoptères du Mexique. 5e Fascicule*. Strasbourg: G. Silbermann, [48] pp. DP: 10 October 1835 (Bousquet 2016).

Chevrolat, L. A. A. 1836: [new taxa]. In: Dejean, P. F. M. A.: *Catalogue des coléoptères de la collection de M. le Comte Dejean*. [1833–1836], Paris: Méquignon-Marvis Père et Fils, 443 pp. DP (pp. 361–443): [31 December] 1836 (Bousquet 2016).

Chevrolat, L. A. A. 1837: [new taxon]. In: Dejean, P. F. M. A.: *Catalogue des coléoptères de la collection de M. le Comte Dejean. Troisième édition, revue, corrigée et augmentée*. [1836–1837], Paris: Méquignon-Marvis Père et Fils, pp. xiv + 503 pp. DP (pp. 385–503): 27 May 1837 (Bousquet 2016).

Chevrolat, L. A. A. 1840: Description de quelques coléoptères de la Galice et du Portugal provenant d'envois de M. Deyrolles fils. *Revue Zoologique, par la Société Cuvierienne* **3** (1): 8–18. DP: 3 February 1840 (*Compt. Rend. Hebdom. Séances Acad. Sci.* 1840, 10: 206).

Chevrolat, L. A. A. 1842: Bothrionopa. P. 669. In: D'Orbigny, C. V. D. (ed.): *Dictionnaire universel d'histoire naturelle résumant et complétant tous les faits présentés par les encyclopédies, les anciens dictionnaires scientifiques, les oeuvres complètes de Buffon, et les meilleurs traités spéciaux sur les diverses branches des sciences naturelles; donnant la description des êtres et des divers phénomènes de la nature, l'étymologie et la définition des noms scientifiques, les principales applications des corps organiques et inorganiques, à l'agriculture, à la médecine, aux arts industriels, etc.; dirigé par M. Charles d'Orbigny, et enrichi d'un magnifique atlas de 288 planches gravées sur acier. Tome deuxième.* Paris: Renard, Martinet et Cie., 795 + [1] pp. DP (pp. 641–688): 20 June 1842 (Evenhuis 2019b).

Chevrolat, L. A. A. 1844: [new taxa]. In: Guérin-Méneville, F. E.: *Iconographie du Règne Animal de G. Cuvier, ou représentation d'après nature de l'une des espèces les plus remarquables, et souvent non encore figurées, de chaque genre d'animaux. Avec un texte descriptif mis au courant de la science. Ouvrage pouvant servir d'atlas à tous les traités de zoologie. Insectes.* [1829–1838], Paris & Londres: J. B. Baillière, 576 pp. DP: 26 August 1844 (Bousquet 2016).

Chevrolat, L. A. A. 1845: Galérucites. Pp. 4–6. In: D'Orbigny, C. V. D. (ed.): *Dictionnaire universel d'histoire naturelle résumant et complétant tous les faits présentés par les encyclopédies, les anciens dictionnaires scientifiques, les oeuvres complètes de Buffon, et les meilleurs traités spéciaux sur les diverses branches des sciences naturelles; donnant la description des êtres et des divers phénomènes de la nature, l'étymologie et la définition des noms scientifiques, les principales applications des corps organiques et inorganiques, à l'agriculture, à la médecine, aux arts industriels, etc.; dirigé par M. Charles d'Orbigny, et enrichi d'un magnifique atlas de 288 planches gravées sur acier. Tome sixième.* Paris: Renard, Martinet et C., 776 pp. DP (pp. 1–64): 28 April 1845 (Evenhuis 2019b).

Chevrolat, L. A. A. 1847: Plectroscelis. P. 267. In: D'Orbigny, C. V. D. (ed.): *Dictionnaire universel d'histoire naturelle résumant et complétant tous les faits présentés par les encyclopédies, les anciens dictionnaires scientifiques, les oeuvres complètes de Buffon, et les meilleurs traités spéciaux sur les diverses branches des sciences naturelles; donnant la description des êtres et des divers phénomènes de la nature, l'étymologie et la définition des noms scientifiques, les principales applications des corps organiques et inorganiques, à l'agriculture, à la médecine, aux arts industriels, etc.; dirigé par M. Charles d'Orbigny, et enrichi d'un magnifique atlas de 288 planches gravées sur acier. Tome dixième.* Paris: Renard, Martinet et C., 760 pp. DP (pp. 257–384): 11 October 1847 (Evenhuis 2019b).

Chevrolat, L. A. A. 1860: Description de coléoptères nouveaux d'Algérie. *Revue et Magasin Zoologique Pure et Appliquée, 2e Série* **12** (10): 448–459. DP: [30] November 1860 (Dickinson et al. 2011).

Chevrolat, L. A. A. 1863: [new taxa]. Grenier, A.: *Catalogue des coléoptères de France et matériaux pour servir à la faune des coléoptères français.* Paris: A. Grenier, iv + 3–79 + 135 pp. DP: 6 August 1863 (book cover, Grenier 1864, see also the comments in the Dating of publications chapter).

Chevrolat, L. A. A. 1872a: Description de six coléoptères exotiques éclos a Paris. *Annales de la Société Entomologique de Belgique* **14** [1870–1871]: 5–8, pl. 1. DP: 2 March 1872 (*Compt.-Rend. Séances Soc. Ent. Belg.*, 1871–1872: xl).

Chevrolat, L. A. A. 1872b: Description de trois nouvelles espèces d'Espagne. Nouvelles et faits divers. No. 32. *L'Abeille, Mémoires d'Entomologie* 8 [1871–1872]: cxxxi–cxxxii. DP: [30] June 1872 (noticed on p. cxxix).

Chevrolat, L. A. A. 1873: Description de quelques espèces nouvelles de coléoptères de Syrie. *Annales de la Société Entomologique de France, Cinquième Série* **3** (2): 203–206. DP: 20 October 1873 (Lefèvre 1885).

Chevrolat, L. A. A. 1877: [Les diagnoses de nouvelles espèces de bruchides]. *Bulletin de la Société Entomologique de France* **1877** (10): 116–117. DP: 31 May 1877 (Evenhuis 2002).

Chevrolat, L. A. A. 1882: [Notes suivantes]. *Bulletin de la Société Entomologique de France* **1882** (19): 212–213. DP: 20 October 1882 (Evenhuis 2002).

Chittenden, F. H. 1927: The species of Phyllotreta North of Mexico. *Entomologica Americana, New Series* **8** [1927–1928] (1): 1–62, 2 pls. DP: 22 September 1927 (issue cover).

Cho, H.-W. 2016: *Revision and classification of the genus Gonioctena Chevrolat, 1837 (Coleoptera: Chrysomelidae: Chrysomelinae)*. Ph.D. Thesis, University of Wrocław, Faculty of Biological Sciences, 403 pp. DP: 24 February 2016 (handwritten date on the Thesis abstract available at Wroclaw university website).

Cho, H.-W. 2017a: Two new species of the subgenus Brachyphytodecta Bechyně, 1948 (Coleoptera, Chrysomelidae, Gonioctena) from Sichuan, Southwest China. *Zootaxa* **4272** (3): 446–450. DP: 30 May 2017 (footer on the paper title page).

Cho, H.-W. 2017b: Two new species of the Gonioctena mauroi species-group from China (Coleoptera: Chrysomelidae: Chrysomelinae). *Acta Entomologica Musei Nationalis Pragae* **57** (1): 173–181. DP: 30 June 2017 (paper title page).

Cho, H.-W. 2021: Definition of the Gonioctena subgeminata species group (Coleoptera, Chrysomelidae, Chrysomelinae), with descriptions of two new species from China and Vietnam. *ZooKeys* **1032**: 79–90. DP: 16 April 2021 (paper title page).

Cho, H.-W. 2022: Two new species of the nominotypical subgenus Gonioctena Chevrolat, 1836, from China and Laos, and a proposed new species-group (Coleoptera: Chrysomelidae: Chrysomelinae). *Zootaxa* **5150** (1): 111–120. DP: 3 June 1922 (footer on the paper title page).

Cho, H.-W. & Borowiec, L. 2013: A new species of the genus Ambrostoma Motschulsky (Coleoptera, Chrysomelidae, Chrysomelinae) from South Korea, with larval descriptions and biological notes. *ZooKeys* **321**: 1–13. DP: 2 August 2013 (paper title page).

Cho, H.-W. & Borowiec, L. 2014: Two new species of the genus Gonioctena Chevrolat (Coleoptera, Chrysomelidae, Chrysomelinae) from Sichuan, China. *Zootaxa* **3765** (3): 295–300. DP: 18 February 2014 (footer on the paper title page).

Cho, H.-W. & Borowiec, L. 2016a: On the genus Gonioctena Chevrolat (Coleoptera: Chrysomelidae: Chrysomelinae), with descriptions of seven new species from the Oriental region and Palaearctic China. *Zootaxa* **4067** (2): 168–184. DP: 21 January 2016 (footer on the paper title page).

Cho, H.-W. & Borowiec, L. 2016b: [new taxa]. In: Cho, H.-W., Kippenberg, H. & Borowiec, L.: Revision of the Gonioctena nivosa species-group (Coleoptera, Chrysomelidae, Chrysomelinae) in the Holarctic region, with descriptions of two new species. *ZooKeys* **596**: 87–128. DP: 8 June 2016 (paper title page).

Cho, H.-W. & Borowiec, L. 2016c: Revision of the Gonioctena flavoplagiata species-group (Coleoptera: Chrysomelidae: Chrysomelinae), with descriptions of two new species from China and Laos. *Acta Entomologica Musei Nationalis Pragae* **56** (2): 755–768. DP: 15 November 2016 (paper title page).

Cho, H.-W. & Kim, S.-K. 2021: An integrative approach reveals a new species of flightless leaf beetle (Chrysomelidae: Suinzona) from South Korea. *Scientific Reports* **11**: 8595. DP: 21 April 2021 (noticed on p. 11).

Cho, H.-W., Kippenberg, H. & Borowiec, L. 2016: Revision of the Gonioctena nivosa species-group (Coleoptera, Chrysomelidae, Chrysomelinae) in the Holarctic region, with descriptions of two new species. *ZooKeys* **596**: 87–128. DP: 3 June 2016 (paper title page).

Cho, H.-W. & Lee, J. E. 2010: Gonioctena koryeoensis (Coleoptera: Chrysomelidae: Chrysomelinae), a new species from Korea, with a description of immature stages. *Zootaxa* **2438**: 52–60. DP: 27 April 2010 (footer on the paper title page).

Cho, H.-W., Takizawa, H. & Borowiec, L. 2016: Notes on Gonioctena tredecimmaculata (Jacoby, 1888), with descriptions of two new species from Taiwan (Coleoptera: Chrysomelidae: Chrysomelinae). *Annales Zoologici* (Warszawa) **66** (3): 357–369. DP: 1 September 2016 (verso of the issue cover).

Chobaut, A. 1896a: Description du Pachybrachis Theryi n. sp. (Col.). *Bulletin de la Société Entomologique de France* **1896** (5): 151–152. DP: 18 May 1896 (*Bibl. Zool.* 1896, 1: 272).

Chobaut, A. 1896b: Descriptions de deux coléoptères nouveaux du Sud Oranais. *Bulletin de la Société Entomologique de France* **1896** (17): 388–390. DP: [31 December] 1896 (footer on p. 381).

Chobaut, A. 1898a: Description de quelques especes et variétés nouvelles de coléoptères Algériens. *Revue d'Entomologie* **17** (3): 74–88. DP: [31] March 1898 (footer on pp. 61 and 77).

Chobaut, A. 1898b: Description d'un eumolpide nouveau du Sahara algérien (Col.). *Bulletin de la Société Entomologique de France* **1898** (16): 308–309. DP: [31 December] 1898 (issue title page).

Chobaut, A. 1899: Description de deux espèces et d'une variété nouvelles de Pachybrachis de la Tunisie méridionale (Col.). *Bulletin de la Société Entomologique de France* **1899** (19): 388–390. DP: [31 December] 1899 (issue title page).

Chobaut, A. 1902: A propos des variétés du Cryptocephalus crassus Ol. (Col.). *Bulletin de la Société Entomologique de France* **1902** (8): 165–166. DP: 13 May 1902 (*dates d'apparition* at the end of volume).

Chobaut, A. 1907: Les sous-variétés de Crioceris macilenta Weise (Col.). *Bulletin de la Société Entomologique de France* **1907** (11): 177–179. DP: 28 July 1907 (*dates d'apparition* at the end of volume).

Chûjô, M. 1932a: Studies on the Chrysomelidae in the Japanese Empire (I). *Transactions of the Natural History Society of Formosa* **22** (121): 310–317. DP: 31 August 1932 (Chûjô 1972).

Chûjô, M. 1932b: Studies on the Chrysomelidae in the Japanese Empire (II). *Transactions of the Natural History Society of Formosa* **22** (122): 337–345. DP: 31 October 1932 (Chûjô 1972).

Chûjô, M. 1933a: Studies on the Chrysomelidae in the Japanese Empire (III). *Sylvia* **4** (1): 19–56. DP: 20 March 1932 (Chûjô 1972).

Chûjô, M. 1933b: Studies on the Chrysomelidae in the Japanese Empire (IV). *Transactions of the Natural History Society of Formosa* **23** (128/129): 304–334, pl. 1. DP: 1 December 1933 (Chûjô 1972).

Chûjô, M. 1934a: Studies on the Chrysomelidae in the Japanese Empire (VI). *Transactions of the Natural History Society of Formosa* **24** (133): 242–247. DP: 9 August 1934 (Chûjô 1972).

Chûjô, M. 1934b: Addition and revision of the Chrysomelidae from the Japanese Empire (I). *Transactions of the Natural History Society of Formosa* **24** (133): 248–250. DP: 9 August 1934 (Chûjô 1972).

Chûjô, M. 1934c: Studies on the Chrysomelidae in the Japanese Empire (V). *Sylvia* **5** (3): 145–182. DP: 22 September 1934 (Chûjô 1972).

Chûjô, M. 1934d: Description of a new Temnaspis-species from Corea (Col.: Chrysomelidae). *Journal of Chosen Natural History Society* **19**: 34, 1 pl. DP: 30 September 1934 (Chûjô 1972).

Chûjô, M. 1934e: Descriptions of some new Cryptocephalus-species (Col. Chrysom.) from Formosa. *Journal of Society of Tropical Agriculture* **6** (3): 513–524. DP: 15 November 1934 (Chûjô 1972).

Chûjô, M. 1934f: H. Sauter's Formosa-Ausbeute: Subfamilies Criocerinae, Clytrinae und Cryptocephalinae (Coleoptera: Chrysomelidae). *Arbeiten über Morphologische und Taxonomische Entomologie aus Berlin-Dahlem* **1** (4): 281–292. DP: 5 December 1934 (volume title page).

Chûjô, M. 1934g: Studies on the Chrysomelidae in the Japanese Empire (VII). *Transactions of the Natural History Society of Formosa* **24** (135): 518–537. DP: 30 December 1934 (Chûjô 1972).

Chûjô, M. 1935a: Chrysomelidae of Loo-Choo Archipelago (I). *Transactions of the Natural History Society of Formosa* **25** (136/139): 69–89. DP: 1 April 1935 (Chûjô 1972).

Chûjô, M. 1935b: Chrysomelidae of Loo-Choo Archipelago (II). *Transactions of the Natural History Society of Formosa* **25** (142): 203–211. DP: 1 July 1935 (Chûjô 1972).

Chûjô, M. 1935c: Description of a new chrysomelid from the Tonkin. *Revue Française d'Entomologie* **2** (2): 94–95. DP: 1 August 1935 (footer on p. 81).

Chûjô, M. 1935d: H. Sauter's Formosa-Ausbeute: Subfamily Galerucinae (Coleoptera: Chrysomelidae). *Arbeiten über Morphologische und Taxonomische Entomologie aus Berlin-Dahlem* **2** (3): 160–174. DP: 15 August 1935 (volume title page).

Chûjô, M. 1935e: Descriptions of a new genus and two new species of Chrysomelidae (Col.). *Arbeiten über Morphologische und Taxonomische Entomologie aus Berlin-Dahlem* **2** (3): 219–220. DP: 15 August 1935 (volume title page).

Chûjô, M. 1935f: Studies on the Chrysomelidae in the Japanese Empire (VIII) Subfamily Halticinae (1). *Transactions of the Natural History Society of Formosa* **25** (145): 354–369. DP: 1 October 1935 (Chûjô 1972).

Chûjô, M. 1935g: Studies on the Chrysomelidae in the Japanese Empire (VIII) Subfamily Halticinae (2). *Transactions of the Natural History Society of Formosa* **25** (146): 392–400. DP: 1 November 1935 (Chûjô 1972).

Chûjô, M. 1935h: Studies on the Chrysomelidae in the Japanese Empire (VIII) Subfamily Halticinae (3). *Transactions of the Natural History Society of Formosa* 25 (147): 459–476. DP: 1 December 1935 (Chûjô 1972).

Chûjô, M. 1936a: Studies on the Chrysomelidae in the Japanese Empire (VIII) Subfamily Halticinae (4). *Transactions of the Natural History Society of Formosa* 26 (148): 15–30. DP: 1 January 1936 (Chûjô 1972).

Chûjô, M. 1936b: Studies on the Chrysomelidae in the Japanese Empire (VIII) Subfamily Halticinae (5). *Transactions of the Natural History Society of Formosa* 26 (149): 84–92. DP: 1 February 1936 (Chûjô 1972).

Chûjô, M. 1936c: Studies on the Chrysomelidae in the Japanese Empire (VIII) Subfamily Halticinae (6). *Transactions of the Natural History Society of Formosa* 26 (150): 108–114. DP: 1 March 1936 (Chûjô 1972).

Chûjô, M. 1936d: Chrysomelid beetles from Corea, collected by Messrs. A. Umeno and K. Yamauchi. *Bulletin of Umeno Entomologial Laboratory* (Kurume) 3: 7–13. DP: 25 June 1936 (Chûjô 1972).

Chûjô, M. 1937a: Studies on the Chrysomelidae in the Japanese Empire (VIII) Subfamily Halticinae (7). *Transactions of the Natural History Society of Formosa* 27 (161): 35–48. DP: 1 February 1937 (Chûjô 1972).

Chûjô, M. 1937b: Studies on the Chrysomelidae in the Japanese Empire (VIII) Subfamily Halticinae (8). *Transactions of the Natural History Society of Formosa* 27 (162): 52–58. DP: 1 March 1937 (Chûjô 1972).

Chûjô, M. 1937c: Studies on the Chrysomelidae in the Japanese Empire (VIII) Subfamily Halticinae (9). *Transactions of the Natural History Society of Formosa* 27 (163): 95–104. DP: 1 April 1937 (Chûjô 1972).

Chûjô, M. 1937d: Descriptions of four new species of the genus Pedrillia Westwood, with a list of the species known from the world (Col. Chrysomelidae). *Transactions of the Natural History Society of Formosa* 27 (165): 146–151. DP: 1 June 1937 (Chûjô 1972).

Chûjô, M. 1937e: Some additions and revisions of Bruchidae (Coleoptera) from the Japanese Empire. *Transactions of the Natural History Society of Formosa* 27 (167/168): 189–201. DP: 1 September 1937 (Chûjô 1972).

Chûjô, M. 1937f: Descriptions of two hispid-beetles, belonging to a new genus, from the South Sea Islands under the Japanese mandate (Coleoptera: Chrysomelidae). *Transactions of the Natural History Society of Formosa* 27 (169): 222–228. DP: 1 October 1937 (Chûjô 1972).

Chûjô, M. 1938a: Description of a new Longitarsus-species from Hokkaido, Japan (Coleoptera: Chrysomelidae). *Transactions of the Natural History Society of Formosa* 28 (172): 15–16. DP: 1 January 1938 (Chûjô 1972).

Chûjô, M. 1938b: H. Sauter's Formosa collection: Subfamily Eumolpinae (Coleoptera: Chrysomelidae II). *Arbeiten über Morphologische und Taxonomische Entomologie aus Berlin-Dahlem* 5 (1): 25–36. DP: 5 February 1938 (volume title page).

Chûjô, M. 1938c: H. Sauter's Formosa-collection: Subfamily Galerucinae (Coleoptera: Chrysomelidae). Second part. *Arbeiten über Morphologische und Taxonomische Entomologie aus Berlin-Dahlem* 5 (2): 135–152. DP: 30 May 1938 (volume title page).

Chûjô, M. 1938d: Beitrag zur Kenntnis der Chrysomeliden-Fauna Chinas (Coleoptera). *Mushi* 11 (2): 158–169. DP: [31] August 1938 (issue cover).

Chûjô, M. 1938e: H. Sauter's Formosa collection: Subfamily Cryptocephalinae. Coleoptera-Chrysomelidae III. *Arbeiten über Morphologische und Taxonomische Entomologie aus Berlin-Dahlem* 5 (3): 294. DP: 11 August 1938 (volume title page).

Chûjô, M. 1938f: Beitrag zur Kenntnis der Chrysomeliden-Fauna von Kyûshû, Japan (2). *Bulletin of the Umeno Entomological Laboratory* (Kurume) 6: 5–13. DP: 30 November 1938 (Chûjô 1972).

Chûjô, M. 1940a: Beitrag zur Chrysomeliden-Fauna der Insel Shikoku (Japan) I. *Kontyû* 14 (2): 65–82. DP: 20 April 1940 (Chûjô 1972).

Chûjô, M. 1940b: Beitrag zur Chrysomeliden-Fauna der Insel Shikoku (Japan) II. *Kontyû* 14 (3): 106–125. DP: 30 May 1940 (Chûjô 1972).

Chûjô, M. 1940c: Fulcidacinae of Formosa (Col. Chrysomelidae). *Transactions of the Natural History Society of Formosa* 30 (202): 265–293. DP: 26 July 1940 (Chûjô 1972).

Chûjô, M. 1940d: Chrysomelid-beetles from Korea (I). *Transactions of the Natural History Society of Formosa* 30 (204): 349–362. DP: 1 September 1940 (Chûjô 1972).

Chûjô, M. 1940e: Chrysomelid-beetles from Borodino Isls. (Daitô-Zima), Loo-Choo, collected by Mr. M. Yanagiwara. *Transactions of the Natural History Society of Formosa* 30 (204): 363–365. DP: 1 September 1940 (Chûjô 1972).

Chûjô, M. 1940f: Chrysomelid-beetles trom Korea (II). *Transactions of the Natural History Society of Formosa* 30 (205): 383–398. DP: 1 October 1940 (Chûjô 1972).

Chûjô, M. 1941a: Chrysomelid-beetles from Korea (III). *Transactions of the Natural History Society of Formosa* 31 (209): 61–75. DP: 1 February 1941 (Chûjô 1972).

Chûjô, M. 1941b: Chrysomelid-beetles from Korea (IV). *Transactions of the Natural History Society of Formosa* 31 (211): 155–174. DP: 13 April 1941 (Chûjô 1972).

Chûjô, M. 1941c: First supplement to the fauna of Korean chrysomelid-beetles (I). *Transactions of the Natural History Society of Formosa* 31 (219): 451–462. DP: 1 December 1941 (Chûjô 1972).

Chûjô, M. 1941d: Description of new chrysomelid-beetles from Formosa (I). *Transactions of the Natural History Society of Formosa* 31 (219): 463–467. DP: 1 December 1941 (Chûjô 1972).

Chûjô, M. 1942a: First supplement to the fauna of Korean chrysomelid-beetles (II). *Transactions of the Natural History Society of Formosa* 32 (220): 31–43. DP: 1 January 1942 (Chûjô 1972).

Chûjô, M. 1942b: Fulcidacinae of Japan proper and Loo-choo (Coleoptera-Chrysomelidae). *Transactions of the Natural History Society of Formosa* 32 (221): 77–100. DP: 25 February 1942 (Chûjô 1972).

Chûjô, M. 1942c: Chrysomelid-beetles from Kwantung-Province. *Mushi* 14 [1941–1942] (2): 51–65. DP: 31 March 1942 (verso of the volume title page).

Chûjô, M. 1942d: Descriptions of two new Orsodacne-species from Japan proper (Col. Chrysomelidae). *Kontyû* 16 (1): 33–36. DP: 30 April 1942 (Chûjô 1972).

Chûjô, M. 1944 Description of new chrysomelid beetle from Formosa (II). *Transactions of the Natural History Society of Taiwan* 33 [1943] (242/243): 570–572. DP: 5 February 1944 (Chûjô 1972).

Chûjô, M. 1949a: On the Hebdomecosta Spaeth with description of a new species from Japan (Coleoptera-Chrysomelidae-Cassidinae). *Transactions of the Kansai Entomological Society* 14 (2): 7–13. DP: 7 November 1949 (Chûjô 1972).

Chûjô, M. 1949b: Descriptions of the Orsodacne-species from Japan (Coleoptera: Chrysomelidae). *Entomological Review of Japan* 1 (2): 22–26. DP: 28 November 1949 (Chûjô 1972).

Chûjô, M. 1950: Description of a new chrysomelid-beetle from Japan. *Kontyû* 18 (5): 114–116. DP: 30 August 1950 (Chûjô 1972).

Chûjô, M. 1951a: A taxonomic study on the Chrysomelidae (Insecta-Coleoptera) of Formosa. Part I subfamily Criocerinae. *Technical Bulletin of the Kagawa Agricultural College* 2 (2): 71–120. DP: 15 January 1951 (Chûjô 1972).

Chûjô, M. 1951b: Chrysomelid-beetles of Shikoku, Japan (II) (Coleoptera). *Transactions of the Shikoku Entomological Society* 2 (3): 31–48. DP: 30 June 1951 (Chûjô 1972).

Chûjô, M. 1951c: Description of a new chrysomelid-beetle from Japan. *Insecta Matsumurana* 17 (3/4): 119–120. DP: 30 June 1951 (Chûjô 1972).

Chûjô, M. 1951d: A taxonomic study on the Chrysomelidae (Insecta-Coleoptera) from Formosa. Part III subfamily Megalopodinae. *Technical Bulletin of the Kagawa Agricultural College* 3 (2): 57–71. DP: 30 September 1951 (Chûjô 1972).

Chûjô, M. 1952a: A taxonomic study on the Chrysomelidae (Insecta-Coleoptera) from Formosa. Part V subfamily Clytrinae (1). *Technical Bulletin of the Kagawa Agricultural College* 4 (1): 32–49. DP: 30 July 1952 (Chûjô 1972).

Chûjô, M. 1952b: A taxonomic study on the Chrysomelidae (Insecta-Coleoptera) from Formosa. Part V subfamily Clytrinae (2). *Technical Bulletin of the Kagawa Agricultural College* 4 (2): 73–91. DP: 30 July 1952 (Chûjô 1972).

Chûjô, M. 1953a: Description d'une espèce nouvelle de Zeugophora du Japon (Coleoptera, Chrysomelidae). *Bulletin d'Institut Royal des Sciences Naturelles de Belgique* 29 (9): 1–3. DP: [31] January 1953 (paper title page and Chûjô 1972).

Chûjô, M. 1953b: A taxonomic study on the Chrysomelidae (Insecta-Coleoptera) from Formosa. Part VI subfamily Lamprosominae. *Technical Bulletin of the Kagawa Agricultural College* 4 (3): 237–249. DP: 31 March 1953 (Chûjô 1972).

Chûjô, M. 1954a: Description of a new chrysomelid-beetle belonging to the galerucine genus Atysa Baly from Japan. *Mushi* 26 (1): 1–4. DP: 15 March 1954 (paper title page and Chûjô 1972).

Chûjô, M. 1954b: Descriptions of a new species and a new subspecies of the genus Lypesthes Baly from Japan, including some notices on its two other known-species (Coleoptera: Chrysomelidae, Eumolpinae). *Insecta Matsumurana* 18 (3/4): 103–108. DP: 30 September 1954 (Chûjô 1972).

Chûjô, M. 1954c: Chrysomelid-beetles from Shikoku, Japan (III). *Transactions of the Shikoku Entomological Society* 4 [1954–1956] (4): 51–62. DP: 1 November 1954 (Chûjô 1972).

Chûjô, M. 1954d: A taxonomic study on the Chrysomelidae (Insecta-Coleoptera) from Formosa. Part VII. Subfamily Cryptocephalinae. *Quarterly Journal of the Taiwan Museum* 7 (3/4): 138–247. DP: [31] December 1954 (Chûjô 1972).

Chûjô, M. 1956a: A taxonomic study on the Chrysomelidae (Insecta-Coleoptera) from Formosa. IX Subfamily Chlamisinae (2). *Technical Bulletin of the Kagawa Agricultural College* 7 [1955–1956] (2): 153–167. DP: 25 February 1956 (reprint title page and Chûjô 1972).

Chûjô, M. 1956b: A taxonomic study on the Chrysomelidae (Insecta: Coleoptera) from Formosa. Part VIII. Subfamily Eumolpinae. *Philippine Journal of Science* **85** (1): 1–180. DP: [31] March 1956 (paper title page and Chûjô 1972).

Chûjô, M. 1956c: Contribution to the fauna of Chrysomelidae (Coleoptera) in Japan (1). *Memories of the Faculty of Liberal Arts & Education, Kagawa University, Part II* **31**: 1–20. DP: 5 April 1956 (Chûjô 1972).

Chûjô, M. 1957a: Chrysomelid-beetles of Loo-Choo Archipelago (III). *Kontyû* **25** (1): 13–20. DP: 1 February 1957 (Chûjô 1972).

Chûjô, M. 1957b: Beitrag zur Kenntnis der Chrysomelidenfauna Chinas (Coleoptera) II. *Memories of the Faculty of Liberal Arts & Education, Kagawa University, Part II* **47**: 1–5. DP: 17 June 1957 (Chûjô 1972).

Chûjô, M. 1957c: Chrysomelid-beetles of Loo-Choo Archipelago (IV). *Memories of the Faculty of Liberal Arts & Education, Kagawa University, Part II* **52**: 1–8. DP: 28 December 1957 (Chûjô 1972).

Chûjô, M. 1958a: Chrysomelid-beetles from the islands Iki and Tsushima, Japan, collected by Dr. K. Baba in 1957. *Memories of the Faculty of Liberal Arts & Education, Kagawa University, Part II* **58**: 1–10. DP: [31] March 1958 (reprint title page).

Chûjô, M. 1958b: A taxonomic study on the Chrysomelidae (Insecta, Coleoptera) from Formosa. Part X. Subfamily Chrysomelinae. *Quarterly Journal of the Taiwan Museum* **11** (1/2): 1–85 + 8 pls. DP: [30] June 1958 (Chûjô 1972).

Chûjô, M. 1958c: Chrysomelid-beetles of Loo-Choo Archipelago (V). *Memories of the Faculty of Liberal Arts & Education, Kagawa University, Part II* **64**: 1–19. DP: 28 December 1958 (Chûjô 1972).

Chûjô, M. 1959: Contribution to the fauna of Chrysomelidae (Coleoptera) in Japan (III). *Memories of the Faculty of Liberal Arts & Education, Kagawa University, Part II* **81**: 1–16. DP: 31 May 1959 (Chûjô 1972).

Chûjô, M. 1961: Chrysomelid-beetles of Loochoo Archipelago (VI). *Publication of Entomological Laboratory, University of Osaka Prefecture* **6**: 83–91. DP: 30 October 1961 (footer on the paper title page and Chûjô 1972).

Chûjô, M. 1962a: Coleoptera from Southeast Asia, 24. Family Chrysomelidae. Pp. 103–106. In: Kira, T. & Umesao, T. (eds.): *Nature and life in Southeast Asia. Volume II*. Kyoto: Fauna and Flora Research Society, 276 pp. DP: 30 March 1962 (Chûjô 1972).

Chûjô, M. 1962b: A taxonomic study on the Chrysomelidae (Insecta: Coleoptera) from Formosa. Part XI. Subfamily Galerucinae. *Philippine Journal of Science* **91** (1/2): 1–239. DP: [30] April 1962 (Chûjô 1972).

Chûjô, M. 1962c: Description of a new chrysomelid-beetle from Japan. Contribution to the fauna of Chrysomelidae (Coleoptera) in Japan (IV). *Niponius, Acta Coleopterologica* 1 (18): 1–3. DP: 13 September 1962 (paper title page).

Chûjô, M. 1963a: Chrysomelid-beetles from Formosa (Taiwan) collected by Dr. K. Baba in 1962. *Niponius, Acta Coleopterologica* 2 (5): 21–30. DP: 13 September 1963 (paper title page and Chûjô 1972).

Chûjô, M. 1963b: Chrysomelid-beetles from Formosa, preserved in the Hungarian Natural History Museum, Budapest. *Annales Historico-Naturales Musei Nationalis Hungarici* 55: 379–402. DP: 31 December 1963 (Merkl et al. 2008).

Chûjô, M. 1964: Coleoptera from Southeast Asia (III). Pp. 161–315. In: Kira, T. & Umesao, T. (eds.): *Nature and life in Southeast Asia. Volume III*. Kyoto: Fauna and Flora Research Society, 466 pp. DP: [31] March 1964 (Chûjô 1972).

Chûjô, M. 1965a: Some chrysomelid beetles from East Asia in the California Academy of Sciences (Coleoptera: Chrysomelidae). *Pan-Pacific Entomologist* 41 (1): 54–58. DP: 4 March 1965 (verso of the issue cover).

Chûjô, M. 1965b: Chrysomelid-beetles of Formosa (I). *Special Bulletin of the Lepidopterological Society of Japan* 1: 88–104. DP: 30 December 1965 (volume title page, Chûjô 1972).

Chûjô, M. 1966a: Coleoptera of East Nepal: 14. Family Chrysomelidae. *Journal of the College of Arts and Sciences, Chiba University* 4 (4): 553–557. DP: [30] November 1966 (paper title page and Chûjô 1972).

Chûjô, M. 1966b: Chrysomelid-beetles from Northeast Nepal. *Memories of the Faculty of Education Kagawa University, Part II* 145: 1–35. DP: 10 December 1966 (reprint title page and Chûjô 1972).

Chûjô, M. & Goecke, H. 1956: Contribution to the fauna of Chrysomelidae (Coleoptera) in Japan (II). *Akitu* 5 (3): 60–62. DP: 7 July 1956 (Chûjô 1972).

Chûjô, M. & Kimoto, S. 1960: Descriptions of three genera and a new species of chrysomelid-beetles from Japan, with some notes on the Japanese species. *Niponius, Acta Coleopterologica* 1 (4): 1–10. DP: 3 March 1960 (paper title page and Chûjô 1972).

Chûjô, M. & Kimoto, S. 1961: Systematic catalogue of Japanese Chrysomelidae (Coleoptera). *Pacific Insects* 3 (1): 117–202. DP: 20 April 1961 (paper title page).

Chûjô, M. & Ohno, M. 1961: Revision of the genus Horaia Chûjô and description of a new Monolepta-species from Japan (Coleoptera: Chrysomelidae). *Memories of the Faculty of Liberal Arts & Education, Kagawa University, Part II* 106: 1–12. DP: 10 October 1961 (footer on the paper title page and Chûjô 1972).

Chûjô, M. & Ohno, M. 1964: [new taxon]. In: Ohno, M.: Revision of Sphaeroderma-species occurring in Japan and the Loo-Choos (Coleoptera, Chrysomelidae, Alticinae). *Journal of the Toyo University, General Education, Natural Science* 5: 29–60. DP: [31] May 1964 (reprint cover).

Chûjô, M. & Ohno, M. 1965: A revision of Luperomorpha-species occurring in Japan and the Loo-Choo Islands (Coleoptera, Chrysomelidae, Alticinae). *Memoirs of the Faculty of Liberal Arts & Education, Kagawa University, Part II* 131: 1–16 DP: 31 January 1965 (footer on p. 1 and Chûjô 1972).

Chûjô, M. & Shirôzu, T. 1955: Chrysomelid beetles from the islands Yakushima and Tanegashima, Kyushu, Japan. *Sieboldia* 1 (3): 237–244. DP: 15 June 1955 (Chûjô 1972).

Čížek, P. 2003: Phyllotreta fornuseki sp. n. (Coleoptera: Chrysomelidae) aus Mähren und aus der Slowakei. *Klapalekiana* 39 (1/3): 63–65. DP: 26 November 2003 (paper title page).

Čížek, P. 2006: *Dřepčíci (Coleoptera: Chrysomelidae: Alticinae) Česka a Slovenska.* [*The flea beetles (Coleoptera: Chrysomelidae: Alticinae) of the Czech and Slovak Republics*]. Nové Město nad Metují: Městské muzeum Nové Město nad Metují, 76 pp. DP: [31 December] 2006 (book cover).

Čížek, P. 2007: Dibolia oudai sp. n. (Coleoptera: Chrysomelidae) aus der Slowakei. *Klapalekiana* 43 (1/2): 15–17. DP: 31 July 2007 (paper title page).

Čížek, P. 2019: Aphthona tatrica sp. nov. (Coleoptera: Chrysomelidae: Galerucinae: Alticini) aus der Slowakei. *Klapalekiana* 55 (3/4): 183–186. DP: 27 December 2019 (paper title page).

Čížek, P. & Doguet, S. 2008: *Klíč k určování dřepčíků (Coleoptera: Chrysomelidae: Alticinae) Česka a Slovenska.* [*Key to identification of flea beetles (Coleoptera: Chrysomelidae: Alticinae) of the Czech and Slovak Republics*]. Nové Město nad Metují: Městské muzeum Nové Město nad Metují, 232 pp. DP: [31 December] 2008 (book cover).

Clark, H. 1861: *Catalogue of Halticidae in the collection of the British Museum. Physapodes and Oedipodes. Part 1.* [1860], London: Trustees of the British Museum, xii + 301 pp., 9 pls. DP: 12 January 1861 (Sherborn 1934).

Clark, H. 1864: Descriptions of new Australian phytophaga. *Journal of Entomology* 2 [1863–1866] (11): 247–263, pl. XII. DP: [31] December 1864 (issue title page).

Clark, H. 1865a: Descriptions of species of phytophaga received from Pulo Penang or its neighbourhood. *Annals and Magazine of Natural History, Third Series* 15 (86): 139–148. DP: 1 February 1865 (Evenhuis 2003).

Clark, H. 1865b: An examination of the Dejeanian genus Coelomera (Coleoptera Phytophaga) and its affinities. *Annals and Magazine of Natural History, Third Series* 16 (94): 256–268. DP: 1 October 1865 (Evenhuis 2003).

Clark, H. 1866a: *A catalogue of Phytophaga. (Coleoptera, Pseudotetramera). Part I. I. Sagridae. II. Donaciidae. III. Crioceridae. IV. Megalopidae. With an appendix, containing descriptions of new species.* London & Edingburg: Williams and Norgate; Paris: A. Deyrolle & Son, 50 + 88 pp. DP: 14 April 1866 (Bousquet 2016).

Clark, H. 1866b: [new taxa]. In: Bates, H. W. & Clark, H.: *Descriptions of new species of Phytophaga beeing the appendix to the catalogue of Phytophaga.* Pp. 1–88. In: Clark, H.: *A catalogue of Phytophaga. (Coleoptera, Pseudotetramera). Part I. I. Sagridae. II. Donaciidae. III. Crioceridae. IV. Megalopidae. With an appendix, containing descriptions of new species.* London & Edingburg: Williams and Norgate; Paris: A. Deyrolle & Son, 50 + 88 pp. DP: 14 April 1866 (Bousquet 2016).

Clark, S. M., LeDoux, D. G., Seeno, T. N., Riley, E. G., Gilbert, A. J. & Sullivan, J. M. 2004: *Host plants of leaf beetle species occurring in the United States and Canada (Coleoptera: Orsodacnidae, Megalopodidae, Chrysomelidae exclusive of Bruchinae).* Special publication No. 2. Sacramento: The Coleopterists Society, 476 pp. DP: 20 April 2004 (colophon).

Clavareau, C. H. 1905: [new taxon]. In: Jacoby, M. & Clavareau, H.: Coleoptera Phytophaga. Fam. Crioceridae. In: Wytsman, P. (ed.): *Genera Insectorum. Fascicule 23*. Bruxelles: V. Verteneuil & L. Desmet, 40 pp., 5 pls. DP: 7 January 1905 (Evenhuis 1994).

Clavareau, C. H. 1906: [new taxon]. In: Jacoby, M. & Clavareau, H.: Coleoptera Phytophaga, Fam. Chrysomelidae, Subfam. Clytrinae. In: Wytsman, P. (ed.): *Genera Insectorum. Fascicule 49*. [1906–1907], Bruxelles: V. Verteneuil & L. Desmet, 87 pp., 5 pls. DP: 10 November 1906 (Evenhuis 1994).

Clavareau, C. H. 1913: Chrysomelidae: 5. Megascelinae, 6. Megalopodinae, 7. Clytrinae, 8. Cryptocephalinae, 9. Chlamydinae, 10. Lamprosominae. In: Schenkling, S. (ed.): *Coleopterorum Catalogus. Pars 53*. Berlin: W. Junk, 278 pp. DP: 28 August 1913 (book cover).

Clavareau, C. H. 1914: Chrysomelidae: 11. Eumolpinae. In: Schenkling, S. (ed.). *Coleopterorum Catalogus. Pars 59*. Berlin: W. Junk, 215 pp. DP: 22 June 1914 (book cover).

Cobos, A. 1952: Nuevas especies de coleopteros Phytophagoidea de la provincia de Almería (Sudeste Español). *Bollettino dell'Associazione Romana di Entomologia* 7 (1): 1–8. DP: 29 January 1952 (last page of the paper).

Cobos, A. 1953: Especies nuevas de coleópteros de la provincia de Almeria, I. *Archivos del Instituto de Aclimatación* (Almería) 1: 127–134. DP: [31] August 1953 (handwritten reception date on the Alonso-Zarazaga's personal copy).

Cobos, A. 1954a: Coleopteros de Sierra Nevada (Familias Cerambycidae y Chrysomelidae). *Archivos del Instituto de Aclimatación* (Almería) 2: 139–155, pl. I. DP: 21 December 1954 (Alonso-Zarazaga 2001).

Cobos, A. 1954b: Una breve campaña entomológica por las Sierras de Tejeda y Almijara (Provincia de Málaga): ins. coleópteros. *Archivos del Instituto de Aclimatación* (Almería) 3: 29–39. DP: 21 December 1954 (Alonso-Zarazaga 2001).

Cobos, A. 1954c: Nota sobre el genero Cyrtonus Latr. (Coleoptera Chrysomelidae) y diagnosis de formas nuevas ibero-marroquies. *Tamuda* 2 (2): 318–324. DP: [31 December] 1954 (title page of reprint, see also comment in Alonso-Zarazaga 2001).

Cobos, A. 1955: Dos nuevas formas de coleópteros Phytophagoidea de Andalucía y notas complementarias. *Archivos del Instituto de Aclimatación* (Almería) 4: 59–62. DP: [31 December] 1955 (Alonso-Zarazaga 2001).

Cobos, A. 1956: Estudios sobre Clytritae (Col. Chrysomelidae) del norte de Africa. *Tamuda* 4 (2): 169–182. DP: [31 December] 1956 (Alonso-Zarazaga 2001).

Cobos, A. 1957: Las Gynandrophthalma Lac. del subgénero Otiocephala Lef. (Coleoptera, Chrysomelidae). *Archivos del Instituto de Aclimatación* (Almería) 6: 163–190, 1 pl. DP: [31 December] 1957 (Alonso-Zarazaga 2001).

Cobos, A. 1958: Mision cientifica J. Mateu al Sahara frances. Coleopteros: Cicindelidae, Malachiidae, Dasytidae, Elateridae, Meloidae, Rhipiphoridae, Alleculidae, Tenebrionidae, Cerambycidae y Chrysomelidae. *Tamuda* 6 (2): 341–354. DP: [31 December] 1958 (Alonso-Zarazaga 2001).

Cobos, A. 1961: Nota sobre Chrysomelidae ibericos (Coleoptera). *Graellsia* **19** (1/3): 27–43. DP: [31 December] 1961 (issue cover).

Cobos, A. 1969: Nuevos materiales para el estudio del género Gynandrophthalma Lac. (Col. Chrysomelidae). *Graellsia* **24** [1968]: 67–77. DP: 30 June 1969 (issue cover).

Codina Padilla, F. 1958: Las Lachnaea Redt. ibero-marroquies (Col. Chrysomelidae). *Graellsia* **16** (1/3): 33–44. DP: [31 December] 1958 (issue cover).

Codina Padilla, F. 1960: Chrysomela vagecincta Fairmaire ssp. obscuriventris nov. (Col. Chrysomelidae). *Comptes Rendus des Séances Mensuelles de la Société des Sciences Naturelles et Physiques du Maroc* **26** (4): 66–68. DP: 25 July 1960 (noticed on p. 76).

Codina Padilla, F. 1961a: Apuntes sobre Chrysomelidae ibero-marroquies (Coleoptera). *Graellsia* **19** (1/3): 45–57. DP: [31 December] 1961 (issue cover).

Codina Padilla, F. 1961b: Nota sobre los Luperus marroquies del subgenero Calomicrus Stephens (Col. Chrysomelidae). *Graellsia* **19** (1/3): 69–79. DP: [31 December] 1961 (issue cover).

Codina Padilla, F. 1962: Nota sobre las Chrysomela ibero-marroquies del subgenero Polysticta Hope (Col. Chrysomelidae). *Bulletin de la Société des Sciences Naturelles et Physiques du Maroc* **41** [1961] (1): 69–83. DP: [30] November 1962 (last page of the issue).

Codina Padilla, F. 1963a: Apuntes sobre Chrysomelidae ibero-marroquies (Coleoptera). *Graellsia* **20** (1/3): 69–77. DP: [31 December] 1963 (issue cover).

Codina Padilla, F. 1963b: Un nuevo Phaedon Latr., marroqui (Col. Chrysomelidae). *Graellsia* **20** (1/3): 79–82. DP: [31 December] 1963 (issue cover).

Codina Padilla, F. 1963c: Nuevas formas de Luperus ibéricos y norteafricanos (Col. Chrysomelidae). *Eos, Revista Española de Entomología* **39** (3/4): 367–378. DP: 31 December 1963 (volume contents).

Codina Padilla, F. 1963d: Nota sobre Chrysomelidae (Coleoptera) ibéricos. Material colectado por A. Cobos en Sierra Harana (Granada) y Sierra Cazorla (Jaén). *Archivos del Instituto de Aclimatación* (Almería) **12**: 43–47. DP: [31 December] 1963 (volume cover, see also Alonso-Zarazaga 2001).

Comolli, A. 1837: *De coleopteris novis ac rarioribus minusve cognitis provinciae Novocomi*. Ticini: Fusi et Socii, 54 pp. DP: 6 November 1837 (Bousquet 2016).

Compayo, L. 1862: *Histoire naturelle du département des Pyrénées-Orientales. Tome troisième.* [1863], Perpignan: J.-B. Alzine, 942 pp. + 1 pl. DP: 20 September 1862 (Bousquet 2016).

Cong, S. & Yu, P.-Y. 1997: The identity and distribution of Donacia fukiensis Goecke with notes on three related species from China (Coleopterea: Chrysomelidae: Donaciinae). *Coleopterists Bulletin* **51** (3): 203–207. DP: 19 September 1997 (verso of the issue cover).

Contarini, N. B. 1847: Notizie sulla fauna terrestre e particolarmente sulla ornitologia del Veneto estuario con cenno sul passaggio degli uccelli e sulla caccia. Pp. 157–259. In: *Venezia e le sue lagune. Volume secondo*. Venezia: Antonelli, 594 pp. DP: [31 December] 1847 (book title page).

Coral Şahin, D. & Özdikmen, H. 2019: A new species of Hydrothassa C. G. Thomson, 1859 from Turkey (Chrysomelidae: Chrysomelinae). *Munis Entomology & Zoology* **14** (2): 326–343. DP:

1 June 2019 (headings throughtout the paper, see also the comments in the Dating of publications chapter).

Cornelius, C. 1851: Zur Entwickelungs- und Ernährungs-Geschichte einiger Schildkäferarten. *Entomologische Zeitung* (Stettin) 12 (3): 88–94. DP: [31] March 1851 (issue title page).

Cornelius, C. 1857: Ernährung und Entwicklung einiger Blattkäfer. *Entomologische Zeitung* (Stettin) 18 (10/12): 392–405. DP: [31] December 1857 (issue title page).

Costa, A. 1842: Secondo rapporto sulla successiva apparizione degli entomati ne' contorni di Palermo. *Bullettino dell'Accademia degli Aspiranti Naturalisti* (Napoli) 1 (11): 124–126. DP: [31] December 1842 (paper title page).

Costa, A. 1884: Notizie ed osservazioni sulla geo-fauna sarda. Memoria terza. Risultamento delle ricerche fatte in Sardegna nella estate del 1883. *Atti della Reale Accademia delle Scienze Fisiche e Matematiche, Serie Seconda* (Napoli) 1 [1888] (9): 1–64. DP: 4 July 1884 (last page of the paper).

Costa, A. 1886: Notizie ed osservazioni sulla geo-fauna sarda. Memoria sesta. Risultamente di ricerche fatte in Sardegna nella state del 1885. *Atti della Reale Accademia delle Scienze Fisiche e Matematiche, Serie Seconda* (Napoli) 2 [1388] (8): 1–40. DP: 16 June 1886 (last page of the paper).

Costessèque, R. 2008: Une espèce nouvelle de cryptocephale: Cryptocephalus halleri sp. n. (Coleoptera, Chrysomelidae). *Revue de l'Association Roussillonnaise d'Entomologie* 17 (2): 49–53. DP: 4 September 2008 (verso of the issue cover).

Cox, M. L. 1994: Diapause in the Chrysomelidae. Pp. 469–502. In: Jolivet, P. H., Cox, M. L. & Petitpierre, E. (eds.): *Novel aspects of the biology of Chrysomelidae*. Dordrecht: Kluwer, 582 pp. DP: [31 December] 1994 (colophon).

Cox, M. L. 2007: *Atlas of the seed and leaf beetles of Britain and Ireland (Coleoptera: Bruchidae, Chrysomelidae, Megalopodidae and Orsodacnidae)*. Oxford: Pisces Publications, 336 pp. DP: [31 December] 2007 (colophon).

Creutzer, C. 1799: *Entomologische Versuche*. Wien: K. Schaumburg und Comp., 142 + [10] pp., 3 pls. DP: 17 October 1799 (Bousquet 2016).

Crişan, A. 1995: Cercetări asupra familiei Chrysomelidae (Coleoptera) în zona rezervaţiei biosferei "Delta Dunării", cu referire specială la Stylosomus tamaricis H.-S. şi Cryptocephalus gamma H.-S. *Buletinul de Informare Entomologică al Societăţii Lepidopterologice Române* 6 (1/2): 145–149. DP: [31 December] 1995 (paper title page).

Crotch, G. R. 1867: On the Coleoptera of the Azores. *Proceedings of the Zoological Society of London* 1867 (2): 359–391, pl. XXIII. DP: [31] October 1867 (Duncan 1937).

Crotch, G. R. 1871: [new names]. In: Sharp, D.: *Catalogue of British Coleoptera*. London: E. W. Janson, [2] + 37 pp. DP: 1 July 1871 (Bousquet 2016).

Crotch, G. R. 1872: Part 2. List of the Coleoptera found during the progress of the survey. Pp. 263–268. In: Wilson, C. W. & Palmer, H. S.: *Ordnance survey to the Peninsula of Sinai.*

Part 1. [1869], Southampton: Ordnance Survey Office, iv + 323 pp., 20 pls. DP: 23 April 1872 (Bousquet 2016).

Crotch, G. R. 1873: Materials for the study of the Phytophaga of the United States. *Proceedings of the Academy of Natural Sciences of Philadelphia* 25: 19–83. DP (pp. 9–56): 4 March 1873, (pp. 57–72): 11 March 1873, (pp. 73–83): 18 March 1873 (verso of the volume title page).

Crotch, G. R. 1874: Descriptions of new species of Coleoptera from the Pacific coast of the United States. *Transactions of the American Entomological Society* 5 [1874–1876] (1): 73–80. DP: [31 December] 1874 (Brown 1964).

Crowson, R. A. 1946: A revision of the genera of the chrysomelid group Sagrinae (Coleoptera). *Transactions of the Royal Entomological Society of London* 97 (4): 75–115. DP: 22 July 1946 (issue cover).

Csiki, E. 1899: Hazánk Orsodacne-féléi. (Die Orsodacne-Arten Ungarns). *Rovartani Lapok* 6 (5): 92–95. DP: [31] May 1899 (issue cover).

Csiki, E. 1900: Hispidae tres novae. *Természetrajzi Füzetek* 23 (1/2): 196. DP: 1 May 1900 (table of contents).

Csiki, E. 1901a: Bogarak-Coleopteren. Pp. 75–120. In: Horváth, G. (ed.): *Zichy Jenő gróf harmadik ázsiai utazása. II. kötet. Zichy Jenő gróf harmadik ázsiai utazásának állattani eredményei. Dritte asiatische Forschungsreise des Grafen Eugen Zichy. Band II. Zoologische Ergebnisse der dritten asiatischen Forschungsreise des Grafen Eugen Zichy.* Budapest: Victor Hornyánszky and Leipzig: Karl W. Hiersemann, xli + 427 pp., xxviii pls. DP: 24 May 1901 (*Bibl. Zool.* 1901, 6: 222).

Csiki, E. 1901b: Coleopterologiai jegyzetek. (Coleopterologische Notizen). *Rovartani Lapok* 8 (5): 99–103. DP: [31] May 1901 (issue cover).

Csiki, E. 1903: Psylliodes Wachsmanni, egy új levélbogár a magyar tengermellékről. (Psylliodes Wachsmanni n. sp.). *Rovartani Lapok* 10 (2): 40–42. DP: [28] February 1903 (issue cover).

Csiki, E. 1916: Bogarak Szibériából és Mongolországból. (Coleopteren aus Sibirien und der Mongolei). *Rovartani Lapok* 23 (9/12): 125–127. DP: 31 December 1916 (footer on the paper title page).

Csiki, E. 1939: [new names]. In: Heikertinger, F. & Csiki, E.: Chrysomelidae: Halticinae I. In: Schenkling, S. (ed.): *Coleopterorum Catalogus. Pars. 166.* s-Gravenhage: W. Junk, 1–336 pp. DP: 14 October 1939 (book cover).

Csiki, E. 1940a: [new names]. In: Heikertinger, F. & Csiki, E.: Chrysomelidae: Halticinae II. In: Schenkling, S. (ed.): *Coleopterorum Catalogus. Pars 169.* s-Gravenhage: W. Junk, 337–635 pp. DP: 15 August 1940 (book cover).

Csiki, E. 1940b: Bogarak. Coleopteren. Pp. 208–288. In: Teleki, P. & Csiki, E. (eds.): *Csiki Ernő állattani kutatásai Albániában. Explorationes zoologicae ab E. Csiki in Albania peracte. Pars 2.* Budapest: Magyar Tudományos Akadémia, vii + 177–334 pp. DP: [31 December] 1940 (volume cover and table of contents).

Csiki, E. 1940c: Coleoptera nova ex Hungaria. *Matematikai és Természettudományi Értesitő* 59: 916–920. DP: [31 December] 1940 (volume cover).

Csiki, E. 1941: Additamenta ad faunam Coleopterorum Hungariae. *Fragmenta Faunistica Hungarica* 4 (3): 54–57. DP: 1 July 1941 (issue cover).

Csiki, E. 1942: Adatok a Görgényi-hegység bogarfaunájának ismeretéhez. (Beiträge zur Kenntnis der Käferfauna des Görgény-Gebirges). *Matematikai és Természettudományi Értesitő* 61: 369–384. DP: [31 December] 1942 (volume cover).

Csiki, E. 1953: Über neue und bekannte Coleopteren aus Ungarn und den angrenzenden Ländern. *Annales Historico-Naturales Musei Nationalis Hungarici, Series Nova* 3 [1952]: 115–135. DP: 31 July 1953 (colophon at the end of the volume).

Csiki, E. & Heikertinger, F. 1940: [new name]. In: Heikertinger, F. & Csiki, E.: Chrysomelidae: Halticinae II. In: Schenkling, S. (ed.): *Coleopterorum Catalogus. Pars 169*. s-Gravenhage: W. Junk, 337–635 pp. DP: 15 August 1940 (book cover).

Curtis, J. 1826: *British entomology; being illustrations and descriptions of the genera of insects found in Great Britain and Ireland: containing coloured figures from nature of the most rare and beautiful species, and in many instances of the plants upon which they are found. Vol. III*. London: J. Curtis, [8] pp. + 99–146 pls. and sheets. DP (pl. 127): 1 August 1826 (Bousquet 2016).

Curtis, J. 1830: *British entomology; being illustrations and descriptions of the genera of insects found in Great Britain and Ireland: containing coloured figures from nature of the most rare and beautiful species, and in many instances of the plants upon which they are found. Vol. VII*. London: J. Curtis, [8] pp. + 290–337 pls. and sheets. DP (pl. 323): 1 September 1830 (Bousquet 2016).

Curtis, J. 1831: *British entomology; being illustrations and descriptions of the genera of insects found in Great Britain and Ireland: containing coloured figures from nature of the most rare and beautiful species, and in many instances of the plants upon which they are found. Vol. VIII*. London: J. Curtis, [8] pp. + 338–385 pls. and sheets. DP (pl. 347): 1 March 1831 (Bousquet 2016).

Curtis, J. 1833: *British entomology; being illustrations and descriptions of the genera of insects found in Great Britain and Ireland: containing coloured figures from nature of the most rare and beautiful species, and in many instances of the plants upon which they are found. Vol. X*. London: J. Curtis, [8] pp. + 434–481 pls. and sheets. DP (pl. 435): 1 January 1833 (Bousquet 2016).

Curtis, J. 1834: *British entomology; being illustrations and descriptions of the genera of insects found in Great Britain and Ireland: containing coloured figures from nature of the most rare and beautiful species, and in many instances of the plants upon which they are found. Vol. XI*. London: J. Curtis, [8] pp. + 482–529 pls. and sheets. DP (pl. 486): 1 February 1834 (Bousquet 2016).

Curtis, J. 1837: *British entomology; being illustrations and descriptions of the genera of insects found in Great Britain and Ireland: containing coloured figures from nature of the most rare and beautiful species, and in many instances of the plants upon which they are found. Vol. XIV*. London: J. Curtis, [8] pp. + 626–673 pls. and sheets. DP (pl. 630): 1 February 1837 (Bousquet 2016).

Curtis, J. 1840: Descriptions, &c. of some rare or interesting indigenous insects. *Annals of Natural History* 5 (31): 274–282. DP: 1 June 1840 (Evenhuis 2003).

Czwalina, G. 1885: Sammelbericht aus Ost- und West-Preussen. *Deutsche Entomologische Zeitschrift* 29 (2): 251–256. DP: [31] December 1885 (table of contents).

Daccordi, M. 1972: Descrizione di una nuova specie e di una nuova sottospecie di Chrysomela appartenenti al sottogenere Palaeosticta Bechyné (Coleoptera, Chrysomelidae). *Memorie del*

Museo Civico di Storia Naturale di Verona **19** [1971]: 511–518. DP: 30 November 1972 (table of contents).

Daccordi, M. 1975: Una nuova specie di Oreomela della Persia (Coleoptera Chrysomelidae). *Bollettino della Società Entomologica Italiana* **107** (6/8): 146–148. DP: 20 October 1975 (Poggi 2008).

Daccordi, M. 1976a: Note sinonimiche su alcuni Entomoscelini, con osservazioni sul sottogenere Synerga e descrizione di due nuove specie di Semenowia (Coleoptera Chrysomelidae). *Bollettino della Società Entomologica Italiana* **108** (1/2): 31–39. DP: 20 February 1976 (Poggi 2008).

Daccordi, M. 1976b: Considerazioni sulle Crosita del sottogenere Bittotaenia Motsch. con descrizione di una nuova specie del Monte Sinai (Coleoptera Chrysomelidae). *Atti della Società Italiana di Scienze Naturali e del Museo Civile di Storia Naturale di Milano* **117** (1/2): 79–84. DP: 15 June 1976 (paper title page).

Daccordi, M. 1976c: Una nuova specie di Pachybrachis Chevr. della penisola iberica (Coleoptera Chrysomelidae). *Bollettino del Museo Civico di Storia Naturale di Verona* **3**: 595–597. DP: 31 December 1976 (table of contents).

Daccordi, M. 1977a: Ergebnisse der Bhutan-Expedition 1972 des Naturhistorischen Museums in Basel. Coleoptera: Fam. Chrysomelidae Subfam. Chrysomelinae. *Entomologica Basiliensia* **2**: 343–350. DP: 7 January 1977 (table of contents).

Daccordi, M. 1977b: Descrizione di Ambrostoma shuteae n. sp. del Nepal e considerazioni su Ambrostoma mahesa (Hope) (Coleoptera Chrysomelidae Chrysomelinae). *Bollettino della Società Entomologica Italiana* **109** (7/8): 133–138. DP: 20 October 1977 (Poggi 2008).

Daccordi, M. 1977c: Coleotteri crisomelidi dell'isola di Creta e descrizione di una nuova specie del genere Pachnephorus Redt. (Coleoptera-Chrysomelidae). *Bollettino del Museo Civico di Storia Naturale di Verona* **4**: 81–93. DP: 31 December 1977 (table of contents).

Daccordi, M. 1978a: Nuove specie di Chrysolina delle Regione Etiopica (Coleoptera Chrysomelidae). *Revue de Zoologie Africaine* **92** (3): 745–752. DP: 29 September 1978 (paper title page).

Daccordi, M. 1978b: Una nuova specie di Phaedon del Sikkim con brevi note su Phaedonia indica Chen (Coleoptera Chrysomelidae, Chrysomelinae). *Bulletin et Annales de la Société Royale Belge d'Entomologie* **114** (7/9): 209–211. DP: 31 October 1978 (issue cover).

Daccordi, M. 1978c: Una nuova specie di "Macrocoma" ("Coleoptera", "Chrysomelidae"). Pp. 155–158. In: García-Talavera, F. (ed.): *Contribución al estudio de la historia natural de las Islas Salvajes. Resultados de la expedicion cientifica "Agamenon 76" (23 de febrero–3 de marzo de 1976)*. Santa Cruz de Tenerife: Museo de Ciencias Naturales del Cabildo Insular de Santa Cruz de Tenerife, Aula de Cultura de Tenerife, 209 pp. DP: [31 December] 1978 (book cover).

Daccordi, M. 1979a: Insects of Saudi Arabia. Coleoptera: Fam. Chrysomelidae Subfam. Eumolpinae e Chlamisinae. *Fauna of Saudi Arabia* **1**: 304–307. DP: 5 March 1979 (table of contents).

Daccordi, M. 1979b: Nuove specie di crisomeline della regione Orientale (Coleoptera: Chrysomelidae Subf. Chrysomelinae). *Entomologica Basiliensia* **4**: 443–461. DP: 10 October 1979 (table of contents).

Daccordi, M. 1980: I sottogeneri afrotropicali di Chrysolina con descrizione di una nuova specie (Coleoptera Chrysomelidae). *Revue Zoologique Africaine* **94** (2): 299–310. DP: 30 June 1980 (paper title page).

Daccordi, M. 1982a: Crisomeline orientali nuove o poco note del Museo di Storia Naturale di Basilea (Svizzera) (Col. Chrysomelidae). *Entomologica Basiliensia* **7**: 394–405. DP: 30 November 1982 (table of contents).

Daccordi, M. 1982b: Descrizione di una nuova specie e di una nuova specie di Crosita (Subgen. Bittotaenia) dell'Oman con osservazioni e nuove sinonimie sui generi Chrysolina e Crosita (Col. Chrysomelidae). *Entomologica Basiliensia* **7**: 406–412. DP: 30 November 1982 (table of contents).

Daccordi, M. 1984: Contributo alla conoscenza delle crisomeline della regione Orientale (Coleoptera: Chrysomelidae). *Revue Suisse de Zoologie* **91** (2): 323–328. DP: [30] June 1984 (paper title page).

Daccordi, M. 1987: Nuove Plagiodera della regione neotropicale (Coleoptera Chrysomelidae, Chrysomelinae). *Bollettino del Museo Civico di Storia Naturale di Verona* **13** [1986]: 443–454. DP: 15 July 1987 (last page of the volume).

Daccordi, M. 1994: Notes for phylogenetic study of Chrysomelinae, with descriptions of new taxa and a list of all the known genera (Coleoptera: Chrysomelidae, Chrysomelinae). Pp. 60–84. In: Furth, D. G. (ed.): *Proceedings of the Third International Symposium on the Chrysomelidae Beijing, 1992*. Leiden: Backhuys Publishers, 150 pp. DP: [31 December] 1994 (book title page).

Daccordi, M. 2001: A new species of Agrosteomela Gistel, 1857 from Nepal (Coleoptera, Chrysomelidae, Chrysomelinae). *Entomologica Basiliensia* **23**: 89–92. DP: 1 September 2001 (table of contents).

Daccordi, M., Bollino, M. & Vela, J. M. 2020: Some techniques for the study of useful characters in the taxonomy of the genus Timarcha Samouelle, 1819 (Coleoptera, Chrysomelidae). *European Journal of Taxonomy* **630**: 1–14. DP: 20 April 2020 (last page of the paper).

Daccordi, M. & Ge, S.-Q. 2007: [new taxa]. In: Ge, S.-Q., Daccordi, M. & Yang, X.-K.: Two new species of the genus Gonioctena Chevrolat from China (Coleoptera: Chrysomelidae: Chrysomelinae). *Genus* **18** (4): 579–587. DP: 28 December 2007 (paper title page).

Daccordi, M. & Ge, S.-Q. 2009a: [new taxon]. In: Ge, S.-Q., Daccordi, M., Wang, S.-Y. & Yang, X.-K.: Study of the genus Entomoscelis Chevrolat (Coleoptera: Chrysomelidae: Chrysomelinae) from China. *Proceedings of the Entomological Society of Washington* **111** (2): 410–425. DP: 15 May 2009 (verso of the issue cover).

Daccordi, M. & Ge, S.-Q. 2009b: [new taxon]. In: Ge, S.-Q., Daccordi, M. & Yang, X.-K. 2009: Two new species of Chrysolina Motschulsky from China, with redescription of C. kozlovi Lopatin and notes on the genus Doeberlia Warchałowski (Coleoptera: Chrysomelidae: Chrysomelinae). *Acta Entomologica Musei Nationalis Pragae* **49** (2): 805–818. DP: 15 December 2009 (paper title page).

Daccordi, M. & Ge, S.-Q. 2011a: [new taxa]. In: Ge, S.-Q., Daccordi, M., Li, W.-Z. & Yang, X.-K.: Seven new species of the genus Chrysolina Motschulsky from China (Coleoptera: Chrysomelidae: Chrysomelinae). *Zootaxa* **2736**: 31–43. DP: 11 January 2011 (footer on the paper title page).

Daccordi, M. & Ge, S.-Q. 2011b: [new taxa]. In: Daccordi, M., Ge, S.-Q., Cui, J.-Z. & Yang, X.-K.: New species of Chrysolina Motschulsky, 1860 from Southwest China (Coleoptera, Chrysomelidae, Chrysomelinae). *Nouvelle Revue d'Entomologie, Nouvelle Série* **26** [2009] (4): 291–316. DP: 17 February 2011 (last page of the cover).

Daccordi, M. & Ge, S.-Q. 2011c: [new taxa]. In: Ge, S.-Q., Daccordi, M., Beutel, R. G., Li, W.-Z. & Yang, X.-K.: Revision of the chrysomeline genera Potaninia, Suinzona and Taipinus (Coleoptera) from eastern Asia, with a biogeographic scenario for the Hengduan mountain region in south-western China. *Systematic Entomology* **36** (4): 644–671. DP: 26 September 2011 (volume contents).

Daccordi, M. & Ge, S.-Q. 2012a: [new taxa]. In: Ge, S.-Q., Daccordi, M., Beutel, R. G., Ren, J., Cui, J.-Z., Li, W.-Z. & Yang, X.-K.: Revision of the Eastern Asian genera Ambrostoma Motschulsky and Parambrostoma Chen (Coleoptera: Chrysomelidae: Chrysomelinae). *Systematic Entomology* **37** (2): 332–345. DP: 29 March 2012 (volume contents).

Daccordi, M. & Ge, S.-Q. 2012b: [new taxa]. In: Ge, S.-Q., Daccordi, M., Lopatin, I. K., Cui, J.-Z., Li, W.-Z. & Yang, X.-K.: A new genus and two new species of Chrysomelinae (Coleoptera: Chrysomelidae) from China. *Entomotaxonomia* **34** (4): 633–640. DP: 25 December 2012 (last page of the cover).

Daccordi, M. & Ge, S.-Q. 2013a: [new taxa]. In: Ge, S.-Q., Daccordi, M., Ren, J., Cui, J.-Z., Li, W.-Z. & Yang, X.-K.: Odontoedon, a new genus from China with descriptions of nine new species (Coleoptera: Chrysomelidae: Chrysomelinae). *Stuttgarter Beiträge zur Naturkunde A, Neue Serie* **6**: 199–222. DP: 30 April 2013 (paper title page).

Daccordi, M. & Ge, S.-Q. 2013b: [new taxa]. In: Ge, S.-Q., Daccordi, M., Cui, J.-Z., Ren, J. & Yang, X.-K.: Revision of genus Sclerophaedon Weise from China (Coleoptera, Chrysomelidae, Chrysomelinae). *Annali del Museo Civico di Storia Naturale "G. Doria"* **105**: 81–102. DP: 24 July 2013 (table of contents).

Daccordi, M. & Ge, S.-Q. 2014: [unavailable names]. In: Yang, X.-K., Ge, S.-Q., Wang, S.-Y., Li, W.-Z. & Cui J.-Z.: *Coleoptera Chrysomelidae Chrysomelinae. Fauna Sinica. Insecta Vol. 61*. Beijing: Science Press, xvi + 641 pp., 8 pls. DP: [30] June 2014 (colophon).

Daccordi, M. & Ge, S.-Q. 2015: [new taxa]. In: Ge, S.-Q., Daccordi, M., Ren, J. & Yang, X.-K.: Revision of Phaedon Latreille from China (Coleoptera: Chrysomelidae). *Zoological Systematics* **40** (1): 1–30. DP: 24 January 2015 (journal website).

Daccordi, M. & Gruev, B. 1991: A new species of Prasocuris from China (Coleoptera, Chrysomelidae, Chrysomelinae). *Bollettino del Museo Civico di Storia Naturale di Verona* **15** [1988]: 379–381. DP: 15 April 1991 (last page of the paper).

Daccordi, M. & Medvedev, L. N. 1996: [new taxa]. In: Medvedev, L. N.: The Chrysomelidae of Arabia. *Fauna of Saudi Arabia* **15**: 211–263. DP: 15 April 1996 (table of contents).

Daccordi, M. & Medvedev, L. N. 2000a: Yunnaedon pankui a new genus and species of Chrysomelinae from China (Coleoptera, Chrysomelidae). *Nouvelle Revue d'Entomologie, Nouvelle Série* 16 [1999] (4): 297–301. DP: 30 June 2000 (last page of the cover).

Daccordi, M. & Medvedev, L. N. 2000b: A study of the species of Sclerophaedon from the Himalaya (Coleoptera Chrysomelidae). *Bollettino della Società Entomologica Italiana* **132** (3): 219–231. DP: 31 October 2000 (paper title page).

Daccordi, M. & Petitpierre, E. 1977: Coleópteros crisomélidos de la Sierra de Cazorla (Jaén) y descripción de una nueva especie de Clytra Laich. (Coleoptera, Chrysomelidae). *Miscelanea Zoologica* (Barcelona) **4** (1): 225–236. DP: [31 December] 1977 (issue title page).

Daccordi, M. & Petitpierre, E. 2013: [new taxon]. In: Petitpierre, E. & Daccordi, M.: Chrysomelidae (Coleoptera) de las sierras del Altiplano de Granada (Granada, Andalucía). *Zoologica Baetica* **24**: 53–78. DP: [31 December] 2013 (paper title page).

Daccordi, M. & Ruffo, S. 1971: Coleotteri crisomelidi raccolti nelle Isole Ponziane e descrizione di una nuova specie del genere Pachybrachis Redt. *Fragmenta Entomologica* 8 [1971–1973] (1): 41–48. DP: 15 September 1971 (reprint title page).

Daccordi, M. & Ruffo, S. 1975: Coleotteri crisomelidi delle Isole Egadi e descrizione di una nuova specie del genere Pachybrachis Chevr. *Bollettino del Museo Civico di Storia Naturale di Verona* 1 [1974]: 427–437. DP: 30 April 1975 (table of contents).

Daccordi, M. & Ruffo, S. 1976: Le specie Appenniniche del genere Oreina (Coleoptera Chrysomelidae). *Bollettino del Museo Civico di Storia Naturale di Verona* **3**: 379–411. DP: 31 December 1976 (table of contents).

Daccordi, M. & Ruffo, S. 1977: Una nuova Oreina del M. Tremalzo (Alpi di Ledro) Oreina liturata subsp. tomasii (Coleoptera Chrysomelidae). *Studi Trentini di Scienze Naturali, Acta Biologica* **54**: 99–102. DP: [31 December] 1977 (paper title page).

Daccordi, M. & Ruffo, S. 1979: Le Chrysolina italiane del sottogenere Threnosoma Motsch. (Coleoptera, Chrysomelidae). *Bollettino del Museo Civico di Storia Naturale di Verona* **6**: 305–332. DP: 31 December 1979 (table of contents).

Daccordi, M. & Ruffo, S. 1980: Una nuova Chrysolina della Grecia appartenente al sottogenere Bechynia (Coleoptera, Chrysomelidae). *Fragmenta Entomologica* **15** [1979–1980] (2): 357–363. DP: 30 July 1980 (table of contents in the cover).

Daccordi, M. & Ruffo, S. 1987: Due nuove sottospecie Appenniniche di Oreina elongata (Suffrian) (Coleoptera, Chrysomelidae). *Bollettino del Museo Civico di Storia Naturale di Verona* **13** [1986]: 13–18. DP: 15 July 1987 (last page of the volume).

Daccordi, M. & Ruffo, S. 1988: Considerazioni sistematiche e biogeografiche sulle Timarcha italiane (Coleoptera, Chrysomelidae). *Atti XV Congresso Nazionale Italiano Entomologia, l'Aquila* **1988**: 343–351. DP: [31 December] 1988 (issue cover).

Daccordi, M. & Ruffo, S. 1990: Una nuova specie di Timarcha delle Alpi Apuane (Coleoptera, Chrysomelidae, Chrysomelinae). *Fragmenta Entomologica* **22** (1): 103–107. DP: 15 September 1990 (table of contents in the cover).

Daccordi, M. & Ruffo, S. 2005: Considerazioni biogeografiche sulle Chrysolina delle province appenninica e sicula con descrizione di Chrysolina (Stichoptera) bourdonnei n. sp (Coleoptera, Chrysomelidae). *Studi Trentini di Scienze Naturali, Acta Biologica* **81** [2004]: 113–127. DP: [31] October 2005 (colophon).

Daccordi, M. & Yang, X.-K. 2009a: [new taxon]. In: Ge, S.-Q., Daccordi, M., Wang, S.-Y. & Yang, X.-K.: Study of the genus Entomoscelis Chevrolat (Coleoptera: Chrysomelidae: Chrysomelinae) from China. *Proceedings of the Entomological Society of Washington* **111** (2): 410–425. DP: 15 May 2009 (verso of the issue cover).

Daccordi, M. & Yang, X.-K. 2009b: [new taxa]. In: Ge, S.-Q., Daccordi, M. & Yang, X.-K.: Two new species of Chrysolina Motschulsky from China, with redescription of C. kozlovi Lopatin and notes on the genus Doeberlia Warchałowski (Coleoptera: Chrysomelidae: Chrysomelinae). *Acta Entomologica Musei Nationalis Pragae* **49** (2): 805–818. DP: 15 December 2009 (paper title page).

Daccordi, M. & Yang, X.-K. 2011a: [new taxon]. In: Ge, S.-Q., Daccordi, M., Li, W.-Z. & Yang, X.-K.: Seven new species of the genus Chrysolina Motschulsky from China (Coleoptera: Chrysomelidae: Chrysomelinae). *Zootaxa* **2736**: 31–43. DP: 11 January 2011 (footer on the paper title page).

Daccordi, M. & Yang, X.-K. 2011b: [new taxa]. In: Daccordi, M., Ge, S.-Q., Cui, J.-Z. & Yang, X.-K.: New species of Chrysolina Motschulsky, 1860 from Southwest China (Coleoptera, Chrysomelidae, Chrysomelinae). *Nouvelle Revue d'Entomologie, Nouvelle Série* **26** [2009] (4): 291–316. DP: 17 February 2011 (last page of the cover).

Daccordi, M. & Yang, X.-K. 2011c: [new taxa]. In: Ge, S.-Q., Daccordi, M., Beutel, R. G., Li, W.-Z. & Yang, X.-K.: Revision of the chrysomeline genera Potaninia, Suinzona and Taipinus (Coleoptera) from eastern Asia, with a biogeographic scenario for the Hengduan mountain region in south-western China. *Systematic Entomology* **36** (4): 644–671. DP: 26 September 2011 (volume contents).

Daccordi, M. & Yang, X.-K. 2013: [new taxon]. In: Ge, S.-Q., Daccordi, M., Ren, J., Cui, J.-Z., Li, W.-Z. & Yang, X.-K.: Odontoedon, a new genus from China with descriptions of nine new species (Coleoptera: Chrysomelidae: Chrysomelinae). *Stuttgarter Beiträge zur Naturkunde A, Neue Serie* **6**: 199–222. DP: 30 April 2013 (paper title page).

Dahl, G. 1823: *Coleoptera und Lepidoptera. Ein systematisches Verzeichniss, mit beygesetzten Preisen der Vorräthe*. Wien: J. E. Akkermann, iv + 105 pp. DP: [31 December] 1823 (Bousquet 2016).

Dahlgren, G. 1984: Chrysomela und Crosita Studien. *Entomologische Blätter* **80** (1): 35–52. DP: 15 May 1984 (issue cover).

Dalman, J. W. 1823: *Analecta entomologica*. Holmiae: Lindhianus, vii + 104 pp., 4 pls. DP: 7 September 1823 (Bousquet 2016).

Dalman, J. W. 1824: *Ephemerides entomologicae. I*. Holmiae: P. A. Nordstedt, 33 pp. DP: [31 December] 1824 (Bousquet 2016).

Dalman, J. W. 1833: [new taxon]. In: Schoenherr, C. J.: *Genera et species curculionidum, cum synonymia hujus familiae. Tomus primus. Pars prima.* Paris: A. Roret, xv + 381 pp. DP: [30] June 1833 (Bousquet 2016).

Damaška, A. F. & Aston, P. 2019: Leaf litter and moss-inhabiting flea beetles of Hong Kong (Coleoptera: Chrysomelidae: Alticini). *Acta Entomologica Musei Nationalis Pragae* **59** (1): 151–161. DP: 8 April 2019 (paper title page).

Damaška, A. F., Ruan, Y.-Y. & Fikáček, M. 2022: The genus Cangshanaltica (Coleoptera: Chrysomelidae: Alticinae): overview, new species, and notes on species complexes. *Zootaxa* **5219** (1): 49–64. DP: 7 December 2022 (paper title page).

Daniel, J. 1903a: Neue, alpine Formen der Gattung Chrysochloa Hoppe (Orina Chevr.). *Münchener Koleopterologische Zeitschrift* **1** [1902–1903] (2): 180–184. DP: 25 January 1903 (table of contents).

Daniel, J. 1903b: [new taxa]. In: Daniel, K. & Daniel, J.: Kleinere Abhandlungen, kritische und synonymische Bemerkungen. *Münchener Koleopterologische Zeitschrift* **1** [1902–1903] (2): 249–255. DP: 25 January 1903 (table of contents).

Daniel, J. 1904a: [new taxa]. In: Daniel, K. & Daniel, J.: Neue paläarktische Koleopteren. *Münchener Koleopterologische Zeitschrift* **2** [1904–1906] (1): 76–93. DP: 16 February 1904 (table of contents).

Daniel, J. 1904b: Revision der paläarktischen Crepidodera-Arten. *Münchener Koleopterologische Zeitschrift* **2** [1904–1906] (2): 237–297. DP: 29 December 1904 (table of contents).

Daniel, J. 1914: *Der Formenkreis der Phytodecta variabilis Oliv.* München: C. Wolf & Sohn, 238–262 pp. DP: [31 December] 1914 (see the comments in Dating of publications) [published only as a reprint; respective volume 4 of the *Münchener Koleopterologische Zeitschrift* was never published].

Daniel, K. 1900: Vorläufige Diagnosen. *Societas Entomologica* **15** [1900–1901] (18): 139–140. DP: 15 December 1900 (issue title page).

Daniel, K. 1904: [new taxon]. In: Daniel, K. & Daniel, J.: Neue paläarktische Koleopteren. *Münchener Koleopterologische Zeitschrift* **2** [1904–1906] (1): 76–93. DP: 16 February 1904 (table of contents).

Daniel, K. 1906: [new taxon]. In: Daniel, K. & Daniel, J.: Neue paläarktische Koleopteren. *Münchener Koleopterologische Zeitschrift* **3** [1906–1908] (2): 170–187. DP: 1 October 1906 (table of contents).

Daniel, K. 1908: [new taxon]. In: Peyerimhoff, P. M. de: Liste des coléoptères du Sinaï. *L'Abeille, Journal d'Entomologie* **31** [1907–1923] (1): 1–48. DP: [31] January 1908 (printed at the end of 1907, distributed in January 1908, see *annonces de la société* on the cover of *Bull. Soc. Ent. Fr.*, 1907, No. 19).

Daniel, K. & Daniel, J. 1898: Zwanzig neue Arten aus dem paläearktischen Faunengebiete. Pp. 61–82. In: *Coleopteren-Studien II*. München: Kgl. Hof-und Universitäts-Buchdruckerei von Dr. C. Wolf & Sohn, [2] + 88 pp. DP: 28 December 1898 (book cover).

David, P. 1952: Notes sur divers Chrysochloa français (Col. Chrysomelidae). *Bulletin de la Société Entomologique de France* **57** (7): 109–111. DP: 3 November 1952 (*dates de publication* on p. 160).

David, P. 1953a: Un nouveau Chrysochloa de France (Col. Chrysomelidae). *Bulletin de la Société Entomologique de France* 58 (1): 13–14. DP: 5 March 1953 (*dates de publication* on p. 168).

David, P. 1953b: Un nouveau sous-genre de Chrysochloa (Col. Chrysomelidae). *Revue Française d'Entomologie* 20 (3): 184–186. DP: 15 October 1953 (footer on p. 177).

David, P. 1954: Sur un nouveau Chrysochloa français (Col. Chrysomelidae). *Bulletin de la Société Entomologique de France* 59 (7/8): 102–104. DP: 3 December 1954 (*dates de publication* on p. 160).

De La Rosa, J. J., Moreno Tamurejo, J. A., García Villanueva, V. & Díaz Rodríguez, E. 2012: Pachybrachis (Pachybrachis) alcantarensis sp. n. del Parque Natural del Tajo Internacional (Cáceres), oeste de España (Coleoptera: Chrysomelidae: Cryptocephalinae). *Arquivos Entomolóxicos* 6: 87–90. DP: 30 April 2012 (verso of volume cover, see also the comments in the Dating of publications chapter).

Debreuil, M. 2014: Chrysomelinae. Pp. 593–601. In: Tronquet, M. (ed.): Catalogue des coléoptères de France. *Revue de l'Association Roussillonnaise d'Entomologie* 23 (supplément): 1–1052. DP: [31] May 2014 (colophon).

Decaux, C. 1894: Métamorphoses du Caryoborus tamarindi coléoptère de la famille des bruchides. *Le Naturaliste* 16 (174): 129–131. DP: 1 June 1894 (issue title page).

Decelle, J. 1951: Contribution à l'étude des Bruchidae du Congo belge (Col. Phytophaga). *Revue de Zoologie et de Botanique Africaines* 45 (1/2): 172–192. DP: 22 December 1951 (paper title page).

Decelle, J. 1969: Le Parc national du Niokolo-Koba (Sénégal). Fascicule III. XVII. Coleoptera Bruchidae. *Mémoires de l'Institut Fondamental de l'Afrique Noire* 84: 287–296. DP: [31] March 1969 (volume cover and last page).

Decelle, J. 1970: Coleoptera: Bruchidae. *South African Animal Life* 14: 256–266. DP: 4 May 1970 (Evenhuis 2022).

Decelle, J. 1971a: Trois nouvelles espèces de Kytorhinus (Col. Bruchidae) du Tibet et du Bhutan. *Bulletin et Annales de la Société Royale d'Entomologie de Belge* 107 (4/6): 105–115. DP: 31 August 1971 (issue cover).

Decelle, J. 1971b: Bruchides (Col.) récoltés en Abyssinie Centrale par l'expédition H. Scott et J. Omer Cooper (IX.1926–I.1927). *Bulletin et Annales de la Société Royale d'Entomologie de Belge* 107 (7/9): 243–259. DP: 31 October 1971 (issue cover).

Decelle, J. 1975a: Les Coléoptères Bruchides d'Angola. *Publicações Culturais, Companhia de Diamantes de Angola* (Lisboa) 89 [1977]: 15–32. DP: 6 January 1975 (reprint title page).

Decelle, J. 1975b: Coleoptera: Bruchidae de Ceylan. *Entomologica Scandinavica* **Supplementum** 4 [1973–1975]: 179–194. DP: 7 May 1975 (verso of issue title page).

Decelle, J. 1975c: Les Bruchidae (Coleoptera) des Iles Canaries. *Bulletin et Annales de la Société Royale Belge d'Entomologie* 111 (4/6): 109–142. DP: 30 June 1975 (issue cover).

Decelle, J. 1977: Ergebnisse der Bhutan-Expedition 1972 des Naturhistorischen Museums in Basel. Coleoptera: Fam. Bruchidae. *Entomologica Basiliensia* 2: 393–409. DP: 7 January 1977 (table of contents).

Decelle, J. 1979a: Insects of Saudi Arabia. Coleoptera: Fam. Bruchidae. *Fauna of Saudi Arabia* 1: 318–330. DP: 5 March 1979 (table of contents).

Decelle, J. 1979b: Une nouvelle espèce de Bruchus (Coleoptera, Bruchidae) des Iles Canaries. *Vieraea* 8 [1978] (1): 143–146. DP: [31] August 1979 (paper title page).

Decelle, J. 1982a: Une nouvelle espece de Spermohagus [sic!] Schonnerr [sic!] (Coleoptera: Bruchidae: Amblycerinae) de Turquie. *Bulletin & Annales de la Société Royale Belge d'Entomologie* 118 (1/3): 31–34. DP: 30 June 1982 (issue table of contents).

Decelle, J. 1982b: Bruchidius medaniensis, nouvelle espèce du Soudan (Coleoptera: Bruchidae: Bruchinae). *Revue de Zoologie Africaine* 96 (2): 282–285. DP: 30 June 1982 (paper title page).

Decelle, J. 1983: Une espece de Coleoptera Bruchidae du Proche-Orient infeodee aux Salvia (Dicotylédones, Tubiflores, Lamiaceae). *Bulletin & Annales de la Société Royale Belge d'Entomologie* 118 [1982] (7/12): 243–248. DP: 28 February 1983 (issue table of contents).

Decelle, J. & Lodos, N. 1989: Contribution to the study of legume weevils of Turkey (Coleoptera: Bruchidae). *Bulletin & Annales de la Société Royale Belge d'Entomologie* 125 (7/9): 163–212. DP: 30 September 1989 (last page of the issue cover).

Dedyukhin, S. V. 2016a: Diversity of phytophagous beetles (Coleoptera: Chrysomeloidea, Curculionoidea) in steppe communities in the forest-steppe of the High Trans-Volga region. *Entomological Review* 95 [2015] (8): 1070–1087. DP: 23 January 2016 (journal website).

Dedyukhin, S. V. 2016b: Consortial associations of phytophagous beetles (Coleoptera: Chrysomeloidea, Curculionoidea) with plants in the East of the Russian plain. *Entomological Review* 96 (6): 679–700. DP: 19 October 2016 (journal website).

Dedyukhin, S. V. 2016c: Species richness and zonal features of partial faunas of phytophagous beetles (Coleoptera: Chrysomeloidea, Curculionoidea) on grassy slopes in the East of the Russian plain and the Cis-Ural-Region. *Entomological Review* 96 (7): 853–865. DP: 24 November 2016 (journal website).

DeGeer, C. 1775: *Mémoires pour servir à l'histoire des insectes. Tome cinquième.* Stockholm: Pierre Hesselberg, vii + 448 pp., 16 pls. DP: 5 April 1775 (Bousquet 2016).

Dejean, P. F. M. A. 1821: *Catalogue de la collection de coléoptères de M. le Baron Dejean.* Paris: Crevot, viii + 136 pp. DP: 12 May 1821 (Bousquet 2016).

Dejean, P. F. M. A. 1835: *Catalogue des coléoptères de la collection de M. le Comte Dejean.* [1833–1836], Paris: Méquignon-Marvis Père et Fils, 443 pp. DP (livraison 4, pp. 257–360): 22 August 1835 (Bousquet 2016).

Dejean, P. F. M. A. 1836: *Catalogue des coléoptères de la collection de M. le Comte Dejean.* [1833–1836], Paris: Méquignon-Marvis Père et Fils, 443 pp. DP (livraison 5, pp. 361–443): [31 December] 1836 (Bousquet 2016).

Delahon, P. 1913: Nachträge zum Schilskyschen Verzeichnis der Käfer Deutschlands von 1909 und einige sonstige Bemerkungen über deutsche Käfer. (Col.). *Deutsche Entomologische Zeitschrift* 1913 (5): 529–539. DP: 30 September 1913 (table of contents).

Delherm de Larcenne, E. 1885: [Description de deux nouveaux coléoptères français]. *Le Naturaliste* 7 (15): 119–120. DP: 1 August 1885 (issue title page).

Della Beffa, G. 1909: Note coleotterologiche. *Rivista Coleotterologica Italiana* 7 (8/11): 192–201. DP: 15 November 1909 (issue title page).

Delobel, A. 2007: [new taxon]. In: Kergoat, G. J., Delobel, P. & Delobel, A.: Phylogenetic relationships of a new species of seed-beetle infesting Cercis siliquastrum L. in China and in Europe (Coleoptera: Chrysomelidae: Bruchinae: Bruchini). *Annales de la Société Entomologique de France, Nouvelle Série* **43** (3): 265–271. DP: 1 November 2007 (table of contents).

Delobel, A. 2010: A new seed beetle associated with Indigofera zollingeriana in Vietnam, with a note on the Bruchidius japonicus (Harold) species group (Coleoptera: Chrysomelidae: Bruchinae). *Genus* **21** (2): 249–255. DP: 30 July 2010 (paper title page).

Delobel, A. 2011: Various species of Spermophagus (Coleoptera, Chrysomelidae: Bruchinae) in the Hungarian Natural History Museum, Budapest. *Annales Historico-Naturales Musei Nationalis Hungarici* **103**: 107–116. DP: 30 November 2011 (verso of the volume title page).

Delobel, A. & Anton, K.-W. 2004: [new taxon]. In: Delobel, A., Anton, K.-W. & Kergoat, G.: New data on European Astragalus-feeding Bruchidius, with the description of a new species from southern Italy (Coleoptera: Bruchidae: Bruchinae). *Genus* **15** (2): 173–185. DP: 30 June 2004 (issue title page).

Delobel, A., Anton, K.-W., Le Ru, B. & Kergoat, G. J. 2013: Morphology, biology and phylogeny of African seed beetles belonging to the Bruchidius ituriensis species group (Coleoptera: Chrysomelidae: Bruchinae). *Genus* **24** (1): 39–63. DP: 30 March 2013 (issue title page).

Delobel, A. & Delobel, B. 2003: Les plantes hôtes des bruches (Coleoptera Bruchidae) de la faune de France, une analyse critique. *Bulletin Mensuel de la Société Linnéenne de Lyon* **72** (6): 199–221. DP: [30] June 2003 (issue cover).

Delobel, A. & Delobel, B. 2007: Contribution to the knowledge of Bulgarian seed beetles (Coleoptera: Bruchidae). *Russian Entomological Journal* **16** (2): 213–218. DP: 1 June 2007 (journal website).

Delobel, A., Le Ru, B., Genson, G., Musyoka, B. K. & Kergoat, G. J. 2015: Molecular phylogenetics, systematics and host-plant associations of the Bruchidius albosparsus (Fåhraeus) species group (Coleoptera, Chrysomelidae, Bruchinae) with the description of four new species. *Zootaxa* **3931** (4): 451–482. DP: 16 March 2015 (footer on the paper title page).

Delobel, A. & Legalov, A. 2009: A new species of Kytorhinus Fisch.-Wald. from Western Pamir (Coleoptera: Bruchinae). *Genus* **20** (1): 71–76. DP: 30 March 2009 (issue title page).

Delobel, A. & Sadeghi, S. E. 2013: Remarks on the Spermophagus rufipes species group, with the description of a new species from Iran (Coleoptera: Chrysomelidae: Bruchinae: Amblycerinae). *Genus* **24** (3/4): 309–314. DP: 28 December 2013 (issue title page).

Delobel, A. & Sadeghi, S. E. 2014: Two new Bruchus species from the Iranian highlands, with biological data (Chrysomelidae: Bruchinae). *Genus* **25** (3): 433–440. DP: 30 September 2014 (issue title page).

Delobel, B. & Delobel, A. 2006: Dietary specialization in European species groups of seed beetles (Coleoptera: Bruchidae: Bruchini). *Oecologia* **149**: 428–443. DP: 23 June 2006 (paper title page).

Demaison, C. 1891: [Description d'un coléoptère nouveau de la famille des Gallerucidae]. *Bulletin de la Société Entomologique de France* **1891** (20): cxciv–cxcv. DP: 31 December 1891 (Evenhuis 2002).

Demaison, C. 1896: Diagnoses de quelques espèces de chrysomélides et note sur les synonymies de deux Clytrides (Col.). *Bulletin de la Société Entomologique de France* **1896** (1): 12–13. DP: 18 May 1896 (*Bibl. Zool.* 1896, 1: 270).

Demaison, C. 1902: Description de trois coléoptères phytophages nouveaux d'Europe et d'Égypte. *Bulletin de la Société Entomologique de France* **1902** (2): 24–25. DP: 10 February 1902 (*dates d'apparition* at the end of volume).

Demaison, C. 1905: Description d'une variété et d'une espèce nouvelles d'Asie Mineure (Col.). *Bulletin de la Société Entomologique de France* [**1904**] (19): 286–287. DP: 7 January 1905 (*dates d'apparition* at the end of volume).

Demaux, J. E. A. 1964: Cryptocephales espagnols (Col., Chrysomelidae). *Bulletin Mensuel de la Société Linnéenne de Lyon* **33** (5): 178–180. DP: [31] May 1964 (issue title page).

De Monte, T. 1948: Una nuova razza del Chrysochus asclepiadeus Pallas. (Coleopt. Chrysomelidae). *Atti del Museo di Civico di Storia Naturale di Trieste* **17** [1948–1949] (1): 57–60. DP: 31 December 1948 (footer on the paper title page).

Depoli, G. 1915: Neue Käferformen aus dem Liburnischen und Istrianer Karst. *Wiener Entomologische Zeitung* **34** (3/4): 109–110. DP: 30 April 1915 (footer on the paper title page).

Depoli, G. 1926a: Su alcuni coleotteri della Carnia. *Bollettino de la Società Entomologica Italiana* **58** (1): 14–15. DP: 29 January 1926 (issue cover and Poggi 2008).

Depoli, G. 1926b: [new taxa]. In: Depoli, G. & Goidanich, A.: I coleotteri della Liburnia. VI. Phytophaga. *Fiume, Rivista Semestrale della "Societá di studi Fiumani" in Fiume* **4** (1): 62–113. DP: [31 December] 1926 (issue cover).

Depoli, G. 1932: Revisione dei coleotteri della collezione Leoni. IV. Chilotoma Redt. *Bollettino del Laboratorio di Entomologia del R. Istituto Superiore Agrario di Bologna* **5**: 86–87. DP: 10 November 1932 (reprint cover).

Depoli, G. 1936: Materiali per le fauna coleotterologica del Gran Sasso d'Italia. *Bollettino de la Società Entomologica Italiana* **68** (8): 133–141. DP: 27 October 1936 (Poggi 2008).

Depoli, G. 1940: I coleotteri della Liburnia. Parte VII: Rhyncophora e supplemento alla parte VI Phytophaga. *Fiume, Rivista della Società di Studi Fiumani in Fiume* **15–16** [1937–1938]: 212–338. DP: 20 July 1940 (last page of the volume).

Depoli, G. & Goidanich, A. 1926: I coleotteri della Liburnia. VI. Phytophaga. *Fiume, Rivista Semestrale della "Societá di studi Fiumani" in Fiume* **4** (1): 62–113. DP: [31 December] 1926 (issue cover).

Derenne, E. 1949: Variété nouvelle de Chrysomela marginata L. *Bulletin & Annales de la Société Entomologique de Belgique* 85 (7/8): 170–171. DP: 15 September 1949 (issue cover).

Desbrochers des Loges, J. 1866: Description d'une nouvelle espèce algérienne du genre Clythra (s. g. Labidostomis). *Bulletin de l'Académie d'Hippone* **2**: 42–44. DP: [31 December] 1866 (volume cover).

Desbrochers des Loges, J. 1869: Description de Polydrosus et autres curculionides nouveaux et d'un genre nouveau de la même famille. *Annales de la Société Entomologique de France, Quatrième Série* **9** (3): 389–400. DP: 8 December 1869 (Lefèvre 1885).

Desbrochers des Loges, J. 1870a: Descriptions de coléoptères nouveaux d'Europe et confins. *L'Abeille, Mémoires d'Entomologie* **7** [1869–1870]: 97–135. DP: [31] March 1870 (footer on p. 129).

Desbrochers des Loges, J. 1870b: [new taxon]. In: Heyden, L. F. J. D. von: Entomologische Reise nach dem südlichen Spanien, der Sierra Guadarrama und Sierra Morena, Portugal und den Cantabrischen Gebirge. *Berliner Entomologische Zeitschrift* **14 (Beiheft)**: [1] + 218 pp., 2 pls. DP: [31 December] 1870 (Bousquet 2016).

Desbrochers des Loges, J. 1870c: Description de coléoptères nouveaux algériens et remarques diverses. *Bulletin de l'Académie d'Hippone* **9**: 68–82. DP: [31 December] 1870 (volume cover).

Desbrochers des Loges, J. 1871: Description de coléoptères nouveaux d'Europe et confins et remarques diverses. *Mittheilungen der Schweizerischen Entomologischen Gesellschaft* **3** [1869–1872] (7): 337–360. DP: [31] January 1871 (issue title page).

Desbrochers de Loges, J. 1873: Notes synonymiques – remarques diverses – description de coléoptères nouveaux. *Annales de la Société Entomologique de France, Cinquième Série* **2** [1872] (4): 420–432. DP: 27 April 1873 (Lefèvre 1885).

Desbrochers des Loges, J. 1875: Notes synonymiques sur quelques coléoptères. *Bulletin de la Société Entomologique de France* **1875** (13): 148–149. DP: 23 July 1875 (Evenhuis 2002).

Desbrochers des Loges, J. 1884: Diagnoses de coléoptères nouveaux algériens. *Bulletin de l'Académie d'Hippone* **19**: 168–171. DP: 9 June 1884 (mentioned in the dated letter in the *Bull. Acad. Hippone* 1885, 20: 119).

Desbrochers des Loges, J. 1891: Monographie des Cassida de France suivie d'observations sur le classement des espèces de ce genre au Catalogus Coleopterorum Europae, Caucasi et Armeniae rossicae. *Le Frelon* [1891–1892] **1**: 1–48. DP (issue 3, pp. 1–8): 5 June 1891, (issue 4, pp. 9–16): 15 July 1891, (issue 5, pp. 17–24): 15 August 1891, (issue 6, pp. 25–36): 15 September 1891, (issue 7, pp. 37–48): 15 October 1891 (Gaedike & Behne 2008).

Desbrochers des Loges, J. 1898: Quelques matériaux pour la faune des coléoptères de Barbarie. II. *Le Frelon* **7** [1898–1899]: 1–51. DP (issue 1, pp. 1–16): 25 August 1898, (issue 2, pp. 17–32): 25 October 1898, (issue 3, pp. 33–48): 30 November 1898, (issue 4, pp. 49–51): 30 December 1898 (Gaedike & Behne 2008).

Dietrich, J. K. 1857: Einiges aus dem Gebiete der schweizerischen Käferfauna. *Entomologische Zeitung* (Stettin) **18** (4/6): 117–138. DP: [30] Juni 1857 (issue title page).

Dillwyn, L. W. 1829: *Memoranda relating to coleopterous insects found in the neighbourhood of Swansea*. Swansea: W. C. Murray and D. Rees, 75 pp. DP: [31 December] 1829 (Bousquet 2016).

Döberl, M. 1991a: Eine neue Alticinae aus Marokko: Psylliodes belarbii nov. spec. (Coleoptera, Chrysomelidae). *Nouvelle Revue d'Entomologie, Nouvelle Série* **7** (3): 337–339. DP: 7 May 1991 (last page of the cover).

Döberl, M. 1991b: Alticinae (Coleoptera, Chrysomelidae) aus Nepal. *Revue Suisse de Zoologie* **98** (3): 613–635. DP: [30] September 1991 (paper title page).

Döberl, M. 1994a: Bemerkenswerte Alticinenfunde aus Westeuropa (Col., Chrysomelidae). *Entomologische Nachrichten und Berichte* **38** (3): 179–182. DP: 31 October 1994 (verso of the issue cover).

Döberl, M. 1994b: Eine neue Alticine vom Kaspischen Meer: Batophila dogueti nov. spec. (Coleoptera, Chrysomelidae, Alticinae). *Nouvelle Revue d'Entomologie, Nouvelle Série* **11** (2): 149–150. DP: 10 November 1994 (last page of the cover).

Döberl, M. 1995a: Eine neue Galerucine aus China: Parexosoma beeneni sp. n. (Col. Chrysomelidae). *Acta Coleopterologica* **11** (2): 51–53. DP: 1 April 1995 (paper title page).

Döberl, M. 1995b: Der heutige Alticinen-Artenbestand der Schweiz (Coleoptera, Chrysomelidae, Alticinae). *Mitteilungen der Entomologischen Gesellschaft Basel, Neue Folge* **45** (2): 42–96. DP: 10 June 1995 (issue title page).

Döberl, M. 1995c: A new Psylliodes from Madeira: Psylliodes erberi nov. spec. (Coleoptera: Chrysomelidae: Alticinae). *Bocagiana* **178**: 1–4. DP: 29 November 1995 (paper title page).

Döberl, M. 1998: Zwei neue Alticinen aus Griechenland und Nord-Spanien (Col., Chrysomelidae, Alticinae). *Entomologische Nachrichten und Berichte* **42** (1/2): 41–43. DP: 30 July 1998 (verso of the issue cover).

Döberl, M. 2000: Beitrag zur Kenntnis der Gattung Epitrix Foudras, 1860 in der Paläarktis (Coleoptera: Chrysomelidae: Alticinae). *Mitteilungen des Internationalen Entomologischen Vereins e. V.* (Frankfurt am Main) **25** (1/2): 1–23. DP: 31 July 2000 (paper title page).

Döberl, M. 2001: Nomenclatorial changes in some Alticinae (Coleoptera: Chrysomelidae) from the Palaearctic and Oriental regions. *Serangga* **6** (2): 379–386. DP: [30] November 2001 (issue cover).

Döberl, M. 2003a: Zwei neue Alticinen aus der Paläarktischen Region, nebst Bemerkungen zu Phyllotreta lubischevi Lopatin, 1992 (Col., Chrysomelidae, Alticinae). *Entomologische Nachrichten und Berichte* 46 [2002] (4): 243–245. DP: 24 February 2003 (verso of the issue cover).

Döberl, M. 2003b: Beitrag zur Kenntnis der Batophila-Arten Nepals (Coleoptera: Chrysomelidae: Alticinae). *Veröffentlichungen des Naturkundemuseums Erfurt* **22**: 149–152. DP: 12 December 2003 (editor of journal M. Hartmann, 2019, pers. comm.).

Döberl, M. 2004: Eine neue westeuropäische Longitarsus-Art: Longitarsus cizeki n. sp. (Col., Chrysomelidae, Alticinae). *Entomologische Nachrichten und Berichte* **48** (1): 15–17. DP: 22 April 2004 (verso of the issue cover).

Döberl, M. 2005: Contribution to the knowledge of the genus Aphthonoides Jacoby, 1885 (Coleoptera: Chrysomelidae: Alticinae). Pp. 53–80. In: Konstantinov, A., Tishechkin, A. & Penev, L. (eds.): *Contributions to systematics and biology of beetles. Papers celebrating the 80th birthday of Igor Konstantinovich Lopatin.* Sofia-Moscow: Pensoft, 388 pp. DP: [30] September 2005 (colophon).

Döberl, M. 2007a: Beitrag zur Kenntnis der Gattung Hemipyxis Dejean, 1836 – die Arten der indo-malayischen Region mit Einschluß der ost-paläarktischen Arten (Coleoptera: Chrysomelidae: Alticinae). 1 Teil. *Russian Entomological Journal* **16** (1): 79–99. DP: 31 March 2007 (journal website).

Döberl, M. 2007b: Beitrag zur Kenntnis der Gattung Minota Kutschera, 1859 in der Ostpaläarktis und in der Orientalis (Coleoptera: Chrysomelidae: Alticinae). *Russian Entomological Journal* 16 (3): 329–332. DP: 4 December 2007 (journal website).

Döberl, M. 2009a: Nomenclatorial notes on Palaearctic Coleoptera (Curculionidae, Cerambycidae, Chrysomelidae). *Entomologische Blätter für Biologie und Systematik der Käfer* 105: 19–23. DP: 15 November 2009 (paper title page).

Döberl, M. 2009b: Contribution to the knowledge of Alticinae from Nepal (Insecta: Coleoptera: Chrysomelidae). *Vernate* 28: 413–420. DP: 29 December 2009 (editor of journal M. Hartmann, 2019, pers. comm.).

Döberl, M. 2010a: Two new Alticinae from Taiwan (Coleoptera: Chrysomelidae: Alticinae). *Acta Coleopterologica* 26 (1): 65–67. DP: 15 January 2010 (paper title page).

Döberl, M. 2010b: Chrysomelidae: Alticinae. Pp. 491–663. In: Löbl, I. & Smetana, A. (eds.): *Catalogue of Palaearctic Coleoptera. Volume 6. Chrysomeloidea.* Stenstrup: Apollo Books, 924 pp. DP: 22 February 2010 (colophon).

Döberl, M. 2010c: Contribution to the knowledge of the alticines from Iran, with description of a new Phyllotreta species (Col.: Chrysomelidae: Alticinae). *Journal of Entomological Society of Iran* 30 (1): 41–54. DP: [30] September 2010 (issue cover).

Döberl, M. 2011: New Alticinae from China and southeastern Asia (Coleoptera: Chrysomelidae). *Genus* 22 (2): 271–283. DP: 31 July 2011 (paper title page).

Döberl, M. 2012: Alticinae (Coleoptera: Chrysomelidae) of Socotra Island. Pp. 429–447. In: Hájek, J. & Bezděk, J. (eds.): Insect biodiversity of the Socotra Archipelago. *Acta Entomologica Musei Nationalis Pragae* 52 (supplementum 2): i–vi + 1–557. DP: 17 December 2012 (colophon and paper title page).

Döberl, M. & Konstantinov, A. 2003: A new genus of flea beetles from Nepal (Coleoptera: Chrysomelidae). *Coleopterists Bulletin* 57 (2): 205–218. DP: 20 June 2003 (last page of the paper).

Dodero, A. 1925: Missione zoologica del Dr. E. Festa in Cirenaica. XIV. *Bollettino del Musei di Zoologia ed Anatomia Comparata della R. Università di Torino* 39: 1–31. DP: 31 March 1925 (last page of the paper).

Doguet, S. 1972: Notes sur les altises d'Algérie (1re note) (Col. Chrysomelidae). Remarques sur quelques Dibolia et Altica d'Algérie. Description de deux espèces nouvelles. *Bulletin de la Société Entomologique de France* 77 (5/6): 133–135. DP: 21 December 1972 (*dates de publication* on p. 320).

Doguet, S. 1973: Description d'une nouvelle espèce de Longitarsus des Pyrénées (Col. Chrysomelidae Alticinae). *L'Entomologiste* 29 (3): 125–127. DP: [30] June 1973 (issue title page).

Doguet, S. 1974: Contribution à l'étude des altises de la faune paléarctique. Notes diverses et description de deux espèces nouvelles (Col. Chrysomelidae Alticinae). *L'Entomologiste* 30 (3): 121–128. DP: [30] June 1974 (issue title page).

Doguet, S. 1975: Contribution à l'étude des Dibolia d'Afrique du Nord et de France (Col. Chrysomelidae). *Nouvelle Revue d'Entomologie* 5 (2): 197–210. DP: 15 June 1975 (issue cover).

Doguet, S. 1976: Alticinae recoltes en Turquie et en Iran par le Dr. W. Wittmer (Col. Chrysomelidae). *Mitteilungen der Entomologischen Gesellschaft Basel, Neue Folge* **26** (3): 91–94. DP: [30] September 1976 (heading on p. 65).

Doguet, S. 1977: Description d'une nouvelle espèce de Phyllotreta d'Algérie (Col. Chrysomelidae). *L'Entomologiste* **33** (4/5): 185–187. DP: [30] September 1977 (DL on last page of the issue cover).

Doguet, S. 1979a: Insects of Saudi Arabia. Coleoptera: Fam. Chrysomelidae, Subfam. Halticinae. *Fauna of Saudi Arabia* **1**: 308–316. DP: 5 March 1979 (table of contents).

Doguet, S. 1979b: Notes systématiques et écologiques sur divers Chrysomelidae paléarctiques. Description de deux espèces nouvelles. *L'Entomologiste* **35** (2): 49–55. DP: [31] March 1979 (DL on last page of the issue cover).

Doguet, S. 1984: Contribution à l'étude des espèces d'Afrique du Nord du genre Phyllotreta (Coleoptera, Chrysomelidae). *Nouvelle Revue d'Entomologie, Nouvelle Série* **1** (3): 243–265. DP: 30 December 1984 (issue cover).

Doguet, S. 1985: Insects of Saudi Arabia. Coleoptera: Fam. Chrysomelidae, Subfam. Alticinae (Part 2). *Fauna of Saudi Arabia* **6** [1984]: 361–366. DP: 28 February 1985 (table of contents).

Doguet, S. 1987: Description d'une espèce nouvelle de Longitarsus d'Algérie (Coleoptera, Chrysomelidae). *Elytron* **1**: 55–57. DP: [31 December] 1987 (volume cover).

Doguet, S. 1988a: Notes faunistiques et taxonomiques sur quelques Aphthona de la faune française avec description d'une espèce nouvelle (Coleoptera, Chrysomelidae). *Nouvelle Revue d'Entomologie, Nouvelle Série* **5** (2): 145–153. DP: 30 June 1988 (last page of the cover).

Doguet, S. 1988b: Contribution à la connaissance des Chrysomelidae de Grèce et de Crète et description de deux sous-espèces nouvelles appartenant aux genres Colaspidea Laporte et Aphthona Chevrolat (Col.). *Bulletin de la Société Entomologique de France* **92** [1987] (9/10): 301–306. DP: 1 July 1988 (*dates de la publication* on p. 328).

Doguet, S. 1990: Une nouvelle espèce du genre Orestia de Corse (Coleoptera, Chrysomelidae). *Nouvelle Revue d'Entomologie, Nouvelle Série* **7** (1): 111–114. DP: 31 October 1990 (last page of the cover).

Doguet, S. 1993: Description d'une nouvelle espèce méditerranéenne de Psylliodes (Coleoptera, Chrysomelidae). *Nouvelle Revue d'Entomologie, Nouvelle Série* **9** [1992] (4): 357–364. DP: 30 April 1993 (last page of the cover).

Doguet, S. 1994: *Coléoptères Chrysomelidae. Volume 2. Alticinae. Faune de France 80*. Paris: Fédération Française des Sociétés de Sciences Naturelles, 697 pp. DP: [31] December 1994 (DL on last page of the book).

Doguet, S. 2013: Contribution à la connaissance des Altises de Libye (Coleoptera, Chrysomelidae, Alticinae). *Nouvelle Revue d'Entomologie, Nouvelle Série* **29** (3/4): 173–177. DP: 31 December 2013 (last page of the cover).

Doguet, S. 2014: Alticinae. Pp. 606–618. In: Tronquet, M. (ed.): Catalogue des coléoptères de France. *Revue de l'Association Roussillonnaise d'Entomologie* **23** (supplément): 1–1052. DP: [31] May 2014 (colophon).

Doguet, S. & Bergeal, M. 2000: Contribution à la connaissance des altises de Grèce et de Turquie (Coleoptera, Chrysomelidae, Alticinae). *Nouvelle Revue d'Entomologie, Nouvelle Série* 17 (2): 123–135. DP: 15 December 2000 (last page of the cover).

Doguet, S. & Bergeal, M. 2002: Étude de diverses altises paléarctiques (Espagne, Maroc, Grèce, Turquie) avec les descriptions de deux nouvelles espèces de Longitarsus (Coleoptera, Chrysomelidae, Alticinae). *Nouvelle Revue d'Entomologie, Nouvelle Série* 18 [2001] (4): 369–374. DP: 5 June 2002 (last page of the cover).

Doguet, S. & Bergeal, M. 2007: Contribution à l'étude des Chrysomelidae du Maroc et d'Espagne, avec description d'une espèce nouvelle de Longitarsus (Coleoptera). *Nouvelle Revue d'Entomologie, Nouvelle Série* 23 [2006] (2): 99–113. DP: 15 January 2007 (last page of the cover).

Doguet, S. & Gruev, B. 1988: Etude des Longitarsus du groupe de L. obliteratus (Rosenhauer) de la faune de France et description d'une espèce nouvelle (Col. Chrysomelidae). *L'Entomologiste* 44 (5): 261–266. DP: [31] October 1988 (issue title page).

Doguet, S. & Petitpierre, E. 1986: Deux nouvelles espèces d'Aphthona d'Espagne (Coleoptera, Chrysomelidae). *Revue Française d'Entomologie, Nouvelle Série* 8 (4): 145–148. DP: 16 December 1986 (last page of the cover).

Dohrn, C. A. 1859: Ein neuer Cryptocephalus der europäischen Fauna. *Entomologische Zeitung* (Stettin) 20 (10/12): 426–428. DP: [31] December 1859 (issue title page).

Dohrn, C. A. 1883: Exotisches. *Entomologische Zeitung* (Stettin) 45 [1884] (1/3): 39–46. DP: [30] November 1883 (noticed on p. 112).

Doi, K. 1928: [Descriptions of some new and unrecorded species of Coleoptera from Japan]. *Dobutsugaku Zasshi* 40 (479): 371–376 (in Japanese). DP: 15 September 1928 (issue cover).

Dollman, H. C. 1912: Longitarsus plantago-maritimus, sp. nov. A coleopteron new to science. *Entomologist's Record and Journal of Variation* 24 (7/8): 187–188. DP: 15 August 1912 (issue cover).

Donckier de Donceel, H. 1900: Catalogue systématique des Hispides. *Annales de la Société Entomologique de France* 68 [1899] (3): 540–608 & (4): 609–615. DP (pp. 540–608): 28 February 1900, (pp. 609–615): 27 June 1900 (Evenhuis 2022).

Donisthorpe, H. 1913: [new taxon]. In: Fowler, W. W. & Donisthorpe, H.: *The Coleoptera of the British Islands. A descriptive account of the families, genera, and species indigenous to Great Britain and Ireland with notes as to localities, habitats, etc. Vol. VI (supplement)*. London: Lovell Reeve and Co. Ltd., xiii + 351 pp., 20 pls. DP: 17 May 1913 (*Athenaeum* 1913, 4464: 547).

Donisthorpe, H. 1934: Some new and unrecorded aberrations of British Coleoptera. *Entomologist's Monthly Magazine, Third Series* 70 (845): 230–231. DP: 5 October 1934 (registered at NHMUK library).

Donisthorpe, H. 1940: Psylliodes napi, Koch, var. hignetti, var. nov. (Col. Chrysomelidae), a variety new to science. *Entomologist's Record and Journal of Variation, New Series* 52 (11): 120. DP: 15 November 1940 (heading on p. 120).

Donisthorpe, H. 1944: Phyllotreta hintoni sp. n. (Col., Halticidae), a species of Coleoptera new to Britain. *Entomologist's Monthly Magazine, Fourth Series* 80 (965): 227–228. DP: [30] October 1944 (issue cover).

Donovan, E. 1793: *The natural history of British Insects; explaining them in their several states, with the periods of their transformations, their food, oeconomy, &c. together with the history of such minute insects as require investigation by the microscope. The whole illustrated by coloured figures, designed and executed from living specimens. Vol. II*. London: F. and C. Rivington, 96 + [6] pp., pls. 37–72. DP: [31 December] 1793 (Bousquet 2016).

Donovan, E. 1794: *The natural history of British Insects; explaining them in their several states, with the periods of their transformations, their food, oeconomy, &c. together with the history of such minute insects as require investigation by the microscope. The whole illustrated by coloured figures, designed and executed from living specimens. Vol. III*. London: F. and C. Rivington, 98 + [7] pp., pls. 73–108. DP: [31 December] 1794 (Bousquet 2016).

Donovan, E. 1799: *The natural history of British Insects; explaining them in their several states, with the periods of their transformations, their food, oeconomy, &c. together with the history of such minute insects as require investigation by the microscope. The whole illustrated by coloured figures, designed and executed from living specimens. Vol. VIII*. London: F. and C. Rivington, 88 + [4] pp., pls. 253–288. DP: [31 December] 1799 (Bousquet 2016).

Donovan, E. 1801: *The natural history of British Insects; explaining them in their several states, with the periods of their transformations, their food, oeconomy, &c. together with the history of such minute insects as require investigation by the microscope. The whole illustrated by coloured figures, designed and executed from living specimens. Vol. X*. London: F. and C. Rivington, 95 + [7] pp., 35 pls. DP: [31 December] 1801 (Bousquet 2016).

Douglas, H. B., Konstantinov, A. S., Brunke, A. J., Moseyko, A. G., Chapados, J. T., Eyres, J., Richter, R., Savard, K., Sears, E., Prathapan, K. D., Ruan, Y.-Y. & Dettman, J. R. 2023: Phylogeny of the flea beetles (Galerucinae: Alticini) and the position of Aulacothorax elucidated through anchored phylogenomics (Coleoptera: Chrysomelidae: Alticini). *Systematic Entomology* 48 (3): 361–386. DP: 3 March 2023 (journal website).

Drapiez, P. A. J. 1819a: Description de huit espèces d'insectes nouveaux. *Annales Générales des Sciences Physiques* (Bruxelles) 1: 45–55, pl. IV. DP: [31 December] 1819 (volume title page).

Drapiez, P. A. J. 1819b: Description de huit espèces d'insectes nouveaux. *Annales Générales des Sciences Physiques* (Bruxelles) 1: 290–298, pl. XI. DP: [31 December] 1819 (volume title page).

Drapiez, P. A. J. 1819c: Description de huit espèces d'insectes nouveaux. *Annales Générales des Sciences Physiques* (Bruxelles) 2: 42–50, pl. XVI. DP: [31 December] 1819 (volume title page).

Drapiez, P. A. J. 1819d: Description de huit espèces d'insectes nouveaux. *Annales Générales des Sciences Physiques* (Bruxelles) 2: 197–204, pl. XXV. DP: [31 December] 1819 (volume title page).

Drury, D. 1773: *Illustrations of natural history. Wherein are exhibited upwards of two hundred and twenty figures of exotic insects, according to their different genera; very few of which have hitherto been figured by any author, being engraved and coloured from nature, with the greatest accuracy, and under the author's own inspection, on fifty copper-plates. With a particular description of each insect: interspersed with remarks and reflections on the nature and properties of many of them. Vol. II*. London: B. White, vii + 90 + [4] pp., 50 pls. DP: 14 May 1773 (Bousquet 2016).

Duan, W.-Y., Wang, F.-Y. & Zhou, H.-Z. 2021a: Two new species of the genus Melixanthus Suffrian (Coleoptera, Chrysomelidae, Cryptocephalinae) from China. *ZooKeys* **1060**: 111–123. DP: 17 September 2021 (paper title page).

Duan, W.-Y., Wang, F.-Y. & Zhou, H.-Z. 2021b: Taxonomy of the Cryptocephalus heraldicus Group (Coleoptera: Chrysomelidae, Cryptocephalinae) from China. *Diversity* **13** (451): 1–24. DP: 20 September 2021 (paper title page).

Duan, W.-Y., Wang, F.-Y. & Zhou, H.-Z. 2022: Two new species of the genus Smaragdina Chevrolat, 1836 (Coleoptera, Chrysomelidae, Cryptocephalinae) from China. *ZooKeys* **1082**: 1–26. DP: 18 January 2022 (paper title page).

Duan, W.-Y. & Zhou, H.-Z. 2021: Revision of the subgenus Burlinius Lopatin (Coleoptera, Chrysomelidae, Cryptocephalinae) from China and description of four new species. *Diversity* **13** (523): 1–40. DP: 23 October 2021 (paper title page).

Duan, W.-Y. & Zhou, H.-Z. 2022: Revision of the genus Adiscus Gistel, 1857 (Coleoptera: Chrysomelidae, Cryptocephalinae) from mainland China. *Zootaxa* **5096** (1): 1–80. DP: 9 February 2022 (paper title page).

Dubois, A. 1887: Tableau des Phytodecta d'Europe (1). *L'Échange, Revue Linnéenne* **3** (25) Supplement: 1–3. DP: 15 January 1887 (issue title page).

Duckett, C. N., Prathapan, K. D. & Konstantinov, A. S. 2006: Notes on identity, new synonymy and larva of Ivalia Jacoby (Coleoptera: Chrysomelidae) with description of a new species. *Zootaxa* **1363**: 49–68. DP: 23 November 2006 (footer on the paper title page).

Dudich, E. 1924: A Cryptocephalus gamma H.-Schäff. változatai (Coleopt., Chrysomelidae). (Aberrationen des Cryptocephalus gamma H.-Schäff.). *Folia Entomologica Hungarica* **1** [1923–1930] (2): 22–24. DP: 15 August 1924 (issue cover).

Dufour, G. 1902: Notes sr [sic] une variété de Chrysomela cerealis. *Feuille des Jeunes Naturalistes, IVe Série* **32** [1901–1902] (376): 83–84. DP: 1 February 1902 (issue cover).

Dufour, L. J. M. 1820: Description de dix espèces nouvelles ou peu connues d'insectes recueillis en Espagne. *Annales Générales des Sciences Physiques* (Bruxelles) **6**: 307–317. DP: [31 December] 1820 (volume title page).

Dufour, L. J. M. 1843: *Excursion entomologique dans les montagnes de la Vallée d'Ossau*. Pau: Veronese, 118 pp. DP: 28 August 1843 (Bousquet 2016) [also issued in *Bulletin des Sciences, Lettres et Arts de Pau* 1843, 3: 5–118].

Dufour, L. J. M. 1848: [Note suivante sur Cyrtonus dufourii]. *Bulletin de la Société entomologique de France* [**1847**] (4): ciii–civ. DP: 14 June 1848 (Evenhuis 2022).

Dufour, L. J. M. 1851: Des zones entomologiques dans nos Pyrénées occidentales, et désignation des insectes qui les habitent. *Actes de la Société Linnéenne de Bordeaux* **17**: 304–364 DP: [31] December 1851 (reprint mentioned in the *Compt. Rend. Hebdom. Séances Acad. Sci.* 1852, 34: 449).

Duftschmid, C. 1825: *Fauna Austriae, oder Beschreibung der österreichischen Insecten, für angehende Freunde der Entomologie. Dritter Theil*. Linz: Verlag der k. k. priv. akademischen Kunst-, Musik- und Buchhandlung, 289 pp. DP: [30] June 1825 (Bousquet 2016).

Duhaldeborde, F. 1999: Description de Cryptocephalus (s. str.) bameuli n. sp., nouvelle espèce paléarctique à large répartition géographique (Coleoptera, Chrysomelidae). *Nouvelle Revue d'Entomologie, Nouvelle Série* **16** (2): 123–135. DP: 31 December 1999 (last page of the cover).

Duhaldeborde, F. 2018: A new Cryptocephalus from Turkey of the C. flavipes Fabricius, 1781, species-group (Coleoptera, Chrysomelidae, Cryptocephalini). *Bulletin de la Société Entomologique de France* **123** (2): 259–268. DP: 22 June 2018 (paper title page).

Duméril, A. M. C. 1816: Altise, Altica. Pp. 520–533. In: Cuvier, F. G. (ed.): *Dictionnaire des sciences naturelles, dans lequel on traite méthodiquement des différents êtres de la nature, considérés soit en eux-mêmes, d'après l'état actuel de nos connoissances, soit relativement à l'utilité qu'en peuvent retirer la médecine, l'agriculture, le commerce et les arts. Suivi d'une biographie des plus célébres naturalistes. Ouvrage destiné aux médecins, aux agriculteurs, aux commerçans, aux artistes, aux manufacturiers, et à tous ceux qui ont intérêt à connoitre les productions de la nature, leurs caractères génériques et spécifiques, leur lieu natal, leurs propriétés et leurs usages. Par plusieurs professeurs du Jardin du roi, et des principales écoles de Paris. Tome premier.* Strasbourg: F. G. Levrault, Paris: Le Normant, lxii + 560 + 138 (Supplément) pp. DP: 12 October 1816 (Bousquet 2016).

Duvivier, A. 1884a: Cinq espèces nouvelles du genre Aulacophora, Chevr. (coléoptères phytophages). *Notes from the Leyden Museum* **6** (2): 119–125. DP: [30] April 1884 (Dickinson 2005).

Duvivier, A. 1884b: Descriptions de quelques phytophages nouveaux. *Bulletin ou Comptes-Rendus des Séances de la Société Entomologique de Belgique, Série III* **1884** (44): cxxxii–cxxxix. DP: 14 May 1884 (*Bull. Soc. Ent. Fr.* 1884: 126).

Duvivier, A. 1884c: Sur quelques espèces du genre Oïdes, Weber (Galerucinae) du Musée de Leyde. *Notes from the Leyden Museum* **6** (4): 236–240. DP: [31] October 1884 (Dickinson 2005).

Duvivier, A. 1884d: Description de quelques phytophages nouveaux et notes rectificatives. *Bulletin ou Comptes-Rendus des Séances de la Société Entomologique de Belgique, Série III* **1884** (50): cccx–cccxx. DP: 10 December 1884 (DP refers to the issue 51 mentioned in the *Bull. Soc. Ent. Fr.* 1884: 231).

Duvivier, A. 1885a: Phytophages exotiques. *Entomologische Zeitung* (Stettin) **46** (4/6): 241–250. DP: [31] May 1885 (noticed on p. 288).

Duvivier, A. 1885b: Phytophages exotiques. *Entomologische Zeitung* (Stettin) **46** (10/12): 385–400. DP: [31] October 1885 (noticed on p. 411).

Duvivier, A. 1885c: Quatre Phytophages nouveaux. *Bulletin ou Comptes-Rendus des Séances de la Société Entomologique de Belgique, Série III* **1885** (64): cxvi–cxix. DP: 9 December 1885 (DP refers to issue 65 mentioned in *Bull. Soc. Ent. Fr.* 1885: ccxx).

Duvivier, A. 1887: Description de trois galérucides nouvelles. *Bulletin ou Comptes-Rendus des Séances de la Société Entomologique de Belgique, Série III* **1887** (87): xlvii–l. DP: 13 July 1887 (*Bull. Soc. Ent. Fr.* 1887: cxxiv).

Duvivier, A. 1890: Liste des coléoptères phytophages recueillis par M. le Dr Plateeuw dans l'île de Sumatra. *Bulletin ou Comptes-Rendus des Séances de la Société Entomologique de Belgique, Série IV* **1890** (3): xxxii–xxxvii. DP: 26 March 1890 (*Bull. Soc. Ent. Fr.* 1890: li).

Duvivier, A. 1891a: Les phytophages du Chota-Nagpore (1re note). *Bulletin ou Comptes-Rendus des Séances de la Société Entomologique de Belgique, Série IV* **1891** (15): xxiv–li. DP: 11 February 1891 (*Bull. Soc. Ent. Fr.* 1891: xxix).

Duvivier, A. 1891b: Mélanges entomologiques. *Bulletin ou Comptes-Rendus des Séances de la Société Entomologique de Belgique, Série IV* **1891** (17): cxlv–clvi. DP: 8 April 1891 (*Bull. Soc. Ent. Fr.* 1891: lxvii).

Duvivier, A. 1892a: Mélanges entomologiques. X. Diagnoses de coléoptères nouveaux du Congo. *Annales de la Société Entomologique de Belgique* **36** (2): 56–60. DP: 9 March 1892 (*Bull. Soc. Ent. Fr.* 1892: lxxviii).

Duvivier, A. 1892b: Les phytophages du Chotah-Nagpore (2e note). *Annales de la Société Entomologique de Belgique* **36** (8): 396–449. DP: 12 October 1892 (*Bull. Soc. Ent. Fr.* 1892: ccxviii).

Egorov, A. B. 1989: Obzor vidov zernovok (Coleoptera, Bruchidae) otnesennykh k rodu Acanthoscelides Schilsky v "Faune SSSR" (Lukyanovich i Ter-Minasyan, 1957). (A review of species of bruchids (Coleoptera, Bruchidae), placed to the genus Acanthoscelides Schilsky in the "Fauna of the USSR" (Luk'yanovich and Ter-Minassian, 1957)). *Entomologicheskoe Obozrenie* **68** (4): 748–758. SFP: 8 December 1989, DP: [31 December] 1989 (issue cover).

Egorov, A. B. 1996: 106. Sem. Bruchidae – Zernovki. Pp. 140–158. In: Ler, P. A. (ed.): *Opredelitel nasekomykh Dalnego Vostoka Rossii v shesti tomakh. Tom 3. Zhestkokrylye, ili zhuki. Chast 3.* (*Key to the insects of Russian Far East. Vol. III. Coleoptera. Pt 3.*). Vladivostok: Dal'nauka, 556 pp. SFP: 4 October 1996, DP: [31 December] 1996 (title page).

Eichler, W. 1924: Nouvelles espèces et variétés de coléoptères du Caucase et de l'Asie-mineure. *Polskie Pismo Entomologiczne* **3** (3): 61–68. DP: 29 November 1924 (stamp on the BHL copy).

Ekiz, A. N., Şen, İ., Aslan, E. G. & Gök, A. 2013: Checklist of leaf beetles (Coleoptera: Chrysomelidae) of Turkey, excluding Bruchinae. *Journal of Natural History* **47** (33/34): 2213–2287. DP: 7 May 2013 (journal website).

El Aalaoui, M., El Bouhssini, M., Bouharroud, R. & Sbaghi, M. 2021: First record of Orsodacne altra [sic !] (Coleoptera: Orsodacnidae) and Cyrtosus cyanipennis (Coleoptera: Malachiidae) associated to Dactylopius opuntiae (Hemiptera: Dactylopiidae) in Morocco. *Polish Journal of Entomology* 90 (3): 145–151. DP: 30 September 2021 (journal website).

El Torkey, A. M. & Al Dhafer, H. M. 2015: Amphimela raydahensis sp. nov. from the Kingdom of Saudi Arabia (Coleoptera: Chrysomelidae: Galerucinae: Alticini). *Zootaxa* **4028** (3): 430–440. DP: 9 October 2015 (footer on the paper title page).

Endrődi, S. 1957: Über die Variabilität der Chrysomeliden-Art Phytodecta fornicata Brüggm. (Coleoptera). *Folia Entomologica Hungarica, Series Nova* 10 (2): 341–348. DP: [31 December] 1957 (issue cover).

Erber, D. & Medvedev, L. N. 1999: Zur Taxonomie der Gattung Smaragdina Chevrolat (Coleoptera: Chrysomelidae: Clytrinae), mit Beschreibungen neuer Taxa aus Asien. *Stuttgarter Beiträge zur Naturkunde, Serie A (Biologie)* **601**: 1–10. DP: 1 December 1999 (paper title page).

Erichson, W. F. 1834: Coleoptera. In: Meyen, F.: Beiträge zur Zoologie, gesammelt auf einer Reise um die Erde. *Novorum Actorum Academiae Caesareae Leopoldino Carolinae Naturae Curiosorum* **16** (supplement 1): 219–276, pls. XXXVII–XXXIX. DP: [31 December] 1834 (issue title page).

Erichson, W. F. 1841: Ueber die Insecten von Algier mit besonders Berücksichtigung ihrer geographischen Verbreitung. Pp. 140–194. In: Wagner, M. (ed.): *Reise in des Regentsschaft Algier in den Jahren 1836, 1837 und 1838. Nebst einem naturhistorischen Anhang und einem Kupferatlas. Dritter Band.* Leipzig: L. Voss, xviii + [2] + 296 pp., 17 pls. DP: 6 February 1841 (Bousquet 2016).

Erichson, W. F. 1842a: Beitrag zur Insecten-Fauna von Vandiemensland, mit besonderer Berücksichtigung der geographischen Verbreitung der Insecten. *Archiv für Naturgeschichte* 8 [1] (1): 83–128 & (2/3): 129–287, pls. IV–V. DP (pp. 83–128): 6 May 1842 (*Allg. Bibl. Deutsch.* 1842, 6: 185), (pp. 129–287): 28 September 1842 (*Lit. Zeit.* 1842 (39): 905).

Erichson, W. F. 1842b: [new taxon]. In: Klug, J. C. F. & Erichson, W. F.: *Verzeichniss der verkäuflichen Doubletten des Berliner Museum.* Berlin, 15 pp. DP: [31 December] 1842 (Bousquet 2016).

Erichson, W. F. 1843: Beitrag zur Insecten-Fauna von Angola, in besonderer Beziehung zur geographischen Verbreitung der Insecten in Afrika. *Archiv für Naturgeschichte* 9 [1] (3): 199–267. DP: [31 December] 1843 (issue title page).

Erichson, W. F. 1847: Conspectus insectorum coleopterorum, quae in Republica Peruana observata sunt. *Archiv für Naturgeschichte* **13** [1] (1): 67–112 & (2) 113–185. DP (pp. 67–112): 14 July 1847 (*Lit. Zeit.* 1847 (57): 914), DP (pp. 113–185): [31 December] 1847 (issue cover).

Erichson, W. F. 1859: [new taxon]. In: Klug, J. C. F. & Erichson, W. F.: Einige Diagnosen. *Entomologische Zeitung* (Stettin) **20** (1/3): 83–87. DP: [31] March 1859 (issue title page).

Eroshkina, G. A. 1992: Obzor rodov Parascela, Lophea, Aulacolepis, Heterotrichus, Hemiplatys, Abirus, Osnaparis i Chalcolema (Coleoptera, Chrysomelidae, Eumolpinae) fauny Vietnama. (Review of the genera Parascela, Lophea, Aulacolepis, Heterotrichus, Hemiplatys, Abirus, Osnaparis and Chalcolema (Coleoptera, Chrysomelidae) in the fauna of Vietnam). Pp. 77–82. In: Medvedev, L. N. (ed.): *Sistematika i ekologia nasekomykh Vietnama.* (*Systematization and ecology of insects of Vietnam*). Moskva, 264 pp. SFP: 12 February 1992, DP: 7 October 1992 (registered at ZIN library).

Escalera, M. M. de la 1914: Los coleópteros de Marruecos. *Trabajos del Museo Nacional de Ciencias Naturales, Serie Zoologica* (Madrid) **11**: 1–553. DP: 10 November 1914 (volume title page).

Escalera, M. M. de la 1928: Formas nuevas de Gynandrophthalma Lac., de España y Marruecos (Col. Chrysomelidae). *Boletín de la Real Sociedad Española de Historia Natural* **28**: 431–433. DP: [31] October 1928 (footer on pp. 417 and 433).

Escalera, M. M. de la 1931: Los Labidostomis Redt. metálicos del Mediterráneo occidental (Col. Chrysom.). *Eos, Revista Española de Entomología* 7 (2): 107–130, 1 pl. DP: 30 July 1931 (issue cover).

Eschscholtz, J. F. G. von 1818: Decades tres eleutheratorum novorum. *Mémoires de l'Académie Impériale des Sciences de Pétersbourg, 5ᵉ Série* **6** [1813–1814]: 451–484. DP: [31 December] 1818 (volume title page).

Evenhuis, N. L., Pape, T. & Pont, A. C. 2008: The problems of subsequent typification in genus-group names and use of the Zoological Record: a study of selected post-1930 Diptera genus-group names without type species designations. *Zootaxa* **1912**: 1–44. DP: 22 October 2008 (paper title page).

Everts, E. 1901: *Coleoptera Neerlandica. De schildvleugelige insecten van Nederland en het aangrenzend gebied. Tweede deel. Eerste gedeelte.* 's Gravenhage: Martinus Nijhoff, 400 pp. DP: 21 December 1901 (Bousquet 2016).

Everts, E. 1919a: Nieuwe vondsten voor de nederlandsche coleopterenfauna XXIII. *Entomologische Berichten* 5 [1917–1921] (108): 160–164. DP: 1 July 1919 (issue title page).

Everts, E. 1919b: Nieuwe vondsten voor de nederlandsche coleopteren-fauna XXIV. *Entomologische Berichten* 5 [1917–1921] (109): 172–175. DP: 1 September 1919 (issue title page).

Everts, E. 1920a: Nieuwe vondsten voor de nederlandsche coleopterenfauna, XXVII. *Entomologische Berichten* 5 [1917–1921] (112): 226–231. DP: 1 March 1920 (issue title page).

Everts, E. 1920b: Nieuwe vondsten voor de Nederlandsche Coleopterenfauna XXXI. *Entomologische Berichten* 5 [1917–1921] (116): 286–292. DP: 1 November 1920 (issue title page).

Everts, E. 1921a: Nieuwe vondsten voor de Nederlandsche Coleopterenfauna XXXII. *Entomologische Berichten* 5 [1917–1921] (117): 302–309. DP: 1 January 1921 (issue title page).

Everts, E. 1921b: Nieuwe vondsten voor de nederlandsche coleopterenfauna XXXVII. *Entomologische Berichten* 6 [1921–1925] (122): 21–26. DP: 1 November 1921 (issue title page).

Fabricius, J. C. 1775: *Systema entomologiae, sistens insectorum classes, ordines, genera, species, adiectis synonymis, locis, descriptionibus, observationibus.* Flensburgi et Lipsiae: Libraria Kortii, [32] + 832 pp. DP: 17 April 1775 (Bousquet 2016).

Fabricius, J. C. 1777: *Genera insectorum eorumque characteres naturales secundum numerum, fuguram, situm et proportionem omnium partium oris adiecta mantissa specierum nuper detectarum.* Chilonii: M. F. Barschii, [16] + 310 pp. DP: 5 October 1777 (Bousquet 2017).

Fabricius, J. C. 1781: *Species insectorum exhibentes eorum differentias specificas, synonyma auctorum, loca natalia, metamorphosin adiectis observationibus, descriptionibus. Tom. I.* Hamburgi et Kilonii: C. E. Bohnii, viii + 552 pp. DP: [31] March 1781 (Bousquet 2016).

Fabricius, J. C. 1782: Appendix specierum nuper detectarum. Pp. 495–510. In: Fabricius, J. C.: *Species insectorum exhibentes eorum differentias specificas, synonyma auctorum, loca natalia, metamorphosin adiectis observationibus, descriptionibus. Tom. II.* Hamburgi et Kilonii: C. E. Bohnii, 517 pp. DP: 28 September 1782 (Bousquet 2016).

Fabricius, J. C. 1787: *Mantissa insectorum sistens eorum species nuper detectas adiectis characteribus genericis, differentiis specificis, emendationibus, observationibus. Tom. I.* Hafniae: C. G. Proft, xvi + 348 pp. DP: 29 April 1787 (Bousquet 2017).

Fabricius, J. C. 1792a: *Entomologia systematica emendata et aucta, secundum classes, ordines, genera, species, adjectis, synonimis, locis, observationibus, descriptionibus. Tom. I. Pars I.* Hafniae: C. G. Proft, xx + 330 pp. DP: 30 September 1792 (Bousquet 2017).

Fabricius, J. C. 1792b: *Entomologia systematica emendata et aucta, secundum classes, ordines, genera, species, adjectis, synonimis, locis, observationibus, descriptionibus. Tom. I. Pars II.* Hafniae: C. G. Proft, 538 pp. DP: 30 September 1792 (Bousquet 2017).

Fabricius, J. C. 1794: *Entomologia systematica emendata et aucta, secundum classes, ordines, genera, species, adjectis, synonimis, locis, observationibus, descriptionibus. Tom. IV.* Hafniae: C. G. Proft, [6] + 472 + [5] pp. DP: 5 October 1794 (Bousquet 2017).

Fabricius, J. C. 1798: *Supplementum entomologiae systematicae.* Hafniae: Proft et Storch, [4] + 572 pp. DP: 29 April 1798 (Bousquet 2017).

Fabricius, J. C. 1801a: *Systema eleutheratorum secundum ordines, genera, species adiectis synonymis, locis, observationibus, descriptionibus. Tomus I.* Kiliae: Bibliopolii Academici Novi, xxiv + 506 pp. DP: 26 April 1801 (Bousquet 2017).

Fabricius, J. C. 1801b: *Systema eleutheratorum secundum ordines, genera, species: adiectis synonymis, locis, observationibus, descriptionibus. Tomus II.* Kiliae: Bibliopoli Academici Novi, 687 pp. DP: [31 December] 1801 (Bousquet 2016, 2017).

Fabricius, J. C. 1803: *Index alphabeticus in I. C. Fabricii Systema Eleutheratorum genera et species continens.* Helmstadii: C. G. Fleckeisen, iv + 93 + [2] pp. DP: [31 December] 1803 (Bousquet 2016).

Fåhraeus, O. I. 1839: [new taxa]. In: Schoenherr, C. J.: *Genera et species curculionidum, cum synonymia hujus familiae; species novae aut hactenus minus cognitae, descriptionibus Dom. L. Gyllenhal, C. H. Boheman, O. J. Fahraeus et entomologis aliis illustratae. Tomus quintus. Pars prima. Supplementum continens.* Paris: A. Roret, Lipsiae: A. F. Fleischer, viii + 456 pp. DP: 15 June 1839 (Bousquet 2016).

Fairmaire, L. 1851a: Monographie du genre Cyrtonus. *Annales de la Société Entomologique de France, Deuxième Série* 8 [1850] (4): 535–548. DP: [31] January 1851 (Evenhuis 2022).

Fairmaire, L. 1851b: Description de sept coléoptères nouveaux provenant du Midi de l'Europe et du Maroc. *Revue et Magasin de Zoologie Pure et Appliquée, 2ᵉ Série* 3 (11): 527–528. DP: [31] December 1851 (Dickinson et al. 2011).

Fairmaire, L. 1856: Coleoptera Maroccana. *Revue et Magasin de Zoologie Pure et Appliquée, 2ᵉ Série* 8 (11): 530–531. DP: [31] December 1851 (Dickinson et al. 2011).

Fairmaire, L. 1857a: Miscellanea entomologica. Première partie. *Annales de la Société Entomologique de France, Troisième Série* 4 [1856] (3): 517–548. DP: 28 January 1857 (Evenhuis 2022).

Fairmaire, L. 1857b: Rapport sur la session extraordinaire tenue a Montpellier au mois de Juin 1857. *Annales de la Société Entomologique de France, Troisième Série* 5 (3): 619–647. DP: 11 November 1857 (Evenhuis 2022).

Fairmaire, L. 1858: Miscellanea entomologica. Deuxième partie. *Annales de la Société Entomologique de France, Troisième Série* 5 [1857] (4): 725–745. DP: 24 February 1858 (Evenhuis 2022).

Fairmaire, L. 1859a: Miscellanea Entomologica. Troisième partie. *Annales de la Société Entomologique de France, Troisième Série* 7 (1): 21–64. DP: 25 May 1859 (Evenhuis 2022).

Fairmaire, L. 1859b: Notice nécrologique sur Ph. Lareynie suivie d'observations sur les coléoptères de l'ile de Corse. *Annales de la Société Entomologique de France, Troisième Série* 7 (2): 261–283. DP: 24 August 1859 (Evenhuis 2022).

Fairmaire, L. 1859c: Descriptions de quelques espèces nouvelles de coléoptères recoltés par M. Gougelet pendant son voyage en Galice dans le courant de l'anné 1859. *Bulletin de la Société Entomologique de France* **1859** (3): cl–clii. DP: 9 November 1859 (Evenhuis 2022).

Fairmaire, L. 1860: [Diagnoses suivantes de quelques coléoptères européens]. *Bulletin de la Société Entomologique de France* [1859] (4): ccxv–ccxvii. DP: 1 March 1860 (Evenhuis 2022).

Fairmaire, L. 1861: [Diagnoses latines suivantes de deux nouvelles espèces de Coléoptères, propres à la faune française]. *Bulletin de la Société Entomologique de France* **1861** (1): xi. DP: 14 August 1861 (Lefèvre 1885).

Fairmaire, L. 1862a: Miscellanea entomologica. Quatrième partie. *Annales de la Société Entomologique de France, Quatrième Série* 1 [1861] (4): 577–596. DP: 26 March 1862 (Lefèvre 1885).

Fairmaire, L. 1862b: Coléoptères nouveaux recuellis pendant l'excursion en Savoie. *Annales de la Société Entomologique de France, Quatrième Série* **1** [1861] (4): 647–650. DP: 26 March 1862 (Lefèvre 1885).

Fairmaire, L. 1864: Coléoptères d'Algérie rapportés par M. Ogier de Baulny. *Annales de la Société Entomologique de France, Quatrième Série* **3** [1863] (4): 637–648. DP: 23 March 1864 (Lefèvre 1885).

Fairmaire, L. 1865: Monographie des chrysomèles de Suffrian. Suite. Traduction. *Annales de la Société Entomologique de France, Quatrième Série* **5** (1): 37–82. DP: 23 August 1865 (Lefèvre 1885).

Fairmaire, L. 1866a: [new taxa]. In: Fairmaire, L. & Coquerel, C.: Essai sur les coléoptères de Barbarie. Quatrième partie. *Annales de la Société Entomologique de France, Quatrième Série* **6** (1): 17–74. DP: 22 August 1866 (Lefèvre 1885).

Fairmaire, L. 1866b: Notice sur les coléoptères récoltés par M. J. Lédérer sur le Bosz-Dagh (Asie Mineure). *Annales de la Société Entomologique de France, Quatrième Série* **6** (2): 249–280. DP: 24 October 1866 (Lefèvre 1885).

Fairmaire, L. 1867: Essai sur les coléoptères de Barbarie. Cinquième partie. *Annales de la Société Entomologique de France, Quatrième Série* **7** (3): 387–416. DP: 11 December 1867 (Lefèvre 1885).

Fairmaire, L. 1868a: *Genera des coléoptères d'Europe comprenant leur classification en famille naturelle, la description de tous les genres, des tableaux dichotomiques destinés à faciliter l'étude, le catalogue de toutes les espèces, de nombreux dessins au trait de caractères par Jacquelin du Val (Camille) et par M. L. Fairmaire et près de seize cents types représentant un ou plusieurs insectes de chaque genre dessinés et peints d'apres nature avec le plus grand soin par M. Jules Migneaux et par M. Théophile Deyrolle. Tome quatrième.* [1864–1868]. Paris: Deyrolle fils, pp. 97–295. DP (pp. 205–?268): 7 December 1868 (for dating of all parts see Bousquet 2016).

Fairmaire, L. 1868b: Essai sur les coléoptères de Barbarie. Sixième partie. *Annales de la Société Entomologique de France, Quatrième Série* **8** (3): 471–502. DP: 23 December 1868 (Lefèvre 1885).

Fairmaire, L. 1869: Notes sur les coléoptères recueillis par Charles Coquerel à Madagascar et sur les côtes d'Afrique. *Annales de la Société Entomologique de France, Quatrième Série* 9 (2): 179–260. DP: 22 September 1869 (Lefèvre 1885).

Fairmaire, L. 1873a: [new taxa]. In: Fairmaire, L. & Raffray, A.: Coléoptères du nord de l'Afrique. *Revue et Magasin de Zoologie Pure et Appliquée, 3ᵉ Série* 1 (7): Pl. 16. DP: 15 July 1873 (*Pet. Nouv. Ent.* 1873, 5 (80): 321) [text published in issue 9].

Fairmaire, L. 1873b: Diagnoses de divers coléoptères nouveaux découverts en Algérie. *Annales de la Société Entomologique de France, Cinquième Série* 3 (3): 391–392. DP: 24 December 1873 (Lefèvre 1885).

Fairmaire, L. 1873c: [new taxa]. In: Fairmaire, L. & Raffray, A.: Coléoptères du nord de l'Afrique. *Revue et Magasin de Zoologie Pure et Appliquée, 3ᵉ Série* 1 (9): 331–385, pls. 15–16. DP: [31 December] 1873 (issue cover) [plate 16 published in issue 7, plate 15 in issue 8].

Fairmaire, L. 1874a: Diagnoses de coléoptères nouveaux du nord de l'Afrique. *Petites Nouvelles Entomologiques* 1 [1869–1875] (6) (97): 388–389. DP: 1 April 1874 (issue title page).

Fairmaire, L. 1874b: Notes rectificatives et complémentaires sur les Timarcha. *Annales de la Société Entomologique de France, Cinquième Série* 3 [1873] (4): 525–528. DP: 22 April 1874 (Lefèvre 1885).

Fairmaire, L. 1874c: Description d'un nouveau genre de la famille des chrysomélides. *Annales de la Société Entomologique de France, Cinquième Série* 4 (1): 42–44. DP: 8 July 1874 (Lefèvre 1885).

Fairmaire, L. 1875a: Diagnose de quelques coléoptères nouveaux d'Algérie. *Petites Nouvelles Entomologiques* 1 [1869–1875] (7) (124): 495–496. DP: 15 May 1875 (issue title page).

Fairmaire, L. 1875b: Coléoptères de la Tunisie récoltés par Mr Abdul Kerim. *Annali di Museo Civico di Storia Naturali di Genova* 7: 475–540. DP: [30] November 1875 (Poggi 2010).

Fairmaire, L. 1876a: Description de trois coléoptères nouveaux. *Petites Nouvelles Entomologiques* 2 [1876–1879] (8) (145): 25–26. DP: 1 April 1876 (issue title page).

Fairmaire, L. 1876b: Diagnoses de coléoptères du nord de l'Afrique. *Petites Nouvelles Entomologiques* 2 [1876–1879] (8) (151): 49–50. DP: 1 July 1876 (issue title page).

Fairmaire, L. 1876c: Diagnoses de coléoptères nouveaux du nord de l'Afrique. *Petites Nouvelles Entomologiques* 2 [1876–1879] (8) (162): 93–94. DP: 15 December 1876 (issue title page).

Fairmaire, L. 1877a: Description de coléoptères nouveaux du nord de l'Afrique. *Petites Nouvelles Entomologiques* 2 [1876–1879] (9) (163): 98. DP: 1 January 1877 (issue title page).

Fairmaire, L. 1877b: Diagnoses de coléoptères de la Nouvelle-Bretagne. *Petites Nouvelles Entomologiques* 2 [1876–1879] (9) (185): 185–186. DP: 1 December 1877 (issue title page).

Fairmaire, L. 1878a: [Description qui suit d'une nouvelle espèce française de coleoptère]. *Bulletin de la Société Entomologique de France* [1877]: clxxix–clxxx. DP: 10 April 1878 (Evenhuis 2002) [published only in the quarterly version of the *Bull. Soc. Ent. Fr.*].

Fairmaire, L. 1878b: [new taxa]. In: Deyrolle, H. & Fairmaire, L.: Descriptions de coléoptères recueillis par M. l'abbé David dans la Chine centrale. *Annales de la Société Entomologique de France, Cinquième Série* 8 (1): 87–112 & (2): 113–140. DP (pp. 87–112): 26 June 1878, (pp. 113–140): 14 August 1878 (Lefèvre 1885).

Fairmaire, L. 1880a: [Deux nouvelles espèces de coléoptères]. *Bulletin de la Société Entomologique de France* 1880 (3): 38–39. DP: 20 February 1880 (Evenhuis 2002).

Fairmaire, L. 1880b: [new taxa]. In: Gestro, R.: Crociera del violante comandato dal Capitano Armatore Enrico D'Albertis durante l'anno 1877. II. Risultati scientifici. Appunti sull' entomologia tunisina. *Annali del Museo Civico di Storia Naturale di Genova* **15**: 405–424. DP: 12 April 1880 (Poggi 2010).

Fairmaire, L. 1880c: Diagnoses de coléoptères de Madagascar. *Le Naturaliste* **2** (40): 316–317. DP: 15 November 1880 (issue title page).

Fairmaire, L. 1880d: Descriptions de coléoptères d'Espagne et de Turquie. *Annales de la Société Entomologique de France, Cinquième Série* 10 (3): 237–244. DP: 8 December 1880 (Lefèvre 1885).

Fairmaire, L. 1880e: Diagnoses de nouvelles espèces du g. Timarcha. *Le Naturaliste* **2** (42): 332–333. DP: 15 December 1880 (issue title page).

Fairmaire, L. 1881: Coléoptères du Nord de l'Afrique. *Revue et Magasin de Zoologie Pure et Appliquée, 3e Série* 7 [1879] (2): 178–218. DP: [28] February 1881 (*Natur. Novit.* 1881, 3 (4): 35).

Fairmaire, L. 1882: Diagnoses de trois coléoptères (Cyrtonus) nouveaux. *Le Naturaliste* **4** (3): 19. DP: 1 February 1882 (issue title page).

Fairmaire, L. 1883: Notes complémentaires sur le genre Cyrtonus. *Anales de la Sociedad Española de Historia Natural* **12** (2): 251–271. DP: 30 September 1883 (issue cover).

Fairmaire, L. 1884a: Descriptions de coléoptères recueillis par le Baron Bonnaire en Algérie. *Bulletin ou Comptes-Rendu des Séances de la Société Entomologique de Belgique, Série III* **1884** (42): lix–lxx. DP: 27 February 1884 (*Bull. Soc. Ent. Fr.* 1884: 64).

Fairmaire, L. 1884b: Liste complémentaire des espèces du g. Timarcha. *Anales de la Sociedad Española de Historia Natural* **13** (1): 75–112. DP: 7 May 1884 (issue cover).

Fairmaire, L. 1884c: [new taxa]. In: Heyden, L. F. J. D. von.: Coleopterologische Ausbeute einer Excursion nach der Sierra d'Espuña. *Deutsche Entomologische Zeitschrift* **28** (2): 355–361. DP: [30] November 1884 (table of contents).

Fairmaire, L. 1884d: Liste des coléoptères recueillis par M. l'abbé David à Akbès (Asie-Mineure) et descriptions des espèces nouvelles. *Annales de la Société Entomologique de France, 6e Série* **4** (3): 165–180. DP: 24 December 1884 (Lefèvre 1895).

Fairmaire, L. 1885a: [new taxon]. *Bulletin de la Société Entomologique de France* **1885** (6): lxiv & (7): lxv. DP (p. lxiv): 4 April 1885, (p. lxv): 18 April 1885 (Evenhuis 2002).

Fairmaire, L. 1885b: Notes sur les coléoptères recueillis par M. Ach. Raffray à Madagascar et descriptions des espèces nouvelles. 1re partie. *Annales de la Société Entomologique de France, 6e Série* **4** [1884] (4): 225–242. DP: 13 May 1885 (Lefèvre 1895).

Fairmaire, L. 1886a: Note sur les coléoptères recueillis par M. Laligant à Obock. *Annales de la Société Entomologique de France, 6e Série* **5** [1885] (4): 435–462. DP: 28 April 1886 (Lefèvre 1895).

Fairmaire, L. 1886b: Descriptions de coléoptères de l'intérieur de la Chine. *Annales de la Société Entomologique de France, 6e Série* **6** (2): 303–304 & (3): 305–356. DP (pp. 303–304): 15 October 1886, (pp. 305–356): 30 December 1886 (Lefèvre 1895).

Fairmaire, L. 1887a: [Descriptions de deux nouvelles espèces de coléoptères]. *Bulletin de la Société Entomologique de France* 1887 (22): cxcii & (23): cxciii. DP (p. cxcii): 3 December 1887, (p. cxciii): 24 December 1887 (Evenhuis 2002).

Fairmaire, L. 1887b: Coléoptères de l'intérieur de la Chine. *Annales de la Société Entomologique de Belgique* **31**: 87–136. DP: 14 December 1887 (reprint mentioned in the *Bull. Soc. Ent. Fr.* 1887: ccviii).

Fairmaire, L. 1887c: [Descriptions de deux nouvelles espèces de coléoptères]. *Bulletin de la Société Entomologique de France* 1887 (22): cxcii & (23): cxciii. DP (p. cxcii): 3 December 1887, (p. cxciii): 24 December 1887 (Evenhuis 2002).

Fairmaire, L. 1887d: Notes sur les coléoptères des environs de Pékin (1re partie). *Revue d'Entomologie* 6 (11): 312–316 & (12): 317–335. DP (pp. 312–316): [30] November 1887, DP (pp. 317–335): [31] December 1887 (footers on pp. 309 and 317).

Fairmaire, L. 1888a: Coléoptères des voyages de M. G. Révoil chez les Somâlis et dans l'intérieur du Zanguebar. *Annales de la Société Entomologique de France, 6e Série* 7 [1887] (3): 277–320 & (4): 321–368. DP (pp. 277–320): 28 December 1887, (pp. 321–368): 11 April 1888 (Lefèvre 1895).

Fairmaire, L. 1888b: Notes sur les coléoptères des environs de Pékin (2e partie). *Revue d'Entomologie* 7 (3): 111–116 & (4): 117–148 & (5): 149–160. DP (pp. 111–116): [31] March 1888, (pp. 117–148): [30] April 1888, (pp. 149–160): [31] May 1888 (footers on pp. 109, 117 and 149).

Fairmaire, L. 1888c: Coléoptères de l'intérieur de la Chine. *Annales de la Société Entomologique de Belgique* **32**: 7–46. DP: [31] August 1888 (reprint mentioned in the *Natur. Novit.* 1888, 10 (17): 268).

Fairmaire, L. 1888d: Énumération des coléoptères recueillis par M. le Dr Hans Schinz dans le sud de l'Afrique. *Annales de la Société Entomologique de France, 6e Série* 8 (2): 173–202. DP: 10 October 1888 (Lefèvre 1895).

Fairmaire, L. 1889a: Descriptions de coléoptères de l'Indo-Chine. *Annales de la Société Entomologique de France, 6e Série* 8 [1888] (4): 333–378. DP: 10 June 1889 (Lefèvre 1895).

Fairmaire, L. 1889b: Coléoptères de l'intérieur de la Chine. 5e partie. *Annales de la Société Entomologique de France, 6e Série* 9 (1): 5–84. DP: 11 September 1889 (Lefèvre 1895).

Fairmaire, L. 1891: Descriptions de coléoptères des montagnes de Kashmir. *Bulletin ou Compte-Rendus des Séance de la Société Entomologique de Belgique, Série IV* **1890** (16): lxxxviii–ciii. DP: 8 April 1891 (*Bull. Soc. Ent. Fr.* 1891: lxvii).

Fairmaire, L. 1894a: Quelques coléoptères du Thibet. *Annales de la Société Entomologique de Belgique* **38** (5): 216–225. DP: 13 June 1894 (*Bull. Soc. Ent. Fr.* 1894: clxvii).

Fairmaire, L. 1894b: Descriptions de coléoptères d'Algérie. *Annales de la Société Entomologique de Belgique* **38** (6): 310–313. DP: 26 July 1894 (*Zool. Anz., Litteratur* 1894, 17: 321).

Fairmaire, L. 1894c: Coléoptères de l'Afrique intertropicale et australe. *Annales de la Société Entomologique de Belgique* **38** (6): 314–335. DP: 26 July 1894 (*Zool. Anz., Litteratur* 1894, 17: 322).

Fairmaire, L. 1898a: Description d'un genre nouveau d'eumolpide (Col.). *Bulletin de la Société Entomologique de France* **1898** (2): 19–20. DP: 18 April 1898 (*Bibl. Zool.* 1898, 3: 187).

Fairmaire, L. 1898b: Matériaux pour la faune coléoptérique de la région malgache. 5e note. *Annales de la Société Entomologique de Belgique* **42** (6): 222–260. DP: 25 June 1898 (footers on pp. 237 and 253).

Fairmaire, L. 1901a: Descriptions de coléoptères nouveaux, de Madagascar. *Bulletin de la Société Entomologique de France* **1901** (5): 126–128. DP: 12 October 1901 (*Bibl. Zool.* 1901, 6: 502).

Fairmaire, L. 1901b: Matériaux pour la faune coléoptérique de la région Malgache (11e Note). *Revue d'Entomologie* **20**: 101–248. DP (issue 5/6, pp. 101–124): [30] June 1901, (issue 7, pp. 125–156): [31] July 1901, (issue 8, pp. 157–188): [31] August 1901, (issue 9, pp. 189–220): [30] September 1901, (issue 10, pp. 221–236): [31] October 1901, (issue 11, pp. 237–248): [30] November 1901 (footers on pp. 93, 109, 125, 141, 157, 173, 189, 205, 221, 237).

Fairmaire, L. 1902: Matériaux pour la faune coléoptérique Malgache. 13e note. *Annales de la Société Entomologique de Belgique* **46** (5/6): 236–271. DP: 4 June 1902 (footers on pp. 239 and 255).

Fairmaire, L. 1904: Matériaux pour la faune coléoptérique Malgache. 18e note. *Annales de la Société Entomologique de Belgique* **48** (7): 225–276. DP: 4 August 1904 (footers on pp. 239, 255 and 271).

Fairmaire, L. & Allard E. 1873: Révision du genre Timarcha. *Annales de la Société Entomologique de France, Cinquième Série* **3** (1): 143–144 & (2): 145–202. DP (pp. 143–144): 10 August 1873, (pp. 145–202): 20 October 1873 (Lefèvre 1885).

Faldermann, F. 1833: Species novae Coleopterorum Mongoliae et Sibiriae incolarum. *Bulletin de la Société Impériale des Naturalistes de Moscou* **6**: 46–72, 1 pl. DP: [31 December] 1833 (Bousquet & Davies 2021).

Faldermann, F. 1835: Coleopterorum ab illustrissimo Bungio in China boreali, Mongolia, et Montibus Altaicis collectorum, nec non ab ill. Turczaninoffio et Stchukino e provincia Irkutzk missorum illustrationes. *Mémoires présentés à l'Académie Impériale des Sciences de Saint-Pétersbourg par Divers Savans et lus dans ses assemblées* **2** (4/5): 337–464, pls. I–V. DP: 12 April 1835 (Dickinson & Daneliya 2011).

Faldermann, F. 1837: Fauna entomologica transcaucasica. Pars II. *Nouveaux Mémoires de la Société des Naturalistes de Moscou* **5**: 1–433, 15 pls. DP: 8 July 1837 (*Bull. Sci. Acad. Imp. Sci. St. Petersbourg* 1837, 2 (18): 288) [also issued separately as: *Fauna entomologica Trans-Caucasica. Coleoptera. Pars II*. Moscou: Auguste Semen, 433 pp., 15 pls.].

Fall, H. C. 1910: Miscellaneous notes and descriptions of North American Coleoptera. *Transactions of the American Entomological Society* **36** (2): 89–197. DP (pp. 89–112): [31] May 1910, (pp. 113–160): [30] June 1910, (pp. 161–197): [31] July 1910 (footers throughout the paper).

Fall, H. C. 1926: A list of the Coleoptera taken in Alaska and adjacent parts of the Yukon territory in the summer of 1924. *Pan-Pacific Entomologist* **2** [1925–1926] (4): 191–208. DP: 14 July 1926 (noticed on p. 216).

Farina, L. 2015: I crisomelidi (Coleoptera; Chrysomelidae) del Parco di Montevecchia e della Valle del Curone (Italia, Lombardia, provincia di Lecco). *Natural History Sciences, Atti della*

Società Italiana di Scienze Naturali e del Museo Civile di Storia Naturale di Milano **2** (2): 61–88. DP: 9 November 2015 (journal website).

Farina, L. 2021: Notes on Longitarsus aphthonoides Weise 1887 and on the species groups of L. lateripunctatus (Rosenhauer 1856) and L. curtus (Allard 1860) (Coleoptera, Chrysomelidae). *Natural History Sciences, Atti della Società Italiana di Scienze Naturali e del Museo Civile di Storia Naturale di Milano* 8 (2): 11–34. DP: 29 October 2021 (paper title page).

Farina, L. 2023: New data on Longitarsus rubellus species group with description of a new species from Western Balkans (Coleoptera, Chrysomelidae, Alticini). *Natural History Sciences, Atti della Società Italiana di Scienze Naturali e del Museo Civile di Storia Naturale di Milano* 10 (Supplement 1): 17–26. DP: 13 December 2023 (journal website).

Fassati, M. 1961: Systematische, faunistische und biologische Bemerkungen über einige tschechoslowakische Chrysomeliden (Coleoptera). *Acta Entomologica Musei Nationalis Pragae* **34**: 417–425. DP: 30 June 1961 (colophon).

Fauvel, A. 1862: Coléoptères de la Nouvelle-Calédonie, recueillis par M. E. Déplanche, chirurgien de la marine impériale (1858–59–60). *Bulletin de la Société Linnéenne de Normandie* 7 [1861–1862]: 120–185, pls. IX–X bis. DP: [31 December] 1862 (volume title page).

Fauvel, A. 1888: Rectifications au Catalogus Coleopterorum Europae et Caucasi (suite). *Revue d'Entomologie* 7 (7): 229–251. DP: [31] July 1888 (footers on pp. 229 and 245).

Fauvel, A. 1892: Deux coléoptères nouveaux de France. *Revue d'Entomologie* **11** (12): 314–315. DP: [31] December 1892 (footer on p. 309).

Feng, C., Yang, X.-K. & Li, Z.-Q. 2022: [new taxon]. In: Feng, C., Yang, X.-K., Li, Z.-Q. & Liu, Y.: Revision of Sphenoraia Clark, 1865 (Coleoptera, Chrysomelidae, Galerucinae) from China, with descriptions of two new species. *ZooKeys* **1132**: 51–83. DP: 25 November 2022 (paper title page).

Feng, C., Yang, X.-K. & Liu, Y. 2022: [new taxon]. In: Feng, C., Yang, X.-K., Li, Z.-Q. & Liu, Y.: Revision of Sphenoraia Clark, 1865 (Coleoptera, Chrysomelidae, Galerucinae) from China, with descriptions of two new species. *ZooKeys* **1132**: 51–83. DP: 25 November 2022 (paper title page).

Feng, C., Yang, X.-K., Liu, Y. & Li, Z.-Q. 2023: Revision of Aplosonyx Chevrolat, 1836 (Coleoptera, Chrysomelidae, Galerucinae) from China, with descriptions of three new species. *ZooKeys* **1154**: 159–222. DP: 21 March 2023 (paper title page).

Fiori, A. 1906: Una breve escursione in Sicilia. *Il Naturalista Siciliano* **18** [1905–1906] (9/10): 200–211. DP: 14 November 1906 (*Bull. Soc. Ent. Fr.* 1906: 236).

Fischer von Waldheim, G. 1809: Sur deux genres nouveaux de coléoptères. *Mémoires de la Société Impériale des Naturalistes de Moscou* **2**: 293–304, pl. 18. DP: [31 December] 1809 (Bousquet & Davies 2021).

Fischer von Waldheim, G. 1823: *Entomographie de la Russie, et genres des insectes. Entomographia Imperii Rossici, sue Caesareae Majestati Alexandro I dicata. Volume II*. Mosquae: Augusti Semen Typographi, Plates XVIII–L [Coleoptera] + VI–XI [Lepidoptera]. DP: [31 December] 1823 (Bousquet 2016).

Fischer von Waldheim, G. 1824: *Entomographie de la Russie, et genres des insectes. Entomographia Imperii Rossici, sue Caesareae Majestati Alexandro I dicata. Volume II*. Mosquae: Augusti Semen Typographi, xx + 264 pp. DP: [31 December] 1824 (Bousquet 2016).

Fischer von Waldheim, G. 1842: *Catalogus coleopterorum in Sibiria orientali a cel. Gregorio Silide Karelin collectorum*. Moscou, 28 pp. DP: 11 August 1842 (Bousquet 2016).

Fitch, A. 1861: The bean-weevil. *Transactions of the Rhode Island Society for the Encouragement of Domestic Industry* [1860]: 62–64. [31 December] 1861 (volume title page).

Fleischer, A. 1892: Eine neue Chrysomela-Varietät. *Wiener Entomologische Zeitung* **11** (5): 141. DP: 31 May 1892 (footers throughout the issue).

Fleischer, A. 1909a: Bemerkenswerte Koloritaberrationen. *Wiener Entomologische Zeitung* **28** (1): 32. DP: 1 February 1909 (footers throughout the issue).

Fleischer, A. 1909b: Ein neuer Stylosomus aus Turkestan. *Wiener Entomologische Zeitung* **28** (7/8): 245. DP: 25 September 1909 (footers throughout the issue).

Fleischer, A. 1909c: Eine neue Varietät des Phaedon pyritosus Rossi aus Turkestan. *Wiener Entomologische Zeitung* **28** (7/8): 271. DP: 25 September 1909 (footers throughout the issue).

Fleischer, A. 1910: Eine neue Coloritaberration des Chrysochus asclepiadeus Pall. *Wiener Entomologische Zeitung* **29** (5/6): 200. DP: 10 July 1910 (footers throughout the issue).

Fleischer, A. 1914: Coleopterologische Notizen. *Wiener Entomologische Zeitung* **33** (1/2): 65. DP: 1 January 1914 (footers throughout the issue).

Fleischer, A. 1916a: Bemerkungen über einige Chrysochloa-Aberrationen. *Wiener Entomologische Zeitung* **35** (3/4): 100. DP: 25 April 1916 (footers throughout the issue).

Fleischer, A. 1916b: Neue Chrysomeliden aus Japan. *Wiener Entomologische Zeitung* **35** (5/7): 222–223. DP: 12 September 1916 (footers throughout the issue).

Fleischer, A. 1917a: Eine neue Phyllotreta aus Kroatien. *Wiener Entomologische Zeitung* **36** (1/2): 17. DP: 30 March 1917 (footers throughout the issue).

Fleischer, A. 1917b: Bemerkenswerte Aberrationen einiger Coleopteren-Arten. *Wiener Entomologische Zeitung* **36** (3/5): 121–122. DP: 10 July 1917 (footers throughout the issue).

Fleischer, A. 1918a: Neue Aberrationen mährischer Coleopteren. *Wiener Entomologische Zeitung* **37** (1/3): 76. DP: 25 March 1918 (footers throughout the issue).

Fleischer, A. 1918b: Eine neue Aberration der Coptocephala scopolina Lin. (Col. Chrysom.). *Wiener Entomologische Zeitung* **37** (4/7): 106. DP: 15 August 1918 (footers throughout the issue).

Fleury, J. & Vitali, F. 2019: Sur la taxonomie de Chrysolina interstincta depressa (Fairmaire, 1854) et sa présence dans le Loiret (Coleoptera Chrysomelidae). *L'Entomologiste* **74** [2018] (5/6): 301–303. DP: [31] January 2019 (editor of journal J.-D. Chapelin-Viscardi, 2023, pers. comm.).

Fogato, W. 1980: Note sui Luperus bicolori italiani e descrizione di L. leonardii n. sp. (Coleoptera Chrysomelidae). *Memorie della Società Entomologica Italiana* **57** [1979]: 46–64. DP: [31 December] 1980 (Poggi 2008).

Formánek, R. 1900: Eine neue Lochmaea Wse. (Col., Chrysom.). *Wiener Entomologische Zeitung* **19** (4/5): 127. DP: 1 May 1900 (footers throughout the issue).

Forsberg, C. P. 1821: Monographia Clythrae. *Nova Acta Regiae Societatis Scientiarum Upsaliensis* 8: 258–292. DP: [31 December] 1821 (volume title page).

Forskål, P. 1775: [new taxon]. In: Forskål, P. & Niebuhr, C.: *Descriptiones animalium avium, amphibiorum, piscium, insectorum, vermium; quae in itinere orientali observavit Petrus Forskål. Post mortem auctoris edidit Carsten Niebuhr. Adjuncta est materia medica kahirina atque tabula maris rubri geographica.* Hauniae: ex officina Mölleri, 19 + [3] + iii – xxxiv + [2] + 164 pp. DP: 7 May 1775 (Bousquet 2017).

Förster, A. 1849: Uebersicht der Käferfauna der Rheinprovinz. Erster Nachtrag. *Verhandlungen des Naturhistorischen Vereins der Preussischen Rheinlande und Westfalens* 6 (supplement): 1–45. DP: [31 December] 1849 (volume cover).

Forster, J. R. 1771: *Novae species insectorum. Centuria I.* Londini: T. Davies et B. White, viii + 100 pp. DP: [30] November 1771 (Bousquet 2016).

Foudras, A. C. M. E. 1860a: Altisides. *Annales de la Société Linnéenne de Lyon, Nouvelle Série* 6 [1859]: 137–384. DP: [31] January 1860 (volume title page).

Foudras, A. C. M. E. 1860b: *Histoire naturelle des Coléoptères de France, par E. Mulsant. Altisides.* Magnin, Blanchard et Cie, Paris, 384 pp. DP: [31] March 1860 (Bousquet 2016) [also issued in the *Annales de la Société Linnéenne de Lyon, Nouvelle série*].

Foudras, A. C. M. E. 1861: Altisides. *Annales de la Société Linnéenne de Lyon, Nouvelle Série* 7 [1860]: 17–128. DP: [28] February 1861 (volume title page).

Fowler, W. W. 1890: *The Coleoptera of the British Islands. A descriptive account of the families, genera, and species indigenous to Great Britain and Ireland, with notes as to localities, habitats, etc. Vol. IV. Lamellicornia – Serricornia – Longicornia – Phytophaga.* London: L. Reeve and Co., 411 pp., pls. 99–142. DP: [31] July 1890 (Bousquet 2016).

Fowler, W. W. 1910: Description of a new species of Galerucella. *Entomologist's Monthly Magazine* 46 (557): 228–229. DP: 4 October 1910 (registered at NHMUK library).

Franz, H. 1938a: Revision der Verwandtschaftsgruppe der Chrysomela gypsophilae Küst. (Coleopt. Chrysomel.). *Entomologische Blätter* 34 (4): 190–210. DP: 15 August 1938 (issue cover).

Franz, H. 1938b: Revision der Verwandtschaftsgruppe der Chrysomela gypsophilae Küst. (Coleopt. Chrysomel.). *Entomologische Blätter* 34 (5): 249–273. DP: 5 November 1938 (issue cover).

Franz, H. 1949a: Zur Kenntnis der Rassenbildung bei Käfern der ostalpinen Fauna. *Zentralblatt für das Gesamtgebiet der Entomologie* 3: 3–23. DP: [31] March 1949 (reprint cover).

Franz, H. 1949b: Zur Kenntnis der Rassenbildung bei einigen Arten der Gattung Cryptocephalus (Coleopt., Chrysom.). *Portugaliae Acta Biologica (Série B), vol. Júlio Henriques*: 165–195. DP: [31 December] 1949 (volume cover).

Franz, H. 1952: 2. Beitrag zur Kenntnis der Gattung Chrysomela (Coleopt., Chrysomel.). Bemerkungen zum Subgenus Colaphoptera Motsch. *Entomologisches Nachrichtenblatt Österreichischer und Schweizer Entomologen* 4 (1/3): 2–6. DP: [31] March 1952 (issue cover).

Franz, H. 1958: Neue und ungenügend bekannte Coleopterenarten aus Spanien. *Eos, Revista Española de Entomología* 34 (2): 117–130. DP: 30 June 1958 (issue cover).

Franz, H. 1974: *Die Nordost-Alpen im Spiegel ihrer Landtierwelt. Band IV. Coleoptera 2. Teil.* Innsbruck-München: Universitätsverlag Wagner, 707 pp. DP: [30 December] 1974 (book cover).

Franz, H. 1982: Zur Kenntnis der kanarischen Cryptocephalus-Arten (Chrysomelidae, Col.). *Koleopterologische Rundschau* **56**: 3–8. DP: [30] December 1982 (issue cover).

Franz, H. & Palmén, E. 1950: Beitrag zur Kenntnis der Untergattung Goniomena Kirby der Gattung Phytodecta (Col., Chrysomelidae). *Annales Entomologici Fennici* **16** (1): 14–18. DP: 11 May 1950 (verso of the volume title page).

Fricken, V. 1888: Berichtigungen und Ergänzungen zu Schilsky's Verzeichniss der deutschen Käfer. *Deutsche Entomologische Zeitschrift* **1888** (2): 321–328. DP: 28 November 1888 (*Bull. Soc. Ent. Fr.* 1888: clxxxv).

Fritzlar, F. 2001: Aphthona warchalowskii n. sp., Aphthona bergeali n. sp., Longitarsus angelikae n. sp. und Longitarsus meridionalis Weise, 1888 n. stat. – 4 Blattkäfer (Coleoptera, Chrysomelidae, Alticinae) aus der südwestlichen Türkei. *Entomologische Blätter* **97** (2/3): 199–224. DP: 28 December 2001 (paper title page).

Fritzlar, F. 2004: Aphthona alanyensis n. sp., Aphthona bergealoides n. sp. und Longitarsus manfredi n. sp. – drei Blattkäfer (Coleoptera, Chrysomelidae, Alticinae) aus der südwestlichen Türkei. *Entomologische Blätter* **100** (2): 105–126. DP: 30 December 2004 (paper title page).

Fritzlar, F., Schöller, M. & Sprick, P. 2022: Rote Liste und Gesamtartenliste der Blatt-, Samen- und Resedakäfer (Coleoptera: Chrysomelidae, Bruchidae; Urodontinae) Deutschlands. In: Ries, M., Balzer, S., Gruttke, H., Haupt, H., Hofbauer, N., Ludwig, G. & Matzke-Hajek, G. (eds.): Rote Liste gefährdeter Tiere, Pflanzen und Pilze Deutschlands, Band 5: Wirbellose Tiere (Teil 3). *Naturschutz und Biologische Vielfalt* **70** (5): 293–331. DP: 28 March 2022 (the date M. Schöller received the book).

Fritzsche, K. & Delobel, A. 2012: Megabruchidius dorsalis (Fåhraeus, 1839), bruche nouvelle pour la faune française (Col., Chrysomelidae, Bruchinae). *Bulletin de la Société Entomologique de France* **117** (3): 381–389. DP: 12 September 2012 (*dates de la publication du Bulletin* on p. 535).

Frivaldszky, J. 1876: Adatok Temes és Krassó megyék faunájához. Data ad fauna Hungariae meridionalis comitatuum Temes et Krassó. *Mathematikai és Természettudományi Közlemények* **13** [1875–1876]: 285–378, 1 pl. DP: [31 December] 1876 (reprint cover) [journal issued 1877].

Frivaldszky, J. 1877: Coleoptera nova ex Hungaria meridionali. *Természetrajzi Füzetek* **1** (4): 228–231. DP: 1 December 1877 (*Entomol. Nachr.* 1877, 3 (12): 192).

Frivaldszky, J. 1881: Coleoptera nova in Europa orientali et Asia minore a D. Eduardo Merkl detecta. *Természetrajzi Füzetek* **4** [1880] (4): 260–266. DP: [31 December] 1881 (volume cover, editorial note on p. 329 dated January 1881).

Frivaldszky, J. 1883: Coleoptera nova ex Hungaria. *Természetrajzi Füzetek* **7**: 9–18. DP: [31 December] 1883 (volume cover).

Frivaldszky, J. 1884: Coleoptera nova ex Asia minore. *Természetrajzi Füzetek* **8** (1): 1–8. DP: [31] March 1884 (issue cover).

Frivaldszky, J. 1892: Coleoptera in expeditione D. Comitis Belae Széchenyi in China, praecipue boreali, a Dominis Gustavo Kreitner et Ludovico Lóczy anno 1879 collecta. Pars secunda. *Természetrajzi Füzetek* 15 (3): 114–125. DP: 31 October 1892 (issue cover).

Frölich, J. A. 1792: Bemerkungen über einige seltene Käfer aus der Insectensammlung des Herrn Hofr. und Prof. Rudolph in Erlangen. *Der Naturforscher* 26: 68–165. DP: 30 September 1792 (*Allg. Verz. Bücher, Michaelismesse* 1792: 240 and Evenhuis 2014).

Frölich, J. A. 1799: Einige neue Gattungen und Arten von Käfern. *Der Naturforscher* 28: 1–65, pl. 1. DP: 29 April 1799 (*Allg. Verz. Bücher, Ostermesse* 1799: 109 and Evenhuis 2014).

Fuente, J. M. de la 1901: Dos coleópteros nuevos. *Butlletí de la Institució Catalana d'Historia Natural* 1 (6): 44–45. DP: [31] December 1901 (issue title page).

Fuente, J. M. de la 1904: Datos para la fauna de la provincia de Ciudad Real. XVII. *Boletín de la Real Sociedad Española de Historia Natural* 4: 381–390. DP: [30] November 1904 (footer on p. 385).

Fuente, J. M. de la 1907: Datos para la fauna de la provincia de Ciudad Real. XIX. *Boletín de la Real Sociedad Española de Historia Natural* 7: 317–323. DP: [31] October 1907 (footer on p. 313).

Fuente, J. M. de la 1908a: Nueva variedad del « Crioceris macilenta ». *Boletín de la Real Sociedad Española de Historia Natural* 8: 121. DP: [29] February 1908 (footer on p. 109).

Fuente, J. M. de la 1908b: Más sobre « Crioceris macilenta » Ws. *Boletín de la Real Sociedad Española de Historia Natural* 8: 388–390. DP: [31] October 1908 (footer on p. 385).

Fuente, J. M. de la 1909a: Datos para la fauna de la provincia de Ciudad Real. XX. *Boletín de la Real Sociedad Española de Historia Natural* 9: 306–310. DP: [30] June 1909 (footer on p. 297).

Fuente, J. M. de la 1909b: [new taxon]. In: Marcet, A. F. 1909: Excursión al Moncayo (11–13 Octubre 1908). *Boletín de la Sociedad Aragonesa de Ciencias Naturales* 8 (6/7): 135–143. DP: [31] July 1909 (issue title page).

Fuente, J. M. de la 1910a: Contribución á la fauna coleopterológica de España. *Boletín de la Real Sociedad Española de Historia Natural* 10: 181–182. DP: [31] March 1910 (footer on p. 169).

Fuente, J. M. de la 1910b: Datos para la fauna de la provincia de Ciudad Real. XXI. Coléopteros. *Boletín de la Real Sociedad Española de Historia Natural* 10: 442–449. DP: [31] December 1910 (footer on p. 445).

Fuente, J. M. de la 1913a: Datos para la fauna de la provincia de Ciudad Real. XXIII. *Boletín de la Real Sociedad Española de Historia Natural* 13: 473–475. DP: [30] November 1913 (footer on p. 469).

Fuente, J. M. de la 1913b: Descripción de algunos Coleópteros nuevos de España. *Boletín de la Real Sociedad Española de Historia Natural* 13: 476–477. DP: [30] November 1913 (footer on p. 469).

Fuente, J. M. de la 1918: Descripción de coleópteros nuevos españoles. *Boletín de la Sociedad Entomológica de España* 1 (1/2): 44–46. DP: [31] January 1918 (issue cover).

Fuente, J. M. de la 1920: Nota sobre coleópteros españoles. *Boletín de la Sociedad Ibérica de Ciencias Naturales* (Zaragoza) 19 (5/7): 149–151. DP: 18 September 1920 (last page of the issue).

Fuente, J. M. de la 1927: Crioceris 12-punctata a. Baguenai n. a. *Boletín de la Sociedad Entomológica de España* 10 (3): 33. DP: 30 June 1927 (last page of the issue).

Fuessly [= Fuesslin, Füessli], J. C. 1775: *Verzeichnis der ihm bekannten schweizerischen Inseckten mit einer ausgemahlten Kupfertafel: nebst der Ankündigung eines neuen Insecten Werks.* Zürich: Füesslin, xii + 62 pp., 1 pl. DP: 7 May 1775 (Bousquet 2017).

Furth, D. G. 1979: Zoogeography and host plant ecology of the Alticinae of Israel, especially Phyllotreta; with descriptions of three new species (Coleoptera: Chrysomelidae). *Israel Journal of Zoology* **28** (1): 1–37. DP: [30] November 1979 (verso of the issue cover).

Furth, D. G. 1980: Zoogeography and host plants of Longitarsus in Israel, with descriptions of six new species (Coleoptera: Chrysomelidae). *Israel Journal of Entomology* **13** [1979]: 79–124. DP: 30 March 1980 (verso of the issue cover).

Furth, D. G. 1984: Alticinae of Israel: Psylliodes (Coleoptera: Chrysomelidae). *Israel Journal of Entomology* **17** [1983]: 37–58. DP: 15 April 1984 (verso of the issue cover).

Fuss, C. 1856: Die siebenbürgischen Chrysomelen und Oreinen. *Verhandlungen und Mittheilungen des Siebenbürgischen Vereins für Naturwissenschaften zu Hermannstadt* 7 (2): 25–32. DP: [29] February 1856 (issue title page).

Fuss, C. 1861a: Beiträge zur siebenbürgischen Käferfauna. *Verhandlungen und Mittheilungen des Siebenbürgischen Vereins für Naturwissenschaften zu Hermannstadt* **12** (9): 151–163. DP: [30] September 1861 (issue title page).

Fuss, C. 1861b: Beiträge zur siebenbürgischen Käferfauna. *Verhandlungen und Mittheilungen des Siebenbürgischen Vereins für Naturwissenschaften zu Hermannstadt* **12** (10): 169–175. DP: [31] October 1861 (issue title page).

Fuss, C. 1864: Berichtigungen und Beiträge zur siebenbürgischen Käferfauna. *Verhandlungen und Mittheilungen des Siebenbürgischen Vereins für Naturwissenschaften zu Hermannstadt* **15** (8): 146–150. DP: [31] August 1864 (issue title page).

Gabdullina, A. U. 2016: Novye vidy zhestkokrylykh nasekomykh (Insecta, Coleoptera) dlya fauny Kazakhstana. (New records of beetle species (Insecta, Coleoptera) from Kazakhstan). *Euroasian Entomological Journal* **15** (1): 95–97. DP: 29 February 2016 (journal website).

Gabriel, J. J. C. 1902: [new taxon]. In: Gerhard, J.: Neuheiten der schlesischen Käferfauna aus dem Jahre 1901. *Deutsche Entomologische Zeitschrift* **1902** (1): 59–60. DP: 1 July 1902 (issue cover).

Gabriel, J. J. C. 1918: [new aberration]. In: Kolbe, W.: Beiträge zur schlesischen Käferfauna. *Entomologische Mitteilungen* 7 (10/12): 200–211. DP: 15 November 1918 (issue cover).

Gabriel, J. J. C. 1928: [new aberration]. In: Kolbe, W.: Beiträge zur schlesischen Käferfauna. *Zeitschrift für Entomologie* (Breslau) **16** (2): 1–10. DP: 1 September 1928 (issue title page).

Gahan, C. J. 1891: Descriptions of new species of the coleopterous genus Oïdes (Galerucidae). *Annals and Magazine of Natural History, Sixth Series* 7 (41): 453–460. DP: 1 May 1891 (Evenhuis 2003).

Gahan, C. J. 1895: On the Coleoptera obtained by Dr. Anderson's collector during Mr. Bent's Expedition to the Hadramaut, South Arabia. *Journal of the Linnean Society, Zoology* **25** [1894–1896] (161): 285–291. DP: 31 July 1895 (verso of the volume title page).

Gahan, C. J. 1896: On Coleoptera from Aden and Somaliland. *Annals and Magazine of Natural History, Sixth Series* **18** (108): 448–460. DP: 1 December 1896 (Evenhuis 2003).

Gahan, C. J. 1903: [new taxon]. In: Forbes, H. O (ed.): The natural history of Sokotra and Abd-el-Kuri being the report upon the results of the conjoint expedition to these island in 1898–9, by Mr. W. R. Ogilvie-Grant, of the British Museum, and Dr. H. O. Forbes, of the Liverpool Museums, together with information from other available sources forming a monograph of the islands. *Special Bulletin of the Liverpool Museums* v–xlvii + 1–598 pp., 27 pls. DP: 7 November 1903 (*Athenaeum* 3967: 599).

Ganglbauer, L. 1897: Einige neue Coleopteren des mitteleuropäischen Faunengebietes. *Verhandlungen der Kaiserlich-Königlichen Zoologisch-Botanischen Gesellschaft in Wien* 47 (8): 565–573. DP: 10 November 1897 (Evenhuis 2023).

Ganglbauer, L. 1905: Coleoptera. Ergebnisse einer naturwissenschaftlichen Reise zum Erdschias-Dagh (Kleinasien). Ausgeführt von Dr. Arnold Penther und Dr. Emerich Zederbauer auf Kosten der "Gesellschaft zur Förderung des naturhistorischen Erforschung des Orients in Wien" (nunmehr "Naturwissenschaftlicher Orientverein in Wien") im Jahre 1902. *Annalen des K. K. Naturhistorischen Hofmuseums* 20 (2/3): 246–290. DP: [31 December] 1905 (issue cover).

Gangola, L. 1969: Two new Galerucinae from Nainital District (Coleoptera-Chrysomelidae). *Entomologist* 102 (1270): 55–59. DP: [31] March 1969 (issue cover).

Gauthier, J., Borer, M., Toussaint, E. F. A., Bilat, J., Kippenberg, H. & Alvarez, N. 2023: Museomics reveals evolutionary history of Oreina alpine leaf beetles (Coleoptera: Chrysomelidae). *Systematic Entomology* 48 (4): 658–671. DP: 10 June 2023 (journal website).

Gautier des Cottes, C. 1861: Description de coléoptères nouveaux propres à la faune française. *Annales de la Société Entomologique de France, Quatrième Série* 1 (2): 193–194. DP: 23 October 1861 (Lefèvre 1885).

Gavrilović, B. D. & Ćurčić, S. B. 2011: Diversity of species of the family Chrysomellidae (Insecta, Coleoptera) in Serbia, with an overview of previous researches. *Acta Zoologica Bulgarica* 63 (3): 231–244. DP: [31] December 2011 (issue cover).

Ge, D.-Y., Wang, S.-Y., Li, W.-Z. & Yang, X.-K. 2008: Study of Phygasia (Coleoptera: Chrysomelidae) from China, with descriptions of eight new species. *Biologia* (Bratislava) 63 (4): 553–565. DP: [31] August 2008 (issue cover).

Ge, D.-Y., Wang, S.-Y. & Yang, X.-K. 2009: A new species of Philopona Weise (Coleoptera: Chrysomelidae) from China, and a key to the known species of China. *Proceedings of the Entomological Society of Washington* 111 (1): 27–32. DP: 27 January 2009 (verso of the issue cover).

Ge, D.-Y., Wang, S.-Y. & Yang, X.-K. 2010: A new species of the genus Phygasia (Coleoptera: Chrysomelidae: Alticinae) from Taiwan of China. *Biologia* (Bratislava) 65 (2): 325–329. DP: [30] April 2010 (issue cover).

Ge, S.-Q. 2011: [new taxon]. In: Daccordi, M., Ge, S.-Q., Cui, J.-Z. & Yang, X.-K.: New species of Chrysolina Motschulsky, 1860 from Southwest China (Coleoptera, Chrysomelidae, Chrysomelinae). *Nouvelle Revue d'Entomologie, Nouvelle Série* 26 [2009] (4): 291–316. DP: 17 February 2011 (last page of the cover).

Ge, S.-Q. & Daccordi, M. 2011a: [new taxa]. In: Ge, S.-Q., Daccordi, M., Li, W.-Z. & Yang, X.-K.: Seven new species of the genus Chrysolina Motschulsky from China (Coleoptera: Chrysomelidae: Chrysomelinae). *Zootaxa* **2736**: 31–43. DP: 11 January 2011 (footer on the paper title page).

Ge, S.-Q. & Daccordi, M. 2011b: [new taxa]. In: Daccordi, M., Ge, S.-Q., Cui, J.-Z. & Yang, X.-K.: New species of Chrysolina Motschulsky, 1860 from Southwest China (Coleoptera, Chrysomelidae, Chrysomelinae). *Nouvelle Revue d'Entomologie, Nouvelle Série* **26** [2009] (4): 291–316. DP: 17 February 2011 (last page of the cover).

Ge, S.-Q. & Daccordi, M. 2011c: [new taxa]. In: Ge, S.-Q., Daccordi, M., Beutel, R. G., Li, W.-Z. & Yang, X.-K.: Revision of the chrysomeline genera Potaninia, Suinzona and Taipinus (Coleoptera) from eastern Asia, with a biogeographic scenario for the Hengduan mountain region in south-western China. *Systematic Entomology* **36** (4): 644–671. DP: 26 September 2011 (volume contents).

Ge, S.-Q. & Daccordi, M. 2013a: [new taxon]. In: Ge, S.-Q., Daccordi, M., Ren, J., Cui, J.-Z. & Yang, X.-K.: The genus Neophaedon Latreille (Coleoptera: Chrysomelidae: Chrysomelinae) from China. *Lavori della Societa Veneziana di Scienze Naturali* **38**: 135–140. DP: 31 January 2013 (footers throughtout the paper).

Ge, S.-Q. & Daccordi, M. 2013b: [new taxa]. In: Ge, S.-Q., Daccordi, M., Ren, J., Cui, J.-Z., Li, W.-Z. & Yang, X.-K.: Odontoedon, a new genus from China with descriptions of nine new species (Coleoptera: Chrysomelidae: Chrysomelinae). *Stuttgarter Beiträge zur Naturkunde A, Neue Serie* **6**: 199–222. DP: 30 April 2013 (paper title page).

Ge, S.-Q. & Daccordi, M. 2014: [unavailable name]. In: Yang, X.-K., Ge, S.-Q., Wang, S.-Y., Li, W.-Z. & Cui, J.-Z.: *Coleoptera Chrysomelidae Chrysomelinae. Fauna Sinica. Insecta Vol. 61*. Beijing: Science Press, xvi + 641 pp., 8 pls. DP: [30] June 2014 (colophon).

Ge, S.-Q. & Daccordi, M. 2015: [new taxon]. In: Ge, S.-Q., Daccordi, M., Ren, J. & Yang, X.-K.: Revision of Phaedon Latreille from China (Coleoptera: Chrysomelidae). *Zoological Systematics* 40 (1): 1–30. DP: 24 January 2015 (journal website).

Ge, S.-Q., Daccordi, M., Beutel, R. G., Li, W.-Z. & Yang, X.-K. 2011: Revision of the chrysomeline genera Potaninia, Suinzona and Taipinus (Coleoptera) from eastern Asia, with a biogeographic scenario for the Hengduan mountain region in south-western China. *Systematic Entomology* **36** (4): 644–671. DP: 26 September 2011 (volume contents).

Ge, S.-Q., Daccordi, M., Beutel, R. G., Ren, J., Cui, J.-Z., Li, W.-Z. & Yang, X.-K. 2012: Revision of the Eastern Asian genera Ambrostoma Motschulsky and Parambrostoma Chen (Coleoptera: Chrysomelidae: Chrysomelinae). *Systematic Entomology* **37** (2): 332–345. DP: 29 March 2012 (volume contents).

Ge, S.-Q. & Wang, S.-Y. 2004a: [new taxa]. In: Ge, S.-Q., Yang, X.-K., Wang, S.-Y., Li, W.-Z. & Cui, J.-Z.: Two new species of Phratora Chevrolat from China (Coleoptera: Chrysomelidae: Chrysomelinae). *Coleopterists Bulletin* **58** (1): 133–137. DP: 2 April 2004 (last page of the paper).

Ge, S.-Q. & Wang, S.-Y. 2004b: [new taxon]. In: Ge, S.-Q., Yang, X.-K. & Cui, J.-Z.: A key to the genus Phaedon (Coleoptera: Chrysomelidae: Chrysomelinae) from China and the description of a new species. *Entomological News* **114** [2003] (2): 75–80. DP: 23 June 2004 (issue cover).

Ge, S.-Q. & Wang, S.-Y. 2014: [unavailable name]. In: Yang, X.-K., Ge, S.-Q., Wang, S.-Y., Li, W.-Z. & Cui, J.-Z.: *Coleoptera Chrysomelidae Chrysomelinae. Fauna Sinica. Insecta Vol. 61*. Beijing: Science Press, xvi + 641 pp., 8 pls. DP: [30] June 2014 (colophon).

Ge, S.-Q., Wang, S.-Y. & Yang, X.-K. 2002a: Notes on the genus Phaedon Latreille of China (Coleoptera: Chrysomelidae: Chrysomelinae). *Acta Zootaxonomica Sinica* 27 (2): 316–325. DP: [30] April 2002 (paper title page).

Ge, S.-Q., Wang, S.-Y. & Yang, X.-K. 2002b: Four new species of genus Phratora from China (Coleoptera: Chrysomelidae: Chrysomelinae). *Acta Zootaxonomica Sinica* 27 (2): 326–329. DP: [30] April 2002 (paper title page).

Ge, S.-Q., Wang, S.-Y., Yang, X.-K. & Li, W.-Z. 2002: A revision of the genus Agrosteella Medvedev (Chrysomelidae: Chrysomelinae). *Pan-Pacific Entomologist* 78 (2): 80–87. DP: 3 July 2002 (verso of the issue cover).

Ge, S.-Q. & Yang, X.-K. 2005: [new taxa]. In: Ge, S.-Q., Li, W.-Z. & Yang, X.-K.: Two new species of Phratora Chevrolat from Qinling Mountains, China (Coleoptera: Chrysomelidae: Chrysomelinae). *Pan-Pacific Entomologist* 81 (1/2): 1–5. DP: 29 September 2005 (verso of the issue cover).

Ge, S.-Q. & Yang, X.-K. 2011: [new taxon]. In: Daccordi, M., Ge, S.-Q., Cui, J.-Z. & Yang, X.-K.: New species of Chrysolina Motschulsky, 1860 from Southwest China (Coleoptera, Chrysomelidae, Chrysomelinae). *Nouvelle Revue d'Entomologie, Nouvelle Série* 26 [2009] (4): 291–316. DP: 17 February 2011 (last page of the cover).

Ge, S.-Q. & Yang, X.-K. 2013: [new taxon]. In: Ge, S.-Q., Daccordi, M., Ren, J., Cui, J.-Z., Li, W.-Z. & Yang, X.-K.: Odontoedon, a new genus from China with descriptions of nine new species (Coleoptera: Chrysomelidae: Chrysomelinae). *Stuttgarter Beiträge zur Naturkunde A, Neue Serie* 6: 199–222. DP: 30 April 2013 (paper title page).

Ge, S.-Q., Yang, X.-K., Wang, S.-Y. & Li, W.-Z. 2004: Revision of the genus Agrosteomela Gistl (Coleoptera: Chrysomelidae: Chrysomelinae). *Coleopterists Bulletin* 58 (2): 273–284. DP: 13 July 2004 (last page of the paper).

Gebler, F. A. von 1817: Insecta Sibiriae rariora. *Mémoires de la Société Impériale des Naturalistes de Moscou* 5: 315–333. DP: [31 December] 1817 (Bousquet & Davies 2021).

Gebler, F. A. von 1823: Chrysomelae Sibiriae rariores. *Mémoires de la Société Impériale des Naturalistes de Moscou* 6: 117–126. DP: [31 December] 1823 (Bousquet & Davies 2021).

Gebler, F. A. von 1825: Coleopterorum Sibiriae species novae. Pp. 42–57. In: Hummel, A. D.: *Essais Entomologiques No. IV. Insectes de 1824. Novae species*. St. Petersbourg: Chancellerie privée du Ministère de l'Intérieur, 71 pp. + index. DP: 1 May 1825 (Bousquet 2016).

Gebler, F. A. von 1830: Bemerkungen über die Insekten Sibiriens, vorzüglich des Altai. (Part III). Pp. 1–228. In: Ledebour, C. F. (ed.): *Reise durch das Altai-Gebirge und die soongorische Kirgisen-Steppe. Auf Kosten der Kaiserlichen Universität Dorpat unternommen im Jahre 1826 in Begleitung der Herren D. Carl Anton Meyer und D. Alexander von Bunge R. K. Collegien-Assessors. Zweiter Theil.* Berlin: G. Reimer, iv + 522 + [2] pp. DP: 4 May 1830 (Bousquet 2016, 2017; see also Schmitt 2017 and the comments in Dating of publications).

Gebler, F. A. von 1832: Notice sur les coléoptères qui se trouvent dans le district des mines de Nertchinsk, dans la Sibérie orientale, avec la description de quelques espèces nouvelles. *Nouveaux Mémoires de la Société Impériale des Naturalistes de Moscou* 2: 23–78. DP: 10 April 1832 (Bousquet & Davies 2021).

Gebler, F. A. von 1833: Notae et additamenta ad Catalogum Coleopterorum Sibiriae occidentalis et confinis Tatariae Operis, C. F. von Ledebours Reise in das Altaigebirge und die soongarische Kirgisensteppe (Zwyter Theil. Berlin 1830). *Bulletin de la Société Impériale des Naturalistes de Moscou* 6: 262–309. DP: [31 December] 1833 (Bousquet & Davies 2021).

Gebler, F. A. von 1836: Lettre de M. le Conseiller d'Etat et Chevalier de Gebler, contenant un rapport d'un voyage dans les hautes montagnes catouniennes jusqu'à la frontière de la Chine; et description de trois nouvelles espèces des coléoptères. *Bulletin de la Société Impériale des Naturalistes de Moscou* 9: 329–345, pl. V. DP: [31] August 1836 (Bousquet & Davies 2021).

Gebler, F. A. von 1839: [new taxon]. In: Schoenherr, C. J.: *Genera et species curculionidum, cum synonymia hujus familiae; species novae aut hactenus minus cognitae, descriptionibus Dom. L. Gyllenhal, C. H. Boheman, O. J. Fahraeus et entomologis aliis illustratae. Tomus quintus. Pars prima. Supplementum continens.* Paris: A. Roret, Lipsiae: A. F. Fleischer, viii + 456 pp. DP: 15 June 1839 (Bousquet 2016).

Gebler, F. A. von 1841a: Charakteristik mehrerer neuen sibirischen Coleopteren. *Bulletin Scientifique del'Académie Impériale des Sciences de Saint-Pétersbourg* 8 [1840–1841] (24): 369–376. DP: 12 May 1841 (last page of the issue).

Gebler, F. A. von 1841b: Notae et additamenta ad Catalogum Coleopterorum Sibiriae occidentalis et confinis Tatariae operis: (C. F. von Ledebour's Reise in das Altaigebirge und die soongarische Kirgisensteppe. Zweyter Theil. Berlin. 1830.). Fasciculus secundus. *Bulletin de la Société Impériale des Naturalistes de Moscou* 14 (4): 577–625. DP: 27 October 1841 (Bousquet & Davies 2021).

Gebler, F. A. von 1842: Charakteristik der vom Hn. Dr. Schrenk im Jahre 1841 in den Steppen und Gebirgen der Songarei gefundenen neuen Coleopteren-Arten. *Bulletin de la Classe Physico-Mathématique de l'Académie Impériale des Sciences de Saint-Pétersbourg* 1 [1842–1843] (1/3): 36–40. DP: 13 September 1842 (last page of the issue).

Gebler, F. A. von 1844: Charakteristik der von Hn. Dr. Schrenk in den Jahren 1842 und 1843 in den Steppen der Dsungarei gefundenen neuen Coleopteren-Arten. *Bulletin de la Classe Physico-Mathématique de l'Académie Impériale des Sciences de Saint-Pétersbourg* 3 [1844–1845] (7): 97–106. DP: 15 June 1844 (last page of the issue).

Gebler, F. A. von 1848: Verzeichniss der im Kolywano-Woskresenskischen Hüttenbezirke Süd-West Sibiriens beobachteten Kaefer mit Bemerkungen und Beschreibungen. *Bulletin de la Société Impériale des Naturalistes de Moscou* 21 (3): 3–85. DP: 28 September 1848 (Bousquet & Davies 2021).

Geiser, E. 2001: *Die Käfer des Landes Salzburg. Faunistische Bestandserfassung und tiergeographische Interpretation. Monographs on Coleoptera, Vol. 2.* Wien: Zoologische-Botanische Gesellschaft in Österreich und Wiener Coleopterologenverein, 706 pp. DP: [31] October 2001 (book title page).

Geiser, E. 2019: To be or not to be a synonym – revision of the Donacia clavareaui-fukiensis complex (Coleoptera, Chrysomelidae, Donaciinae). *ZooKeys* **856**: 27–50. DP: 17 June 2019 (paper title page).

Geiser, E. 2023: Revision of the Palaearctic species of the genus Plateumaris C. G. Thomson, 1859 (Coleoptera, Chrysomelidae, Donaciinae). *ZooKeys* **1177**: 167–233. DP: 30 August 2023 (paper title page).

Geiser, E. & Geiser, R. 2023: Original descriptions of Palaearctic species of the genus Plateumaris C. G. Thomson, 1859 (Coleoptera, Chrysomelidae, Donaciinae) and their translations. *ZooKeys* **1177**: 235–258. DP: 30 August 2023 (paper title page).

Geiser, E. & Jäch, M. 2021: Explanatory notes on the updates concerning the genus Donacia Fabricius, 1775 in the second edition of the Catalogue of Palaearctic Coleoptera, Vol. 6/2 (Coleoptera: Chrysomelidae). *Koleopterologische Rundschau* **91**: 155–178. DP: [31] October 2021 (paper title page).

Gené, J. [C. G.] 1839: De quibusdam insectis sardiniae novis aut minus cognitis. Fasciculus II. *Memorie della Reale Accademia delle Scienze di Torino, Serie Seconda* **1**: 43–84, pls. I–II. DP: [31 December] 1839 (volume title page).

Gentner, L. G. 1926: New North American Halticinae (Coleoptera) with notes on other species. *Canadian Entomologist* **58** (6): 149–154. DP: [30] June 1926 (issue cover).

Geoffroy, E. L. 1762: *Histoire abregée des insectes qui se trouvent aux environs de Paris; dans laquelle ces animaux sont rangés suivant un ordre méthodique. Tome premier*. Paris: Durand, [4] + xxxviii + 523 + [1] pp., 10 pls. DP: [31 December] 1762 (Bousquet 2016) [work rejected for nomenclatorial purposes (ICZN 1954, Opinion 228), but some genus-level names subsequently made available (ICZN 1994, Opinion 1754)].

Geoffroy, E. L. 1785: [new taxa]. In: Fourcroy, A. F. de.: *Entomologia parisiensis; sive Catalogus Insectorum quae in Agro Parisiensi reperiuntur; Secundum methodum Geoffraeanam in sectiones, genera et species distributus: cui addita sunt nomina trivialia et fere trecentae novae Species. Pars prima*. Parisiis: Privilegio Academiae, vii + [1] + 231 pp. DP: 1 August 1785 (Bousquet 2016).

Gerhardt, J. 1891: Zugänge zur schlesischen Koleopteren-Fauna. *Zeitschrift für Entomologie, Neue Folge* (Breslau) **16**: 26–29. DP: [30] November 1891 (*Natur. Novit.* 1891, 13 (21): 434).

Gerhardt, J. 1904: Neuheiten der schlesischen Koleopterenfauna aus dem Jahre 1903. *Zeitschrift für Entomologie, Neue Folge* (Breslau) **29**: 77–78. DP: [31] December 1904 (*Natur. Novit.* 1904, 26 (24): 675).

Gerhardt, J. 1906: Neuheiten der schlesischen Koleopterenfauna aus dem Jahre 1905. *Deutsche Entomologische Zeitschrift* **1906** (1): 238. DP: [28] February 1906 (table of contents).

Gerhardt, J. 1909: Neuheiten der schlesischen Käferfauna aus dem Jahre 1908, geordnet nach dem Katalog europäischer Käfer von 1906. *Deutsche Entomologische Zeitschrift* **1909** (3): 415–422. DP: 1 May 1909 (table of contents).

Gerhardt, J. 1910: Neuheiten der schlesischen Käferfauna aus dem Jahre 1909. (Col.). *Deutsche Entomologische Zeitschrift* **1910** (5): 554–557. DP: 1 September 1910 (table of contents).

Gerhardt, E. 1911: Neuheiten der schlesischen Käferfauna aus dem Jahre 1910. (Col.). *Deutsche Entomologische Zeitschrift* **1911** (3): 337–338. DP: 4 May 1911 (table of contents).

Germain, P. 1903: Lista de los insectos traidos por los Señores Reiche i Machado de un viaje á Lebu y á La Mocha. Enero i febrero 1902. *Anales del Museo Nacional de Chile* **16**: 49–63. DP: [31 December] 1903 (volume title page).

Germar, E. F. 1810: [new taxon]. In: Ahrens, A.: Beiträge zu einer Monographie der Rohrkäfer. *Neue Schriften der Naturforschenden Gesellschaft zu Halle* 1 (3): 9–48. DP: 13 May 1810 (*Allg. Verz. Bücher, Ostermesse* 1810: 154 and Evenhuis 2014).

Germar, E. F. 1811: Nachträge zu Ahrens Monographie der Rohrkäfer. *Neue Schriften der Naturforschenden Gesellschaft zu Halle* **1** (6): 25–38. DP: [31 December] 1811 (issue title page).

Germar, E. F. 1813: Neue Insekten. *Magazin der Entomologie* **1** (1): 114–133. DP: 9 May 1813 (*Allg. Verz. Bücher, Ostermesse* 1813: 40 and Evenhuis 2014).

Germar, E. F. 1817: *Reise nach Dalmatien und in das Gebiet von Ragusa.* Leipzig und Altenberg: F. A. Brockhaus, xii + 323 pp., 9 pls, 2 maps. DP: [31] August 1817 (Bousquet 2016).

Germar, E. F. 1821: [new taxon]. In: Wiedemann, C. R. W. & Germar, E. F.: Neue exotische Käfer. *Magazin der Entomologie* **4**: 107–183. DP: [31 December] 1821 (see the comments in Dating of publications).

Germar, E. F. 1822a: *Augusti Ahrensii Fauna Insectorum Europae. Fasciculus V.* Halae: C. A. Kümmel, 25 nrs. DP: [30] June 1822 (Bousquet 2016).

Germar, E. F. 1822b: *Fauna Insectorum Europae. Fasciculus septimus.* Halae: C. A. Kümmel, 25 nrs. DP: [30] June 1822 (Bousquet 2016).

Germar, E. F. 1823: *Coleopterorum species novae aut minus cognitae, descriptionibus illustratae. Volumen primum. Coleoptera.* [1824], Halae: J. C. Hendelii & Filii, xxiv + 624 pp., 2 pls. DP: 5 October 1823 (Bousquet 2017).

Germar, E. F. 1831: *Fauna Insectorum Europae. Fasciculus quartusdecimus.* Halae: Car. Aug. Kümmelii, 25 nrs. DP: 2 October 1831 (Bousquet 2017).

Germar, E. F. 1834: *Fauna Insectorum Europae. Fasciculus sextusdecimus.* Halae: C. A. Kümmel, 25 nrs. DP: 20 April 1834 (Bousquet 2017).

Germar, E. F. 1842: *Fauna Insectorum Europae. Fasciculus vicesimus secundus.* Halae: Car. Aug. Kümmelii, 25 nrs. DP: 4 June 1842 (Bousquet 2016).

Gerstaecker, C. E. A. 1855: Diagnosen neuer Coleopteren und Lepidopteren. *Bericht über die zur Bekanntmachung Geeigneten Verhandlungen der Königl. Preuss. Akademie der Wissenschaften zu Berlin* **1855**: 636–643. DP: [31 December] 1855 (volume title page).

Gerstaecker, C. E. A. 1862: Coleoptera. Käfer. Pp. 268–348, pls. XV–XX. In: Peters, W. C. H.: *Naturwissenschaftliche Reise nach Mossambique auf Befehl seiner Majestät des Königs Friedrich Wilhelm IV in den Jahren 1842 bis 1848 ausgeführt von Wilhelm C. H. Peters. Zoologie. V. Insecten und Myriopoden. Bearbeiten in Verbingung mit Klug, Loew, Schaum, Hagen, Gerstaecker, Hopffer.* Berlin: Georg Reimer, 566 pp. + 25 pls. DP: 31 July 1862 (Bousquet 2016).

Gerstaecker, C. E. A. 1871: Beitrag zur Insekten-Fauna von Zanzibar. III. Coleoptera. *Archiv für Naturgeschichte* **37** [1] (1): 42–86. DP: 14 October 1871 (*Lit. Centralbl. Deutschl.* 1871 (41): 1038).

Gestro, R. 1885: Note entomologiche. II. Materiali per lo studio delle Hispidae Malesi e Papuane. *Annali del Museo Civico di Storia Naturale di Genova* **22**: 153–178, pl. IV. DP: 14 July 1885 (Poggi 2010).

Gestro, R. 1888a: Viaggio di Leonardo Fea in Birmania e regioni vicine. IV. Nuove specie di coleotteri. Decade I e II. *Annali del Museo Civico di Storia Naturali di Genova* **26**: 87–132. DP: 14 March 1888 (Poggi 2010).

Gestro, R. 1888b: Viaggio di Leonardo Fea in Birmania e regioni vicine. VI. Nuove specie di coleotteri. Decade III. *Annali del Museo Civico di Storia Naturali di Genova* **26**: 171–184. DP: 2 May 1888 (Poggi 2010).

Gestro, R. 1889: Viaggio ad Assab nel Mar Rosso dei Signori G. Doria ed O. Beccari con il R. Avviso "Esploratore" dal 16 Novembre 1879 al 26 Febbraio 1880. IV. Coleotteri. *Annali del Museo Civico di Storia Naturali di Genova* **27**: 5–72. DP: 19 April 1889 (Poggi 2010).

Gestro, R. 1890: Viaggio di Leonardo Fea in Birmania e regioni vicine. XXIX. Enumerazione delle Hispidae. *Annali del Museo Civico di Storia Naturali di Genova* **30** [1890–1891]: 225–268. DP: 17 November 1890 (Poggi 2010).

Gestro, R. 1892: Viagio di Lamberto Loria nella Papuasia Orientale. VI. Hispidae. *Annali del Museo Civico di Storia Naturali di Genova* **30**: 1015–1021. DP: 27 September 1892 (Poggi 2010).

Gestro, R. 1897a: Materiali per lo studio delle Hispidae. I. Enumerazione delle Hispidae di Sumatra. *Annali del Museo Civico di Storia Naturale di Genova* **38**: 37–110. DP: 14 June 1897 (Poggi 2010).

Gestro, R. 1897b: Materiali per lo studio delle Hispidae. II. Elementi per lo studio del genere Platypria. *Annali del Museo Civico di Storia Naturale di Genova* **38**: 110–116. DP: 14 June 1897 (Poggi 2010).

Gestro, R. 1897c: Materiali per lo studio delle Hispidae. III. Descrizioni di nuove species di Hispidae. *Annali del Museo Civico di Storia Naturale di Genova* **38**: 116–135. DP: 14 June 1897 (Poggi 2010).

Gestro, R. 1897d: Quelques espéces nouvelles d'hispides de Sumatra appartenant au Musée de Leyde. *Notes from the Leyden Museum* **19** (1/2): 67–72. DP: 15 July 1897 (verso of the volume title page).

Gestro, R. 1898: Tre nuove specie di Hispidae appartenenti al Museo Nazionale di Budapest. *Természetrajzi Füzetek* **21** (1/2): 260–263. DP: 10 March 1898 (issue cover).

Gestro, R. 1899a: Nuove forme del gruppo delle Platypria. *Annali del Museo Civico di Storia Naturale di Genova* **40** [1899–1901]: 172–176. DP: 17 July 1899 (Poggi 2010).

Gestro, R. 1899b: Contribution à l'étude de la faune de Sumatra (Côte ouest – vice-résidence de Païnan). Relevé des hispides recueillies à Sumatra par M. J.-L. Weyers. *Annales de la Société Entomologique de Belgique* **43** (7): 315–326. DP: 29 July 1899 (footer on p. 311).

Gestro, R. 1899c: Materiali per lo studio delle Hispidae. V. Appunti sul genere Downesia. *Annali del Museo Civico di Storia Naturale di Genova* **40** [1899–1901]: 218–224. DP: 3 November 1899 (Poggi 2010).

Gestro, R. 1899d: Materiali per lo studio delle Hispidae VII. Intorno ad alcune Hispa platiacante. *Annali del Museo Civico di Storia Naturale di Genova* **40** [1899–1901]: 227–229. DP: 3 November 1899 (Poggi 2010).

Gestro, R. 1906: Materiali per lo studio delle Hispidae. XXVIII. Descrizioni di alcune Hispidae inedite. *Annali del Museo Civico di Storia Naturale di Genova* **42** [1905]: 468–500. DP: 15 November 1906 (Poggi 2010).

Gestro, R. 1908: Materiali per lo studio delle Hispidae. XXXII. Descrizioni di alcune specie nuove. *Bullettino della Società Entomologica Italiana* **38** [1906] (3/4): 176–193. DP: 1 April 1908 (Poggi 2008).

Gestro, R. 1911a: Collezioni zoologiche fatte nell'Uganda dal Dott. E. Bayon. X. Nota sulle Hispidae. *Annali del Museo Civico di Storia Naturale di Genova* **45**: 23–33. DP: 20 January 1911 (Poggi 2010).

Gestro, R. 1911b: Materiali per lo studio delle Hispidae. XLII. Secondo saggio sulle Hispidae di Borneo. *Bullettino della Società Entomologica Italiana* **41** [1909]: 122–150. DP: 28 February 1911 (Poggi 2008).

Gestro, R. 1914: Materiali per lo studio delle Hispidae. *Annali del Museo Civico di Storia Naturale di Genova* **46**: 75–78. DP: 31 January 1914 (Poggi 2010).

Gestro, R. 1920: Materiali per lo studio delle Hispidae LIII. Hispidae Indocinesi raccolte dal signor R. Vitalis de Salvaza. *Annali del Museo Civico di Storia Naturale Giacomo Doria* **48** [1918–1919]: 385–403. DP: 2 February 1920 (Poggi 2010).

Gestro, R. 1923: Materiali per lo studio delle Hispidae. LIV. Contributi alla sistematica della tribù e descrizione di specia nove. *Annali del Museo Civico di Storia Naturale Giacomo Doria* **51**: 5–22. DP: 10 March 1923 (Poggi 2010).

Ghahari, H. & Beenen, R. 2020: An annotated catalogue of the Iranian Galerucinae sensu stricto (Coleoptera: Chrysomelidae) with taxonomic remarks. *Oriental Insects* **55** (1): 10–45. DP: 10 February 2020 (footer on the paper title page).

Ghahari, H. & Borowiec, L. 2017: A checklist of seed-beetles (Coleoptera: Chrysomelidae: Bruchinae) from Iran. *Zootaxa* **4268** (2): 215–237. DP: 16 May 2017 (footer on the paper title page).

Gillerfors, G. 1986: Two new species of Tarphius Erichson (Coleoptera, Colydiidae) and a new species of Mniophilosoma Wollaston (Coleoptera, Chrysomelidae) from the Azores. *Bocagiana* **98**: 1–10. DP: 15 September 1986 (paper title page).

Gistel, J. N. F. X. 1831: Entomologische Fragmente. *Isis von Oken* **1831** (3): 301–310. DP: [31] March 1831 (published monthly).

Gistel, J. N. F. X. 1832: Entomologische Notizen. *Faunus, Zeitschrift für Zoologie und Vergleichende Anatomie* **1**: 128–151. DP: [31 December] 1832 (volume title page).

Gistel, J. N. F. X. 1848a: Abtheilung I. Thierreich. Pp. 1–626. In: Gistel, J. N. F. X. & Bromme, F.: *Handbuch der Naturgeschichte aller drei Reiche, für Lehrer und Lernende, für Schule und Haus.* [1850], Hoffmann, Stuttgart, 1037 pp., 48 pl. DP (pp. 1–160 + 8 pls.): 29 February 1848; (pp. 161–480 + 16 pls.): [31] May 1848; (pp. 481–640 + 8 pls.): [31] July 1848 (Bousquet 2016, Bezděk 2020).

Gistel, J. N. F. X. 1848b: *Naturgeschichte des Thierreichs. Für höhere Schulen*. Stuttgart: Hoffmann'sche Verlags-Buchhandlung, xvi + 216 + 4 pp., 32 pls. DP: [30] June 1848 (Bousquet 2016).

Gistel, J. N. F. X. 1848c: Faunula monacensis cantharologica. (Fortsetzung). *Isis von Oken* **1848** (9): [unnumbered pp. "8–9"]. DP: [30] September 1848 (published monthly).

Gistel, J. N. F. X. 1848d: Faunula monacensis cantharologica. (Fortsetzung). *Isis von Oken* **1848** (10): [unnumbered pp. "10–11"]. DP: [31] October 1848 (published monthly).

Gistel, J. N. F. X. 1856: *Die Mysterien der europäischen Insectenwelt. Ein geheimer Schlüssel für Sammler aller Insecten-Ordnungen und Stände, behufs des Fangs, des Aufenthalts-Orts, der Wohnung, Tag- und Jahreszeit u. s. w., oder autoptische Darstellung des Insectenstaats in seinem zusammenhange zum Bestehen des Naturhaushaltes überhaupt und insbesondere in seinem Einflusse auf die phanerogamische und cryptogamische Pflanzenbevölkerung Europa's. Zum ersten Male nach fünfundzwanzigjährigen eigenen Erfahrungen zusammengestellt und herausgegeben.* Kempten: Tobias Dannheimer. xii + 532 pp. DP: 18 February 1856 (Bousquet 2016).

Gistel, J. N. F. X. 1857a: Emendationes in Fabricii Systema Eleutheratorum. Pp. 138–143. In: *Vacuna oder die Geheimnisse aus der organischen und leblosen Welt. Zweiter Band.* Straubing: Schorner, 1031 pp. DP: [31] October 1857 (Bousquet 2016).

Gistel, J. N. F. X. 1857b: Achthundert und zwanzig neue oder unbeschriebene wirbellose Thiere. Pp. 513–606. In: *Vacuna oder die Geheimnisse aus der organischen und leblosen Welt. Zweiter Band.* Straubing: Schorner, 1031 pp. DP: [31] October 1857 (Bousquet 2016) [reprint has pagination 1–92; earliest evidence of existence [30] November 1857 – *Ztschr. Ges. Naturwiss.* 1857, 10 (10/11): 453)].

Gmelin, J. F. 1790: *Caroli a Linné, systema naturae per regna tria naturae, secundum classes, ordines, genera, species, cum characteribus, differentiis, synonymis, locis. Editio decima tertia, aucta, reformata. Tom I.* Pars *IV. Classis V. Insecta.* Lipsiae: Georg Emanuel Beer, pp. 1517–2224. DP: 25 April 1790 (Bousquet 2017).

Goecke, H. 1934: Revision asiatischer Donaciinen (Col. Chrys.). I. (3. Beitrag zur Kenntnis der Donaciinen). *Koleopterologische Rundschau* 20 (6): 215–230. DP: 31 December 1934 (issue cover).

Goecke, H. 1936: Revision asiatischer Donaciinen (Col. Chrys.) II. 8. Beitrag zur Kenntnis der Donaciinen. *Entomologische Blätter* **32** (5): 220–228. DP: 31 October 1936 (issue cover).

Goecke, H. 1940: Donacia Freyi nov. sp. (Col. Chrys.). 10. Beitrag zur Kenntnis der Donaciinen. *Mitteilungen der Münchner Entomologischen Gesellschaft* 30 (3): 852–856. DP: 15 September 1940 (volume table of contents).

Goecke, H. 1944: Revision asiatischer Donaciinen (Col. Chrys.) III. 14. Beitrag zur Kenntnis der Donaciinen. *Entomologische Blätter* 40 (1/2): 7–14. DP: 30 April 1944 (issue cover).

Goecke, H. 1957: Plateumaris roscida. *Entomologische Blätter* **53** (1/2): 124. DP: 1 November 1957 (issue cover).

Goecke, H. 1960: Monographie der Schilfkäfer III. Die Gattungen und Arten der Donaciinae (Col. Chrys.) und ihre Verbreitung. *Entomologische Blätter* **56** (1): 1–19. DP: 1 August 1960 (issue cover).

Goeze, J. A. E. 1777: *Entomologische Beyträge zu des Ritter Linné zwölften Ausgabe des Natursystems. Erster Theil.* Leipzig: Weidmanns Erben und Reich, xvi + 736 pp. DP: 13 April 1777 (Bousquet 2016).

Gök, A. 2005a: A new species of the genus Phyllotreta Chevrolat (Coleoptera: Chrysomelidae) from southwestern Turkey. *Zootaxa* **923**: 1–5. DP: 29 March 2005 (footer on the paper title page).

Gök, A. 2005b: Psylliodes yalvacensis sp. n. (Coleoptera, Chrysomelidae, Alticinae) from Turkey. *Biologia* (Bratislava) **60** (2): 133–135. DP: [31] March 2005 (issue cover).

Gök, A. & Aslan, E. G. 2007: A new species of Psylliodes Latreille (Coleoptera: Chrysomelidae) from Turkey. *Entomological News* **118** (4): 371–376. DP: 14 November 2007 (issue cover).

Gök, A. & Çilbiroğlu, E. G. 2004: A new species of the genus Psylliodes Latreille (Coleoptera: Chrysomelidae) from Turkey. *Zootaxa* **440**: 1–6. DP: 25 February 2004 (footer on the paper title page).

Gök, A., Doguet, S. & Çilbiroğlu, E. G. 2003: Psylliodes cerenae sp. nov., a new Alticinae species from Southwest Turkey (Coleoptera: Chrysomelidae). *Annales Zoologici* (Warszawa) **53** (2): 201–202. DP: 1 May 2003 (verso of the issue cover).

Gök, A. & Sassi, D. 2002: A new species of genus Cryptocephalus Müller, 1764 from Southwest Turkey (Coleoptera: Chrysomelidae). *Annales Zoologici* (Warszawa) **52** (1): 155–156. DP: 29 March 2002 (verso of the issue cover).

Gómez-Zurita, J. 2004: Molecular systematics and time-scale for the evolution of Timarcha, a leaf-beetle genus with a disjunct Holarctic distribution. *Molecular Phylogenetics and Evolution* **32** (2): 647–665. DP: [31] August 2004 (journal website).

Gómez-Zurita, J. 2008: Species and speciation in Timarcha. Pp. 17–39. In: Jolivet, P., Santiago-Blay, J. & Schmitt, M. (eds.): *Research on Chrysomelidae. Volume 1*. Leiden: Brill, 430 pp. DP: 10 July 2008 (publisher website).

Gómez-Zurita, J. & Cardoso, A. 2021: Molecular systematics, higher-rank classification and Gondwanan origins of Cryptocephalinae leaf beetles. *Zoologica Scripta* **50** (5): 592–615. DP: [30] September 2021 (issue cover).

Gómez-Zurita, J., Hunt, T. & Vogler, A. P. 2008: Multilocus ribosomal RNA phylogeny of the leaf beetles (Chrysomelidae). *Cladistics* **24** (1): 34–50. DP: [29] February 2008 (journal website).

Gómez-Zurita, J., Juan, C. & Petitpierre, E. 2000a: The evolutionary history of the genus Timarcha (Coleoptera, Chrysomelidae) inferred from mitochondrial CoII gene and partial 16S rDNA sequences. *Molecular Phylogenetics and Evolution* **14** (2): 304–317. DP: [29] February 2000 (journal website).

Gómez-Zurita, J., Juan, C. & Petitpierre, E. 2000b: Sequence, secondary structure and phylogenetic analyses of the ribosomal internal transcribed spacer 2 (ITS2) in the Timarcha leaf beetles (Coleoptera: Chrysomelidae). *Insect Molecular Biology* **9** (6): 591–604. DP: [31] December 2000 (journal website).

Gómez-Zurita, J. & Kippenberg, H. 2010: Subfamily, Chrysomelinae, genus Timarcha. Pp. 437–443. In: Löbl, I. & Smetana, A. (eds.): *Catalogue of Palaearctic Coleoptera. Volume 6. Chrysomeloidea*. Stenstrup: Apollo Books, 924 pp. DP: 22 February 2010 (colophon).

Gómez-Zurita, J., Pons, J. & Petitpierre, E. 2004: The evolutionary origin of a novel karyotype in Timarcha (Coleoptera, Chrysomelidae) and general trends of chromosome evolution in the genus. *Journal of Zoological Systematics and Evolutionary Research* **42** (4): 332–341. DP: [30] November 2004 (journal website).

Gorham, H. S. 1885: Revision of the phytophagous Coleoptera of the Japanese fauna: subfamilies Cassidinae and Hispinae. *Proceedings of the Zoological Society of London* **1885** (2): 280–286. DP: [31] August 1885 (Duncan 1937).

Gortani, M. 1906: Saggio sulla distribuzione geografica dei coleotteri in Friuli. *In Alto, Cronaca della Societa Alpina Friulana* **16/17**: 3–93. DP: [31 December] 1906 (reprint cover).

Gory, H. L. 1833: [new taxon, pl. 49]. In: Guérin-Méneville, F. E.: *Iconographie du règne animal de G. Cuvier, ou représentation d'après nature de l'une des espèces les plus remarquables et souvent non encore figurées, de chaque genre d'animaux. Avec un texte descriptif mis au courant de la science. Ouvrage pouvant servir d'atlas à tous les traités de zoologie. Tome II. Planches des animaux invertébrés.* [1829–1844], Paris & Londres: J. B. Bailière, 450 pls. DP (pls. 47–49): [30] September 1833 (Bousquet 2016).

Gory, H. L. 1844: [new taxon]. In: Guérin-Méneville, F. E.: *Iconographie du Règne Animal de G. Cuvier, ou représentation d'après nature de l'une des espèces les plus remarquables, et souvent non encore figurées, de chaque genre d'animaux. Avec un texte descriptif mis au courant de la science. Ouvrage pouvant servir d'atlas à tous les traités de zoologie. Insectes.* [1829–1838], Paris & Londres: J. B. Baillière, 576 pp. DP: 26 August 1844 (Bousquet 2016).

Gozis, M. des 1881: Quelques rectifications synonymiques touchant différente genres et espèces de coléoptères français (4e partie). *Bulletin de la Société Entomologique de France* **1881** (20): 180–181. DP: 31 October 1881 (Evenhuis 2002).

Gozis, M. des 1882: Notes et remarques pour le futur catalogue des coléoptères gallo-rhénans. *Revue d'Entomologie* **1** (9): 193–207. DP: [30] September 1882 (footer on p. 193).

Gozis, M. des 1886: *Recherche de l'espèce typique de quelques anciens genres. Rectifications synonymiques et notes diverses.* Montluçon: Herbin, 36 pp. DP: 10 March 1886 (Bousquet 2016).

Gozis, M. des 1907: Tableaux analytiques pour déterminer les coléoptères de France. Genre Cryptocephalus Geoffroy. *Revue Scientifique du Bourbonnais et du Centre de la France* **20** (2): 66–80 & (3): 81–104 & (4/1): 144–175. DP (pp. 66–80): 26 June 1907 (*Bull. Soc. Ent. Fr.* 1907: 202); (pp. 81–104): 4 October 1907 (stamp on p. 81 of BHL copy); (pp. 144–175): [31 December] 1907 (issue cover).

Gozis, M. des 1908: Tableaux analytiques pour déterminer les Coléoptéres de France (suite). *Revue Scientifique du Bourbonnais et du Centre de la France* **21** (1): 31–36 & (2): 37–49. DP: (pp. 31–36): 22 April 1908, (pp. 37–49): 22 July 1908 (*Bull. Soc. Ent. Fr.* 1908: 159, 256).

Gradl, H. 1881: Aus der Fauna des Egerlands. Neue Beschreibungen von Insekten. *Entomologische Nachrichten* **7** (20): 294–300 & (21): 301–309. DP (pp. 294–300): 15 October 1881, (pp. 301–309): 1 November 1881 (headings throughout the paper).

Gradl, H. 1882: Aus der Fauna des Egerlands. Zweite Folge. *Entomologische Nachrichten* **8** (24): 323–332. DP: 15 December 1882 (headings throughout the paper).

Graells, M. de la Paz 1851: Decription de algunos insectos nuevos pertenencientes a la fauna central de España. *Memorias de la Real Academia de Ciencias de Madrid* **1** (2): 109–163, 3 pls. DP: [31 December] 1851 (see the comments in Dating of publications).

Graells, M. de la Paz 1858a: Insectos nuevos de España. Pp. 36–111. In: Luján, F. de: *Memorias de la Comisión del Mapa Geológico de España, año de 1855. Parte zoológica*. Madrid: Imprenta Nacional, 111 + [1] pp., 7 pls. DP: [30] September 1858 (Bousquet 2016).

Graells, M. de la Paz 1858b: Insectos nuevos de España. *Memoria que comprende los trabajos verificados en el año de 1855 por las diferentes secciones de la Comision encargada de formar el mapa geológico de la provincia de Madrid y el general del reino* [1855]: 74–149, 6 pls. DP: [31 December] 1858 (volume title page).

Grasso, D. 1958: Osservazioni su Lachnaea italica Weise e L. longipes Fabr. con descrizione di una nuova specie (Col. Chrysomelidae). *Bollettino della Società Entomologica Italiana* 88 (3/4): 57–60. DP: 20 April 1958 (issue cover and Poggi 2008).

Grasso, D. 1963: Una nuova sottospecie di Lachnaea italica Weise (Coleoptera Chrysomelidae). *Bollettino della Società Entomologica Italiana* 93 (5/6): 70–72. DP: 20 June 1963 (issue cover and Poggi 2008).

Gravenhorst, J. L. C. 1807: *Vergleichende Uebersicht des Linneischen und einiger neuern zoologischen Systeme nebst dem eingeschalteten Verzeichnisse der zoologischen Sammlung des Verfassers und den Beschreibungen neuer Thierarten, die in derselben vorhanden sind*. Goettingen: Dieterich, [20] + 476 pp. DP: 19 March 1807 (Bousquet 2016).

Gredler, I. V. M. 1857: Die Käfer von Passeier. II. Heft. *Zeitschrift des Ferdinandeums für Tirol und Vorarlberg, Dritte Folge* 6: 105–160. DP: [31 December] 1857 (volume cover).

Gredler, I. V. M. 1866: *Die Käfer von Tirol nach ihrer horizontalen und vertikalen Verbreitung. II. Hälfte: Dascillidae – Schluss. Mit mehren diagnosirten Novitäten*. Bozen: G. Ferrari, vormals Eberle, [2] + 235–491 pp. DP: 19 April 1866 (Bousquet 2016).

Gredler, I. V. M. 1868: Nachlese zu den Käfern von Tirol. *Coleopterologische Hefte* 3: 56–79. DP: 5 September 1868 (*Ann. Soc. Ent. Belg.* 12: ii (*Comptes rendus*)).

Grenier, A. 1863: *Catalogue des coléoptères de France et matériaux pour servir à la faune des coléoptères français*. Paris: A. Grenier, iv + 3–79 + 135 pp. DP: 6 August 1863 (book cover, Grenier 1864, see also the comments in the Dating of publications chapter).

Gressitt, J. L. 1938a: Some tortoise beetles from Hainan Island (Coleoptera: Chrysomelidae: Cassidinae). *Lingnan Science Journal* 17 (2): 185–193. DP: 16 June 1938 (reprint cover).

Gressitt, J. L. 1938b: Some hispine beetles from Hainan Island (Coleoptera: Chrysomelidae: Hispinae). *Lingnan Science Journal* 17 (3): 321–336, pl. 11. DP: 1 July 1938 (reprint cover).

Gressitt, J. L. 1938c: Some tortoise beetles from southeastern China (Coleoptera: Chrysomelidae: Cassidinae). *Lingnan Science Journal* 17 (3): 383–388. DP: 1 July 1938 (reprint cover).

Gressitt, J. L. 1938d: Tortoise beetles in the collection of the Lingnan Natural History Survey and Museum (Coleoptera: Chrysomelidae: Cassidinae). *Lingnan Science Journal* 17 (4): 571–590. DP: 8 October 1938 (reprint cover).

Gressitt, J. L. 1939a: Hispine beetles in the collection of the Lingnan Natural History Survey and Museum (Coleoptera: Chrysomelidae: Hispinae). *Lingnan Science Journal* 18 (2): 161–186. DP: 13 May 1939 (reprint cover).

Gressitt, J. L. 1939b: Some hispine beetles from Southeastern China (Coleoptera: Chrysomelidae: Hispinae). *Lingnan Science Journal* 18 (2): 197–208, pl. 7. DP: 13 May 1939 (reprint cover).

Gressitt, J. L. 1939c: East Asian Hispinae and Cassidinae in the collection of the California Academy of Sciences (Coleoptera, Chrysomelidae). *Pan-Pacific Entomologist* 15 (3): 132–143. DP: 9 August 1939 (volume table of contents).

Gressitt, J. L. 1942a: Plant-beetles from South and West China. I. Sagrinae, Donaciinae, Orsodacninae and Megalopodinae (Coleoptera). *Lingnan Science Journal* 20 (2/4): 271–293, pls. 11–14. DP: 27 August 1942 (issue cover).

Gressitt, J. L. 1942b: Plant-beetles from South and West China. II. Criocerinae (Chrysomelidae). *Lingnan Science Journal* 20 (2/4): 295–323, pls. 15–18. DP: 27 August 1942 (issue cover).

Gressitt, J. L. 1942c: Plant-beetles from South and West China. III. Clytrinae, Cryptocephalinae and Chlamydinae (Coleoptera). *Lingnan Science Journal* 20 (2/4): 325–376, pls. 19–22. DP: 27 August 1942 (issue cover).

Gressitt, J. L. 1942d: New tortoise beetles from China (Chrysomelidae: Cassidinae). *Special Publication Lingnan Natural History Survey and Museum* 5: 1–4. DP: 12 October 1942 (Evenhuis 2015).

Gressitt, J. L. 1945a: On some genera of Oriental Orsodacninae and Eumolpinae (Col. Chrysom.). *Lingnan Science Journal* 21 (1/4): 135–146. DP: 15 August 1945 (issue cover).

Gressitt, J. L. 1945b: A new tortoise-beetle name (Col.: Chrys.: Cassid.). *Lingnan Science Journal* 21 (1/4): 147. DP: 15 August 1945 (issue cover).

Gressitt, J. L. 1946: Chinese chrysomelid beetles of the subfamily Chlamisinae. *Annals of the Entomological Society of America* 39 (1): 84–100, 1 pl. DP: 29 March 1946 (*mailing dates* at the end of volume).

Gressitt, J. L. 1950: The hispine beetles of China (Coleoptera: Chrysomelidae). *Lingnan Science Journal* 23 (1/2): 53–142, pls. 4–8. DP: 23 December 1950 (Evenhuis 2015).

Gressitt, J. L. 1952: The tortoise beetles of China (Chrysomelidae: Cassidinae). *Proceedings of the California Academy of Sciences, Fourth Series* 27 (17): 433–592, pls. 27–36, 4 maps. DP: 8 December 1952 (reprint cover).

Gressitt, J. L. 1953: Supplement to "The hispine beetles of China" (Coleoptera: Chrysomelidae). *Pan-Pacific Entomologist* 29 (3): 121–126. DP: 25 August 1953 (volume table of contents).

Gressitt, J. L. 1955: Insects of Micronesia. Coleoptera: Chrysomelidae. *Insects of Micronesia* 17 (1): 1–60, 2 pls. DP: 15 July 1955 (verso of issue title page).

Gressitt, J. L. 1959: Longicorn beetles from New Guinea, I (Cerambycidae). *Pacific Insects* 1 (1): 59–171. DP: 17 July 1959 (journal website).

Gressitt, J. L. & Kimoto, S. 1961: The Chrysomelidae (Coleopt.) of China and Korea, part 1. *Pacific Insects Monograph* 1A: 1–299. DP: 16 February 1961 (journal website).

Gressitt, J. L. & Kimoto, S. 1963a: The Chrysomelidae (Coleopt.) of China and Korea, part 2. *Pacific Insects Monograph* 1B: 301–1026. DP: 25 February 1963 (paper title page).

Gressitt, J. L. & Kimoto, S. 1963b: Supplement to "The Chrysomelidae (Coleopt.) of China and Korea". *Pacific Insects* 5 (4): 921–932. DP: 30 December 1963 (paper title page).

Gressitt, J. L. & Kimoto, S. 1965: Second supplement to "The Chrysomelidae (Coleopt.) of China and Korea". *Pacific Insects* 7 (4): 799–806. DP: 30 December 1965 (paper title page).

Grimmer, K. H. B. 1841: *Steiermark's Coleoptern mit Einhundert sechs neu beschriebenen Spezies*. Grätz: C. Tanzer, iv + 5–50 pp. DP: 26 May 1841 (Bousquet 2016).

Gröndal, C. F. 1808: [new taxa]. In: Schoenherr, C. J.: *Synonymia insectorum, oder: Versuch einer Synonymie aller bisher bekannten Insecten; nach Fabricii Systema Eleutheratorum &c. geordnet. Erster Band. Eleutherata oder Käfer. Zweiter Theil. Spercheus – Cryptocephalus*. Stockholm: C. F. Marquard, ix + 424 pp., 1 pl. DP: [31 December] 1808 (Bousquet 2016).

Gruev, B. 1973a: [new taxon]. In: Tomov, V. & Gruev, B. 1973: Neue Blattkäfer für die Fauna Bulgariens mit Notizen über einige mitgeteilten Arten (Coleoptera, Chrysomelidae). *Travaux Scientifiques, Université de Plovdiv "Paissi Hilendarski"* 11 (5 – Biologie): 91–102. DP: [31 December] 1973 (issue title page).

Gruev, B. 1973b: Neue Angaben über die Systematik und Verbreitung einiger paläarktischen Arten der Gattung Longitarsus Latr. mit Beschreibung einer neuen Art aus Bulgarien (Coleoptera, Chrysomelidae). *Travaux Scientifiques, Université de Plovdiv "Paissi Hilendarski"* 11 (5 – Biologie): 103–124. DP: [31 December] 1973 (issue title page).

Gruev, B. 1973c: Über die Einheit der Longitarsus lycopi-Gruppe und der Longitarsus obliteratus-Gruppe mit Notizen über Verbreitung, Nomenklatur und diagnostische Merkmale einiger Arten, sowie Beschreibung von zwei neuen Arten (Coleoptera, Chrysomelidae). *Travaux Scientifiques, Université de Plovdiv "Paissi Hilendarski"* 11 (5 – Biologie): 125–145. DP: [31 December] 1973 (issue title page).

Gruev, B. 1974a: Zwei neue Unterarten von Derocrepis serbica (Kutschera, 1860) aus Bulgarien (Col. Chrysomelidae). *Entomologische Arbeiten aus dem Museum G. Frey* 25: 121–130. DP: 1 October 1974 (table of contents).

Gruev, B. 1974b: Eine neue Unterart von Chrysochloa speciosissima Scop. aus Bulgarien (Col. Chrysomelidae). *Comptes Rendus de l'Académie Bulgare des Sciences* 27 (7): 969–970. DP: [31 December] 1974 (issue title page).

Gruev, B. 1975a: Zwei neue Arten der Unterfamilie Halticinae und taxonomische Notizen über Chalcoides nigricoxis (Allard, 1878) (Coleoptera, Chrysomelidae). *Acta Zoologica Bulgarica* 1: 89–95. DP: [31] January 1975 (paper title page).

Gruev, B. 1975b: Eine neue Longitarsus-Art aus Kroatien (Col., Chrysomelidae). *Wissenschaftliche Mitteilungen des Bosnisch-herzegovinischen Landesmuseums* (Sarajevo) 4/5 [1974–1975] (C – Naturwissenschaft): 139–140. DP: [31 December] 1975 (reprint cover).

Gruev, B. 1975c: Eine neue Epithrix-Art aus Asien und neue Angaben über Epithrix ermischi Mohr, 1968 (Col. Chrysomelidae). *Travaux Scientifiques, Université de Plovdiv "Paissi Hilendarski"* 13 (4 – Biologie): 171–175. DP: [31 December] 1975 (paper title page).

Gruev, B. 1976: Eine neue Chrysomela-Art aus der Türkei und Bemerkungen über Chrysomela fallaciosa Müller, 1948 (Coleoptera, Chrysomelidae). *Acta Zoologica Bulgarica* 5: 75–78. DP: [30] November 1976 (paper title page).

Gruev, B. 1977: Neue Angaben über einige paläarktischen Blattkäfer und Beschreibung von zwei neuen Alticinen aus Bulgarien (Coleoptera, Chrysomelidae). *Acta Zoologica Bulgarica* 8: 65–73. DP: [31] October 1977 (paper title page).

Gruev, B. 1978: Eine neue Cassida-Art aus Jugoslawien (Coleoptera, Chrysomelidae). *Deutsche Entomologische Zeitschrift, Neue Folge* 25 (1/3): 171–172. DP: 1 June 1978 (table of contents).

Gruev, B. 1979: Chrysomelidae (Coleoptera) Jugoslawiens (Unterfamilien: Lamprosomatinae, Eumolpinae, Chrysomelinae, Alticinae, Hispinae, Cassidinae). *Deutsche Entomologische Zeitschrift, Neue Folge* 26 (1/3): 113–152. DP: 1 June 1979 (table of contents).

Gruev, B. 1980a: A contribution to the knowledge of the Chrysomelidae of Korea, III. (Coleoptera). *Entomological Review of Japan* 34 (1/2): 29–38. DP: 30 May 1980 (General index 1978–1981).

Gruev, B. 1980b: A new subspecies of Ambrostoma quadriimpressum Motschulsky, 1845 from East China (Col., Chrysomelidae: Chrysomelinae). *Entomological Review of Japan* 34 (1/2): 83–84. DP: 30 May 1980 (General index 1978–1981).

Gruev, B. 1981a: A new species of Epitrix Foudras from the Azores (Insecta, Coleoptera, Chrysomelidae). *Boletim da Sociedade Portuguesa de Entomologia* 10: 1–2. DP: 10 March 1981 (paper title page).

Gruev, B. 1981b: New data about some chrysomelid beetles from China with a description of a new species of Psylliodes Latreille (Col., Chrysomelidae). *Entomological Review of Japan* 35 (1/2): 71–76. DP: 30 March 1981 (General index 1978–1981).

Gruev, B. 1981c: Bemerkungen über die Arten der Gattung Angulaphthona Bechyné, 1960, mit Beschreibung einer neuen Art aus Jemen (JAR) (Coleoptera, Chrysomelidae, Alticinae). *Reichenbachia* 19: 55–58. DP: 15 December 1981 (paper title page).

Gruev, B. 1981d: Chrysomelidae (Coleoptera) collected by Mr P. M. Hammond in China. II. Subfamily Alticinae. *Entomologist's Monthly Magazine* 117 (1404/1407): 147–155. DP: 21 December 1981 (issue cover).

Gruev, B. 1982a: Eine neue Phyllotreta-Art aus der Balkanhalbinsel (Coleoptera, Chrysomelidae). *Deutsche Entomologische Zeitschrift, Neue Folge* 29 (1/3): 99–100. DP: 15 March 1982 (table of contents).

Gruev, B. 1982b: Neue Angaben über einige Blattkäfer aus der Alten Welt (Insecta, Coleoptera, Chrysomelidae). *Faunistische Abhandlungen Staatliches Museum für Tierkunde in Dresden* 9 [1981] (1): 109–114. DP: 28 June 1982 (paper title page).

Gruev, B. 1982c: Beitrag zur Kenntnis der Longitarsus obliteratus-Gruppe (Col., Chrysomelidae, Alticinae). *Deutsche Entomologische Zeitschrift, Neue Folge* 29 (4/5): 469–473. DP: 1 November 1982 (contens pages).

Gruev, B. 1985a: Leaf-beetles collected by P. Beron and S. Andreev from the National Museum (Natural History) – Sofia in India, Sri Lanka, Nepal, Burma and Thailand in 1981 and 1984 (Coleoptera, Chrysomelidae). *Entomological Review of Japan* 40 (1): 35–47. DP: 26 June 1985 (General index 1981–1985).

Gruev, B. 1985b: Two new taxa of Alticinae from China (Coleoptera, Chrysomelidae). *Entomological Review of Japan* 40 (2): 125–127. DP: 20 December 1985 (General index 1981–1985).

Gruev, B. 1988a: A new species of Longitarsus Latreille from Nepal (Coleoptera, Chrysomelidae: Alticinae). *Entomological Review of Japan* 43 (2): 101–102. DP: 20 December 1988 (verso of the issue cover).

Gruev, B. 1988b: Check-Liste der Arten von Eumolpinae, Chrysomelinae, Alticinae, Hispinae und Cassidinae in Afghanistan (Coleoptera, Chrysomelidae). *Entomological Review of Japan* 43 (2): 145–170. DP: 20 December 1988 (verso of the issue cover).

Gruev, B. 1989: Two new species of Gonioctena (Brachyphytodecta) Chevrolat (Coleoptera, Chrysomelidae: Chrysomelinae) from South Asia and a note on Gonioctena issikii Chûjô, 1958. *Entomological Review of Japan* 44 (1): 51–54. DP: 30 June 1989 (verso of the issue cover).

Gruev, B. 1990a: Two new species of Alticinae (Coleoptera, Chrysomelidae) from Nepal. *Acta Zoologica Bulgarica* 39: 73–75. DP: [31] March 1990 (paper title page).

Gruev, B. 1990b: Leaf-beetles collected by P. Beron of the National Natural History Museum – Sofia in Nepal (Coleoptera, Chrysomelidae). *Entomological Review of Japan* 45 (1): 59–66. DP: 30 June 1990 (verso of the issue cover).

Gruev, B. 1990c: Chaetocnema imitatrix sp. n. – a new chrysomelid species from Bulgaria (Insecta, Coleoptera, Chrysomelidae: Alticinae). *Reichenbachia* 28 (1): 89–90. DP: 30 November 1990 (paper title page).

Gruev, B. 1990d: On the geographical distribution of some leaf beetles in Korea with a description of Psylliodes takizawai sp. n. (Coleoptera, Chrysomelidae). *Entomological Review of Japan* 45 (2): 119–133. DP: 31 December 1990 (verso of the issue cover).

Gruev, B. 1991: Oreomela lopatini – a new species of Oreomela Jacobson (Coleoptera, Chrysomelidae, Chrysomelinae) from the USSR (Kirghiz SSR). *Acta Zoologica Bulgarica* 41: 67–68. DP: [31] March 1991 (paper title page).

Gruev, B. 1992: *Geographical distribution of the leaf beetle subfamilies Lamprosomatinae, Eumolpinae, Chrysomelinae, Alticinae, Hispinae and Cassidinae (Coleoptera: Chrysomelidae) on the Balkan Peninsula.* Plovdiv: Plovdiv University Press, 512 pp. DP: [31 December] 1992 (colophon).

Gruev, B. 1994: New distributional data about some leafbeetles in the Korean Peninsula and descriptions of four new species (Coleoptera, Chrysomelidae). *Insecta Koreana* 11: 75–84. DP: 15 December 1994 (paper title page).

Gruev, B. 1996: Bibliography of the descriptions and the nomenclatoric changes of the palaearctic Longitarsus species after Csiki & Heikertinger: Chrysomelidae Halticinae, Longitarsus, in Coleopterorum Catalogus, Junk & Schenkling (1939–1940) (Coleoptera Chrysomelidae). *Memorie della Società Entomologica Italiana* 74 [1995]: 33–63. DP: 29 February 1996 (volume cover).

Gruev, B. 1998: New data about some Palaearctic Alticinae from the Zoological museum of the University of Helsinki with description of a new subspecies of Phyllotreta (Coleoptera:

Chrysomelidae). *Acta Zoologica Bulgarica* 50 (2/3): 37–42. DP: [31] December 1998 (issue cover).

Gruev, B. A. 2005: A comparative list of the leaf beetles of the Balkan countries (Coleoptera: Chrysomelidae). *Scientific Studies, University of Plovdiv "Paissi Hilendarski", Animalia* 41: 23–46. DP: [31 December] 2005 (paper title page).

Gruev, B. & Arnold, U. 1989: Eine neue Longitarsus-Art aus Bulgarien (Coleoptera, Chrysomelidae, Alticinae). *Entomologische Nachrichten und Berichte* 33 (4): 175–176. DP: 15 August 1989 (verso of the issue cover).

Gruev B. & Aslan, I. 1998: Four new species of the subfamily Alticinae (Coleoptera, Chrysomelidae) from Turkey. *Türkiye Entomoloji Dergisi* 22 (3): 163–169. DP: [30] September 1998 (table of contents).

Gruev, B. & Döberl, M. 1997: General distribution of the flea beetles in the Palaearctic Subregion (Coleoptera, Chrysomelidae: Alticinae). *Scopolia* 37: 1–496. DP: [31] December 1997 (volume cover).

Gruev, B. & Döberl, M. 2005: *General distribution of the flea beetles in the Palaearctic Subregion (Coleoptera, Chrysomelidae: Alticinae). Supplement.* Sofia – Moscow: Pensoft, 239 pp. DP: [30] June 2005 (colophon).

Gruev, B. & Kasap, H. 1985: A list of some Alticinae from Turkey with descriptions of two new species (Col. Chrysomelidae). *Deutsche Entomologische Zeitschrift, Neue Folge* 32 (1/3): 59–73. DP: 15 March 1985 (issue cover).

Gruev, B., Merkl, O. & Víg, K. 1993: Geographical distribution of Alticinae (Coleoptera, Chrysomelidae) in Romania. *Annales Historico-naturales Musei Nationalis Hungarici* 85: 75–132. DP: 31 December 1993 (Merkl et al. 2008).

Gruev, B. & Petitpierre, E. 1979: New data about some species of Heyrovskya Madar et Madar, 1968 (Col., Chrysomelidae, Alticinae) with descriptions of two new species. *Vieraea* 8 [1978] (1): 125–130. DP: [31] August 1979 (paper title page).

Gruev, B. & Tomov, V. 1968: Novyy rod i vid zhuka listoeda Rhodopaea angelovi, gen. et sp. n. (Coleoptera, Chrysomelidae) iz Bolgarii. (A new genus and species of the leaf-beetle Rhodopaea angelovi, gen. et sp. n. (Coleoptera, Chrysomelidae) from Bulgaria). *Entomologicheskoe Obozrenie* 47 (3): 553–555. SFP: 6 September 1968, DP: 9 October 1968 (registered at ZIN library).

Gruev, B. & Tomov, V. 1979: Zur Kenntnis einiger in der Türkei, Jugoslawien und Griechenland vorkommender Arten der familie Chrysomelidae (Coleoptera) aus der Zoologischen Staatssammlung München. *Spixiana* 2 (3): 259–267. DP: 1 November 1979 (paper title page).

Gruev, B. & Tomov, V. 1987: *Coleoptera, Chrysomelidae. Part II. Chrysomelinae, Galerucinae, Alticinae, Hispinae, Cassidinae. Fauna Bulgarica 16.* [1986], Sofia: Bulgarian Academy of Sciences, 388 pp. DP: 5 January 1987 (colophon).

Gruev, B. A. & Tomov, V. 1998: *Catalogus Faunae Bulgaricae. Vol. 3. Coleoptera: Chrysomelidae.* Sofia: Pensoft, 160 pp. DP: [30] April 1998 (colophon).

Gruev, B. & Tomov, V. 2007: *A distributional atlas and catalogue of the leaf beetles of Bulgaria (Coleoptera: Chrysomelidae)*. Sofia-Moscow: Pensoft, 350 pp. DP: [31] January 2007 (colophon).

Guérin-Méneville, F. E. 1838: Crustacés, arachnides et insectes. In: Lesson, M.: *Voyage autour du Monde, exécuté par ordre du Roi, sur la corvette de sa Majesté, La Coquille, pendant les années 1822, 1823, 1824 et 1825, sous le Ministère et conformément aux instructions de S. E. M. Le Marquis de Clermont-Tonnerre, ministre de la marine; et publié sous les auspices de son Excellence Mgr le Cte de Chabrol, Ministre de la marine et des colonies, par M. L. I. Duperrey. Zoologie. Tome second – 2e partie*. Paris: Arthus Berthand, xii + 9–319 pp. DP: [31 December] 1838 (Bousquet 2016).

Guérin-Méneville, F. E. 1840a: Coléoptères nouveaux du plateau des Neelgheries, dans les Indes orientales, découverts par M. Adolphe Delessert. *Revue Zoologique, par la Société Cuvierienne* 3 (2): 37–42. DP: 2 March 1840 (*Compt. Rend. Hebdom. Séances Acad. Sci.* 1840, 10: 206).

Guérin-Méneville, F. E. 1840b: Description de cinq espèces d'hispes, formant une division distincte dans ce genre. *Revue Zoologique, par la Société Cuvierienne* 3 (5): 139–142. DP: 1 June 1840 (*Compt. Rend. Hebdom. Séances Acad. Sci.* 1840, 10: 877).

Guérin-Méneville, F. E. 1840c: Note monographique sur le genre de coléoptères nommé Alurnus, par Fabricius, et sur quelques groupes voisins. *Revue Zoologique, par la Société Cuvierienne* 3 (11): 330–334. DP: 28 December 1840 (*Compt. Rend. Hebdom. Séances Acad. Sci.* 1840, 11: 1066).

Guérin-Méneville, F. E. 1841: Description de treize espèces du genre Hispa proprement dit. *Revue Zoologique, par la Société Cuvierienne* 4 (1): 6–14. DP: 8 February 1841 (*Compt. Rend. Hebdom. Séances Acad. Sci.* 1841, 12: 315).

Guérin-Méneville, F. E. 1844: *Iconographie du Règne Animal de G. Cuvier, ou représentation d'après nature de l'une des espèces les plus remarquables, et souvent non encore figurées, de chaque genre d'animaux. Avec un texte descriptif mis au courant de la science. Ouvrage pouvant servir d'atlas à tous les traités de zoologie. Insectes*. [1829–1838], Paris & Londres: J. B. Baillière, 576 pp. DP: 26 August 1844 (Bousquet 2016).

Guérin-Méneville, F. E. 1858: Mélanges et nouvelles. Ravages des altises. *Revue et Magasin de Zoologie Pure et Appliquée, 2e Série* 10 (9): 414–416. DP: [31] October 1858 (Dickinson et al. 2011).

Guérin-Méneville, F. E. 1860: [Coléoptères nouveaux de l'oasis Ouargla]. *Bulletin de la Société Entomologique de France* [1859] (4): clxxxvi–cxciii. DP: 1 March 1860 (Evenhuis 2022).

Guillebeau, F. 1885: Descriptions de quelques nouvelles espèces de coléoptères. *L'Échange, Revue Linnéenne* 1 (7): 2–3. DP: 15 July 1885 (issue title page).

Guillebeau, F. 1891: Révision du genre Luperus Geoffroy d'après Weise. Traduite et complétée. *Revue d'Entomologie* 10 (10): 290–304 & (11/12): 305. DP (pp. 290–304): [31] October 1891, (p. 305): [31] December 1891 (footers on pp. 289, 305).

Guillebeau, F. 1894: Description de deux espèces nouvelles de coléoptères et une note sur Anthribus scapularis Gebler. *Bulletin de la Société Entomologique de France* 1894 (14): cxcii–cxciv. DP: 6 October 1894 (Evenhuis 2002).

Guillebeau, F. 1895a: [Note relative à son travail concernant les Phalacridae et la description d'une nouvelle espèce de Phyllotreta (Coléoptères)]. *Bulletin de la Société Entomologique de*

France 1895 (2): xxv–xxvii. DP: 13 March 1895 (DP refers to issue 4 mentioned in *Bull. Soc. Ent. Fr.* 1895 (5): cliii).

Guillebeau, F. 1895b: Étude sur quelques espèces de Longitarsus Latr. et observations sur leur synonymie (Col.). *Bulletin de la Société Entomologique de France* 1895 (19): ccclxxxvi–ccclxxxix. DP: [31] December 1895 (footer on p. ccclxxxi).

Guillebeau, F. 1895c: Description d'une nouvelle espèce de Longitarsus Latr. (Col.). *Bulletin de la Société Entomologique de France* 1895 (19): ccclxxxix. DP: [31] December 1895 (footer on p. ccclxxxi).

Guillebeau, F. 1897a: Description de quelques espèces nouvelles de coléoptères. *Bulletin de la Société Entomologique de France* 1897 (10): 163–166. DP: 11 October 1897 (*Bibl. Zool.* 1897, 2: 536).

Guillebeau, F. 1897b: Descriptions de quelques espèces nouvelles de coléoptères. *Bulletin de la Société Entomologique de France* 1897 (13): 222–227. DP: [31 December] 1897 (issue title page).

Günther, V. 1958: Nový druh a nový podrod čeledi Cassididae (Col.). Eine neue Art und eine neue Untergattung der Familie Cassididae (Col.). *Acta Entomologica Musei Nationalis Pragae* 32: 567–569. DP: 31 December 1958 (colophon).

Guskova, E. V. 2006: Novye dannye po listoedam (Coleoptera, Chrysomelidae) Mongolskogo Altaya. (New data on leaf-beetles (Coleoptera, Chrysomelidae) from the Mongolian Altai). *Euroasian Entomological Journal* 5 (4): 303–306. DP: 29 December 2006 (journal website).

Guskova, E. V. 2010: The leaf-beetles (Coleoptera, Chrysomelidae) of the South Urals. *Entomofauna* 31 (14): 169–228. DP: 19 November 2010 (paper title page).

Guskova, E. V., Efimov, D. A. & Korshunov, A. V. 2016: New for the fauna of the Kemerovo region (Russia, Siberia) leaf beetles (Coleoptera: Chrysomelidae). *Far Eastern Entomologist* 307: 11–16. DP: [31] March 2016 (first page of issue).

Guskova, E. V., Akulov, E. N. & Loshchev, S. M. 2018: The Cryptocephalinae (Coleoptera, Chrysomelidae) of the Krasnoyarsk Krai (except for the Taymyrsky Dolgano-Nentsky District), Siberia, Russia. *Coleopterists Bulletin* 72 (3): 565–576. DP: 20 September 2018 (last page of the paper).

Gyllenhal, L. 1808: [new taxa]. In: Schoenherr, C. J.: *Synonymia insectorum, oder: Versuch einer Synonymie aller bisher bekannten Insecten; nach Fabricii Systema Eleutheratorum &c. geordnet. Erster Band. Eleutherata oder Käfer. Zweiter Theil. Spercheus – Cryptocephalus.* Stockholm: C. F. Marquard, ix + 424 pp., 1 pl. DP: [31 December] 1808 (Bousquet 2016).

Gyllenhal, L. 1813: *Insecta Svecica. Classis I. Coleoptera sive Eleuterata. Tomi 1 Pars III.* Scaris: Leverentz, [4] + 730 + [2] pp. DP: [31 December] 1813 (Bousquet 2016).

Gyllenhal, L. 1817: [new taxa]. In: Schoenherr, C. J.: *Appendix ad C. J. Schönherr Synonymiam insectorum Tom 1. Part 3. sistens descriptiones novarum specierum.* Scaris: Leverentz: 266 pp., pls. 5–6. DP: [31 December] 1817 (Bousquet 2016).

Gyllenhal, L. 1827: *Insecta Svecica. Classis I. Coleoptera sive Eleuterata. Tom I. Pars IV. Cum appendice ad partes priores.* Lipsiae: Fleischer, viii + [2] + 761 + [1] pp. DP: 6 May 1827 (Bousquet 2016).

Gyllenhal, L. 1833: [new taxa]. In: Schoenherr, C. J.: *Genera et species curculionidum, cum synonymia hujus familiae. Tomus primus. Pars prima.* Paris: A. Roret, xv + 381 pp. DP: [30] June 1833 (Bousquet 2016).

Gyllenhal, L. 1839: [new taxa]. In: Schoenherr, C. J.: *Genera et species curculionidum, cum synonymia hujus familiae; species novae aut hactenus minus cognitae, descriptionibus Dom. L. Gyllenhal, C. H. Boheman, O. J. Fahraeus et entomologis aliis illustratae. Tomus quintus. Pars prima. Supplementum continens.* Paris: A. Roret, Lipsiae: A. F. Fleischer, viii + 456 pp. DP: 15 June 1839 (Bousquet 2016).

György, Z. & Merkl, O. 2005: Seed beetles preserved in the Savaria Museum, Hungary, with a national checklist of the family (Coleoptera: Bruchidae). *Praenorica, Folia Historico-Naturalia* 8: 65–78. DP: [31 December] 2005 (volume cover).

Haddad, S., Shin, S., Lemmon, A. R., Moriarty Lemmon, E., Švácha, P., Farrel, B., Ślipiński, A., Windsor, D. & McKenna, D. D. 2018: Anchored hybrid enrichment provides new insights into the phylogeny and evolution of longhorned beetles (Cerambycidae). *Systematic Entomology* **43** (1): 68–89. DP: 18 August 2018 (last page of the paper).

Haldemann, S. S. 1849: Cryptocephalinarum boreali-americae diagnoses cum speciebus novis musei lecontiani. *Journal of the Academy of Natural Sciences of Philadelphia, Second Series* 1 [1847–1850] (3): 245–265. DP: 7 August 1849 (Nolan 1913).

Hagenbach, J. J. 1822: *Symbola faunae insectorum Helvetiae exhibentia vel species novas vel nondum depicta. Fasciculus primus. Cum tabulis 15 color. ad viv. expressis.* Basileae: J. G. Neukirch, vi + 7–48 pp., 15 pls. DP: [31 December] 1822 (Bousquet 2016).

Hagenbach, J. J. 1825: [new taxon]. In: Hoppe, D. H. & Hornschuch, F.: Insecta Coleoptrata, quae in itineribus, suis praesertim Alpinis. *Nova Acta Physico-Medica Academiae Caesareae Leopoldino-Carolinae Naturae Curiosorum* 12 [1824–1825] (2): 477–490, pl. XLV. DP: 23 April 1825 (*Allg. Verz. Bücher, Ostermesse* 1825: 5 and Evenhuis 2014).

Hajóss, J. 1938: Egy új Magyarországi Bruchidius. [One new Hungarian Bruchidius]. *Folia Entomologica Hungarica* 3 (1/4): 128–130. DP: 1 March 1938 (volume cover).

Halbherr, B. 1898: *Elenco sistematico dei coleotteri finora raccolti nella Valle Lagarina. Fascicolo X. Chrysomelidae – Coccinellidae. Epilogo – Bibliografia – Indice alfabetico delle famiglie, dei generi e sottogeneri.* Rovereto: Tipografia Roveretana, 79 pp. DP: 30 November 1898 (Bousquet 2016).

Halbherr, B. 1908: Aggiunte all'elenco sistematico dei coleotteri finora raccolti nella Valle Lagarina. *Publicazione della Società "Museo Civico in Rovereto"* 45: 1–41. DP: [31] December 1908 (*Natur. Novit.* 1908, 30 (24): 626).

Hänel, K. 1911: Beiträge zur Fauna Saxonica. *Entomologische Blätter* 7 (5/6): 124–125. DP: 3 June 1911 (issue title page).

Hänel, K. 1912: Beiträge zur Fauna Saxonica. *Entomologische Blätter* 8 (3): 90–91. DP: 20 March 1912 (issue title page).

Hänel, K. 1934a: Cryptocephalus violaceus a. nov. viridicollis. *Entomologische Blätter* 30 (2): 90. DP: 30 April 1934 (issue cover).

Hänel, K. 1934b: Chrysochloa speciosissima a. nov. Fothi. *Entomologische Blätter* 30 (2): 90. DP: 30 April 1934 (issue cover).

Hänel, K. 1937: Katalog der Chrysochloa-Arten und ihrer Formen nach dem Kennwort-System dargestellt. *Koleopterologische Rundschau* **23** (1/2): 25–34. DP: 31 March 1937 (issue cover).

Harold, E. von 1872: Geänderte Namen. *Coleopterologische Hefte* **9/10**: 254. DP: [31 December] 1872 (volume title page).

Harold, E. von 1874a: Palesida, neue Eumolpiden-Gattung. *Berliner Entomologische Zeitschrift* 18 (1/2): 23–24. DP: [30] April 1874 (volume table of contents).

Harold, E. von 1874b: Geänderte Namen. *Coleopterologische Hefte* **12**: 152. DP: 1 June 1874 (*Proc. Ent. Soc. London* 1874: xiii).

Harold, E. von 1874c: Chrysomelidae (Pars I.). In: Gemminger, M. & Harold, E. von: *Catalogus coleopterorum hucusque descriptorum synonymicus et systematicus. Tom. XI.* Monachii: G. Beck, 3233–3478 + [4] pp. DP: 26 September 1874 (Bousquet 2016).

Harold, E. von 1875a: Correspondenz. *Entomologische Zeitung* (Stettin) **36** (4/6): 176. DP: [30] April 1875 (last page of the issue).

Harold, E. von 1875b: Geänderte Namen. *Coleopterologische Hefte* **13**: 185. DP: 7 June 1875 (*Proc. Ent. Soc. London* 1875: xiv).

Harold, E. von 1875c: Diagnosen neuer Arten. *Coleopterologische Hefte* **14**: 137–140. DP: [31 December] 1875 (volume title page).

Harold, E. von 1876a: Diagnosen neuer Arten. *Coleopterologische Hefte* **15**: 118–124. DP: 1 November 1876 (*Ent. Nachricht.* 1876, 2 (11): 176).

Harold, E. von 1876b: Geänderte Namen. *Coleopterologische Hefte* **15**: 174. DP: 1 November 1876 (*Ent. Nachricht.* 1876, 2 (11): 176).

Harold, E. von 1876c: Chrysomelidae (Pars II.). Pp. 3479–3676. In: Gemminger, M. & Harold, E. von: *Catalogus coleopterorum hucusque descriptorum synonymicus et systematicus. Vol. XII: Chrysomelidae (Pars II), Languriidae, Erotylidae, Endomychidae, Coccinellidae, Corylophidae, Platypsyllidae.* Monachii: Theodor Ackermann, 3479–3676 + [2] + lxxiii pp. DP: 18 November 1876 (Bousquet 2016).

Harold, E. von 1877a: Beiträge zur Käferfauna von Japan. (Zweites Stück). Japanische Käfer des Berliner Königl. Museums. *Deutsche Entomologische Zeitschrift* **21** (2): 337–367. DP: [31] December 1877 (volume table of contents).

Harold, E. von 1877b: Beschreibungen einiger Oedionychis-Arten. *Deutsche Entomologische Zeitschrift* **21** (2): 433–434. DP: [31] December 1877 (volume table of contents).

Harold, E. von 1877c: Coleopterorum species novae. *Mittheilungen des Münchener Entomologischen Vereins* **1** (1): 97–111. DP: [31 December] 1877 (issue title page).

Harold, E. von 1878: Beiträge zur Käferfauna von Japan. (Viertes Stück). Japanische Käfer des Berliner Königl. Museums. *Deutsche Entomologische Zeitschrift* **22** (1): 65–88. DP: [31] May 1878 (volume table of contents).

Harold, E. von 1879a: Neue Coleopteren aus dem Berliner Museum. *Entomologische Zeitung* (Stettin) **40** (7/9): 329–336. DP: [31] July 1879 (last page of the issue).

Harold, E. von 1879b: Nachtrag zu Stück IV. der Beiträge zur Käferfauna von Japan. *Deutsche Entomologische Zeitschrift* **23** (2): 366–368. DP: [30] November 1879 (volume table of contents).

Harold, E. von 1879c: Diagnosen neuer Arten. *Coleopterologische Hefte* **16**: 225–232. DP: [31] August 1879 (*Tijd. Ent.* 1884, 24: xlix).

Harold, E. von 1880a: Ueber ostindische Galeruciden. *Entomologische Zeitung* (Stettin) **41** (4/6): 142–157. DP: [30] April 1880 (last page of the issue).

Harold, E. von 1880b: [new taxon]. In: Leder, H.: Beitrag zur kaukasischen Käfer-Fauna. Unter Mitwirkung hervorragender Fachgenossen. *Verhandlungen der Kaiserlich-Königlichen Zoologisch-Botanischen Gesellschaft in Wien* **29** [1879] (2): 451–488. DP: 15 April 1880 (Evenhuis 2023).

Harrer, G. A. 1784: *Beschreibung derjenigen Insecten welche Herr D. Jacob Christoph Schäffer in CCLXXX ausgemahlten Kupfertafeln unter dem Titel Icones insectorum circa Ratisbonam indigenorum ehemals in drey Theilen herausgegeben hat. I. Theil Hartschaalige Insecten.* Regensburg: Keyser, [30] + 328 pp. DP: 26 April 1784 (Bousquet 2017).

Harrer, G. A. 1791: *Beschreibungen zu des Herrn D. Iacob Christian Schaeffers natürlich ausgemahlten Abbildungen regensburgischer Insecten. Erster Band.* Regensburg: Montag und Weiss, [3] + xii + 144 pp. DP: 15 May 1791 (Bousquet 2016).

Harris, T. W. 1851: Insects on the potato-vine. *Journal of Agriculture* **1**: 99–102. DP: 3 September 1851 (*Bibl. Imp. Contr. Amer. Econ. Ent.* 1895, 4: 126).

Harrison, S. & Harrisson, T. 2022: Zeugophora flavicollis (Marsham) (Megalopodidae) in Cambridgeshire, with comments on the taxonomy of Z. frontalis Suffrian. *Coleopterist* **31** (1): 19–20. DP: [31] May 2022 (footer on p. 20).

Hasegawa, T. 1980: A taxonomic revision of the Chrysolina angusticollis-complex in Japan (Coleoptera, Chrysomelinae). *Insecta Matsumurana, New Series* **21**: 1–17. DP: [31] July 1980 (paper title page).

Hasenkamp, R. & Wagner, T. 2000: Revision of Afromaculepta gen. n., a monophyletic group of Afrotropical galerucine leaf beetles (Coleoptera: Chrysomelidae). *Insect Systematics & Evolution* **31** (1): 3–26. DP: [30] April 2000 (paper title page).

Hatch, M. H. 1935: A new sub-alpine genus of Halticini from North America. (Coleop.: Chrysomelidae). *Entomological News* **46** (10): 276–278. DP: 27 December 1935 (*Ent. News* 1936, 47 (1): 28).

Hatch, M. H. 1939: List of the Coleoptera taken by Mr. George P. Engelhardt in Alaska in 1938. *Bulletin of the Brooklyn Entomological Society, New Series* 34 (1): 45–50. DP: 27 February 1939 (issue cover).

Havelka, J. 1955: I. Beitrag zur Kenntnis der Gattung Galeruca Geoffroy (Col., Chrysomelidae). *Acta Societatis Entomologicae Čechosloveniae* **51** [1954]: 115–118. DP: [31 December] 1955 (volume title page).

Havelka, J. 1956: II. Beitrag zur Kenntnis der Gattung Galeruca Geoffroy (Coleoptera, Galerucidae). *Tijdschrift voor Entomologie* **99** (1/2): 47–54. DP: 12 May 1956 (volume title page).

Havelka, J. 1958: III. Beitrag zur Kenntnis der Gattung Galeruca Geoffr. (Coleoptera, Chrysomelidae). *Annales Historico-Naturales Musei Nationalis Hungarici* 50: 195–208. DP: 31 December 1958 (Merkl et al. 2008).

Havelka, J. 1960: Abhandlung über die Gattung Arima Chapuis (Col., Galerucidae). *Memorie della Società Entomologica Italiana* 38 [1959]: 97–136. DP: 27 January 1960 (table of contents).

Hayashi, M. 2000: Description of Donacia tominagai sp. nov. from Hokkaido, Japan, and taxonomic notes on its allies (Coleoptera: Chrysomelidae: Donaciinae). *Entomological Review of Japan* 55 (1): 5–13. DP: 30 June 2000 (paper title page).

Hayashi, M. 2020: Revision of extant and fossil Donaciinae (Coleoptera: Chrysomelidae) of Japan. *Japanese Journal of Systematic Entomology, Supplementary Series* 2: 61–160. DP: 30 April 2020 (paper title page).

Hayashi, M. & Lee, C.-F. 2007: A new Donacia from Taiwan, with a new record of Donacia frontalis Jacoby (Coleoptera, Chrysomelidae, Donaciinae). *Elytra* 35 (2): 551–557. DP: 3 November 2007 (paper title page).

Hayashi, M. & Lee, C.-F. 2009: Description of a new species, Donacia lungtanensis (Coleoptera: Chrysomelidae: Donaciinae) from Taiwan. *Entomological Review of Japan* 64 (2): 153–159. DP: 30 December 2009 (paper title page).

Hayashi, M. & Pütz, A. 2008: A new species of Donacia Fabricius, 1775 from China (Coleoptera: Chrysomelidae, Donaciinae). *Entomologische Zeitschrift* 118 (6): 243–248. DP: 16 December 2008 (table of contents).

Hayashi, M. & Sota, T. 2013: Quaternary donaciine beetles (Coleoptera, Chrysomelidae) in Japan: colonization and divergence patterns inferred from fossil and molecular data. *Quaternary International* 341 [2014]: 255–266. DP: 16 September 2013 (paper title page).

Heer, O. 1836: *Observationes entomologicae continentes metamorphoses coleopterorum nonnullorum adhuc incognitas*. Turici: Orellii, Fuesslini & Sociorum, 36 pp., 6 pls. DP: 24 April 1836 (Bousquet 2016).

Heer, O. 1845: Ueber die obersten Gränzen des thierischen und pflanzlichen Lebens in den Schweizer-Alpen. *An die Zürcherische Jugend* (Zürich) 47: 1–19, 1 pl. DP: [31 December] 1845 (paper title page).

Heikertinger, F. 1909a: Beschreibungen zweier neuer Halticinen der europäischen Fauna. *Verhandlungen der Kaiserlich-Königlichen Zoologisch-Botanischen Gesellschaft in Wien* 59 (1/2): (9)–(16). DP: 15 March 1909 (volume title page).

Heikertinger, F. 1909b: Diagnosen von Halticinen. *Verhandlungen der Kaiserlich-Königlichen Zoologisch-Botanischen Gesellschaft in Wien* 59 (7/8): (290)–(295). DP: 5 November 1909 (volume title page).

Heikertinger, F. 1910a: Beschreibungen zweier neuer Halticinen der Fauna Europas. *Verhandlungen der Kaiserlich-Königlichen Zoologisch-Botanischen Gesellschaft in Wien* 59 [1909] (10): (361)–(364). DP: 20 January 1910 (volume title page).

Heikertinger, F. 1910b: Beschreibungen eines neuen Subgenus und einer neuen Halticinenspezies der paläarktischen Fauna. *Verhandlungen der Kaiserlich-Königlichen Zoologisch-Botanischen Gesellschaft in Wien* 59 [1909] (10): (369)–(372). DP: 20 January 1910 (volume title page).

Heikertinger, F. 1910c: Die Halticinengattung Orestia Germ. vom systematischen und geographischen Standpunkte. *Verhandlungen der Kaiserlich-Königlichen Zoologisch-Botanischen Gesellschaft in Wien* 60 (2/3): (42)–(48) & (4/5): (49). DP (pp. (42)–(48)): 30 April 1910; (p. (49)): 25 June 1910 (volume title page).

Heikertinger, F. 1910d: Beschreibungen von drei neuen, beziehungsweise wenig gekannten Halticinenformen der Balkanhalbinsel. *Verhandlungen der Kaiserlich-Königlichen Zoologisch-Botanischen Gesellschaft in Wien* 60 (4/5): (52)–(55). DP: 25 June 1910 (volume title page).

Heikertinger, F. 1911a: Beschreibungen von vier neuen Aphthona-Formen aus dem paläarktischen Asien. *Verhandlungen der Kaiserlich-Königlichen Zoologisch-Botanischen Gesellschaft in Wien* 61 (1/2): (4)–(11). DP: 24 March 1911 (volume title page).

Heikertinger, F. 1911b: Beschreibungen einiger neuen Psylliodes-Formen der paläarktischen Fauna. *Verhandlungen der Kaiserlich-Königlichen Zoologisch-Botanischen Gesellschaft in Wien* 61 (1/2): (16)–(24). DP: 24 March 1911 (volume title page).

Heikertinger, F. 1911c: Beobachtungen über geographische Einflüsse auf Form und Färbung bei Halticinen (Col.). Nebst synonymischen Bemerkungen und Neubeschrebungen. I. Kleinasiatische Dibolia-Arten. *Deutsche Entomologische National-Bibliothek* 2 (15): 118–120. DP: 1 August 1911 (issue title page).

Heikertinger, F. 1911d: Die Halticinengattung Ochrosis Foudr. *Archiv für Naturgeschichte* 77 (1) (1. Supplementheft): 19–34. DP: [30] September 1911 (issue cover).

Heikertinger, F. 1911e: Die Halticinengattung Neocrepidodera (nov. gen.). *Archiv für Naturgeschichte* 77 (1) (1. Supplementheft): 34–38. DP: [30] September 1911 (issue cover).

Heikertinger, F. 1911f: Beschreibungen von vier neuen Phyllotreta-Formen aus der paläarktischen Region. *Verhandlungen der Kaiserlich-Königlichen Zoologisch-Botanischen Gesellschaft in Wien* 61 (7/8): (159)–(160) & (9/10): (161)–(168). DP (pp. (159)–(160)): 28 November 1911, (pp. (161)–(168)): 28 February 1912 (volume title page).

Heikertinger, F. 1912a: Notizen zur Halticinengattung Podagrica Foudr. *Societas Entomologica* 27 (4): 17–18. DP: 17 February 1912 (issue tite page).

Heikertinger, F. 1912b: Beschreibungen von vier neuen Phyllotreta-Formen aus der paläarktischen Region. *Verhandlungen der Kaiserlich-Königlichen Zoologisch-Botanischen Gesellschaft in Wien* 61 [1911] (7/8): (159)–(160) & (9/10): (161)–(168). DP (pp. (159)–(160)): 28 November 1911, (pp. (161)–(168)): 28 February 1912 (volume title page).

Heikertinger, F. 1912c: Skizzen zur Systematik und Nomenklatur der paläarktischen Halticinen. *Entomologische Blätter* 8 (3): 65–70. DP: 20 March 1912 (issue title page).

Heikertinger, F. 1912d: Diagnosen neuer paläarktischer Halticinen. *Verhandlungen der Kaiserlich-Königlichen Zoologisch-Botanischen Gesellschaft in Wien* 62 (2): (44)–(48). DP: 30 April 1912 (volume title page).

Heikertinger, F. 1912e: Notizen zur Halticinengattung Minota Kutsch. *Societas Entomologica* 27 (13): 60–61. DP: 29 June 1912 (issue title page).

Heikertinger, F. 1912f: Biogeographische Skizzen über paläarktische Halticinen (Col.). *Entomologische Mitteilungen* 1 (8): 236–241. DP: 1 August 1912 (issue title page).

Heikertinger, F. 1912g: Über dunkle Extremitätenfärbungen bei asiatischen Formen der Halticinengattung Chalcoides Foudr. *Berliner Entomologische Zeitschrift* 57 (1/2): 103–106. DP: [30] November 1912 (verso of the volume title page).

Heikertinger, F. 1912h: Biogeographische Skizzen über paläarktische Halticinen (Col.) (Fortsetzung). *Entomologische Mitteilungen* 1 (12): 383–388. DP: 1 December 1912 (issue title page).

Heikertinger, F. 1912i: Skizzen zur Systematik und Nomenklatur der paläarktischen Halticinen. *Entomologische Blätter* 8 (12): 291–297. DP: 7 December 1912 (issue title page).

Heikertinger, F. 1912j: Die einheimischen Kohlerdflöhe. Eine kritische Darstellung der in Mitteleuropa an gebauten Cruciferen schädlichen Halticinenarten. *Centralblatt für Bakteriologie, Parasitenkunde und Infektionkrankheiten, Zweite Abteilung* **36** [1913] (1/5): 98–127. DP: 14 December 1912 (issue title page).

Heikertinger, F. 1913a: Skizzen zur Systematik und Nomenklatur der paläarktischen Halticinen. *Entomologische Blätter* 9 (1/2): 27–32. DP: 18 January 1913 (issue title page).

Heikertinger, F. 1913b: Skizzen zur Systematik und Nomenklatur der paläarktischen Halticinen. *Entomologische Blätter* 9 (3/4): 67–71. DP: 20 March 1913 (issue title page).

Heikertinger, F. 1913c: [new taxon]. In: Paganetti-Hummler, G.: Beitrag zur Halticinenfauna des nordwestlichen Spanien. *Entomologische Mitteilungen* 2 (5): 139–144. DP: 1 May 1913 (issue title page).

Heikertinger, F. 1913d: Skizzen zur Systematik und Nomenklatur der paläarktischen Halticinen. *Entomologische Blätter* 9 (5/6): 145–149. DP: 10 May 1913 (issue title page).

Heikertinger, F. 1913e: Skizzen zur Systematik und Nomenklatur der paläarktischen Halticinen. *Entomologische Blätter* 9 (7/8): 180–182. DP: 31 July 1913 (issue title page).

Heikertinger, F. 1913f: Unterfamilie: Halticinae. Pp. 143–212. In: Reitter, E.: *Fauna Germanica. Die Käfer des Deutschen Reiches. IV. Band.* [1912], Stuttgart: K. G. Lutz, 236 pp., pls. 129–152. DP: [31] December 1913 (*Natur. Novit.* 1913, 35 (24): 612, see also *Ent. Bl.* 1914, 10: 127).

Heikertinger, F. 1914: Neue oder interessante Psylliodes-Formen der Fauna Niederösterreichs. *Verhandlungen der Kaiserlich-Königlichen Zoologisch-Botanischen Gesellschaft in Wien* **64** (3/4): (95)–(108). DP: 30 April 1914 (volume title page).

Heikertinger, F. 1915: *Monographie der Koleopterengattung Derocrepis Weise (Chrysomelidae, Halticini)*. München: C. Wolf & Sohn, 271–343 pp. DP: [31] January 1915 (see the comments in Dating of publications) [published only as a reprint; respective volume 4 of the *Münchener Koleopterologische Zeitschrift* was never published].

Heikertinger, F. 1916: Zur Kenntnis der Halticinengattung Psylliodes (Col. Chrysomelidae). *Entomologische Blätter* 12 (1/3): 29–47. DP: 31 March 1916 (issue title page).

Heikertinger, F. 1918: Über künstliche Abänderung der Färbungen toter Insekten. *Koleopterologische Rundschau* 7 [1918–1919] (1/2): 5–8. DP: 30 April 1918 (issue cover).

Heikertinger, F. 1921a: Bestimmungstabelle der Halticinengattung Psylliodes aus dem paläarktischen Gebiete mit Ausschluß Japans und der Kanarischen Inseln. I. Die ungeflügelten Arten (52. Beitrag zur Kenntnis der Halticinen). *Koleopterologische Rundschau* 9 (1/3): 39–48 & (4/9): 49–62. DP (pp. 39–48): 30 March 1921, (pp. 49–62): 30 September 1921 (issue covers).

Heikertinger, F. 1921b: Die paläarktischen Arten der Halticinengattung Batophila Foudr. (Glyptina Lec.). (54. Beitrag zur Kenntnis der Halticinen). *Koleopterologische Rundschau* 9 (10/12): 89–98. DP: 30 December 1921 (issue cover).

Heikertinger, F. 1922: *Monographie der Halticinengattung Derocrepis Weise (Coleopt., Chrysomelidae)*. Wien: Selbstverlag des Verfassers, 73 pp. [271–343]. DP: [31 December] 1922 (Heikertinger 1925b, Strouhal 1955).

Heikertinger, F. 1923a: Zur Kenntnis der Halticinengattung Crepidodera (Col. Chrysomel.). I. Die braunen Crepidodera-Formen Europas. (55. Beitrag zur Kenntnis der Halticinen). *Wiener Entomologische Zeitung* 40 (1/4): 121–123. DP: 25 April 1923 (footers throughout the issue).

Heikertinger, F. 1923b: Zur Kenntnis der Halticinengattung Crepidodera (Col. Chrysomel.). II. Systematische und verbreitungsgeographische Bemerkungen. – III. Die Gattung Crepidomorpha und ihre nächsten Verwandten. (55. Beitrag zur Kenntnis der Halticinen). *Wiener Entomologische Zeitung* 40 (5/10): 129–140. DP: 15 December 1923 (footers throughout the issue).

Heikertinger, F. 1924a: Die Halticinengenera der Palaearktis und Nearktis. Bestimmungstabellen. (Monographie der palaearktischen Halticinen: Systematischer Teil. – Zweites Stück.). *Koleopterologische Rundschau* 11 (1/2): 25–48. DP: 10 October 1924 (issue cover).

Heikertinger, F. 1924b: Monographie der Halticinengattung Orestia Germ. (Col., Chrysom.). (Monographie der paläearktischen Halticinen: Systematischer Teil. – Viertes Stück) (Zugleich 57. Beitrag zur Kenntnis der Halticinen). *Verhandlungen der Zoologisch-Botanischen Gesellschaft in Wien* 74/75 [1924–1925]: 56–125. DP: [31] October 1924 (notice on p. 125).

Heikertinger, F. 1925a: Die Halticinengenera der Palaearktis und Nearktis. Bestimmungstabellen. (Monographie der palaearktischen Halticinen: Systematischer Teil. – Zweites Stück.) *Koleopterologische Rundschau* 11 (3/4): 49–70. DP: 30 March 1925 (issue cover).

Heikertinger, F. 1925b: Monographie der Halticinengattung Derocrepis Weise (Coleopt., Chrysomelidae). (Monographie der paläarktischen Halticinen: Systematischer Teil. – Zweite Auflage des Dritten Stücks.) (Zugleich 60. Beitrag zur Kenntnis der Halticinen.). *Wiener Entomologische Zeitung* 42 (4/10): 95–178. DP: 15 November 1925 (footers throughout the issue).

Heikertinger, F. 1926: Bestimmungstabelle der Halticinengattung Psylliodes aus dem paläarktischen Gebiete. II. Die hellfarbigen Arten. (61. Beitrag zur Kenntnis der Halticinen). *Koleopterologische Rundschau* 12 (2): 101–112 & (3/4): 113–138. DP (pp. 101–112): 24 April 1926, (pp. 113–138): 31 July 1926 (issue covers).

Heikertinger, F. 1940: [new names]. In: Heikertinger, F. & Csiki, E: Chrysomelidae: Halticinae. In: Schenkling, S. (ed.): *Coleopterorum Catalogus. Pars 169*. W. Junk: s-Gravenhage, 337–635 pp. DP: 15 August 1940 (book cover).

Heikertinger, F. 1941a: Bestimmungs-Tabellen europäischer Käfer (7. Stück). LXXXII. Fam. Chrysomelidae. 5. Subfam. Halticinae. 1. Gatt. Phyllotreta Steph. Bestimmungstabelle der paläarktischen Phyllotreta-Arten. *Koleopterologische Rundschau* 27 (1/3): 15–64. DP: 15 September 1941 (issue cover).

Heikertinger, F. 1941b: Bestimmungs-Tabellen europäischer Käfer (7. Stück). LXXXII. Fam. Chrysomelidae. 5. Subfam. Halticinae. 1. Gatt. Phyllotreta Steph. Bestimmungstabelle der paläarktischen Phyllotreta-Arten (Teil II). *Koleopterologische Rundschau* 27 (4/6): 69–116. DP: 15 November 1941 (issue cover).

Heikertinger, F. 1942: Die Phyllotreta-Arten des indoaustralischen Faunengebietes (Col. Chrysom. Haltic.). *Entomologische Blätter* 38 (4): 138–144 & (5/6): 145–152. DP (pp. 138–144): 30 September 1942, (pp. 145–152): 31 December 1942 (issue covers).

Heikertinger, F. 1944: Bestimmungstabellen europäischer Käfer (10. Stück). LXXXII. Fam. Chrysomelidae. 5. Subfam. Halticinae. 2. Gatt. Aphthona Chevr. Bestimmungstabelle der paläarktischen Aphthona-Arten. Mit einem Anhang, betreffend die Aphthona-Arten des orientalischen Faunen-gebietes, sowie Ergänzungen zur Bestimmungstabelle der paläarktischen Phyllotreta-Arten. *Koleopterologische Rundschau* 30 (1/3): 37–80 & (4/6): 81–124. DP (pp. 37–80): 31 July 1944, (pp. 81–124): 30 September 1944 (issue covers).

Heikertinger, F. 1948: Bestimmungstabellen europäischer Käfer (11. Stück). LXXXII. Fam. Chrysomelidae. 5. Subfam. Halticinae. Abteilung II. 11.–30. Gattung: Crepidodera-Verwandtschaft weitesten Sinnes. Bestimmungstabelle der paläarktischen Arten der Halticinengattungen: Crepidodera (Chevr.) Steph., Derocrepis Weise (mit Aeschrocnemis Weise), Ochrosis Foudr. (mit Lythraria Bedel und Neocrepidodera Heiktgr.), Orestia Germ., Hippuriphila Foudr., Chalcoides Foudr., Epithrix Foudr., Arrhenocoela Foudr., Minota Kutsch., Cardax Weise, Batophila Foudr., sowie der ostpaläarktischen Arten der asiatischen Gattungen Xuthea Baly, Novofoudrasia Jacobs., Pseudodera Baly, Crepidomorpha Heiktgr., Phygasia Baly (Scallodera Har.), Lipromima Heiktgr. und Manobia Jacoby. Mit Berücksichtigung der Arten der orientalischen Grenzgebiete und des holarktischen Nordamerika. *Koleopterologische Rundschau* 31 [1945–1950] (1/3): 15–80. DP: 15 April 1948 (issue cover).

Heikertinger, F. 1950: Bestimmungstabellen europäischer Käfer (11. Stück). LXXXII. Fam. Chrysomelidae. 5. Subfam. Halticinae. Abteilung II. 11.–30. Gattung: Crepidodera-Verwandtschaft weitesten Sinnes. Bestimmungstabelle der paläarktischen Arten der Halticinengattungen: Crepidodera (Chevr.) Steph., Derocrepis Weise (mit Aeschrocnemis Weise), Ochrosis Foudr. (mit Lythraria Bedel und Neocrepidodera Heiktgr.), Orestia Germ., Hippuriphila Foudr., Chalcoides Foudr., Epithrix Foudr., Arrhenocoela Foudr., Minota Kutsch., Cardax Weise, Batophila Foudr., sowie der ostpaläarktischen Arten der asiatischen Gattungen Xuthea Baly, Novofoudrasia Jacobs., Pseudodera Baly, Crepidomorpha Heiktgr., Phygasia Baly (Scallodera Har.), Lipromima Heiktgr. und Manobia Jacoby. Mit Berücksichtigung der Arten der orientalischen Grenzgebiete und des holarktischen Nordamerika. *Koleopterologische Rundschau* 31 [1945–1950] (4/6): 81–139. DP: 30 June 1950 (issue cover).

Heikertinger, F. 1951: Bestimmungstabellen europäischer Käfer (12. Stück). LXXXII. Fam. Chrysomelidae. 5. Subfam. Halticinae. Abteilung II. Bestimmungstabellen der paläarktischen Arten der Gattungen Podagrica Foudr., Mantura Steph. und Chaetocnema Steph. *Koleopterologische Rundschau* **32** [1951–1954] (1/3): 1–84. DP: 31 August 1951 (issue cover).

Heikertinger, F. & Csiki, E. 1939: Chrysomelidae: Halticinae I. In: Schenkling, S. (ed.): *Coleopterorum Catalogus. Pars. 166.* W. Junk: s-Gravenhage, 1–336 pp. DP: 14 October 1939 (book cover).

Heinze, E. 1927a: 1. Beitrag zur Kenntnis der Criocerinen (Col., Chrysomelid.). Über einige Afrikaner nebst synonymischen Bemerkungen. *Entomologische Mitteilungen* **16** (2): 138–142. DP: 1 March 1927 (issue cover).

Heinze, E. 1927b: Drei neue Criocerinen-Gattungen, drei neue Lema-Arten und einige Bemerkungen über bekannte Lema-Arten von Afrika. (3. Beitrag zur Kenntnis der Criocerinen (Col., Chrysomel.)). *Entomologische Blätter* **23** (4): 161–170. DP: 31 December 1927 (issue cover).

Heinze, E. 1928: Bemerkungen über afrikanische Criocerinen aus Anlass einer Durchsicht des Materials im Berliner Zoologischen Museum, nebst Beschreibung neuer Arten (4. Beitrag zur Kenntnis der Criocerinen (Col., Chrysomel.)). *Mitteilungen aus dem Zoologischen Museum in Berlin* **14** (1): 43–81. DP: [31] May 1928 (issue title page).

Heinze, E. 1929: Übersicht der Arten des afrikanischen Festlandes der Gattung Hapsidolema Heinze. (12. Beitrag zur Kenntnis der Criocerinen (Col. Chrysomel.)). *Deutsche Entomologische Zeitschrift* **1929** (4): 289–297. DP: 30 November 1929 (issue cover).

Heinze, E. 1930: Über afrikanische Criocerinen, vorzugsweise aus dem Kongo-Museum Tervueren. (17. Beitrag zur Kenntnis der Criocerinen (Col. Chrysomelidae). *Revue de Zoologie et de Botanique Africaines* **20** (1): 23–55. DP: 15 November 1930 (paper title page).

Heinze, E. 1931: Über neue und bekannte afrikanische Criocerinen, großenteils aus Londoner Museen. (19. Beitrag zur Kenntnis der Criocerinen, Col., Chrysomel.). *Wiener Entomologische Zeitung* **48** (4): 175–213. DP: 25 December 1931 (footers throughout the paper).

Heinze, E. 1941: Über Bradylema femorata Guér. und quadripunctata Ol. (27. Beitrag zur Kenntnis der Criocerinen (Col.; Chrysomel.)). *Entomologische Blätter* **37** (5/6): 207–209. DP: 31 December 1941 (issue cover).

Heinze, E. 1942: Über alte und neue Criocerinen der Alten Welt (Coleoptera: Chrysomelidae). (28. Beitrag zur Kenntnis der Criocerinae). *Arbeiten über Morphologische und Taxonomische Entomologie aus Berlin-Dahlem* **9** (1): 47–61. DP: 25 March 1942 (volume title page).

Heinze, E. 1943a: Über australische Criocerinen. 30. Beitrag zur Kenntnis der Criocerinen (Col., Chrysomelidae). *Entomologische Blätter* **39** (1/2): 22–28. DP: 30 April 1943 (issue cover).

Heinze, E. 1943b: Über bekannte und neue Criocerinen. (29. Beitrag zur Kenntnis der Criocerinen (Col., Chrysomel.)). *Stettiner Entomologische Zeitung* **104** (1/2): 101–109. DP: 1 May 1943 (issue cover).

Heisswolf, A., Ulmann, S., Obermaier, E., Mitesser, O. & Poethke, H. J. 2007: Host plant finding in the specialised leaf beetle Cassida canaliculata: an analysis of small-scale movement behaviour. *Ecological Entomology* **32**: 194–200. DP: 30 January 2007 (last page of the paper).

Hellén, W. 1922: Koleopterologische Mitteilungen aus Finland II. *Notulae Entomologicae* **1** [1921] (4): 97–103. DP: 18 January 1922 (volume table of contents).

Hellén, W. 1933: Koleopterologische Mitteilungen aus Finnland X. *Notulae Entomologicae* **13** (1/2): 37–40. DP: 8 May 1933 (verso of the volume title page).

Hellén, W. 1937: Über die Haemonia-Arten Finnlands (Col.). *Notulae Entomologicae* **17** (1): 1–4. DP: 12 April 1937 (verso of the volume title page).

Heller, K. M. 1916: Über Hispinae aus Celebes. *Entomologische Blätter* **12** (4/6): 114–117. DP: 6 July 1916 (issue cover).

Heller, K. M. 1923: Die Coleopterenausbeute der Stötznerschen Sze-Tschwan-Expedition (1913–1915). *Entomologische Blätter* **19** (2): 61–79. DP: 30 June 1923 (issue cover).

Helliesen, T. 1912: La grupo Chrysomela sanguinolenta. *Stavanger Museums Aarshefte* **22** [1911]: 1–15. DP: [31 December] 1912 (volume title page).

Helliesen, T. 1913: La norvegaj specoj registrantaj al la subgenero Spartophila de la genro Phytodecta Kirby. *Stavanger Museums Aarshefte* **23** [1912]: 1–9. DP: [31 December] 1913 (volume title page).

Herbst, J. F. W. 1783: Kritisches Verzeichniss meiner Insektensammlung. *Archiv der Insectengeschichte* [J. C. Fuessly] **2** (4): 1–72, pls. 19–23. DP: [31 December] 1783 (issue title page, Griffin 1937).

Herbst, J. F. W. 1784: Kritisches Verzeichniss meiner Insektensammlung. Fortsetzung. *Archiv der Insectengeschichte* [J. C. Fuessly] **2** (5): 73–128, pls. 24–28. DP: 26 April 1784 (*Allg. Verz. Bücher, Ostermesse* 1784: 774 and Evenhuis 2014).

Herbst, J. F. W. 1786: Erste Mantisse zum Verzeichniss der ersten Klasse meiner Insektensammlung. *Archiv der Insectengeschichte* [J. C. Fuessly] **2** (7/8): 153–182, pls. 43–48. DP: 7 May 1786 (*Allg. Verz. Bücher, Ostermesse* 1786: 253 and Evenhuis 2014).

Herbst, J. F. W. 1799: *Natursystem aller bekannten in- und ausländischen Insekten, als eine Fortsetzung der von Buffonschen Naturgeschichte. Der Käfer achter Theil.* Berlin: Geh. Commerzien-Raths Pauli, xiv + 420 pp., pls. 117–137. DP: 14 April 1799 (Bousquet 2017).

Hering, M. 1924: Zur Kenntnis der Blattminenfauna des Banats. I. *Zeitschrift für Wissenschaftliche Insektenbiologie* **19** (1): 31–41. DP: 20 February 1924 (footer on p. 1924).

Herrich-Schäffer, G. A. W. 1830: *Die Fortsetzung von Panzer Faunae insectorum Germaniae initia. Heft 116*. Regensburg: Manz, 24 pls. DP: 3 October 1830 (Nagel & Grieder 2019).

Herrich-Schäffer, G. A. W. 1835a: *Die Fortsetzung von Panzer Faunae insectorum Germaniae initia. Heft 127*. Regensburg: Manz, 24 pls. DP: 14 February 1835 (Nagel & Grieder 2019).

Herrich-Schäffer, G. A. W. 1835b: *Die Fortsetzung von Panzer Faunae insectorum Germaniae initia. Heft 130*. Regensburg: Manz, 24 pls. DP: 6 June 1835 (Nagel & Grieder 2019).

Herrich-Schäffer, G. A. W. 1835c: *Die Fortsetzung von Panzer Faunae insectorum Germaniae initia. Heft 135*. Regensburg: Manz, 24 pls. DP: 6 June 1835 (Nagel & Grieder 2019).

Herrich-Schäffer, G. A. W. 1836: *Die Fortsetzung von Panzer Faunae insectorum Germaniae initia. Heft 143*. Regensburg: Manz, 24 pls. DP: 1 October 1836 (Nagel & Grieder 2019).

Herrich-Schäffer, G. A. W. 1838a: *Die Fortsetzung von Panzer Faunae insectorum Germaniae initia. Heft 156*. Regensburg: Manz, 24 pls. DP: 1 April 1838 (Nagel & Grieder 2019).

Herrich-Schäffer, G. A. W. 1838b: *Die Fortsetzung von Panzers Faunae insectorum Germaniae initia. Heft 157*. Regensburg: Manz, 24 pls. DP: 1 May 1838 (Nagel & Grieder 2019).

Herrich-Schäffer, G. A. W. 1838c: *Die Fortsetzung von Panzers Faunae insectorum Germaniae initia. Heft 164*. Regensburg: Manz, 24 pls. DP: 31 October 1838 (Nagel & Grieder 2019).

Heyden, L. F. J. D. von 1863: Beitrag zur Coleopterenfauna des Ober-Engadins, insbesondere der Umgegend von St. Moritz. *Jahresbericht der Naturforschenden Gesellschaft Graubündens, Neue Folge* **8** [1861–1862]: 1–52. DP: [30] November 1863 (*Ztschr. Ges. Naturwiss.* 1863, 22 (10/11): 443).

Heyden, L. F. J. D. von 1870: Entomologische Reise nach dem südlichen Spanien, der Sierra Guadarrama und Sierra Morena, Portugal und den Cantabrischen Gebirge. *Berliner Entomologische Zeitschrift* **14** (Beiheft): [1] + 218 pp., 2 pls. DP: [31 December] 1870 (Bousquet 2016).

Heyden, L. F. J. D. von 1877: Die Käfer von Nassau und Frankfurt. *Jahrbücher des Nassauischen Vereins für Naturkunde* **29/30** [1876–1877]: 55–412. DP: [31 December] 1877 (volume title page).

Heyden, L. F. J. D. von 1878: [new taxa]. In: Schneider, O. & Leder, H.: *Beiträge zur Kenntniss der kaukasischen Käferfauna*. Brünn: W. Burkart, 359 + [1] pp., 6 pls. DP: [31 December] 1878 (Döberl 2009, see also the comments in the Dating of publications chapter).

Heyden, L. F. J. D. von 1882a: Beitrag zur Coleopteren-Fauna der Sierra de Cordoba. *Deutsche Entomologische Zeitschrift* **26** (1): 43–48. DP: 8 March 1882 (*Bull. Soc. Ent. Fr.* 1882: 92).

Heyden, L. F. J. D. von 1882b: *Catalog der Coleopteren von Sibirien mit Einschluß derjenigen der turanischen Länder, Turkestans und der chinesischen Grenzgebiete. Mit specieller Angabe der einzelnen Fundorte in Sibirien und genauer Citirung der darauf bezüglichen einzelnen Arbeiten nach eigenem Vergleich, sowie mit besonderer Rücksicht auf die geographische Verbreitung der einzelnen Arten über die Grenzländer, namentlich Europa und Deutschland.* [1880–1881], Berlin: A. W. Schade, xxiv + 224 pp. DP (pp. 113–224): [30] March 1882 (Bousquet 2016).

Heyden, L. F. J. D. von 1883a: Phytodecta affinis und Verwandte. *Entomologische Nachrichten* **9** (4): 52–53. DP: 15 February 1883 (headings throughout the issue).

Heyden, L. F. J. D. von 1883b: Phytodecta, Coccinella, Adoxus. *Entomologische Nachrichten* **9** (4): 53. DP: 15 February 1883 (headings throughout the issue).

Heyden, L. F. J. D. von 1883c: Verzeichniss der von Herrn Dr. med. W. Kobelt in Nordafrika und Spanien gesammelten Coleopteren. *Bericht über die Senckenbergische Naturforschende Gesellschaft* **[1882/1883]**: 217–237. DP: [30] September 1883 (volume title page).

Heyden, L. F. J. D. von 1884: Beitrag zur Coleopterenfauna der Insel Askold und anderer Theile des Amurgebietes. *Deutsche Entomologische Zeitschrift* **28** (2): 273–300. DP: [30] November 1884 (table of contents).

Heyden, L. F. J. D. von 1886a: Ueber Phytodecta-Varietäten. *Deutsche Entomologische Zeitschrift* **30** (1): 85–86. DP: [30] June 1886 (table of contents).

Heyden, L. F. J. D. von 1886b: Zusammenstellung der von Herrn Dr. med. W. Kobelt von seiner Reise in der Provinzen Alger und Constantine, sowie von Tunis mitgebrachten Coleopteren. *Bericht über die Senckenbergische naturforschende Gesellschaft in Frankfurt am Main* [1886]: 35–57. DP: [30] September 1886 (volume title page).

Heyden, L. F. J. D. von 1886c: Zur Turkestan-Fauna. *Deutsche Entomologische Zeitschrift* 30 (2): 277. DP: [31] December 1886 (table of contents).

Heyden, L. F. J. D. von 1886d: Neue Käferarten aus Malatia (in südlichen Kleinasian). *Deutsche Entomologische Zeitschrift* 30 (2): 278–280. DP: [31] December 1886 (table of contents).

Heyden, L. F. J. D. von 1887a: Kleine coleopterologische Mittheilungen. *Wiener Entomologische Zeitung* 6 (3): 98. DP: 8 April 1887 (footers throughout the issue).

Heyden, L. F. J. D. von 1887b: Zweiter Beitrag zur Coleopteren-Fauna von Pecking in Nord-China. *Deutsche Entomologische Zeitschrift* **31** (2): 293–295. DP: [30] November 1887 (table of contents).

Heyden, L. F. J. D. von 1887c: XI. Beitrag zur Coleopteren-Fauna von Turkestan. *Deutsche Entomologische Zeitschrift* **31** (2): 305–336. DP: [30] November 1887 (table of contents).

Heyden, L. F. J. D. von 1887d: Verzeichniss der von Herrn Otto Herz auf der chinesischen Halbinsel Korea gesammelten Coleopteren. *Horae Societatis Entomologicae Rossicae* **21** (3/4): 243–273. DP: 27 December 1887 (Kerzhner 1984, Standfuss & Kerzhner 2004).

Heyden, L. F. J. D. von 1888: Neue und interessante Coleopteren aus Malatia in Mesopotamien. *Deutsche Entomologische Zeitschrift* **32** (1): 72–78. DP: [30] April 1888 (table of contents).

Heyden, L. F. J. D. von 1889a: Synonymische Bemerkungen über Coleoptera. *Wiener Entomologische Zeitung* 8 (5): 202–203. DP: 31 May 1889 (footers throughout the issue).

Heyden, L. F. J. D. von 1889b: Descriptiones Coleopterorum novorum Regionis Palaearcticae. *Deutsche Entomologische Zeitschrift* **1889** (2): 325–330. DP: [30] November 1889 (table of contents).

Heyden, L. F. D. J. von 1892a: Bemerkenswerthe Chrysomeliden-Varietäten. *Deutsche Entomologische Zeitschrift* [1891] (2): 368. DP: [31] March 1892 (table of contents).

Heyden, L. F. J. D. von 1892b: XIV. Beitrag zur Coleopteren-Fauna von Turkestan, Turkmenien und Süd-West-Sibirien. *Deutsche Entomologische Zeitschrift* **1892** (1): 99–104. DP: [31] August 1892 (table of contents).

Heyden, L. F. J. D. von 1892c: XV. Beitrag zur Coleopteren-Fauna von Turkestan, Turkmenien und Süd-West-Sibirien. *Deutsche Entomologische Zeitschrift* **1892** (1): 105–110. DP: [31] August 1892 (table of contents).

Heyden, L. F. J. D. von 1894: [new taxon]. In: Hauser, F.: Beiträge zur Coleopteren-Fauna von Transcaspien und Turkestan. *Deutsche Entomologische Zeitschrift* **1894** (1): 17–74. DP: 10 January 1894 (*Bull. Soc. Ent. Fr.* 1894: xiv).

Heyden, L. F. J. D. von 1895: *Catalog der Coleopteren von Siberien, mit Einschluss derjenigen des östlichen Caspi-Gebietes, von Turcmenien, Turkestan, Nord-Thibet und des Amur-Gebietes. Nachtrag I.* [1893–1895]. Berlin: Deutsche entomologische Gesellschaft, 217 pp. DP: 31 October 1895 (Heyden 1896: 2).

Heyden, L. F. J. D. von 1906: Die Varietäten der Crioceris asparagi L. und macilenta Wse. (Coleopt.). *Wiener Entomologische Zeitschrift* **25** (2/4): 123–126. DP: 15 March 1906 (footers throughout the issue).

Hincks, W. D. 1949: Some nomenclatorial notes on Chrysomelidae (Col.). No. 1, Galerucinae. *Annals and Magazine of Natural History, Twelfth Series* **2** (20): 607–622. DP: 7 October 1949 (Evenhuis 2003).

Hincks, W. D. 1950: Some nomenclatorial notes on Chrysomelidae (Col.). No. 3, Cassidinae. *Annals and Magazine of Natural History, Twelfth Series* **3** (30): 506–512. DP: 1 June 1950 (Evenhuis 2003).

Hincks, W. D. 1952: The genera of the Cassidinae (Coleoptera: Chrysomelidae). *Transactions of the Royal Entomological Society of London* **103** (10): 327–358. DP: 31 December 1952 (issue cover).

Hochhut, H. 1847: Enumeration der Rüsselkäfer, welche vom Baron Maximilian von Chaudoir und vom Baron A. v. Gotsch auf ihren Reisen im Kaukasus und in Transkaukasien im Jahre 1845 gesammelt wurden; nebst Beschreibung der neuentdeckten Arten. *Bulletin de la Société Impériale des Naturalistes de Moscou* **20** (2): 448–587. DP: 30 September 1847 (Bousquet & Davies 2021).

Hodge, P. J. 1997: Bruchidius varius (Olivier) (Chrysomelidae) new to the British Isles. *Coleopterist* **5** (3): 65–68. DP: [31] January 1997 (paper title page).

Hoffmann, A. 1938: Tableaux analytiques des Bruchides (Col.) du bassin de la Seine. *Bulletin de la Société des Sciences de Seine-et-Oise, 3e Série* **6** (1/3): 1–13. DP: [31 December] 1938 (paper title page).

Hoffmann, A. 1945: *Coléoptères Bruchides et Anthribides. Faune de France 44.* Paris: P. Chevalier, 184 pp. DP: [31 December] 1945 (book title page).

Hoffmann, A. 1950: Descriptions et observations concernant divers coléoptères phytophages de la région paléarctique. *Revue Française d'Entomologie* **17** (3): 192–199. DP: 15 October 1950 (footer on p. 189).

Hoffmann, A. 1953: Coléoptères phytophages nouveaux ou peu connus, de la faune paléarctique. *Revue Française d'Entomologie* **20** (3): 186–194. DP: 15 October 1953 (footer on p. 177).

Hoffmann, A. 1956: Curculionides nouveaux rapportés par la mission G. Remaudière en Iran. *Revue de Pathologie Végétale et d'Entomologie Agricole de France* **35** (4): 241–249. DP: [31] December 1956 (issue title page).

Hoffmann, A. 1961: Coléoptères phytophages inédits ou critiques de la région paléarctique. *Bulletin de la Société Entomologique de France* **66** (1/2): 32–39. DP: 24 June 1961 (*dates de publication* on p. 250).

Hoffmann, A. 1964: Description d'un Bruchidius nouveau de Crimée (Coleoptera: Bruchidae). *Annales Historico-Naturales Musei Nationalis Hungarici* **56**: 419–420. DP: 31 December 1964 (Merkl et al. 2008).

Hoffmann, A. 1965: Observations sur les Kytorrhinus et description d'une espèce inédite de la Mongolie centrale (Col. Bruchidae). *Annales de la Société Entomologique de France, Nouvelle Série* **1** (1): 63–70. DP: 18 February 1965 (last page of the issue).

Hoffmann, A. 1970: 142. Bruchidae II. Ergebnisse der zoologischen Forschungen von Dr. Z. Kaszab in der Mongolei (Coleoptera). *Reichenbachia* **11** (1): 1–3. DP: 9 September 1968 (paper title page).

Holdhaus, K. 1914: Eine neue Chrysomela aus den Ostalpen. *Verhandlungen der Kaiserlich-Königlichen Zoologisch-Botanischen Gesellschaft in Wien* **64** (5/6): (126)–(127). DP: 5 September 1914 (volume title page).

Hope, F. W. 1831: Synopsis of the new species of Nepaul insects in the collection of Major General Hardwicke. Pp. 21–32. In: Gray, J. E. (ed.): *Zoological Miscellany*. [1831–1844], London: Treuttel, Wurtz & Co., 86 pp., 4 pls. DP (pp. 1–40): 19 February 1831 (Dickinson et al. 2011).

Hope, F. W. 1841: *The coleopterist's manual, part the third, containing various families, genera and species, of beetles, recorded by Linneus and Fabricius. Also, descriptions of newly discovered and unpublished insects*. [1840], London: J. C. Bridgewater and Bowdery and Kerby, [5] + 191 pp., 3 pls. DP: [30] April 1841 (Low 1853, Bousquet 2016).

Hope, F. W. 1842a: On some rare and beautiful coleopterous insects from Silhet, the major part belonging to the collection of Frederic Parry, Esq., of Cheltenham. *Proceedings of the Linnean Society of London* **1** (14): 127–128. DP: 5 April 1842 (Kappel 1896).

Hope, F. W. 1842b: On some rare and beautiful Coleopterous insects from Silhet, the major part belonging to the collection of Frederic Parry, Esq., of Cheltenham. *Annals and Magazine of Natural History* **9** (57): 247–248. DP: 1 May 1842 (Evenhuis 2003).

Hope, F. W. 1843a: Descriptions of the coleopterous insects sent to England by Dr. Cantor from Chusan and Canton, with observations on the entomology of China. *Annals and Magazine of Natural History* **11** (67): 62–66. DP: 1 January 1843 (Evenhuis 2003).

Hope, F. W. 1843b: Descriptions of the coleopterous insects sent to England by Dr. Cantor from Chusan and Canton, with observations on the entomology of China. *Journal of Proceedings of the Entomological Society of London* [**1842**]: 59–64. DP: 1 January 1843 (Wheeler 1912).

Hope, F. W. 1845: On the entomology of China, with description of the new species sent to England by Dr. Cantor from Chusan and Canton. *Transactions of the Entomological Society of London* **4** (1): 4–17. DP: 7 April 1845 (Wheeler 1912).

Hoppe, D. H. 1795: *Enumeratio insectorum elytratorum circa Erlangam indigenarum observationibus iconibusqve illustrata. Dissertatio inauguralis quam summi numinis auspicio ex decreto gratiosae facultatis medicae pro summis in medicina honoribus legitime obtinendis publico eruditorum examini submittit David Henricus Hoppe*. Erlangae: Hilpertianis, 70 pp., 1 pl. DP: [30] April 1795 (Bousquet 2016).

Horion, A. 1939: Zur Käferfauna der Rheinprovinz. Nachtrag XVIII. *Entomologische Blätter* **35** (3): 129–142. DP: 30 June 1939 (table of contents).

Horn, G. H. 1873: Revision of the Bruchidae of the United States. *Transactions of the American Entomological Society* **4** [1872–1873] (4): 311–342. DP: [31 December] 1873 (Brown 1964).

Horn, G. H. 1885: Contribution to the coleopterology of the United States. (No. 4). *Transactions of the American Entomological Society and Proceedings of the Entomological Section of the Academy of Natural Sciences* **12**: 128–162, pls. IV–V. DP (pp. 128–160): [31] March 1885, (pp. 161–162): [30] April 1885 (footers throughout the volume).

Horn, G. H. 1889: A synopsis of the Halticini of boreal America. *Transactions of the American Entomological Society and Proceedings of the Entomological Section of the Academy of Natural Sciences* **16**: 163–320, pls. V–VII. DP (pp. 163–192): [31] May 1889, (pp. 193–264): [30] June 1889, (pp. 265–320): [31] July 1889 (footers throughout the volume).

Horn, G. H. 1893: The Galerucini of boreal America. *Transactions of the American Entomological Society* 20: 57–144, pl. I. DP (pp. 57–72): [30] April 1893, (pp. 73–88) [31] May 1893, (pp. 89–144) [30] June 1893 (footers throughout the volume).

Horn, G. H. 1894: The Coleoptera of Baja California. *Proceedings of the California Academy of Sciences, Second Series* 4 [1893–1894]: 302–449, pls. VII–VIII. DP: 3 August 1894 (footers throughout the volume).

Hornstedt, C. F. 1787: Beschreibung neuer Blatkäferarten. *Schriften der Gesellschaft Naturforschender Freunde zu Berlin* 8 (1): 1–8, 1 pl. DP: 29 April 1787 (*Allg. Verz. Bücher, Ostermesse* 1787: 605 and Evenhuis 2014).

Hubenthal, W. 1898: Ergänzungen zu Kellners Verzeichniß der Käfer Thüringens: 1. Chrysomelidae. *Aus der Heimat. Blätter der Vereinigung für Gothaische Geschichte und Alterthumsforschung* 1 [1897–1898] (6): 89–93. DP: 1 March 1898 (paper title page).

Hubenthal, W. 1911: Beschreibungen neuer europäischer Coleopteren und Bemerkungen zu bekannten Arten. *Entomologische Blätter* 7 (10/11): 185–193. DP: 31 October 1911 (issue title page).

Hummel, A. D. 1825: Novae species variorum. Pp. 58–71. In: Hummel, A. D.: *Essais Entomologiques No. IV*. St. Petersbourg: Chancellerie privée du Ministère de l'Intérieur, 71 pp. + index. DP: 1 May 1825 (Bousquet 2016).

Hummel, A. D. 1827: Insectes de 1826. Pp. 1–20. In: Hummel, A. D.: *Essais Entomologiques No. VI*. St. Petersbourg: Chancellerie privée du Ministère de l'Intérieur, 49 pp. DP: [31 December] 1827 (Bousquet 2016).

Iablokoff-Khnzorian, S. M. 1954: Novyy vid bloshek iz Armyanskoy SSR (Coleoptera, Chrysomelidae) Haltica armeniaca sp. n. [New species of flea-beetles from Armenian SSR (Coleoptera, Chrysomelidae) Haltica armeniaca sp. n.]. *Doklady Akademii Nauk Armyanskoy SSR* **19** (2): 59–60. SFP: 3 December 1954, DP: 27 December 1954 (registered at VKP).

Iablokoff-Khnzorian, S. M. 1956: Chetyre novykh vida zhestkokrylykh iz Armyanskoy SSR (Coleoptera, Insecta). [Four new species of beetles from Armenian SSR (Coleoptera, Insecta)]. *Doklady Akademii Nauk Armyanskoy SSR* **22** (3): 135–139. SFP: 3 May 1956, DP: [31 December] 1956 (issue title page).

Iabkokoff-Khnzorian S. M. 1957: Novye vidy zhestkokrylykh iz Armyanskoy SSR i Nakh. ASSR. [New species of beetles from the Armenian SSR and the Nakhichevan ASSR]. *Zoologicheskiy Sbornik Akademii Nauk Armyanskoy SSR* 10: 153–183. SFP: 26 July 1957, DP: 26 September 1957 (registered at ZIN library).

Iablokoff-Khnzorian, S. M. 1959: Neue Käfer aus Sovjetarmenien (Bruchidae et Curculionidae). *Notulae Entomologicae* 38 [1958] (3): 73–82. DP: 20 February 1959 (verso of the volume title page).

Iabkokoff-Khnzorian, S. M. 1960: Novye vidy zhestkokrylykh (Coleoptera, Insecta) iz Armyanskoy SSR i Nakhichevanskoy ASSR. II. [New species of beetles (Coleoptera, Insecta) from the Armenian SSR and the Nakhichevan ASSR. II]. *Zoologicheskiy Sbornik Akademii Nauk Armyanskoy SSR* 11 [1959]: 65–78. SFP: 12 January 1960, DP: 6 June 1960 (registered at ZIN library).

Iablokoff-Khnzorian, S. M. 1961: Coléoptères nouveaux de l'Arménie Soviétique. *Notulae Entomologicae* 40 [1960] (4): 140–153. DP: 20 January 1961 (verso of the volume title page).

Iablokoff-Khnzorian, S. M. 1962: Novye vidy zhestkokrylykh iz Zakavkazya. (New species of Coleoptera from Transcaucasus (Insecta – Coleoptera)). *Zoologicheskiy Sbornik Akademii Nauk Armyanskoy SSR* 12: 99–124. SFP: 30 March 1962, DP: 9 July 1962 (registered at ZIN library).

Iabkokoff-Khnzorian S. M. 1964: Novye rody i vidy zhestkokrylykh iz Zakavkazya i Sredney Azii. (New genera and species of Coleoptera from Transcaucasus and Middle Asia). *Zoologicheskiy Sbornik Akademii Nauk Armyanskoy SSR* 13: 151–186. SFP: 29 July 1964, DP: 17 November 1964 (registered at ZIN library).

Iablokoff-Khnzorian, S. M. 1965: Novye vidy zhestkokrylykh (Coleoptera: Dytiscidae, Staphylinidae, Chrysomelidae, Bruchidae, Curculionidae) iz Kirgizii. [New species of beetles (Coleoptera: Dytiscidae, Staphylinidae, Chrysomelidae, Bruchidae, Curculionidae) from Kyrgyzstan]. Pp. 16–26. In: Protsenko, A. I. (ed.): *Entomologicheskie issledovaniya v Kirgizii. [Entomological research in Kyrgyzstan]*. Frunze: Izdatel'stvo "ILIM", 121 pp. SFP: 30 June 1965, DP: [31 December] 1965 (book title page).

Iablokoff-Khnzorian, S. M. 1966: Considérations sur l'édéage des Chrysomelidae et son importance phylogénique. *L'Entomologiste* 22 (6): 115–137. DP: [31] December 1966 (issue title page).

Iablokoff-Khnzorian, S. M. 1968: Notes sur les Chrysomelidae de l'Arménie Soviétique (Col.). *Annales de la Société Entomologique de France, Nouvelle Série* 4 (2): 259–277. DP: 2 August 1968 (stamp on the Gallica copy).

Iablokoff-Khnzorian, S. M. 1970: Novye vidy zhestkokrylykh iz Armenii i drugikh chastey SSSR (Insecta, Coleoptera). (New species of beetles from Armenia and other lands of USSR (Insecta, Coleoptera)). *Zoologicheskiy Sbornik Akademii Nauk Armyanskoy SSR* 15: 50–80. SFP: 26 June 1970, DP: 11 December 1970 (registered at ZIN library).

Iablokoff-Khnzorian, S. M. 1974: Novye vidy zhestkokrylykh-zernovok iz SSSR (Coleoptera, Bruchidae). (New species of Coleoptera Bruchidae from USSR). *Doklady Akademii Nauk Armyanskoy SSR* 58 (2): 123–128. SFP: 1 April 1974, DP: [31 December] 1974 (issue title page).

Iablokoff-Khnzorian, S. M. 1978: Tri novykh vida zhestkokrylykh-listoedov iz Sredney Azii (Coleoptera, Chrysomelidae). (Three new species of chrysomelid-beetles from Central Asia (Coleoptera, Chrysomelidae)). *Doklady Akademii Nauk Armyanskoy SSR* 66 (2): 119–123. SFP: 23 May 1978, DP: 22 August 1978 (registered at VKP).

Iablokoff-Khnzorian, S. M. 1983: Novyy listoed iz Tadzhikistana (Coleoptera, Chrysomelidae). (A new species of chrysomelid-beetles from Tadzhikistan). *Doklady Akademii Nauk Armyanskoy SSR* 76 (2): 94–96. SFP: 23 March 1983, DP: [31 December] 1983 (issue title page).

ICZN 1954: Opinion 228. Rejection for nomenclatorial purposes of Geoffroy, 1762, "Histoire abregee des insectes qui se trouvent aux environs de Paris". *Opinions and Declarations Rendered by the International Commission on Zoological Nomenclature* 4 (18): 209–220. DP: 21 April 1954 (issue cover).

ICZN 1955: Opinion 376. Suppression under the Plenary Powers of the generic name Lobacantha Kirby (W.), 1837, for the purpose of rendering the generic name Platypria Guérin-Méneville, 1840 (Class Insecta, Order Coleoptera) the oldest available name for the genus concerned. *Opinions and Declarations Rendered by the International Commission on Zoological Nomenclature* 11 [1955–1956] (26): 391–400. DP: 2 December 1955 (issue cover).

ICZN 1964: Opinion 710. Enhydrus Laporte, 1834 (Insecta, Coleoptera): validated under the plenary powers. *Bulletin of Zoological Nomenclature* **21** (4): 242–245. DP: 16 October 1964 (issue cover).

ICZN 1970: Opinion 908. Crioceris Müller, 1764, and Lema Fabricius, 1798 (Insecta, Coleoptera): designation of type-species under the plenary powers. *Bulletin of Zoological Nomenclature* **27** (1): 12–13. DP: 5 June 1970 (issue cover).

ICZN 1979: Opinion 1130. Lilioceris Reitter, 1912 (Insecta; Coleoptera) correction of type species. *Bulletin of Zoological Nomenclature* **36** (2): 79–81. DP: 1 August 1979 (issue cover).

ICZN 1984a: Opinion 1270. Chrysomela flavicornis and C. tibialis Suffrian, 1851 (Insecta, Coleoptera): conserved. *Bulletin of Zoological Nomenclature* **41** (1): 22–23. DP: 29 March 1984 (issue cover).

ICZN 1984b: Opinion 1273. Anaspis, Luperus, Lampyris and Clerus (Insecta, Coleoptera): determination of authorship and fixation of type species. *Bulletin of Zoological Nomenclature* **41** (1): 28–31. DP: 29 March 1984 (issue cover).

ICZN 1984c: Opinion 1279. Chrysolina Motschulsky, 1860 (Insecta, Coleoptera): conserved. *Bulletin of Zoological Nomenclature* **41** (4): 218–220. DP: 30 November 1984 (issue cover).

ICZN 1985a: *International code of zoological nomenclature. Third edition*. London: International trust for zoological nomenclature, xx + 338 pp. DP: [28] February 1985 (colophon).

ICZN 1985b: Opinion 1290. Leptinotarsa Chevrolat, 1837 (Insecta, Coleoptera): conserved. *Bulletin of Zoological Nomenclature* **42** (1): 21–23. DP: 2 April 1985 (issue cover).

ICZN 1986: Opinion 1382. Zeugophora Kunze, 1818 (Insecta, Coleoptera): conserved. *Bulletin of Zoological Nomenclature* **43** (1): 44–45. DP: 9 April 1986 (issue cover).

ICZN 1991: Opinion 1656. Longitarsus symphyti Heikertinger, 1912 (Insecta, Coleoptera): specific name conserved. *Bulletin of Zoological Nomenclature* 48 (3): 270–271. DP: 30 September 1991 (issue cover).

ICZN 1994: Opinion 1754. Histoire abrégée des insectes qui se trouvent aux environs de Paris (Geoffroy, 1762): some generic names conserved (Crustacea, Insecta). *Bulletin of Zoological Nomenclature* **51** (1): 58–70. DP: 30 March 1994 (issue cover).

ICZN 1995a: Opinion 1809. Bruchus Linnaeus, 1767, Ptinus Linnaeus, 1767 and Mylabris Fabricius, 1775 (Insecta, Coleoptera): conserved. *Bulletin of Zoological Nomenclature* **52** (2): 208–210. DP: 30 June 1995 (issue cover).

ICZN 1995b: Opinion 1820. A.A.H. Lichtenstein's (1796, 1797) Catalogus musei zoologici ... Sectio Tertia. Continens Insecta and D.H. Schneider's (1880) Verzeichniss einer Parthei Insekten ... : supressed, with conservation of some Lichtenstein (1796) names (Insecta and Arachnida). *Bulletin of Zoological Nomenclature* 52 (3): 283–285. DP: 28 September 1995 (issue cover).

ICZN 1999: *International code of zoological nomenclature. Fourth edition.* London: International trust for zoological nomenclature, xxix + 306 pp. DP: [31 December] 1999 (colophon).

ICZN 2002: Opinion 1989 (Case 3103). Orsodacne Latreille, 1802 (Insecta, Coleoptera): Chrysomela cerasi Linnaeus, 1758 designated as the type species. *Bulletin of Zoological Nomenclature* 59 (1): 55. DP: 27 March 2002 (issue cover).

ICZN 2009: Opinion 2227 (Case 3375). Fidia Baly, 1863 and Lypesthes Baly, 1863 (Insecta, Coleoptera): usage not conserved and priority maintained for Fidia Motschulsky, 1860. *Bulletin of Zoological Nomenclature* 66 (2): 198–200. DP: [30] June 2009 (paper title page).

ICZN 2011a: Opinion 2271 (Case 3481). Crioceris quadripunctata Olivier, 1808 (currently Petauristes quadripunctatus; Insecta, Coleoptera): specific name conserved. *Bulletin of Zoological Nomenclature* 68 (2): 145–146. DP: [30] June 2011 (paper title page).

ICZN 2011b: Opinion 2278 (Case 3489). Chrysomela elongata Suffrian, 1851 (currently Oreina elongata; Insecta, Coleoptera): name conserved. *Bulletin of Zoological Nomenclature* 68 (3): 220–221. DP: [30] September 2011 (paper title page).

ICZN 2012a: Opinion 2298 (Case 3519). Eumolpus Weber, 1801, Chrysochus Chevrolat in Dejean, 1836 and Bromius Chevrolat in Dejean, 1836 (Insecta, Coleoptera, Chrysomelidae): usage conserved. *Bulletin of Zoological Nomenclature* 69 (2): 147–149. DP: [30] June 2012 (paper title page).

ICZN 2012b: Amendment of Articles 8, 9, 10, 21 and 78 of the International Code of Zoological Nomenclature to expand and refine methods of publication. *Zootaxa* 3450: 1–7. DP: 4 September 2012 (paper title page).

ICZN 2015a: Opinion 2351 (Case 3575). Haltica undulata Kutschera, 1860 (currently Phyllotreta undulata, Insecta, Coleoptera, Chrysomelidae): precedence given over Haltica bivittata Waterhouse, 1838 (currently Phyllotreta bivittata). *Bulletin of Zoological Nomenclature* 72 (1): 84–85. DP: [31] March 2015 (paper title page).

ICZN 2015b: Opinion 2357 (Case 3607). Haltica schillingii Letzner, 1847 (currently Dibolia schillingii; Insecta, Coleoptera, Chrysomelidae): no precedence given over Petalopus metallicus Motschulsky, 1845 (currently Dibolia metallica). *Bulletin of Zoological Nomenclature* 72 (2): 164–165. DP: [30] June 2015 (paper title page).

ICZN 2019: Opinion 2439 (Case 3725) – Lema decempunctata Gebler, 1829 (Insecta, Coleoptera, Chrysomelidae): spelling of the specific name conserved. *Bulletin of Zoological Nomenclature* 76 (1): 154–154. DP: 30 August 2019 (paper title page).

ICZN 2021: Opinion 2474 (Case 3760) – Chrysomela tenebricosa Fabricius, 1775 (currently Timarcha tenebricosa; Insecta, Coleoptera, Chrysomelidae): specific name conserved. *Bulletin of Zoological Nomenclature* 78 (1): 57–58. DP: 30 April 2021 (paper title page).

Ihering, H. von 1904: [new taxon]. In: Ihering, H. von & Ihering, R. von: Bibliographia 1902–1904. Historia natural e anthropologia do Brazil. *Revista do Museu Paulista* 6: 584–659. DP: [31 December] 1904 (volume cover).

Illiger, J. C. W. 1794: Beschreibung einiger neuen Käferarten aus der Sammlung des Herrn Professors Hellwig in Braunschweig. *Neuestes Magazin für die Liebhaber der Entomologie* 1 (5): 593–620. DP: [31 December] 1794 (see the comments in Dating of publications).

Illiger, J. K. W. 1798: *Verzeichniss der Käfer Preussens. Entworfen von Johann Gottlieb Kugelann Apotheker in Osterode. Ausgearbeitet von Johann Karl Wilhelm Illiger. Mit einer Vorrede des Professors und Pagenhofmeisters Hellwig in Braunschweig, und dem angehängten Versuche einer natürlichen Ordnungs- und Gattungs-Folge der Insekten.* Halle: Johann Jacob Gebauer, xlii + 510 + [1] pp. DP: 19 September 1798 (Bousquet 2016).

Illiger, J. C. W. 1800: Vierzig neue Insekten aus der Hellwigischen Sammlung in Braunschweig. *Archiv für Zoologie und Zootomie* 1 (2): 103–150. DP: 5 October 1800 (*Allg. Verz. Bücher, Michaelismesse* 1800: 392 and Evenhuis 2014).

Illiger, J. C. W. 1801a: Namen der Insekten-Gattungen, ihr Genitiv, ihr grammatisches Geschlecht, ihr Silbenmass, ihre Herleitung; Zugleich mit den Deutschen Benennungen. *Magazin für Insektenkunde* 1 [1801–1802] (1/2): 125–156. DP: 24 July 1801 (*Leipzig. Jahrb. Neuest. Lit.* 1801: 164).

Illiger, J. C. W. 1801b: Neue Insekten. *Magazin für Insektenkunde* 1 [1801–1802] (1/2): 163–208. DP: 24 July 1801 (*Leipzig. Jahrb. Neuest. Lit.* 1801: 164).

Illiger, J. C. W. 1802: Zusätze, Berichtigungen und Bemerkungen zu Fabricii Systema Eleutheratorum. Tomus I. *Magazin für Insektenkunde* 1 [1801–1802] (3/4): 306–425. DP: 9 May 1802 (*Allg. Verz. Bücher, Ostermesse* 1802: 102 and Evenhuis 2014).

Illiger, J. K. W 1805: Zusätze, Berichtigungen und Bemerkungen zu Fabricii Systema Eleutheratorum Tomus II (Beschluss.). *Magazin für Insektenkunde* 4: 69–174. DP: 5 May 1805 (*Allg. Verz. Bücher, Ostermesse* 1805: 107 and Evenhuis 2014).

Illiger, J. C. W. 1807a: Portugiesische Käfer. *Magazin für Insektenkunde* 6: 1–80. DP: [31 December] 1907 (volume title page).

Illiger, J. C. W. 1807b: Verzeichniss der Arten der Flohkäfer, Halticae, in der Hellwig-Hoffmanseggischen Sammlung mit Beschreibung der neuen, und Bezeichnung der übrigen Arten. *Magazin für Insektenkunde* 6: 81–177. DP: [31 December] 1907 (volume title page).

İriboz, N. Ş 1934: *Şekerpancarı hastalıkları.* [*Sugar-beet diseases*]. Istanbul: Resimli Ay Matbaasi, 164 pp. DP: [31 December] 1934 (book title page).

Isono, M. 1990: A revision of the genus Demotina (Coleoptera, Chrysomelidae) from Japan, the Ryukyus, Taiwan and Korea, I. *Japanese Journal of Entomology* 58 (2): 375–382. DP: 25 June 1990 (paper title page).

Israelson, G. 1956: Haltica cirsii n. sp. (Col. Chrysomelidae). *Opuscula Entomologica* 21 (2/3): 129–131. DP: 15 November 1956 (table of contents).

Israelson, G. 1969: A new species of Luperus from the Canary Islands (Col. Chrysomelidae). *Eos, Revista Española de Entomología* 44 [1968] (1/4): 159–164. DP: 1 October 1969 (issue cover).

Israelson, G. 1980: Taxonomical and nomenclatural notes on some Canarian Coleoptera. *Vieraea* 9 [1979] (1/2): 183–210. DP: [30] September 1980 (paper title page).

Israelson, G. 1991: A Longitarsus from the Azores (Coleoptera, Chrysomelidae). *Bocagiana* **144**: 1–4. DP: 18 November 1991 (paper title page).

Israelson, W. 1926: Ueber einige neue deutsche Coleopteren-Aberrationen. *Entomologische Zeitschrift* 39 [1925–1926] (41): 166. DP: 6 February 1926 (issue title page).

Jacobson [= Jacobsohn], G. G. 1892a: Analytische Übersicht der bekannten Donacia- und Plateumaris-Arten der Alten Welt. *Horae Societatis Entomologicae Rossicae* **26** [1891–1892] (3/4): 412–437. DP: 7 July 1892 (Kerzhner 1984, Standfuss & Kerzhner 2004).

Jacobson [= Jacobsohn], G. G. 1892b: Conspectus specierum generis Haltica Geoffr. ex Asia media hucusque notarum. *Horae Societatis Entomologicae Rossicae* **26** [1891–1892] (3/4): 462–467. DP: 7 July 1892 (Kerzhner 1984, Standfuss & Kerzhner 2004).

Jacobson [= Jacobsohn], G. G. 1893a: Über einige Chrysomelidae aus Sibirien. *Horae Societatis Entomologicae Rossicae* 27 [1892–1893] (1/2): 122–126. DP (preprint): 12 February 1893 (Kerzhner 1984, Standfuss & Kerzhner 2004).

Jacobson [= Jacobsohn], G. G. 1893b: Beitrag zur west-turkestanischen Chrysomeliden-Fauna. *Horae Societatis Entomologicae Rossicae* 27 [1892–1893] (1/2): 236–248. DP (preprint): 12 April 1893 (Kerzhner 1984, Standfuss & Kerzhner 2004).

Jacobson [= Jacobsohn], G. G. 1894a: Zur Chrysomeliden-Fauna der Umgegend vom See Issyk-kul. *Deutsche Entomologische Zeitschrift* **1894** (1): 97–107. DP: 10 January 1894 (*Bull. Soc. Ent. Fr.* 1894: xiv).

Jacobson [= Jacobsohn], G. G. 1894b: Materialia ad cognitionem faunae Chrysomelidarum provinciae Astrachanensis. *Horae Societatis Entomologicae Rossicae* 28 [1893–1894] (1/2): 152–160. DP (preprint): 12 February 1894 (Kerzhner 1984, Standfuss & Kerzhner 2004).

Jacobson [= Jacobsohn], G. G. 1894c: Chrysomelidae ab A. Trotzina in provinciis Transcapica et Ferganensi a. 1893 collectae. *Horae Societatis Entomologicae Rossicae* 28 [1893–1894] (1/2): 238–241. DP (preprint): 12 April 1894 (Kerzhner 1984, Standfuss & Kerzhner 2004).

Jacobson [= Jacobsohn], G. G. 1894d: Adnotationes de Chrysomelidis nonnullis novis vel parum cognitis. *Horae Societatis Entomologicae Rossicae* **28** [1893–1894] (1/2): 242–246. DP (preprint): 12 April 1894 (Kerzhner 1984, Standfuss & Kerzhner 2004).

Jacobson [= Jacobsohn], G. G. 1894e: Chrysomelidae palaearcticae novae ac parum cognitae. *Horae Societatis Entomologicae Rossicae* 28 [1893–1894] (1/2): 269–278. DP (preprint): 12 April 1894 (Kerzhner 1984, Standfuss & Kerzhner 2004).

Jacobson, G. G. 1894f: Spisok donaciy Zoologicheskago Muzeya Imperatorskago Moskovskago Universiteta. [List of reed-beetles from the Zoological Museum of Moscow Imperial University]. *Dnevnik Zoologicheskago Otdeleniya Obshchestva i Zoologichskago Muzeya* (Moscow) 2 (1/2): 21–22. DP: [31 December] 1894 (issue cover).

Jacobson, G. G. 1895a: Revisio generis Xenomela Weise. *Horae Societatis Entomologicae Rossicae* **29** [1894–1895] (1/2): 271–281. DP: 12 May 1895 (Kerzhner 1984, Standfuss & Kerzhner 2004).

Jacobson, G. G. 1895b: Corrigenda. *Horae Societatis Entomologicae Rossicae* **29** [1894–1895] (1/2): 294–297. DP: 12 May 1895 (Kerzhner 1984, Standfuss & Kerzhner 2004).

Jacobson, G. G. 1896a: Chrysomelidae palaearcticae novae vel parum cognitae. II. *Horae Societatis Entomologicae Rossicae* 29 [1894–1895] (3/4): 529–558. DP: 1 January 1896 (Kerzhner 1984, Standfuss & Kerzhner 2004).

Jacobson, G. G. 1896b: Materialia ad cognitionem faunae Chrysomelidarum Provinciae Saratowensis. *Horae Societatis Entomologicae Rossicae* 30 [1895–1896] (1/2): 103–121. DP (preprint): 12 March 1896 (Kerzhner 1984, Standfuss & Kerzhner 2004).

Jacobson, G. G. 1897a: Materialia ad cognitionem faunae Chrysomelidarum provinciae Orenburgensis. *Horae Societatis Entomologicae Rossicae* 30 [1895–1896] (3/4): 429–437. DP (preprint): 12 February 1897 (Kerzhner 1984, Standfuss & Kerzhner 2004).

Jacobson, G. G. 1897b: Chrysomelidae palaearctici novi vel parum cogniti. I. *Annuaire du Musée Zoologique de l'Académie Impériale des Sciences de St.-Pétersbourg* 2 (1): 74–77. DP (preprint): 12 April 1897 (last page of preprint, see also Asanovich et al. 2007).

Jacobson, G. G. 1897c: Duo Chrysomelidae novi transcaucasici. *Annuaire du Musée Zoologique de l'Académie Impériale des Sciences de St.-Pétersbourg* 2 (2): 215–217. DP (preprint): 12 July 1897 (last page of preprint, see also Asanovich et al. 2007).

Jacobson, G. G. 1898: Chrysomelidae palaearctici novi vel parum cogniti. II. *Annuaire du Musée Zoologique de l'Académie Impériale des Sciences de St.-Pétersbourg* 3 [1898] (2): 191–202. DP: [31] December 1898 (Asanovich et al. 2007).

Jacobson, G. G. 1899a: De specie nova g. Luperus Geoffr. e Caucaso. *Horae Societatis Entomologicae Rossicae* 33 [1900–1901] (1/2): 141–142. DP (preprint): 12 May 1899 (Kerzhner 1984, Standfuss & Kerzhner 2004).

Jacobson, G. G. 1899b: Chrysomelidae palaearctici novi vel parum cogniti. III. *Annuaire du Musée Zoologique de l'Académie Impériale des Sciences de St.-Pétersbourg* 4 (1): 1–11. DP (preprint): 12 June 1899 (last page of preprint, see also Asanovich et al. 2007).

Jacobson, G. G. 1901a: Chrysomelidae Sibiriae occidentalis. *Horae Societatis Entomologicae Rossicae* 35 (1/2): 73–102. DP (preprint): 13 January 1901 (Kerzhner 1984, Standfuss & Kerzhner 2004).

Jacobson, G. G. 1901b: Symbola ad cognitionem Chrysomelidarum Rossicae asiaticae. *Öfversigt af Finska Vetenkaps-Societetens Förhandlingar* 43 [1900–1901]: 99–147. DP: [31 December] 1901 (volume title page).

Jacobson, G. G. 1903: Spisok zhukov, naydennykh v 1899–1901 gg. shtab-rotmistrom A. N. Gudzenko v yuzhnoy Manchzhurii i na Kvantungskom poluostrov. (Coleoptera Mandshuriae meridionalis et peninsulae Quantungensis, ab A. N. Gudzenko allata). *Annuaire du Musée Zoologique de l'Académie Impériale des Sciences de St. Pétersbourg* 8 (1): 11–16. DP: 13 August 1903 (Asanovich et al. 2007).

Jacobson, G. G. 1905: De Cryptocephali specie nova turkestanica (Coleoptera, Chrysomelidae). *Russkoe Entomologicheskoe Obozrenie* 5 (3/4): 111–113. DP: 13 September 1905 (footers throughout the issue).

Jacobson, G. G. 1906a: Labidostomis sardoa sp. n. *Rivista Coleotterologica Italiana* 4 (2): 30–31. DP: 15 February 1906 (issue cover).

Jacobson, G. G. 1906b: Donacia Clavareaui spec. nova. *Annales de la Société Entomologique de Belgique* 50 (11): 311–312. DP: 30 November 1906 (footer on p. 309).

Jacobson, G. G. 1906c: [Rewiew of:] Jacoby M. et Clavareau H.: Coleoptera Phytophaga (Genera Insectorum publiés par P. Wytsman). *Russkoe Entomologicheskoe Obozrenie* 6 (3/4): 309–312. DP: [31] December 1906 (footers throughout the issue).

Jacobson, G. G. 1907a: O sluchayakh neposledovatelnoy izmenchivosti v okraske zhukov. [On cases of non-compliance variability in beetle coloration]. *Horae Societatis Entomologicae Rossicae* 38 [1907–1908] (1/2): LX–LXV. DP: 3 May 1907 (Kerzhner 1984, Standfuss & Kerzhner 2004).

Jacobson, G. G. 1907b: Donacia semenovi, sp. nov. (Coleoptera, Chrysomelidae). *Russkoe Entomologicheskoe Obozrenie* 7 (1): 5–6. DP: [31] December 1907 (footers throughout the issue).

Jacobson, G. G. 1907c: De duabus novis formis generis Crioceris Geoffr. (Coleoptera, Chrysomelidae), additis annotationibus synonymicis. *Russkoe Entomologicheskoe Obozrenie* 7 (1): 25–26. DP: [31] December 1907 (footers throughout the issue).

Jacobson, G. G. 1908: De tribus generibus novis Chrysomelidarum faunae rossicae. *Horae Societatis Entomologicae Rossicae* 38 [1907–1908] (3/4): 619–627, pl. VII. DP: 14 October 1908 (Kerzhner 1984, Standfuss & Kerzhner 2004).

Jacobson, G. G. 1909: Pl. 55. In: *Zhuki Rossii i zapadnoy Evropy. Rukovodstvo k opredelenyu zhukov.* [1905–1915]. St. Petersburg: Devrien, [1] + 1024 pp., 83 pls. DP (pl. 55): 4 March 1909 (for publication dates of all parts see Griffin 1932).

Jacobson, G. G. 1910a: Ueber die Chrysomelidae, Elateridae und Curculionidae der Sammlung der Russischen Nordpolar-Expedition. Pp. 53–66, pl. 1. In: Poppius, B., Zaitzev, Ph. & Jacobson, G. G.: Beiträge zur Coleopteren-fauna des äußersten Nordens von Sibirien. *Mémoires de l'Académie Impériale des Sciences de St.-Pétersbourg, VIIIe Série* 18 [1905–1910] (9): 1–66, pl. 1. DP: [31] August 1910 (verso of paper title page).

Jacobson, G. G. 1910b: De chrysomelidis palaearcticis. Descriptionum et annotationum series I. *Russkoe Entomologicheskoe Obozrenie* 10 (1/2): 53–60. DP: 14 September 1910 (*répartition des livraisons* on p. xxv).

Jacobson, G. G. 1911: Pl. 59. In: *Zhuki Rossii i zapadnoy Evropy. Rukovodstvo k opredelenyu zhukov.* [1905–1915]. St. Petersburg: Devrien, [1] + 1024 pp., 83 pls. DP (pl. 59): 3 June 1911 (for publication dates of all parts see Griffin 1932).

Jacobson, G. G. 1915a: [new taxon]. In: Schreiner, J. O.: Novyy vid roda Phyllotreta Foudr. (Coleoptera, Chrsomelidae). [Une espèce nouvelle du genre Phyllotreta Foudr. (Coleoptera, Chrysomelidae)]. *Russkoe Entomologicheskoe Obozrenie* 15 (1): 72–73. DP: 31 March 1915 (*répartition des livraisons* on p. xxviii).

Jacobson, G. G. 1915b: Fauna okresnostey g. Novo-Alexandriy, Lyublinskoy gub. II. Zhestkokrylyya (Coleoptera). 1. (Cerambycidae, Chrysomelidae (chastyu), Lucanidae, Scarabaeidae). [Fauna of surroundings of Nova Alexandria, Lublin governorate II. Beetles (Coleoptera). 1. (Cerambycidae, Chrysomelidae (partim), Lucanidae, Scarabaeidae)]. *Zapiski*

Novo-Aleksandriyskago Instituta Selskago Khozyaystva i Lesovodstva (Petrograd) **23** (3): 150–173. DP: [31 December] 1915 (issue cover).

Jacobson, G. G. 1917: De genere Thelyterotarso Weise (Coleoptera, Chrysomelidae). *Russkoe Entomologicheskoe Obozrenie* **16** [1916] (3/4): 266–274. DP: 14 May 1917 (issue cover).

Jacobson, G. G. 1921: K morfologii i sistematike nekotorykh rodov podsem. Chrysomelini. (Contributions a la morphologie et classification des quelques genres de la sous-famille Chrysomelini). *Izvestiya Petrogradskoy Oblastnoy Stancii Zashchity Rasteniy ot Vrediteley* **2**: 20–24. DP: [30] November 1921 (volume cover).

Jacobson, G. G. 1922a: Chrysomelidae palaearctici novi vel parum cogniti (Coleoptera). IV. *Annuaire du Musée Zoologique de l'Académie des Sciences de Russie* **23** [1918–1922] (3/4): 517–534. DP: [31] December 1922 (Asanovich et al. 2007).

Jacobson, G. G. 1922b: [new taxa]. In: Barovskiy, V.: Novye i interesnye dlya fauny Petrogradskoy gubernii nasekomyya. VIII–IX. (Insectes nouveaux ou rares de la faune du gouvernement de Petrograd. VIII–IX). *Russkoe Entomologicheskoe Obozrenie* **18** (1): 53–64. DP: [31 December] 1922 (issue cover).

Jacobson, G. G. 1923: De uno genere novo et specie nova e Chrysomelidarum tribu Clytrinorum. *Entomologische Mitteilungen* **12** (5/6): 233–235. DP: 30 September 1923 (issue cover).

Jacobson, G. G. 1924a: Geograficheskoe rasprostranenie vidov roda Chrysochloa Hope (Coleoptera, Chrysomelidae). (La distribution géographiques des espèces du genre Chrysochloa Hope (Coleoptera, Chrysomelidae)). *Comptes Rendus de l'Académie des Sciences de Russie* **1924** (janvier–mars): 20–21. DP: [29] February 1924 (verso of the issue cover).

Jacobson, G. G. 1924b: De sex speciebus novis generis Chrysomelae L. (Col., Chrysom.). *Wiener Entomologische Zeitung* **41** (4/10): 78–84. DP: 25 September 1924 (footers throughout the issue).

Jacobson, G. G. 1924c: Annotationes synonymicae et systematicae de coleopteris. *Russkoe Entomologicheskoe Obozrenie* **18** (4): 237–244. DP: [31 December] 1924 (issue cover).

Jacobson, G. G. 1925a: Zhuki-listoedy yuzhnogo Eniseya. (Chrysomeliden des Süd Jenissey). *Jahrbuch des Martjanov'schen Staatsmuseums in Minussinsk* **3** (1): 1–52. DP: 30 July 1925 (issue cover).

Jacobson, G. G. 1925b: K sistematike i geograficheskomu rasprostraneniyu nekotorykh rodov triby Galerucini (Coleoptera, Chrysomelidae). (Contributions à la systématique et à la distribution géographique de quelques genres da la tribu de Galerucini (Coleoptera, Chrysomelidae)). *Comptes Rendus de l'Académie des Sciences de Russie* **1925** (avril–juin): 49–52. DP: [31] July 1925 (verso of the issue cover).

Jacobson, G. G. 1925c: De Chrysomelidis palaearcticis. Descriptionum et annotationum series II. *Russkoe Entomologicheskoe Obozrenie* **19** (1): 7–16. DP: 10 September 1925 (registered at ZIN library).

Jacobson, G. G. 1925d: De Chrysomelidis palaearcticis. Descriptionum et annotationum series III. *Russkoe Entomologicheskoe Obozrenie* **19** (2): 143–148. DP: [31 December] 1925 (issue cover).

Jacobson, G. G. 1926a: Chrysomelidae (Coleoptera) palaearctici novi vel parum cogniti. V–VI. *Annuaire du Musée Zoologique de l'Académie des Sciences de l'URSR* **26** [1925] (3/4): 231–276. DP: [30] September 1926 (verso of issue title page).

Jacobson, G. G. 1926b: De Chrysomelidis palaearcticis. Descriptionum et annotationum series IV. *Russkoe Entomologicheskoe Obozrenie* **19** [1925] (3/4): 163–169. DP: 28 October 1926 (registered at ZIN library, 1926 also mentioned on the issue cover).

Jacobson, G. G. 1927: Donacia knipowitschi, spec. nova. (Coleoptera, Chrysomelidae). Pp. 45–46. In: *Sbornik v chest' professora Nikolaya Mikhalovicha Knipovicha 1885–1925. (Festschrift für Prof. N. M. Knipowitsch 1885–1925).* Moskva: Izdanie Narodnogo komissariata zemledelia RSFSR, 447 pp. DP: 27 December 1927 (stamp on the Russian Geographical Society's copy).

Jacobson, G. G. 1931: *Opredelitel zhukov. Izdanie 2-e dopolnennoe D. A. Ogloblinym.* Moskva-Leningrad: Gosudarstvennoe izdatelstvo selskokhozyastvennoy i kolkhozno-kooperativnoy literatury, 454 pp. SFP: 6 October 1931, DP: [31 December] 1931 (book title page).

Jacoby, M. 1880: Descriptions of new species of phytophagous Coleoptera. *Proceedings of the Zoological Society of London* **1879** (4): 773–793. DP: [30] April 1880 (Duncan 1937).

Jacoby, M. 1881: Descriptions of new genera and species of phytophagous Coleoptera. *Proceedings of the Zoological Society of London* **1881** (2): 439–450. DP: [31] August 1881 (Duncan 1937).

Jacoby, M. 1882: [Phytophaga (part)]. Pp. 169–184. In: Godman, F. D. & Salvin, O. (eds.): *Biologia Centrali-Americana. Insecta. Coleoptera. Vol. VI. Part 1.* [1880–1892], London: Taylor and Francis, xx + 625 pp. DP: [30] June 1882 (Bousquet 2016).

Jacoby, M. 1883a: On some new species of phytophagous Coleoptera from the island of Saleyer. *Notes from the Leyden Museum* **5** (3): 197–203. DP: [31] July 1883 (Dickinson 2005).

Jacoby, M. 1883b: Descriptions of some new species of beetles of the family Galerucidae. *Proceedings of the Zoological Society of London* **1883** (3): 399–406, pl. XLV. DP: [31] October 1883 (Duncan 1937).

Jacoby, M. 1884a: Descriptions of new genera and species of phytophagous Coleoptera from the Indo-Malayan and Austro-Malayan subregions, contained in the Genoa Civic Museum. First Part. *Annali del Museo Civico di Storia Naturale di Genova* **20** [1883–1884]: 188–233. DP: 18 January 1884 (Poggi 2010).

Jacoby, M. 1884b: Descriptions of new genera and species of phytophagous Coleoptera from Sumatra. *Notes from the Leyden Museum* **6** (1): 9–70. DP: [31] January 1884 (Dickinson 2005).

Jacoby, M. 1884c: Descriptions of new genera and species of phytophagous Coleoptera collected by Dr. B. Hagen at Serdang (East Sumatra). *Notes from the Leyden Museum* **6** (4): 201–230. DP: [31] October 1884 (Dickinson 2005).

Jacoby, M. 1884d: Description of a new genus and three new species of Malayan Galerucinae. *Notes from the Leyden Museum* **6** (4): 233–235. DP: [31] October 1884 (Dickinson 2005).

Jacoby, M. 1885a: Descriptions of new genera and species of phytophagous Coleoptera from the Indo-Malayan and Austro-Malayan subregions, contained in the Genoa Civic Museum. Second part. *Annali del Museo Civico di Storia Naturale di Genova* **22**: 20–76. DP: 8 June 1885 (Poggi 2010).

Jacoby, M. 1885b: Descriptions of the phytophagous Coleoptera of Japan, obtained by Mr. George Lewis during his second journey, from February 1880 to September 1881. Part I. *Proceedings of the Zoological Society of London* 1885 (1): 190–211, pl. XI. DP: [30] June 1885 (Duncan 1937).

Jacoby, M. 1886a: Descriptions of the phytophagous Coleoptera of Japan obtained by Mr. George Lewis during his second journey, from February 1880 to September 1881. – Part II. Halticinae and Galerucinae. *Proceedings of the Zoological Society of London* [1885] (4): 719–755, pl. XLVI. DP: [30] April 1886 (Duncan 1937).

Jacoby, M. 1886b: Descriptions of some new species and a new genus of phytophagous Coleoptera. *Proceedings of the Zoological Society of London* [1885] (4): 925–929. DP: [30] April 1886 (Duncan 1937).

Jacoby, M. 1886c: Beschreibung einer neuen Oedionychis-Art von der Insel Creta. *Entomologische Zeitung* (Stettin) 47 (4/6): 215–216. DP: [30] April 1886 (last page of the issue).

Jacoby, M. 1886d: Descriptions of new genera and species of phytophagous Coleoptera from the Indo-Malayan and Austro-Malayan subregions, contained in the Genoa Civic Museum. Third Part. *Annali del Museo Civico di Storia Naturale di Genova* 24: 41–121. DP: 25 October 1886 (Poggi 2010).

Jacoby, M. 1886e: Descriptions of some undescribed species of phytophagous Coleoptera from Abyssinia, contained in the Genoa Civic Museum. *Annali del Museo Civico di Storia Naturale di Genova* 24: 122–128. DP: 25 October 1886 (Poggi 2010).

Jacoby, M. 1887a: Descriptions of the phytophagous Coleoptera of Ceylon, obtained by Mr. George Lewis during the years 1881–1882. *Proceedings of the Zoological Society of London* 1887 (1): 65–119, pls. X–XI. DP: [30] June 1887 (Duncan 1937).

Jacoby, M. 1887b: Descriptions of some new genera and species of phytophagous Coleoptera contained in the Leyden Museum. *Notes from the Leyden Museum* 9 (3): 229–243. DP: [31] July 1887 (Dickinson 2005).

Jacoby, M. 1888: Descriptions of new species of phytophagous Coleoptera from Kiukiang (China). *Proceedings of the Zoological Society of London* 1888 (3): 339–351. DP: [31] October 1888 (Duncan 1937).

Jacoby, M. 1889a: Descriptions of some new species of phytophagous Coleoptera. *Entomologist's Monthly Magazine* 25 [1888–1889] (297): 203–206. DP: 1 February 1889 (registered at NHMUK library).

Jacoby, M. 1889b: Viaggio di Leonardo Fea in Birmania e regioni vicine. XVII. List of the phytophagous Coleoptera obtained by signor L. Fea at Burmah and Tenasserim, with descriptions of the new species. *Annali del Museo Civico di Storia Naturale di Genova* 27: 147–237. DP: 11 May 1889 (Poggi 2010).

Jacoby, M. 1889c: List of the phytophagous Coleoptera collected by Signor Modigliani at Nias and Sumatra, with descriptions of the new species. *Annali del Museo Civico di Storia Naturale di Genova* 27: 278–287, pl. IV. DP: 31 May 1889 (Poggi 2010).

Jacoby, M. 1890a: Descriptions of new species of phytophagous Coleoptera received by Mr. J. H. Leech, from Chang-Yang, China. *Entomologist* 23 (322): 84–89. DP: [31] March 1890 (issue cover).

Jacoby, M. 1890b: Descriptions of new species of phytophagous Coleoptera received by Mr. J. H. Leech, from Chang-Yang, China. *Entomologist* 23 (323): 114–118, pl. I–II. DP: [30] April 1890 (issue cover).

Jacoby, M. 1890c: Descriptions of new species of phytophagous Coleoptera received by Mr. J. H. Leech, from Chang-Yang, China. *Entomologist* 23 (324): 161–167. DP: [31] May 1890 (issue cover).

Jacoby, M. 1890d: Descriptions of new species of phytophagous Coleoptera received by Mr. J. H. Leech, from Chang-Yang, China. *Entomologist* 23 (325): 193–197. DP: [30] June 1890 (issue cover).

Jacoby, M. 1890e: Descriptions of new species of phytophagous Coleoptera received by Mr. J. H. Leech, from Chang-Yang, China. *Entomologist* 23 (326): 214–217. DP: [31] July 1890 (issue cover).

Jacoby, M. 1890f: Descriptions of two new species of phytophagous Coleoptera from the East. *Entomologist* 23 (327): 253–254. DP: 25 August 1890 (stamp on the issue cover of the BHL copy).

Jacoby, M. 1891a: Descriptions of some new species of phytophagous Coleoptera from India. *Entomologist* 24 (supplement): 31–34. DP (pp. 31–32): [28] February 1891, (pp. 33–34): [30] April 1891 (footers on pp. 17 and 33).

Jacoby, M. 1891b: On some new species of phytophagous Coleoptera from various regions. *Entomologist* 24 (supplement): 35–41. DP (pp. 35–40): [30] April 1891, (p. 41): [31] May 1891 (footers on pp. 33 and 41).

Jacoby, M. 1891c: Descriptions of some new species of phytophagous Coleoptera. *Entomologist* 24 (supplement): 62–65. DP (pp. 62–64): [30] June 1891, (p. 65): [30] September 1891 (footers on pp. 57 and 65).

Jacoby, M. 1891d: [Phytophaga (part)]. Pp. 281–312. In: Godman, F. D. & Salvin, O. (eds.): *Biologia Centrali-Americana. Insecta. Coleoptera. Vol. VI. Part 1, Supplement.* [1888–1892], London: Taylor and Francis, [2] + 392 pp. + 43 pls. DP: [31] December 1891 (Bousquet 2016).

Jacoby, M. 1892a: Descriptions of some new species of phytophagous Coleoptera, and synonymic notes. *Entomologist* 25 (supplement): 86–88. DP: [30] April 1892 (footer on p. 85).

Jacoby, M. 1892b: Viaggio di Leonardo Fea in Birmania e regioni vicine. LI. Description of the new genera and species of the phytophagous Coleoptera obtained by Sign. L. Fea in Burma. *Annali del Museo Civico di Storia Naturale di Genova* 32: 869–999. DP: 14 December 1892 (Poggi 2010).

Jacoby, M. 1893a: Descriptions of some new genera and new species of Halticidae. *Transactions of the Entomological Society of London* 1893 (2): 145–158. DP: 6 June 1893 (Wheeler 1912).

Jacoby, M. 1893b: Descriptions of some new species of Donaciinae and Criocerinae contained in the Brussels Museum and that of my own. *Annales de la Société Entomologique de Belgique* 37 (5): 261–271. DP: 14 June 1893 (*Bull. Soc. Ent. Fr.* 1893: ccxxxi).

Jacoby, M. 1893c: Descriptions of some new species of phytophagous Coleoptera from the East. *Entomologist* 26 (supplement): 105–111. DP (pp. 105–108): [31] October 1893, (pp. 109–111): [28] February 1895 (footers on pp. 105 and 109).

Jacoby, M. 1894: Descriptions of new genera and species of phytophagous Coleoptera obtained by W. Doherty in the Malayan Archipelago. *Novitates Zoologicae* 1 (2): 267–330. DP: 16 April 1894 (Evenhuis 2022).

Jacoby, M. 1895a: Chrysomeliden von Togo (Bismarckburg). *Deutsche Entomologische Zeitschrift* **1895** (1): 165–188. DP: [28] February 1895 (table of contents).

Jacoby, M. 1895b: Descriptions of the new species of phytophagous Coleoptera obtained by Mr Andrews in India. *Annales de la Société Entomologique de Belgique* **39** (5): 252–288. DP: 27 May 1895 (issue cover).

Jacoby, M. 1895c: Descriptions of new species of phytophagous Coleoptera from the Indo- and Austro-Malayan Regions. *Entomologische Zeitung* (Stettin) **56** (1/6): 52–80. DP: [30] September 1895 (*Zool. Anz., Litteratur* 1895: 373).

Jacoby, M. 1896a: Descriptions of some new species of phytophagous Coleoptera from the Loo-Choo Islands. *Entomologist* **29** (392): 5–8. DP: 8 January 1896 (*Bull. Soc. Ent. Fr.* 1896: 19).

Jacoby, M. 1896b: Descriptions of the new genera and species of phytophagous Coleoptera obtained by Dr. Modigliani in Sumatra. *Annali del Museo Civico di Storia Naturale di Genova* **36**: 377–501. DP: 11 May 1896 (Poggi 2010).

Jacoby, M. 1896c: Descriptions of the new genera and species of phytophagous Coleoptera obtained by Mr Andrewes in India. Part II. Chrysomelinae, Halticinae and Galerucinae. *Annales de la Société Entomologique de Belgique* 40 (6): 250–304. DP: 1 July 1896 (issue cover).

Jacoby, M. 1896d: List and descriptions of the phytophagous Coleoptera obtained by Dr. Modigliani from Mentawi Islands. *Annali del Museo Civico di Storia Naturale di Genova* **37** [1896–1897]: 126–148. DP: 14 December 1896 (Poggi 2010).

Jacoby, M. 1897a: Further contributions to the knowledge of the phytophagous Coleoptera of Africa, including Madagascar. *Proceedings of the Zoological Society of London* **1897** (1): 238–265, pl. XVII. DP: [30] June 1897 (Duncan 1937).

Jacoby, M. 1897b: List of the phytophagous Coleoptera obtained by Mr. H. Raap in the Batu islands with descriptions of the new species. *Annali del Museo Civico di Storia Naturale di Genova* **38**: 405–411. DP: 7 October 1897 (Poggi 2010).

Jacoby, M. 1897c: Further contributions to the knowledge of the phytophagous Coleoptera of Africa, including Madagascar. – Part II. *Proceedings of the Zoological Society of London* **1897** (3): 527–577, pl. XXXIV. DP: [31] October 1897 (Duncan 1937).

Jacoby, M. 1897d: Descriptions of some new species of phytophagous Coleoptera from India. *Annales de la Société Entomologique de Belgique* **41** (11): 420–424. DP: 10 December 1897 (footer on p. 407).

Jacoby, M. 1898a: Descriptions of some new species of Indian phytophagous Coleoptera. *Annales de la Société Entomologique de Belgique* **42** (4): 185–191. DP: 5 May 1898 (footer on p. 183).

Jacoby, M. 1898b: New species of phytophagous Coleoptera from Australia and the Malayan regions. *Annales de la Société Entomologique de Belgique* **42** (9): 350–380. DP: 5 October 1898 (footer on p. 343).

Jacoby, M. 1899a: Some new genera and species of phytophagous Coleoptera collected during Captain Bottego's last expedition. *Annali del Museo Civico di Storia Naturale di Genova* **39** [1898]: 521–535. DP: 2 February 1899 (Poggi 2010).

Jacoby, M. 1899b: Some new genera and species of phytophagous Coleoptera from India and Ceylon. *Entomologist* **32** (430): 67–70. DP: 8 March 1899 (*Bull. Soc. Ent. Fr.* 1899: 107).

Jacoby, M. 1899c: Some new genera and species of phytophagous Coleoptera from India and Ceylon. *Entomologist* **32** (431): 80–84. DP: 12 April 1899 (*Bull. Soc. Ent. Fr.* 1899: 140).

Jacoby, M. 1899d: Additions to the knowledge of the phytophagous Coleoptera of Africa. Part II. *Proceedings of the Zoological Society of London* **1899** (2): 339–380, pl. XXI. DP: [31] August 1899 (Duncan 1937).

Jacoby, M. 1900a: Descriptions of the new species of phytophagous Coleoptera obtained by Dr. Dohrn in Sumatra. *Entomologische Zeitung* (Stettin) **60** [1899] (10/12): 259–313, 1 pl. DP: [30] April 1900 (last page of the issue).

Jacoby, M. 1900b: New species of Indian Phytophaga principally from Mandar in Bengal. *Mémoires de la Société Entomologique de Belgique* **7**: 95–140. DP: 21 July 1900 (*Bibl. Zool.* 1900, 5: 397).

Jacoby, M. 1900c: On new genera and species of phytophagous Coleoptera from South and Central Africa. *Proceedings of the Zoological Society of London* **1900** (2): 203–266, pl. XX. DP: [31] August 1900 (Duncan 1937).

Jacoby, M. 1901a: Descriptions of some new species of phytophagous Coleoptera of the family Chlamydae. *Proceedings of the Zoological Society of London* **1901** [1] (1): 153–164, pl. XIV. DP: [30] June 1901 (Duncan 1937).

Jacoby, M. 1901b: A further contribution to our knowledge of African phytophagous Coleoptera. *Transactions of the Entomological Society of London* **1901** [1901–1902] (3): 209–256, pl. X. DP: 30 September 1901 (verso of the volume title page).

Jacoby, M. 1903a: Descriptions of the new genera and species of phytophagous Coleoptera obtained by Mr H.-L. Andrewes and Mr T.-R.-D. Bell at the Nilgiri Hills and Kanara. *Annales de la Société Entomologique de Belgique* **47** (3): 80–128. DP: 2 April 1903 (footer on p. 79).

Jacoby, M. 1903b: A further contribution to our knowledge of African phytophagous Coleoptera, part II. *Transactions of the Entomological Society of London* **1903** [1903–1904] (1): 1–38. DP: 29 April 1903 (verso of the volume title page).

Jacoby, M. 1904a: Another contribution to the knowledge of Indian phytophagous Coleoptera. *Annales de la Société Entomologique de Belgique* **48** (11): 380–406. DP: 2 December 1904 (footer on p. 379).

Jacoby, M. 1904b: Description of some new species of phytophagous Coleoptera. *Entomologist* **37** (499): 293–296. DP: 13 December 1904 (stamp on the issue cover of the BHL copy).

Jacoby, M. 1905a: Descriptions of new genera and species of phytophagous Coleoptera obtained by Dr. Loria in New Guinea. *Annali del Museo Civico di Storia Naturale di Genova* **41** [1904]: 469–514. DP: 13 February 1905 (Poggi 2010).

Jacoby, M. 1905b: Descriptions of new Malayan and one Bornean species of phytophagous Coleoptera. Pp. i–vii (Appendix). In: Anandale, N. & Robinson, H. C. (eds.): *Fasciculi Malayenses. Anthropological and zoological results of an expedition to Perak and the Siamese Malay states, 1901–1902. Zoology. Part III*. London: University Press of Liverpool, 148 pp + i–vii (Appendix). DP (preprint): [28] February 1905 (preprint title page).

Jacoby, M. 1905c: Descriptions of thirty-two new species of Halticinae (Phytophagous Coleoptera) from South and Central America. *Proceedings of the Zoological Society of London* **1904** [2] (2): 396–413. DP: 18 April 1905 (Duncan 1937).

Jacoby, M. 1906: Descriptions of new genera and species of African Halticinae and Galerucinae. *Transactions of the Entomological Society of London* **1906** [1906–1907] (1): 11–52, pl. III. DP: 29 May 1906 (verso of the volume title page).

Jacoby, M. 1908: Coleoptera, Chrysomelidae. Vol. 1. In: Bingham, C. T. (ed.): *The Fauna of British India, including Ceylon and Burma*. London: Taylor & Francis, xx + 534 pp., 2 pls. DP: 21 March 1908 (*Athenaeum* 1908, 4195: 355).

Jacquelin du Val, P. N. C. 1853: Description de deux genres nouveaux et de plusieurs espèces nouvelles de coléoptères propres à la faune française. *Annales de la Société Entomologique de France, Deuxième Série* 10 [1852] (4): 695–718. DP: 23 March 1853 (Evenhuis 2022).

Jacquet, E. 1886: [La description d'une nouvelle espèce de coléoptère]. *Bulletin de la Société Entomologique de France* **1886** (20): clxxi–clxxii. DP: 20 November 1886 (Evenhuis 2002).

Jacquet, E. 1887: Notes entomologiques. *L'Échange, Revue Linnéenne* **3** (33): 2. DP: 15 September 1887 (issue title page).

Jacquet, E. 1888: Tableaux analytiques pour la détermination des Rhynchophores de la Faune française. *L'Échange, Revue Linnéenne* **4** (43): 3–10 [special pagination]. DP: 15 July 1888 (issue cover).

Jacquet, J. 1935: "Chrysomela" varians Schall., v. Malleti nov. *Bulletin Mensuel de la Société Linnéenne de Lyon* **4** (6): 95–96. DP: [30] June 1935 (issue title page).

Jakob, H. 1952a: Revision des Subg. Protorina Wse. des Gen. Chrysochloa Hope (Chrys. Col.). *Entomologische Arbeiten aus dem Museum Georg Frey* **3** (1): 96–103. DP: 31 August 1952 (table of contents).

Jakob, H. 1952b: Drei neue Subspecies der Melasoma populi L. aus Asien (Chrys. Col). *Entomologische Arbeiten aus dem Museum G. Frey* **3** (1): 104–106. DP: 31 August 1952 (table of contents).

Jakob, H. 1953a: Eine neue Melasoma aus Transbaikalien. Melasoma selengensis nov. spec. (Chrys., Col.). *Entomologisches Nachrichtenblatt Österreichischer und Schweizer Entomologen* **5** (3/4): 35–37. DP: [30] April 1953 (issue cover).

Jakob, H. 1953b: Ergebnisse der Revision der Chrysomeliden des Naturhistorischen Museums (1. Teil). *Annalen des Naturhistorischen Museums in Wien* **59** [1952/1953]: 157–158. DP: [31] December 1953 (volume title page).

Jakob, H. 1953c: Eine neue "Chrysochloa-Art" aus den südlichen Alpen. Pp. 132–135. In: Mandl, K.: Ergebnisse einer koleopterologischen Sammelreise in das südwestliche Adamello-Gebiet. *Studi Trentini di Scienze Naturali* 30 (2): 128–135. DP: [31 December] 1953 (issue title page).

Jakob, H. 1954a: Eine neue Lema-Art aus China (Chrys. Col.). 9. Beitrag zur Kenntnis pal. Chrysomeliden. *Koleopterologische Rundschau* **32** [1951–1954] (4/6): 102–103. DP: 31 July 1954 (issue cover).

Jakob, H. 1954b: Bemerkungen zu Chrysomela cerealis ssp. alternans Panz. (= ssp. plorans Bechyne). (Chrys. Col.). 10. Beitrag zur Kenntnis pal. Chrysomeliden. *Koleopterologische Rundschau* **32** [1951–1954] (4/6): 104. DP: 31 July 1954 (issue cover).

Jakob, H. 1954c: Die Ergebnisse der österreichischen Iran-Expedition 1949/50. Coleoptera III. Teil. *Sitzungsberichte der Österreichischen Akademie der Wissenschaften, Mathematish-naturwissenschaftliche Klasse (Abteilung 1)* **163** (1/2): 45–50. DP: 23 December 1954 (registered at ZIN library).

Jakob, H. 1955: Ergebnisse der Revision der Chrysomelidae des Naturhistorischen Museums (2. Teil). *Annalen des Naturhistorischen Museums in Wien* **60** [1954/1955]: 230–235. DP: [31] December 1955 (volume title page).

Jakob, H. 1958a: Ergebnisse der Revision der Chrysomelidae des Naturhistorischen Museums, Wien (3. Teil). 11. Beitrag zur Kenntnis pal. Chrysomeliden. *Entomologisches Nachrichtenblatt Österreichischen und Schweizer Entomologen* **9** [1957] (3): 33–35. DP: [30] April 1958 (noticed on p. 56).

Jakob, H. 1958b: Eine neue Melasoma aus Zentralasien. 12. Beitrag zur Kenntnis pal. Chrysomeliden. *Entomologisches Nachrichtenblatt Österreichischer und Schweizer Entomologen* **10** (2): 38–39. DP: [30] November 1958 (noticed on p. 47).

Jakob, H. 1960: Ergebnisse der Revision der Chrysomelidae des Naturhistorischen Museums, Wien (4. Teil). 14. Beitrag zur Kenntnis pal. Chrysomeliden. *Zeitschrift der Arbeitsgemeinschaft Österreichischer Entomologen* **12** (1): 26–27. DP: [31] May 1960 (noticed on p. 31).

Jakob, H. 1961: Zwei neue Lilioceris-Arten (Col. Chrysomel.) aus China. 16. Beitrag zur Kenntnis pal. Chrysomeliden. *Zeitschrift der Arbeitsgemeinschaft Österreichischer Entomologen* **13** (2): 61–62. DP: [31] December 1961 (noticed on p. 64).

Jakob, H. 1962: The 3rd Danish Expedition to Central Asia. Zoological Results 28. Chrysomelidae (Insecta) aus Afghanistan. *Videnskabelige Meddelelser fra Dansk Naturhistorisk Forening i København* **124**: 193–198. DP: 14 May 1962 (footer on p. 198).

Jakob, H. 1965: Eine neue Subspecies von Chrysochloa speciosissima Scop. aus Österreich (Chrys. Col.). 19. Beitrag zur Kenntnis pal. Chrysomeliden. *Koleopterologische Rundschau* **42** [1964]: 11–12. DP: 1 June 1965 (volume cover).

Jakob, H. 1979: *Coleoptera, Fam. Chrysomelidae. Catalogus faunae Austriae. Teil XVp.* Wien: Österreichischen Akademie der Wissenschaften, 36 pp. DP: [31 December] 1979 (book cover).

Jankowski, J. 1932: Eine neue Art der Gattung Thelyterotarsus aus Mittelasien (Coleopt., Chrysom., Cryptocephalini). *Wiener Entomologische Zeitung* **49** (3): 168–170. DP: 10 October 1932 (footers througout the issue).

Jansson, A. 1947: Zwei Namenänderungen (Coleoptera). *Opuscula Entomologica* (Lund) **12**: 118. DP: 30 June 1947 (verso of the volume title page).

Japoshvili, G. & Aslan, E. G. 2020: Checklist of leaf beetles (Coleoptera: Chrysomelidae) from Sakartvelo (Georgia). *Annals of Agrarian Science* **18**: 332–358. DP: [31 December] 2020 (paper title page).

Jeanne, C. 1967: Révision des espèces françaises du genre Timarcha Latr. (Coléopt. Chrysomelidae). *Actes de la Société Linnéenne de Bordeaux, Série A* **102** [1965] (7): 1–25. DP: 25 January 1967 (last page of the issue).

Jekel, H. 1855: *Insecta Saundersiana: or characters of undescribed insects in the collection of William Wilson Saunders, Esq., F.R.S., F.L.S., &c. Coleoptera. Curculionides. – Part I*. London: J. v. Voorst, 153 pp., 2 pls. DP: 5 March 1855 (Bousquet 2016).

Jiang, S.-Q. 1988a: Four new species of Galerucinae from Xizang, China (Coleoptera: Chrysomelidae). *Acta Zootaxonomica Sinica* **13** (4): 392–396. DP: [31] October 1988 (paper title page).

Jiang, S.-Q. 1988b: Five new species of the genus Haplosomoides (Coleoptera: Chrysomelidae). *Sinozoologia* **6**: 177–182. DP: [31] December 1988 (paper title page).

Jiang, S.-Q. 1988c: A study on the Chinese Liroetis (Coleoptera: Chrysomelidae). *Sinozoologia* **6**: 183–198. DP: [31] December 1988 (paper title page).

Jiang, S.-Q. 1989: Four new Chinese species of Japonitata (Coleoptera: Chrysomelidae, Galerucinae). *Acta Entomologica Sinica* **32** (2): 221–225. DP: [31] May 1989 (paper title page).

Jiang, S.-Q. 1990a: New Galerucinae from the Mt. Namjagbarwa region (Coleoptera: Chrysomelidae). *Sinozoologia* **7**: 137–139. DP: [30] September 1990 (paper title page).

Jiang, S.-Q. 1990b: On the Chinese species of the Galerucinae genus Zangia (Coleoptera: Chrysomelidae). *Sinozoologia* **7**: 141–144. DP: [30] September 1990 (paper title page).

Jiang, S.-Q. 1990c: A new species of the genus Pseudosepharia (Coleoptera: Chrysomelidae). *Acta Entomologica Sinica* **33** (4): 455–456. DP: [30] November 1990 (paper title page).

Jiang, S.-Q. 1991: Six new species of the genus Pseudadimonia (Coleoptera: Chrysomelidae). *Acta Entomologica Sinica* **34** (1): 83–88. DP: [28] February 1991 (paper title page).

Jiang, S.-Q. 1992: Coleoptera: Chrysomelidae – Galerucinae. Pp. 646–674. In: Chen, S.-H. (ed.): *Insects of Hengduan Mountain Region. Volume I*. Beijing: Science Press, xii + 865 pp. DP: [31] August 1992 (colophon).

Joannis, L. de 1865: Monographie des galérucides d'Europe, du Nord de l'Afrique et de l'Asie. Gallerucides, tribu de la famille des phytophages, ou chrysomélines. *L'Abeille, Mémoires d'Entomologie* **3** [1865–1867]: 1–168. DP: 27 December 1865 (*Bull. Soc. Ent. Fr.* 1865: lxvii).

Jolivet, P. 1943a: Tableau dichotomique pour la détermination des espèces et variétés françaises du genre Timarcha, Latr. (Col., Chrysomelidae). *Miscellanea Entomologica* **40** (7): 67–72. DP: 27 December 1943 (reprint in Mauro Daccordi's library with paper stripe with the date glued on first page).

Jolivet, P. 1943b: Chrysomela menthastri Suffr. et ses variations dans le Département de la Manche. *Notices, Mémoires et Documents Publiés par la Société d'Archéologie et d'Histoire Naturelle du Départment de la Manche* 54: 78–79. DP: [31 December] 1943 (reprint title page).

Jolivet, P. 1944: Notes sur quelques espèces et variétés du genre Timarcha, Latr. (Col., Chrysomelidae). *Miscellanea Entomologica* 41 (7): 96–100. DP: [31] December 1944 (footer on p. 89).

Jolivet, P. 1950a: Une nouvelle espèce de Timarcha (Col. Chrysomelidae) des Pyrénées Orientales. *Vie et Milieu* 1 (2): 235–239. DP: [30] September 1950 (DL on last page of the issue cover).

Jolivet, P. 1950b: Bechyneia spinosa nov. gen., nov. sp. (Col. Chrysomelidae). *Bulletin & Annales de la Société Entomologique de Belgique* 86 (1/2): 36–39. DP: [31 December] 1950 (volume title page) [the issue cover was not examined].

Jolivet, P. 1951a: Contribution à l'étude du genre Gastrophysa Chevrolat (Coleoptera Chrysomelidae) (3me note). *Bulletin d'Institut Royal des Sciences Naturelles de Belgique* 27 (21): 1–47. DP: [31] May 1951 (paper title page).

Jolivet, P. 1951b: Variations géographiques chez deux espèces du genre Chrysolina Motschulsky (Coleoptera Chrysomelidae). *Bulletin d'Institut Royal des Sciences Naturelles de Belgique* 27 (51): 1–7. DP: [31] August 1951 (paper title page).

Jolivet, P. 1951c: Contribution a l'étude des coléoptères Chrysomeloidea de la région Indo-Australienne (1re note). *Bulletin d'Institut Royal des Sciences Naturelles de Belgique* 27 (52): 1–7. DP: [31] August 1951 (paper title page).

Jolivet, P. 1951d: Rectifications de nomenclature. *Bulletin & Annales de la Société Entomologique de Belgique* 87 (9/10): 194. DP: 30 October 1951 (issue cover).

Jolivet, P. 1951e: A propos de l'écologie de Chrysolina menthastri Suffr. (Col. Chrysomelidae). *Vie et Milieu* 2 (4): 470–472. DP: [31 December] 1951 (issue title page).

Jolivet, P. 1966a: Une nouvelle sous-espèce de Timarcha marocaine (Col. Chrysomelidae). *Bulletin de la Société des Sciences Naturelles et Physiques du Maroc* 46 (1/2): 121–125. DP: 9 November 1966 (last page of the issue).

Jolivet, P. 1966b: Notes systematiques et ecologiques sur les chrysomelides Marocains (Coleoptera) (1re note: Chrysomelinae). *Bulletin de la Société des Sciences Naturelles et Physiques du Maroc* 46 (1/2): 127–143. DP: 9 November 1966 (last page of the issue).

Jolivet, P. 1968a: Les races de Timarcha punctella Latreille (Col. Chrysomelidae) au Maroc. *Bulletin Mensuel de la Société Linnéenne de Lyon* 37 (2): 92–95. DP: [29] February 1968 (issue cover).

Jolivet, P. 1968b: Les races de Timarcha punctella Latreille (Col. Chrysomelidae) au Maroc (suite). *Bulletin Mensuel de la Société Linnéenne de Lyon* 37 (3): 113–118. DP: [31] March 1968 (issue cover).

Jolivet, P. 1968c: Les complexe de Timarcha scabripennis Fairmaire au Maroc (Col. Chrysomelidae). *Entomologische Arbeiten aus dem Museum G. Frey* 19: 229–236. DP: 1 June 1968 (table of contents).

Jolivet, P. 1979: Reflexions sur l'ecologie, l'origine et la distribution des chrysomelides (Col.) des Iles Mascareignes, Ocean Indien, avec la description de deux especes nouvelles. *Bulletin Mensuel de la Société Linnéenne de Lyon* 48 (10): 606–608 & 641–649. DP: [31] December 1979 (issue cover).

Jolivet, P. & Hawkeswood, T. J. 1995: *Host-plants of Chrysomelidae of the world. An essay about the relationships between the leaf-beetles and their food-plants*. Leiden: Backhuys Publishers, 281 pp. DP: [31 December] 1995 (book title page).

Jolivet, P. & Petitpierre, E. 1973: Plantes-hôtes connues des Timarcha Latreille (Col. Chrysomelidae). Quelques considérations sur les raisons possibles du trophisme sélectif. *Bulletin de la Societé Entomologique de France* 78 (1/2): 9–25. DP: 8 June 1973 (*dates de la publication du Bulletin* on p. 344).

Jolivet, P., Poinar, G. & Verma, K. K. 2014: Timarcha Latreille: A strange beetle and a living fossil. *Terrestrial Arthropod Reviews* 7 (1): 3–20. DP: 28 April 2014 (journal website).

Johnson, C. D., Southgate, B. J. & Delobel, A. 2004: A revision of the Caryedontini (Coleoptera: Bruchidae: Pachymerinae) of Africa and the Middle East. *Memoirs of the American Entomological Society* 44: 1–120. DP: 29 February 2004 (colophon).

Johnson, W. F. & Halbert, J. N. 1902: A list of the beetles of Ireland. *Proceedings of the Royal Irish Academy, Third Series* 6 (4): 535–827. DP: [31] July 1902 (volume contents).

Jurado-Rivera, J. A. & Petitpierre, E. 2015: New contributions to the molecular systematics and the evolution of host-plant associations in the genus Chrysolina (Coleoptera, Chrysomelidae, Chrysomelinae). *ZooKeys* 547: 165–192. DP: 17 December 2015 (paper title page).

Kacker, R. K. 1969: On a new species of the genus Crepidodera Chevrolat, 1844 (Coleoptera: Chrysomelidae: Halticinae). *Journal of the Zoological Society of India* 20 [1968] (1/2): 55–57. DP: [31] May 1969 (issue title page).

Kantner, F. 2007: Labidostomis warchalowskii sp. nov. from China (Coleoptera: Chrysomelidae: Clytrinae). *Genus* 18 (4): 559–565. DP: 28 December 2007 (paper title page).

Kapp, A. 2007: Eine neue Art der Gattung Orestia Germar aus den Abruzzen und aktuelle Funde von Orestia aus Mittel- und Süditalien (Coleoptera, Chrysomelidae). *Entomologische Blätter* 102 [2006] (1/3): 95–100. DP: 27 March 2007 (paper title page).

Karapetjan, A. P. 1973: Novyy vid zernovok (Coleoptera, Bruchidae) iz SSSR. [A new species of seed-beetles (Coleoptera, Bruchidae) from SSSR]. *Doklady Akademii Nauk Armyanskoy SSR* 57 (1): 44–46. SFP: 19 October 1973, DP: [31 December] 1973 (issue cover).

Karapetjan, A. P. 1976: Novyy vid zhukov zernovok iz Armyanskoy SSR (Coleoptera, Bruchidae). (A new species of Bruchidae from Armenia (Coleoptera, Bruchidae)). *Doklady Akademii Nauk Armyanskoy SSR* 62 (3): 189–191. SFP: 7 July 1976, DP: [31 December] 1976 (issue cover).

Karapetjan, A. P. 1989: Novyy vid zhukov zernovok iz Armyanskoy SSR (Coleoptera, Bruchidae). (A new species of Bruchidae from the Armenian SSR (Coleoptera, Bruchidae)). *Doklady Akademii Nauk Armyanskoy SSR* 83 (4): 178–180. SFP: 4 August 1989, DP: [31 December] 1989 (issue cover).

Karsch, F. 1882: Verzeichniss der von Herrn Stabsarzt Dr. Falkenstein in Westafrika (Chinchoxo) gesammelten Chrysomeliden, Endomychiden, Coccinelliden und Anthotribiden. *Berliner Entomologische Zeitschrift* **26** (2): 395–403. DP: [31] December 1882 (issue title page).

Kasap, H. 1984: A new species of Cheilotoma (Coleoptera: Chrysomelidae) from Turkey with lectotype designation of C. fulvicollis Sahlberg. *Coleopterists Bulletin* 38 (3): 215–219. DP: 9 November 1984 (last page of the issue cover).

Kasatkin, D. G. 2000: Materialy k izucheniyu fauny zhukov-zernovok (Coleoptera: Bruchidae) yuga evropeyskoy chasti Rossii i severnogo Kavkaza. (Materials on studying the fauna of Bruchidae (Coleoptera) from the south of the European part of Russia and the northern Caucasus). *Izvestiya Kharkovskogo Entomologicheskogo Obshchestva* 8 (1): 95–106. DP: [31 December] 2000 (paper title page).

Kasatkin, D. G. 2018: A new species of seed-beetles of the genus Bruchus Linnaeus, 1767 (Coleoptera: Chrysomelidae: Bruchinae) from Western Iran. *Caucasian Entomological Bulletin* 14 (1): 41–42, pls. 3–4. DP: 16 July 2018 (table of contents and journal website).

Kaszab, Z. 1962: Beiträge zur Kenntnis der Chrysomeliden-Fauna des Karpatenbeckens nebst Beschreibung neuer Formen (Coleoptera). *Folia Entomologica Hungarica, Series Nova* **15** (1): 25–93. DP: 20 June 1962 (table of contents).

Kaszab, Z. & Székessy, V. 1953: Bátorliget bogár-faunája, Coleoptera. [The beetle fauna of Bátorliget, Coleoptera]. Pp. 194–285. In: Székessy, V. (ed.): *Bátorliget élővilága*. Budapest: Akademiai Kiadó, 486 pp. DP: 2 September 1953 (colophon).

Kelly, P. G. & Reid, C. A. M. 1999: Designation of type species for the genera of Australian paropsine beetles (Coleoptera: Chrysomelidae). *Memoirs of Museum Victoria* 57 (2): 263–266. DP: 31 July 1999 (issue cover).

Kemner, N. A. 1919: De svenska arterna av släktet Haltica. *Entomologisk Tidskrift* 40 (2/4): 143–165. DP: 29 December 1919 (verso of the volume title page).

Kergoat, G. J. 2011: Évolution de la spécificité des relations entre les bruches et leurs plantes-hôtes et de leur aptitude à devenir des ravageurs. Pp. 25–31, pls. I–IV. In: Huignard, J., Glitho, I. A., Monge, J.-P. & Regnault-Roger, C. (eds.): *Insectes ravageurs des grains de légumineuses. Biologie des Bruchinae et lutte raisonnée en Afrique*. Versailles Cedex: Éditions Quae, 145 pp. [30] November 2011 (last page of the book).

Kergoat, G. J. & Alvarez, N. 2008: Assessing the phylogenetic usefulness of a previously neglected morphological structure through elliptic Fourier analyses: a case study in Bruchus seed-beetles (Coleoptera: Chrysomelidae: Bruchinae). *Systematic Entomology* **33**: 289–300. DP: 2 January 2008 (last page of the paper).

Kergoat, G. J., Delobel, P. & Delobel, A. 2007: Phylogenetic relationships of a new species of seed-beetle infesting Cercis siliquastrum L. in China and in Europe (Coleoptera: Chrysomelidae: Bruchinae: Bruchini). *Annales de la Société Entomologique de France, Nouvelle Série* **43** (3): 265–271. DP: 1 November 2007 (table of contents).

Kergoat, G. J., Delobel, A., Le Ru, B. & Silvain, J.-F. 2008: Seed-beetles in the age of the molecular: recent advances on systematics and host plant association patterns. Pp. 59–96, 376–378. In:

Jolivet, P., Santiago-Blay, J. & Schmitt, M. (eds.): *Research on Chrysomelidae. Volume 1*. Leiden: Brill, 430 pp. DP: 10 July 2008 (publisher website).

Kergoat, G. J., Le Ru, B. P., Genson, G., Cruaud, C., Couloux, A. & Delobel, A. 2011: Phylogenetics, species boundaries and timing of resource tracking in a highly specialized group of seed beetles (Coleoptera: Chrysomelidae: Bruchinae). *Molecular Phylogenetics and Evolution* **59**: 746–760. DP: 21 March 2011 (paper title page).

Kergoat, G. J. & Silvain, F.-J. 2004: Le genre Bruchidius (Coleoptera: Bruchidae) est-il monophylétique? Apports des méthodes de parcimonie, maximum de vraisemblance et inférence bayésienne. *Biosystema* **22**: 113–125. DP: [31 December] 2004 (footer on the paper title page).

Kevan, D. 1967: The British species of the genus Longitarsus Latreille (Col., Chrysomelidae). *Entomologist's Monthly Magazine* **103** (1235/1237): 83–110. DP: 20 October 1967 (issue cover).

Kiesenwetter, E. A. H. von 1850: Funfzig Diagnosen unbeschriebener oder wenig bekannter europäischer Käfer. *Entomologische Zeitung* (Stettin) **11** (7): 217–225. DP: [31] July 1850 (issue title page).

Kiesenwetter, E. A. H. von 1861: Eine entomologische Excursion in das Wallis und nach dem Monte Rosa im Sommer 1861. *Berliner Entomologische Zeitschrift* **5** (3/4): 360–395. DP: 26 December 1861 (Evenhuis 2022).

Kiesenwetter, E. A. H. von 1873: Bemerkungen zur Bearbeitung der Luperus-Arten in der Monographie des Gallerucides von Joannis in Marseul's Abeille T. III Ao. 1866. p. 1 fig. *Berliner Entomologische Zeitschrift* **17** (1/2): 23–31. DP: [31] October 1873 (issue cover).

Kiesenwetter, E. A. H. von 1878: [new taxon]. In: Schneider, O. & Leder, H.: *Beiträge zur Kenntniss der kaukasischen Käferfauna*. Brünn: W. Burkart, 359 + [1] pp., 6 pls. DP: [31 December] 1878 (Döberl 2009, see also the comments in the Dating of publications chapter).

Kiesenwetter, E. A. H. von 1879: Chrysomela Kenderesyi n. sp. *Deutsche Entomologische Zeitschrift* **23** (2): 256. DP: [30] November 1879 (table of contents).

Kimoto, S. 1957: Description of a new species of Hispinae, Leptispa miyamotoi, from Amami-Ôshima. Studies on Chrysomelidae III (Coleoptera, Chrysomelidae). *Insecta Matsumurana* **21** (1/2): 77–79. DP: [31] August 1957 (*dates of publication in numbers* on p. ii).

Kimoto, S. 1963: Descriptions of new species of Chrysomelidae from Japan, with notes on some known species. *Fragmenta Coleopterologica* **1963** (3): 13–14 & (4): 15–18. DP: [31] May 1963 (issues title pages).

Kimoto, S. 1964a: The Chrysomelidae of Japan and the Ryukyu Islands. III. *Journal of the Faculty of Agriculture, Kyushu University* (Fukuoka) **13** [1964–1966] (1): 141–164. DP: 30 March 1964 (paper title page).

Kimoto, S. 1964b: The Chrysomelidae of Japan and the Ryukyu Islands. VI. Subfamily Galerucinae I. *Journal of the Faculty of Agriculture, Kyushu University* (Fukuoka) **13** [1964–1966] (2): 287–308. DP: 30 October 1964 (paper title page).

Kimoto, S. 1965: The Chrysomelidae of Japan and the Ryukyu Islands. VII. Subfamily Galerucinae II. *Journal of the Faculty of Agriculture, Kyushu University* (Fukuoka) **13** [1964–1966] (3): 369–459. DP: 30 May 1965 (paper title page).

Kimoto, S. 1966a: The Chrysomelidae of Japan and the Ryukyu Islands. X. Subfamily Alticinae III. *Journal of the Faculty of Agriculture, Kyushu University* (Fukuoka) **13** [1964–1966] (4): 601–671. DP: 15 February 1966 (paper title page).

Kimoto, S. 1966b: The Chrysomelidae of Japan and the Ryukyu Islands. XI. *Journal of the Faculty of Agriculture, Kyushu University* (Fukuoka) **13** [1964–1966] (4): 635–671. DP: 15 February 1966 (paper title page).

Kimoto, S. 1966c: A list of the chrysomelid specimens of Taiwan preserved in the Zoological Museum, Berlin. *Esakia* **5**: 21–38. DP: 15 November 1966 (paper title page).

Kimoto, S. 1967a: A list of the chrysomelid species from Hong Kong, with descriptions of three new species. *Esakia* **6**: 55–63. DP: 10 November 1967 (paper title page).

Kimoto, S. 1967b: A list of the chrysomelid specimens from the Himalayas and Kashmir, preserved in the Zoological Museum, Berlin. *Esakia* **6**: 65–75. DP: 10 November 1967 (paper title page).

Kimoto, S. 1969: Notes on the Chrysomelidae from Taiwan II. *Esakia* **7**: 1–68. DP: 31 January 1969 (paper title page).

Kimoto, S. 1970a: Notes on the Chrysomelidae from Taiwan III. *Kontyû* **38** (2): 176–183. DP: 30 June 1970 (heading on p. 93).

Kimoto, S. 1970b: Notes on the Chrysomelidae from Taiwan IV. *Kontyû* **38** (3): 205–221. DP: 30 September 1970 (heading on p. 187).

Kimoto, S. 1970c: A list of the Nepalese chrysomelid specimens preserved in Zoologische Sammlung des Bayerischen Staates, München. *Khumbu Himal* **3** (3): 412–421. DP: 1 October 1970 (issue cover).

Kimoto, S. 1970d: Notes on the Chrysomelidae from Taiwan V. *Kontyû* **38** (4): 292–313. DP: 15 December 1970 (heading on p. 281).

Kimoto, S. 1971a: Notes on the Chrysomelidae from Taiwan, VI. *Entomological Review of Japan* **23** (2): 73–87. DP: 30 September 1971 (General index 1968–1973).

Kimoto, S. 1971b: [new taxa]. In: Kimoto, S. & Hiura, I.: A list of the chrysomelid specimens preserved in the Osaka Museum of Natural History, III (Insecta: Coleoptera). *Bulletin of the Osaka Museum of Natural History* **25**: 1–26. DP: 25 November 1971 (reprint cover).

Kimoto, S. 1974a: New or little known Chrysomelidae (Coleoptera) from Japan and its adjacent regions I. *Kontyû* **42** (2): 144–150. DP: 25 June 1974 (paper title page).

Kimoto, S. 1974b: Notes on the Chrysomelidae from Taiwan VII. *Entomological Review of Japan* **26** (1/2): 21–26. DP: 30 June 1974 (General index 1974–1977).

Kimoto, S. 1974c: On some infraspecific variation of chrysomelid beetles (Coleoptera) occuring in the Ryukyu Archipelago. *Kontyû* **42** (3): 270–282. DP: 25 September 1974 (paper title page).

Kimoto, S. 1976: Notes on the Chrysomelidae from Taiwan, VIII. *Entomological Review of Japan* **29** (1/2): 1–9. DP: 31 August 1976 (General index 1974–1977).

Kimoto, S. 1977: Ergebnisse der Bhutan-Expedition 1972 des Naturhistorischen Museums in Basel. Coleoptera: Fam. Chrysomelidae Subfam. Galerucinae. *Entomologica Basiliensia* **2**: 351–392. DP: 7 January 1977 (table of contents).

Kimoto, S. 1978a: Revisional study on the Japanese species of genus Hemipyxis Dejean (Coleoptera, Chrysomelidae, Alticinae). *Entomological Review of Japan* **31** (1/2): 13–21, pls. 1–2. DP: 30 June 1978 (General index 1978–1981).

Kimoto, S. 1978b: Notes on the Chrysomelidae from Taiwan, China, IX. *Entomological Review of Japan* **31** (1/2): 69–74. DP: 30 June 1978 (General index 1978–1981).

Kimoto, S. 1979a: New or little known Chrysomelidae (Coleoptera) from Japan and its adjacent regions, II. *Entomological Review of Japan* **33** (1/2): 41–45. DP: 30 July 1979 (General index 1978–1981).

Kimoto, S. 1979b: The Galerucinae (Coleoptera: Chrysomelidae) of Nepal, Bhutan and northern territories of India, in the Natural History Museum in Basel, I. *Entomologica Basiliensia* **4**: 463–478. DP: 10 October 1979 (table of contents).

Kimoto, S. 1981a: New or little known Japanese Donaciinae (Coleoptera: Chrysomelidae). *Bulletin of the Osaka Museum of Natural History* **34**: 23–26. DP: 31 March 1981 (verso of volume cover).

Kimoto, S. 1981b: Notes on the Chrysomelidae from Taiwan, China. X. *Entomological Review of Japan* **36** (1): 1–4. DP: 30 September 1981 (General index 1981–1985).

Kimoto, S. 1981c: The Cassidinae of Nepal, Bhutan and northern territories of India, in the Natural History Museum in Basel (Coleoptera, Chrysomelidae). *Entomological Review of Japan* **36** (1): 55–62. DP: 30 September 1981 (General index 1981–1985).

Kimoto, S. 1982a: The Galerucinae of Nepal, Bhutan and northern territories of India, in the Natural History Museum in Basel, II. (Coleoptera: Chrysomelidae). *Entomological Review of Japan* **37** (1): 7–24. DP: 30 June 1982 (General index 1981–1985).

Kimoto, S. 1982b: Description of a new species of galerucid beetle from Taiwan, China (Coleoptera: Chrysomelidae). Pp. 151–152. In: Satô, M., Hori, Y., Arita, Y. & Okadome, T. (eds.): *Special issue to the memory of retirement of Emeritus Professor Michio Chûjô*. Nagoya: Nagoya Women's University, [4] + 185 pp. DP: 17 December 1982 (colophon).

Kimoto, S. 1983a: Revisional study on Megalopodinae, Donaciinae and Clytrinae of Japan (Coleoptera: Chrysomelidae). *Entomological Review of Japan* **38** (1): 5–23. DP: 30 June 1983 (General index 1981–1985).

Kimoto, S. 1983b: New or little known Chrysomelidae (Coleoptera) from Japan and its adjacent regions, III. *Entomological Review of Japan* **38** (1): 45–54. DP: 30 June 1983 (General index 1981–1985).

Kimoto, S. 1984: Notes on the Chrysomelidae from Taiwan, China, XI. *Entomological Review of Japan* **39** (1): 39–58. DP: 30 June 1984 (General index 1981–1985).

Kimoto, S. 1986: New or little known Chrysomelidae (Coleoptera) from Japan and its adjacent regions, V. *Entomological Review of Japan* **41** (2): 123–129. DP: 26 December 1986 (verso of the issue cover).

Kimoto, S. 1987: Descriptions of two new species of Chrysomelidae (Coleoptera) collected by the Nagoya University Scientific Expedition to Taiwan in 1986. *Entomological Review of Japan* **42** (2): 95–98. DP: 22 December 1987 (verso of the issue cover).

Kimoto, S. 1989a: Chrysomelidae (Coleoptera) of Thailand, Cambodia, Laos and Vietnam. IV. Galerucinae. *Esakia* 27: 1–241. DP: 25 January 1989 (colophon).

Kimoto, S. 1989b: Descriptions of a new genus and three new species of Taiwanese Chrysomelidae (Coleoptera) collected by Dr. Kintaro Baba, on the occasion of his entomological survey in 1986. *Entomological Review of Japan* **44** (2): 73–78. DP: 20 December 1989 (verso of the issue cover).

Kimoto, S. 1991a: Descriptions of a new genus and six new species of Chrysomelidae (Coleoptera) collected by Dr. Kintaro Baba in Taiwan, China. *Entomological Review of Japan* **46** (1): 13–20. DP: 28 June 1991 (verso of the issue cover).

Kimoto, S. 1991b: Notes on the Chrysomelidae from Taiwan, China, XII. *Entomological Review of Japan* **46** (2): 115–124. DP: 31 December 1991 (verso of the issue cover).

Kimoto, S. 1993: New or little known Chrysomelidae (Coleoptera) from Japan and its adjacent regions, VI. *Entomological Review of Japan* **48** (2): 93–101. DP: 28 December 1993 (verso of the issue cover).

Kimoto, S. 1994: Description of a new galerucid species from Taiwan, China (Coleoptera: Chrysomelidae). *Transactions of the Shikoku Entomological Society* **20** [1993–1994] (3/4): 191–192. DP: 10 March 1994 (issue cover).

Kimoto, S. 1996: Notes on the Chrysomelidae from Taiwan, China, XIII. *Entomological Review of Japan* **51** (1): 27–51. DP: 15 June 1996 (paper title page).

Kimoto, S. 1997: [new names]. In: Kimoto, S. & Takizawa, H.: *Leaf beetles (Chrysomelidae) of Taiwan*. Tokyo: Tokai University Press, xvii + 581 pp., 119 pls. DP: [31] July 1997 (colophon).

Kimoto, S. 2000a: Descriptions of some new genera and species of Chrysomelidae (Coleoptera) from Thailand, Laos, and Vietnam. *Serangga* **5** (1): 1–39. DP: [31] January 2000 (issue cover).

Kimoto, S. 2000b: Chrysomelidae (Coleoptera) of Thailand, Cambodia, Laos and Vietnam. VII. Alticinae. *Bulletin of the Institute of Comparative Studies of International Cultures and Societies* **26**: 103–299. DP: [30] September 2000 (volume cover).

Kimoto, S. 2001: The Chrysomelidae (Insecta: Coleoptera) collected by the Kyushu University Scientific Expedition to the Nepal Himalaya in 1971 and 1972. *Bulletin of the Kitakyushu Museum of Natural History* **20**: 17–80. DP: 31 March 2001 (paper title page).

Kimoto, S. 2004: New or little known Chrysomelidae (Coleoptera) from Nepal, Bhutan and the northern territories of Indian subcontinent. *Bulletin of the Kitakyushu Museum of Natural History and Human History, Series A* **2**: 47–63. DP: 31 March 2004 (paper title page).

Kimoto, S. 2005: Systematic catalogue of the Chrysomelidae (Coleoptera) from Nepal and Bhutan. *Bulletin of the Kitakyushu Museum of Natural History & Human History, Series A* **3**: 13–114. DP: 31 March 2005 (paper title page).

Kimoto, S. & Chu Y.-I. 1996: Systematic catalog of Chrysomelidae of Taiwan (Insecta: Coleoptera). *Bulletin of the Institute of Comparative Studies of International Cultures and Societies* **16**: 1–152. DP: [31] January 1996 (volume cover).

Kimoto, S. & Gressitt, J. L. 1966: The Chrysomelidae of the Ryukyu Archipelago. *Pacific Insects* **8** (2): 467–577. DP: 15 July 1966 (paper title page).

Kimoto, S. & Gressitt, J. L. 1979: Chrysomelidae (Coleoptera) of Thailand, Cambodia, Laos and Vietnam. I. Sagrinae, Donaciinae, Zeugophorinae, Megalopodinae and Criocerinae. *Pacific Insects* 20 (2/3): 191–256. DP: 16 May 1979 (paper title page).

Kimoto, S. & Gressitt, J. L. 1981: Chrysomelidae (Coleoptera) of Thailand, Cambodia, Laos and Vietnam. II. Clytrinae, Cryptocephalinae, Chlamisinae, Lamprosomatinae and Chrysomelinae. *Pacific Insects* 23 (3/4): 286–391. DP: 11 December 1981 (paper title page).

Kimoto, S. & Gressitt, J. L. 1982: Chrysomelidae (Coleoptera) of Thailand, Cambodia, Laos and Vietnam. III. Eumolpinae. *Esakia* 18: 1–141. DP: 25 November 1982 (colophon).

Kimoto, S. & Kawase, E. 1966: A list of some chrysomelid specimens collected in E. Manchuria and N. Korea. *Esakia* 5: 39–48. DP: 15 November 1966 (paper title page).

Kimoto, S. & Osawa, S. 1995: Description of a new species of the genus Ambrostoma Motschulsky from Taiwan, China (Coleoptera, Chrysomelidae, Chrysomelinae). *Entomological Review of Japan* 50 (1): 15–16, pl. 5. DP: 29 July 1995 (verso of the issue cover).

Kimoto, S. & Takahashi, Y. 1992: Description of a new species of Galerucinae (Chrysomelidae, Col.) from Japan. *Entomological Review of Japan* 47 (2): 99–101. DP: 30 December 1992 (verso of the issue cover).

Kimoto, S. & Takizawa, H. 1972: Chrysomelid-beetles of Nepal, collected by the Hokkaido University Scientific Expedition to Nepal Himalaya, 1968. Part I. *Kontyû* 40 (4): 215–223. DP: 25 November 1972 (*dates of publications* at the end of issue 4).

Kimoto, S. & Takizawa, H. 1973: The chrysomelid-beetles of Nepal, collected by the Hokkaido University Scientific Expedition to Nepal Himalaya, 1968. Part II. *Kontyû* 41 (2): 170–180. DP: 25 June 1973 (*dates of publications* at the end of issue 4).

Kimoto, S. & Takizawa, H. 1981: Chrysomelid-beetles of Nepal, collected by the Hokkaido University Scientific Expeditions to Nepal Himalaya, 1968 and 1975. Part III (Coleoptera). *Entomological Review of Japan* 35 (1/2): 51–65, pls. 2–4. DP: 30 March 1981 (General index 1978–1981).

Kimoto, S. & Takizawa, H. 1983: Chrysomelid beetles of Nepal, collected by the Himalaya Expedition 1979 of the National Science Museum, Tokyo (part 1). *Bulletin of the National Science Museum, Tokyo* (A) 9 (2): 83–96. DP: 22 June 1983 (paper title page).

Kimoto, S. & Takizawa, H. 1994: *Leaf beetles (Chrysomelidae) of Japan*. Tokyo: Tokai University Press, xvii + 539 pp., 133 pls. DP: [31 December] 1994 (colophon).

Kimoto, S. & Takizawa, H. 2002: Chrysomelid beetles of Nepal collected by the Himalaya expedition 1979 of the National Science Museum, Tokyo (part 2). *Bulletin of the National Science Museum, Tokyo* (A) 28 (3): 143–149. DP: 22 September 2002 (paper title page).

Kingsolver, J. M. 1973: Description of a new genus and a new species of Bruchidae from South America (Coleoptera). *Journal of the Washington Academy of Sciences* 63 (4): 142–146. DP: [31] December 1973 (issue cover).

Kingsolver, J. M. 1999: A new species of Callosobruchus (Coleoptera: Bruchidae) from Thailand and China. *Entomological News* 110 (4): 209–213. DP: 22 October 1999 (*mailing dates* on p. 320).

Kingsolver, J. M. & Borowiec, L. 1988: The genus Spermophagus in the New World (Coleoptera, Bruchidae). *Elytron* 2: 81–84. DP: [31 December] 1988 (volume cover).

Kippenberg, H. 1974: Pachybrachis seidenstueckeri, eine neue Art aus Kleinasien. *Entomologische Blätter* 70 (1): 47–50. DP: 15 March 1974 (paper title page).

Kippenberg, H. 1975: Eine auffällige Form von Cryptocephalus carinthiacus Suffr. *Nachrichtenblatt der Bayerischen Entomologen* 24 (3): 64. DP: 15 June 1975 (issue title page).

Kippenberg, H. 1994: 88. Familie: Chrysomelidae. Pp. 17–92, 142. In: Lohse, G. A. & Lucht, W. H. (eds.): *Die Käfer Mitteleuropas. Band 14. 3. Supplementband mit Katalogteil.* Krefeld: Goecke & Evers, 403 pp. DP: [31 December] 1994 (colophon).

Kippenberg, H. 1998: 88. Familie: Chrysomelidae. Pp. 313–318. In: Lucht, W. H. & Klausnitzer, B. (eds.): *Die Käfer Mitteleuropas. Band 15. 4. Supplementband.* Krefeld: Goecke & Evers, 398 pp. DP: [31 December] 1998 (colophon).

Kippenberg, H. 2001: Neuordnung der Gonioctena variabilis – Gruppe (Coleoptera, Chrysomelidae). *Entomologische Blätter* 97 (1): 13–34. DP: 10 May 2001 (paper title page).

Kippenberg, H. 2003: Rote Liste gefährdeter Blatt- und Samenkäfer (Coleoptera: Chrysomelidae et Bruchidae) Bayerns. Pp. 154–160. In: *Rote Liste gefährdeter Tiere Bayerns. Schriftenreihe Heft 166.* Augsburg: Bayerisches Landesamt für Umweltschutz, 384 pp. DP: [31 December] 2003 (footer on the paper title page).

Kippenberg, H. 2004: Chrysolina (Taeniosticha) petitpierrei n. sp. aus den Pyrenäen (Coleoptera, Chrysomelidae). *Entomologische Blätter* 100 (1): 19–22. DP: 30 June 2004 (paper title page).

Kippenberg, H. 2008: Revision der Untergattung Protorina Weise der Gattung Oreina Chevrolat (Coleoptera: Chrysomelidae: Chrysomelinae). *Koleopterologische Rundschau* 78: 367–418. DP: [31] July 2008 (paper title page).

Kippenberg, H. 2010a: New acts and comments. Chrysomelidae: Chrysomelinae. Pp. 67–73. In: Löbl, I. & Smetana, A. (eds.): *Catalogue of Palaearctic Coleoptera. Volume 6. Chrysomeloidea.* Stenstrup: Apollo Books, 924 pp. DP: 22 February 2010 (colophon).

Kippenberg, H. 2010b: Chrysomelidae: Chrysomelinae. Pp. 390–437. In: Löbl, I. & Smetana, A. (eds.): *Catalogue of Palaearctic Coleoptera. Volume 6. Chrysomeloidea.* Stenstrup: Apollo Books, 924 pp. DP: 22 February 2010 (colophon).

Kippenberg, H. 2010c: Beitrag zur Kenntnis von Donacia delagrangei Pic und D. kraatzi Weise (Coleoptera: Chrysomelidae: Donaciinae). *Koleopterologische Rundschau* 80: 183–188. DP: [30] September 2010 (paper title page).

Kippenberg, H. 2011: Cryptocephalus messutati sp. n. aus der Türkei (Col., Chrysomelidae). *Mitteilungen des Internationalen Entomologischen Vereins e. V.* (Frankfurt am Main) 36 (1/2): 43–49. DP: 30 June 2011 (paper title page).

Kippenberg, H. 2012a: Eine neue Unterart der iberischen Blattkäferart Chrysolina (Chalcoidea) janbechynei (Cobos). *Mitteilungen des Internationalen Entomologischen Vereins e. V.* (Frankfurt am Main) 37 (1/2): 56–61. DP: 30 April 2012 (paper title page).

Kippenberg, H. 2012b: Lopatinica subg. n., eine neue Untergattung von Chrysolina Motschulsky aus dem Kaukasus und der Türkei (Coleoptera: Chrysomelidae). *Koleopterologische Rundschau* 82: 317–337. DP: [30] September 2012 (paper title page).

Kippenberg, H. 2012c: Überprüfung der ehemals zur Untergattung Colaphoptera Motschulsky gerechneten Chrysolina-Arten Europas (Coleoptera, Chrysomelidae), Teil I: Untergattungen. *Entomologische Blätter und Coleoptera* 108: 141–150. DP: [31] December 2017 (volume cover).

Kippenberg, H. 2015: Strukturierung von artenreichen Chrysomelidae-Gattungen, am Beispiel der Gattung Donacia Fabricius (Coleoptera: Chrysomelidae). *Koleopterologische Rundschau* 85: 249–282. DP: [30] September 2015 (paper title page).

Kippenberg, H. 2017: Überprüfung der ehemals zur Untergattung Colaphoptera Motschulsky gerechneten Chrysolina-Arten Europas (Col., Chrysomelidae), Teil II. *Entomologische Blätter und Coleoptera* 113 (1): 95–178. DP: [30] November 2017 (issue cover).

Kippenberg, H. 2018: Eine übersehene Smaragdina-Art aus dem Iran (Coleoptera: Chrysomelidae). *Koleopterologische Rundschau* 88: 217–220. DP: [30] September 2018 (paper title page).

Kippenberg, H. 2019: Rhodopaea heinzi sp. n. aus Anatolien (Coleoptera: Chrysomelidae: Eumolpinae). *Koleopterologische Rundschau* 89: 321–324. DP: [30] September 2019 (paper title page).

Kippenberg, H. 2020a: [new taxa]. In: Kippenberg, H. & Mikhailov, Y. E.: Contribution to the knowledge of Entomoscelis adonidis (Pallas, 1771) and allied species (Coleoptera: Chrysomelidae: Chrysomelinae). *Koleopterologische Rundschau* 90: 257–290. DP: [30] September 2020 (paper title page).

Kippenberg, H. 2020b: Chrysolina (Stichoptera) latecincta (Demaison, 1896) und ihre Subspezies in den Alpen (Coleoptera, Chrysomelidae). *Mitteilungen der Münchner Entomologischen Gesellschaft* 110: 53–64. DP: 15 November 2020 (paper title page).

Kirby, W. 1797: A history of three species of Cassida. *Transactions of the Linnean Society* (London) 3: 7–11. DP: 25 May 1797 (Raphael 1970).

Kirby, W. 1837: Part the fourth and last. The insects. In: Richardson, J.: *Fauna Boreali-Americana; or the zoology of the northern parts of British America: containing descriptions of the objects of natural history collected on the late Northern Land Expeditions, under command of captain Sir John Franklin, R. N.* Norwich: J. Fletcher, xxxix + 325 + [1] pp., 8 pls. DP: 14 October 1837 (Bousquet 2016).

Kirsch, T. 1866: Rhaphidopalpa signata n. sp. *Berliner Entomologische Zeitschrift* 10 (1/3): 284. DP: [31] July 1866 (table of contents).

Kirsch, T. 1875: Neue Käfer aus Malacca. *Mittheilungen des Königlichen Zoologischen Museums zu Dresden* 1: 25–58. [31 December] 1875 (volume title page).

Kirsch, T. 1880: Adimonia costipennis n. sp. *Entomologische Monatsblätter* 2 (21): 164. DP: [30] November 1880 (issue title page).

Kismali, S. & Sassi, D. 1994: Preliminary list of Chrysomelidae with notes on distribution and importance of species in Turkey, II. Subfamily Cassidinae Spaeth. *Türkiye Entomoloji Dergisi* 18 (3): 141–156. DP: [30] September 1994 (table of contents).

Klug, J. C. F. 1835: Insekten. Pp. 27–50 pp., pls. xv–xvi. In: Erman, A.: *Reise um die Erde durch Nord-Asien und die beide Oceanien in den Jahren 1828, 1829 und 1830. Naturhistorischer Atlas.* Berlin: G. Reimer, vi + 64 pp., 17 pls. DP: 30 December 1835 (Bousquet 2016).

Knoch, A. W. 1801: *Neue Beyträge zur Insectenkunde. Erster Theil.* Leipzig: Schwickert, xii + 208 pp., 9 pls. DP: [31 December] 1801 (Bousquet 2016).

Koch, J. D. W. 1803a: Monographie der von den Verfassern in dem Departemente vom Donnersberge, und den angrenzenden Gegenden der Departemente von der Saar, und von Rhein und Mosel einheimisch beobachteten Flohkäfer. (Haltica.). Pp. 5–90. In: Hoffmann, J. J. J., Koch, J. D. W., Müller, P. W. J. & Linz, J. M.: *Entomologische Hefte enthaltend Beiträge zur weitern Kenntniss und Aufklärung der Insektengeschichte. Eine Vorarbeit zu einer künftigen Faune des Departements vom Donnersberge und den angrenzenden Gegenden der Departemente von der Saar, und von Rhein und Mosel. Ausgearbeitet von einigen Freunden der Naturgeschichte. Zweites Heft.* Frankfurt am Main: Friedrich Esslinger, 130 + [4] pp., 2 pls. DP: [31 December] 1803 (Bousquet 2016).

Koch, J. D. W. 1803b: Nachtrag. Pp. 120–130. In: Hoffmann, J. J. J., Koch, J. D. W., Müller, P. W. J. & Linz, J. M.: *Entomologische Hefte enthaltend Beiträge zur weitern Kenntniss und Aufklärung der Insektengeschichte. Eine Vorarbeit zu einer künftigen Faune des Departements vom Donnersberge und den angrenzenden Gegenden der Departemente von der Saar, und von Rhein und Mosel. Ausgearbeitet von einigen Freunden der Naturgeschichte. Zweites Heft.* Frankfurt am Main: Friedrich Esslinger, 130 + [4] pp., 2 pls. DP: [31 December] 1803 (Bousquet 2016).

Kocher, L. 1953a: Antipa Paykulli Lac et fasciata Lef (Col. Chrysom.). *Comptes Rendus des Séances Mensuelles de la Société de Sciences Naturelles du Maroc* **18** [1952] (7): 117–118. DP: 5 January 1953 (noticed on p. 121).

Kocher, L. 1953b: Notes sur les Timarcha marocaines (Coléopt. chrysomélides). Contribution a l'étude systématique du genre. *Bulletin de la Société des Sciences Naturelles du Maroc* **31** [1951] (2): 313–332. DP: 30 January 1953 (noticed on p. 357).

Kocher, L. 1953c: Localisations nouvelles ou intéressantes de coléoptères marocains. *Travaux de l'Institut Scientifique Chérifien, Série Zoologie* **7**: 6–142. DP: 30 April 1953 (last page of the volume).

Kocher, L. 1955: Prospection entomologique (Coléoptères) de la Moyenne Moulouya. *Bulletin de la Société des Sciences Naturelles et Physiques du Maroc* **34** [1954] (3): 263–286. DP: [31] March 1955 (last page of the issue).

Kocher, L. 1956: Description d'une nouvelle clytrine du Pré-Sahara marocaine (Col. Chrysom.). *Comptes Rendus des Séances Mensuelles de la Société de Sciences Naturelles et Physique du Maroc* **22** (7): 127–128. DP: 31 December 1956 (last page of the issue).

Kocher, L. 1957: Observations complémentaires sur les Timarcha Marocaines (Col. Chrysom.). *Bulletin de la Société des Sciences Naturelles et Physiques du Maroc* **36** [1956] (2): 125–137. DP: 15 March 1957 (last page of the issue).

Kocher, L. 1958: Catalogue commenté des coléoptères du Maroc. Fascicule VIII. Phytophages. *Travaux de l'Institut Scientifique Chérifien, Série Zoologie* **19**: 1–172. DP: 5 November 1958 (last page of the volume).

Kocher, L. 1959: Contribution à l'étude des chrysomélides du Maroc. *Mémoires de la Société des Sciences Naturelles et Physiques du Maroc, Nouvelle série, Zoologie* **5** [1958]: 1–82. DP: 22 June 1959 (last page of the paper).

Kocher, L. 1961: Nouvelles espèces de coléoptères du Maroc. *Bulletin de la Société des Sciences Naturelles et Physiques du Maroc* 40 [1960] (3): 233–241. DP: 15 September 1961 (last page of the issue).

Kocher, L. 1963: Coléoptères nouveaux ou intéressants du Maroc présaharien. *Bulletin de la Société des Sciences Naturelles et Physiques du Maroc* **42** [1960] (3): 195–214. DP: [31 December] 1963 (see the comments in Dating of publications).

Kocher, L. 1965: Coléoptères nouveaux ou intéressants de la région Rifaine. *Bulletin de la Société des Sciences Naturelles et Physiques du Maroc* **43** [1963] (4): 193–199. DP: [31 December] 1965 (*Comptes Rendus Séances Mens. Soc. Sci. Nat. Phys. Maroc* 1965, 31 (1): 2).

Kocher, L. 1967: Note complementaire sur les Macrocoma du Maroc (Col. Chrysom. Eumolp.). *Bulletin de la Société des Sciences Naturelles et Physiques du Maroc* **46** [1966] (3/4): 289–303. DP: 27 October 1967 (last page of the issue).

Kocher, L. 1969a: Description de nouveaux coleopteres du Maroc. *Bulletin de la Société des Sciences Naturelles et Physiques du Maroc* **48** [1968] (1/2): 107–113. DP: 15 May 1969 (last page of the issue).

Kocher, L. 1969b: Catalogue commenté des coléoptères du Maroc. Feuillet rectificatif No. 19 (Fasc. VIII, p. 82 à 85) (Fam. Chrysomelidae). *Travaux de l'Institut Scientifique Chérifien et de la Faculte des Sciences, Série Zoologie* **1969**: 1–6. DP: [31 December] 1969 (volume cover).

Kolbe, H. J. 1886: Beiträge zur Kenntniss der Coleopteren-Fauna Koreas, bearbeitet auf Grund der von Herrn C. Gottsche während der Jahre 1883 und 1884 in Korea veranstalteten Sammlung; nebts Bemerkungen über die zoogeographischen Verhältnisse dieses Faunengebietes und Untersuchungen über einen Sinnesapparat im Gaumen von Misolampidius morio. *Archiv für Naturgeschichte* **52** [1] (2): 139–240, pls. X–XI. DP: 27 September 1886 (*Zool. Anz.* 1886, 9 (233): 560).

Kolenati, F. 1858: Meletemata Entomologica. Fasc. VIII. Curculionina Caucasi et vicinorum. *Bulletin de la Société Impériale des Naturalistes de Moscou* **31** (1): 102–184, pls. I–IV. DP: 7 July 1858 (Bousquet & Davies 2021).

Kollar, V. & Redtenbacher, L. 1844: Aufzählung und Beschreibung der von Freiherrn Carl v. Hügel auf seiner Reise durch Kaschmir und das Himaleyagebirge gesammelten Insecten. Pp. 393–564, 28 pls. In: Hügel, K. F. von (ed.): *Kaschmir und das Reich der Siek. Vierter Band. Zweite Abtheilung*. [1844–1848], Stuttgart: Hallbergerischer Verlag, pp. 247–865. DP (pp. 247–586): [31 December] 1844, (pp. 587–865 + [6]): [31 December] 1848 (Bousquet 2016).

Kolosov, Yu. M. 1925: Novye aberracii zhestkokrylykh Srednego Urala. *Izvestiya Entomologicheskogo i Fitopatologicheskogo Byuro Uralskogo Obshchestva Lyubiteley Estestvoznaniya* (Sverdlovsk) **1925** (1): 3. DP: 5 May 1925 (handwritten date on the J. Bezděk's copy).

Kolov, S. V. 2017: A new species of the leaf-beetle genus Cryptocephalus Geoffroy (Coleoptera, Chrysomelidae: Cryptocephalinae) from southeastern Kazakhstan. *Entomological Review* **96** [2016] (7): 899–901. DP: 22 March 2017 (registered at ZIN library).

Komiya, Y. 1985: Studies on the Trichochrysea-species of Japan, Ryukyu Archipelago, Taiwan and Korea (Coleoptera, Chrysomelidae, Eumolpinae). *Elytra* 12 (2): 11–25. DP: 10 February 1985 (paper title page).

Komiya, Y. 1986: Notes on the Taiwanese Megalopodinae with description of a new species (Coleoptera, Chrysomelidae). *Elytra* 14 (1): 1–9. DP: 10 August 1986 (paper title page).

Komiya, Y. 1988: Studies on the Colaspoides-species of Japan, Ryûkyû Archipelago and Taiwan (Coleoptera: Chrysomelidae-Eumolpinae). Part 1. *Transactions of the Shikoku Entomological Society* 19 [1988–1991] (1/2): 17–24. DP: 30 November 1988 (issue cover).

Komiya, Y. 1991: Studies on the Colaspoides-species of Japan, Ryûkyû Archipelago and Taiwan (Coleoptera: Chrysomelidae-Eumolpinae). Part 2. *Transactions of the Shikoku Entomological Society* 19 [1988–1991] (4): 139–150. DP: 30 September 1991 (issue cover).

Komiya, Y. 2001: Description of a new Donacia species (Coleoptera, Chrysomelidae, Donaciinae) from West Japan. *Elytra* 29 (1): 41–44. DP: 15 June 2001 (paper title page).

Komiya, Y. 2002: Description of a new Notosacantha-species (Coleoptera, Chrysomelidae, Cassidinae) from Okinawa-Hontô Is., Southwest Japan. *Elytra* 30 (1): 215–217. DP: 30 June 2002 (paper title page).

Komiya, Y. 2005: Description of a new species of the genus Galerucella (Coleoptera, Chrysomelidae, Galerucinae) from Hokkaido, Northern Japan. *Elytra* 33 (1): 95–99. DP: 30 May 2005 (paper title page).

Komiya, Y. 2006a: Studies on the genus Lipromorpha (Alticinae, Chrysomelidae, Coleoptera) in Japan, with description of a new species. *Elytra* 34 (1): 199–205. DP: 20 May 2006 (paper title page).

Komiya, Y. 2006b: Description of a new species of the genus Lipromorpha (Coleoptera, Chrysomelidae, Alticinae) from Okinoerabu-jima Island, Southwest Japan. *Elytra* 34 (2): 375–378. DP: 18 November 2006 (paper title page).

Komiya, Y. 2006c: Studies on the genus Pseudoliprus (Coleoptera, Chrysomelidae, Alticinae) in Taiwan, with descriptions of a new genus and four new species. *Japanese Journal of Systematic Entomology* 12 (2): 185–196. DP: 20 December 2006 (paper title page).

Komiya, Y. & Kubota, M. 1987: A new Donacia (Coleoptera, Chrysomelidae, Donaciinae) from Central Japan. *Elytra* 15 (1/2): 41–44. DP: 7 November 1987 (paper title page).

Kôno, H. & Tamanuki, K. 1926: Die Käfer-Ausbeute aus Nordsachalin im Jahre 1922. *Dobutsugaku Zasshi* 38 (455): 276–296. DP: 15 September 1926 (issue title page).

Konstantinov, A. S. 1986: Novyy vid roda Longitarsus (Coleoptera, Chrysomelidae) s severo-zapadnogo Kavkaza. (A new Longitarsus species (Coleoptera, Chrysomelidae) from the north-western Caucasus). *Vestnik Zoologii* 1986 (4): 81–82. SFP: 28 July 1986, DP: 20 September 1986 (registered at ZIN library).

Konstantinov, A. S. 1991: K sistematike listoedov-bloshek roda Asiorestia (Coleoptera, Chrysomelidae, Alticinae). (On taxonomy of Asiorestia (Coleoptera, Chrysomelidae, Alticinae)).

Zoologicheskiy Zhurnal 70 (4): 143–144. SFP: 19 March 1991, DP: 3 July 1991 (registered at ZIN library).

Konstantinov, A. S. 1992a: [new taxa]. In: Konstantinov, A. S. & Lopatin, I. K.: On the taxonomy of Phyllotreta Chevr. of the Palaearctic region (Coleoptera, Chrysomelidae, Alticinae). *Spixiana* 15 (3): 261–267. DP: 1 November 1992 (paper title page).

Konstantinov, A. S. 1992b: [new taxa]. In: Lopatin, I. K. & Konstantinov, A. S.: Materialy k faune i sistematike zhukov-listoedov Kavkaza (Coleoptera, Chrysomelidae). (Materials to the fauna and systematics of the leaf-cutting beetles of the Caucasus). Pp. 31–50. In: Abdurakhmanov, G. M. (ed.): *Nasekomye Kavkaza (Fauna, sistematika, geograficheskoe rasprostranenie). Tom I. Novye i maloizvestnye nasekomye s Kavkaza. (Insects of the Caucasus (Fauna, systematics, geographical spreading). Volume 1. New and less known insects from the Caucasus).* [1991], Makhachkala: Ministry of Education of Russia, Daghestan State Pedagogical Institute, Daghestan Ecological Centre, 52 pp. DP: [31 December] 1992 (see the comments in Dating of publications).

Konstantinov, A. S. 1993: Four new species of Longitarsus Latreille from West Tien Shan (Coleoptera: Chrysomelidae: Alticinae). *Elytron* 6 [1992]: 41–46. DP: [31] October 1993 (last page of the volume).

Konstantinov, A. S. 1994: [new taxa]. In: Lopatin, I. K. & Konstantinov, A. S.: New species of Chrysomelidae (Coleoptera) from Palearctic and Oriental regions. *Lambillionea* 94 (4): 524–530. DP: [31] December 1994 (paper title page).

Konstantinov, A. S. 1995: [new taxon]. In: Lopatin, I. K. & Konstantinov, A. S.: Materialy k faune i sistematike zhukov-listoedov Kavkaza (Coleoptera, Chrysomelidae). (Materials on the systematics of leaf-beetles from the Caucasus (Coleoptera, Chrysomelidae)). Pp. 180–200. In: Lopatin, I. K. (ed.): *Fauna i sistematika: Trudy Zoologicheskogo muzeya Belorusskogo universiteta. Vyp. 1. (Fauna and taxonomy: Proceedings of Zoological Museum of the Byelorussian University. Vol. 1).* Minsk: Navuka i tekhnika, 328 pp. SFP: 7 August 1995, DP: [31 December] 1995 (book title page).

Konstantinov, A. S. 1996: Review of Palearctic species of Crepidodera Chevrolat (Coleoptera, Chrysomelidae, Alticinae). *Spixiana* 19 (1): 21–37. DP: 1 March 1997 (paper title page).

Konstantinov, A. S. 1998a: Revision of the Aphthona crypta group of species and a key to the species groups in Aphthona Chevrolat (Coleoptera: Chrysomelidae: Alticinae). *Coleopterists Bulletin* 52 (2): 134–146. DP: 2 June 1998 (verso of the issue cover).

Konstantinov, A. S. 1998b: Revision of the Palearctic species of Aphthona Chevrolat and cladistic classification of the Aphthonini (Coleoptera: Chrysomelidae: Alticinae). *Memoirs on Entomology, International* 11: i–viii + 1–429. DP: 1 July 1998 (colophon).

Konstantinov, A. S. 2000: [new taxa]. In: Konstantinov, A. S. & Lopatin, I. K.: Review of the Longitarsus asperifoliarum group of species (Coleoptera: Chrysomelidae: Alticinae). *Coleopterists Bulletin* 54 (2): 200–220. DP: 27 July 2000 (verso of the issue cover).

Konstantinov, A. S. 2002: Revision of Paraminota Scherer (Coleoptera: Chrysomelidae). *Journal of the New York Entomological Society* 110 (2): 234–246. DP: 1 August 2002 (journal website).

Konstantinov, A. S. 2005: New species of Middle Asian Longitarsus Latreille with discussion of their subgeneric placement (Coleoptera: Chrysomelidae). *Zootaxa* **1056**: 19–42. DP: 30 September 2005 (footer on the paper title page).

Konstantinov, A. S. 2006: A revision of the Aphthona gracilis species group (Coleoptera: Chrysomelidae). *Proceedings of the Russian Entomological Society* (St. Petersburg) 77: 178–188. SFP: 6 March 2006, DP: 3 April 2006 (registered at ZIN library).

Konstantinov A. S. 2009: [new taxa]. In: Lopatin, I. K. & Konstantinov, A. S.: New genera and species of leaf beetles (Coleoptera: Chrysomelidae) from China and South Korea. *Zootaxa* **2083**: 1–18. DP: 27 April 2009 (footer on the paper title page).

Konstantinov, A. S., Baselga, A., Grebennikov V. V., Prena, J. & Lingafelter, S. W. 2011: *Revision of the Palearctic Chaetocnema species (Coleoptera: Chrysomelidae: Galerucinae: Alticini)*. Pensoft: Sofia-Moscow, 363 pp. DP: [31] January 2011 (colophon).

Konstantinov, A. S., Chamorro, M. L., Prathapan, K. D., Ge S.-Q. & Yang, X.-K. 2013: Moss-inhabiting flea beetles (Coleoptera: Chrysomelidae: Galerucinae: Alticini) with description of a new genus from Cangshan, China. *Journal of Natural History* **47** (37/38): 2459–2477 DP: 3 May 2013 (journal website).

Konstantinov, A. S. & Dorr, M. F. 2023: Longitarsus leonardicarloi (Coleoptera: Chrysomelidae: Galerucinae: Alticini), new species from Turkmenistan. *Natural History Sciences, Atti della Società Italiana di Scienze Naturali e del Museo Civile di Storia Naturale di Milano* 10 **(Supplement 1)**: 45–48. DP: 13 December 2023 (journal website).

Konstantinov, A. S. & Duckett, C. N. 2005: New species of Clavicornaltica Scherer (Coleoptera: Chrysomelidae) from continental Asia. *Zootaxa* **1037**: 49–64. DP: 19 August 2005 (footer on the paper title page).

Konstantinov, A. S. & Lingafelter, S. 2002: *Revision of the Oriental species of Aphthona Chevrolat (Coleoptera: Chrysomelidae)*. Washington: Entomological Society of Washington, 349 pp. DP: 8 July 2002 (colophon).

Konstantinov, A. S. & Moseyko, A. G. 2019: A new species of Phyllotreta Chevrolat, 1836 (Coleoptera: Chrysomelidae: Galerucinae: Alticini) from Omsk Province of Russia with comments on Phyllotreta species diversity in Northeastern Palearctic. *Zootaxa* **4679** (3): 499–510. DP: 3 October 2019 (footer on the paper title page).

Konstantinov, A. S. & Prathapan, K. D. 2006: Kashmirobia, a new genus of flea beetles from the Oriental region (Coleoptera: Chrysomelidae). *Zootaxa* **1377**: 61–68. DP: 11 December 2006 (footer on the paper title page).

Konstantinov, A. S. & Prathapan, K. D. 2008: New generic synonyms in the Oriental flea beetles (Coleoptera: Chrysomelidae). *Coleopterists Bulletin* **62** (3): 381–418. DP: 16 October 2008 (last page of the paper).

Konstantinov, A. S. & Ruan, Y.-Y. 2017a: [new taxa]. In: Ruan, Y.-Y., Konstantinov, A. S., Prathapan, K. D. & Yang, X.-K.: New contributions to the knowledge of Chinese flea beetle fauna (I): Gansuapteris new genus and Primulavorus new genus (Coleoptera: Chrysomelidae: Galerucinae). *Zootaxa* **4282** (1): 111–122. DP: 23 June 2017 (footer on the paper title page).

Konstantinov, A. S. & Ruan, Y.-Y. 2017b: [new taxa]. In: Ruan, Y.-Y., Konstantinov, A. S., Prathapan, K. D. & Yang, X.-K.: Contributions to the knowledge of Chinese flea beetle fauna (II): Baoshanaltica new genus and Sinosphaera new genus (Coleoptera, Chrysomelidae, Galerucinae, Alticini). *ZooKeys* 720: 103–120. DP: 11 December 2017 (paper title page).

Konstantinov, A. S., Ruan, Y.-Y. & Prathapan, K. D. 2018: [new taxa]. In: Ruan, Y.-Y., Konstantinov, A. S., Prathapan, K. D., Zhang, M., Jiang, S.-H. & Yang, X.-K.: New contributions to the knowledge of Chinese flea beetle fauna (III): revision of Meishania Chen & Wang with description of five new species (Coleoptera: Chrysomelidae: Galerucinae). *Zootaxa* 4403 (1): 186–200. DP: 3 April 2018 (footer on the paper title page).

Konstantinov, A. S. & Sprecher-Uebersax, E. 2005: Genus Aphthona Chevrolat in Nepal (Coleoptera, Chrysomelidae). *Mitteilungen der Schweizerischen Entomologischen Gesellschaft* 78 (1/2): 173–219. DP: [30] June 2005 (last page of the cover).

Konstantinov, A. S., Volkovitsh, M. G. & Cristofaro, M. 2001: New data on Palearctic Aphthona (Coleoptera: Chrysomelidae) with description of a new species: taxonomic and faunistic results of biological control exploration. *Entomological News* 112 (1): 31–41. DP: 25 February 2001 (*mailing dates* on p. 340).

Kontkanen, P. 1933: Zur Kenntnis der Gattung Phaedon Latr. (Col., Chrysom.). *Annales Societatis Zoologicae-Botanicae Fennicae Vanamo* 14 (3): 67–74. DP: 14 March 1933 (table of contents).

Kontkanen, P. 1950: [new taxon]. In: Heikertinger, F.: Bestimmungstabellen europäischer Käfer (11. Stück). LXXXII. Fam. Chrysomelidae. 5. Subfam. Halticinae. Abteilung II. 11.–30. Gattung: Crepidodera-Verwandtschaft weitesten Sinnes. Bestimmungstabelle der paläarktischen Arten der Halticinengattungen: Creidodera (Chevr.) Steph., Derocrepis Weise (mit Aeschrocnemis Weise), Ochrosis Foudr. (mit Lythraria Bedel und Neocrepidodera Heiktgr.), Orestia Germ., Hippuriphila Foudr., Chalcoides Foudr., Epithrix Foudr., Arrhenocoela Foudr., Minota Kutsch., Cardax Weise, Batophila Foudr., sowie der ostpaläarktischen Arten der asiatischen Gattungen Xuthea Baly, Novofoudrasia Jacobs., Pseudodera Baly, Crepidomorpha Heiktgr., Phygasia Baly (Scallodera Har.), Lipromima Heiktgr. und Manobia Jacoby. Mit Berücksichtigung der Arten der orientalischen Grenzgebiete und des holarktischen Nordamerika. *Koleopterologische Rundschau* 31 [1945–1950] (4/6): 81–139. DP: 30 June 1950 (issue cover).

Kontkanen, P. 1958: Über Chrysolina montana Gebl. und einige andere Arten der Untergattung Heliostola Motsch. (Col., Chrysomelidae). *Annales Entomologici Fennici* 23 [1957] (4): 208–210. DP: 10 January 1958 (verso of the volume title page).

Kontkanen, P. 1959: Über einige Chrysolina-Arten der Sectio Pleurosticha Motsch. sensu Jacobson 1910 (Col., Chrysomelidae). *Annales Entomologici Fennici* 25 (1): 27–35. DP: 26 February 1959 (verso of the volume title page).

Kovalev, O. V. 2002: Formirovanie solitonopodobnykh voln pri invaziyakh organizmov i v evolyucii biosfery. [The formation of the soliton-like waves under the invasions of living organisms and in the evolution of the biosphere]. Pp. 65–81. In: Stegniy, V. N. (ed.): *Evolyucionnaya biologiya. Tom 2. Materialy II Mezhdunarodnoy konferencii "Problema vida i vidobrazonanie" g. Tomsk, 24–26 oktyabrya 2001 g.* [*Proceedings of the 2nd International conference "Problems of Species and Speciation: Evolutionary Biology"*]. Tomsk: Tomskiy gosudarstvennyy universitet, 412 pp. SFP: 25 September 2002, DP: DP: [31 December] 2002 (book title page).

Kraatz, G. 1859: Ueber einige Oreina-Arten. *Berliner Entomologische Zeitschrift* 3 (4): 275–294. DP: 7 December 1859 (*Verh. Zool.-Bot. Ges. Wien* 1859, 9: 123).

Kraatz, G. 1866: Phyllobrotica elegans nov. sp. *Berliner Entomologische Zeitschrift* 10 (1/3): 285–286. DP: [31] July 1866 (table of contents).

Kraatz, G. 1868: Ueber Bruchus pallidicornis Schh. und Verwandte. *Berliner Entomologische Zeitschrift* 12 (1/2): 313–323. DP: [31] May 1868 (table of contents).

Kraatz, G. 1870a: Ueber deutsche Donacien. *Berliner Entomologische Zeitschrift* 13 [1869] (3/4): 263–272. DP: [31] January 1870 (table of contents).

Kraatz, G. 1870b: Einige Bemerkungen über Allard'sche Bruchus. *Berliner Entomologische Zeitschrift* 13 [1869] (3/4): 331–334. DP: [31] January 1870 (table of contents).

Kraatz, G. 1870c: Eine neue russische Clythra-Art. *Horae Societatis Entomologicae Rossicae* 8 [1871] (1): 28–31. DP: 24 October 1870 (Kerzhner 1984, Standfuss & Kerzhner 2004).

Kraatz, G. 1872a: Eine neue deutsche Criocerinen-Art. *Berliner Entomologische Zeitschrift* 15 [1871] (2/3): 162. DP: [29] February 1872 (table of contents).

Kraatz, G. 1872b: Bemerkungen über europäische Clythriden. *Berliner Entomologische Zeitschrift* 16 (2/4): 193–232. DP: [30] November 1872 (table of contents).

Kraatz, G. 1874a: Verzeichniss andalusischer Cassida-Arten. *Berliner Entomologische Zeitschrift* 18 (1/2): 103–104. DP: [30] April 1874 (table of contents).

Kraatz, G. 1874b: Ueber russische Cassida-Arten. *Horae Societatis Entomologicae Rossicae* 10 [1873–1874] (2/4): 124–128. DP: 13 September 1874 (Kerzhner 1984, Standfuss & Kerzhner 2004).

Kraatz, G. 1876: Ueber andalusische Cryptocephalus. *Deutsche Entomologische Zeitschrift* 20 (1): 137–138. DP: [29] February 1876 (table of contents).

Kraatz, G. 1879a: Neue ostsibirische Arten der Chrysomelinen-Gattung Pedrillia. *Deutsche Entomologische Zeitschrift* 23 (1): 119–120, pl. II. DP: [31] January 1879 (issue cover, see also the comments in the Dating of publications chapter).

Kraatz, G. 1879b: Neue Käfer vom Amur. *Deutsche Entomologische Zeitschrift* 23 (1): 121–144, pl. II. DP: [31] January 1879 (issue cover, see also the comments in the Dating of publications chapter).

Kraatz, G. 1879c: Ueber die Verwandten der Phytodecta viminalis L. *Zeitschrift für Entomologie, Neue Folge* (Breslau) 7: 45–56. DP: [30] November 1879 (*Natur. Novit.* 1879, 32 (22): 211).

Kraatz, G. 1879d: Die Cryptocephalen von Sibirien und Japan und ihre geographische Verbreitung. *Deutsche Entomologische Zeitschrift* 23 (2): 257–265. DP: [30] November 1879 (table of contents).

Kraatz, G. 1879e: Ueber die sibirischen Entomoscelis-Arten (Coleopt., Chrysomelidae, Chrysomelini). *Deutsche Entomologische Zeitschrift* 23 (2): 266. DP: [30] November 1879 (table of contents).

Kraatz, G. 1879f: Die Cassiden von Ost-Sibirien und Japan. *Deutsche Entomologische Zeitschrift* 23 (2): 267–275. DP: [30] November 1879 (table of contents).

Kraatz, G. 1879g: Rhaebus Gebleri Fischer, oder eine neue Rhaebus-Art in Europa einheimisch? (Coleopt., Chrysomelin.? Sagrin.?). *Deutsche Entomologische Zeitschrift* **23** (2): 276–278. DP: [30] November 1879 (table of contents).

Kraatz, G. 1879h: [Zwei neue Käfer-Arten aus dem nördlichen Persien]. *Deutsche Entomologische Zeitschrift* **23** (2): 288. DP: [30] November 1879 (table of contents).

Kraatz, G. 1879i: Vier neue spanische Timarcha-Arten. *Deutsche Entomologische Zeitschrift* **23** (2): 380–384. DP: [30] November 1879 (table of contents).

Kraatz, G. 1880: Die Unterscheidung der Oreinen-Arten nach den männlichen Geschlechtsorganen. *Entomologische Monatsblätter* **2** (13): 33–38. DP: [31] March 1880 (issue title page).

Kraatz, G. 1882a: [new taxa]. In: Heyden, L. F. J. D. von & Kraatz, G.: Käfer um Margelan, gesammelt von Haberhauer. *Deutsche Entomologische Zeitschrift* **26** (1): 99–118. DP: 8 March 1882 (*Bull. Soc. Ent. Fr.* 1882: 92).

Kraatz, G. 1882b: [new taxon]. In: Heyden, L. F. J. D. von & Kraatz, G.: Käfer um Samarkand, gesammelt von Haberhauer. *Deutsche Entomologische Zeitschrift* **26** (2): 297–338. DP: [31] August 1882 (table of contents).

Kraatz, G. 1885: [new taxon]. In: Heyden, L. F. J. D. von & Kraatz, G.: Beiträge zur turkestanischen Coleopteren-Fauna. *Deutsche Entomologische Zeitschrift* **29** (2): 273–298. DP: [31] December 1885 (table of contents).

Kraatz, G. 1888: [new taxon]. In: Heyden, L. F. J. D. von: Neue und interessante Coleopteren aus Malatia in Mesopotamien. *Deutsche Entomologische Zeitschrift* **1888** (1): 72–78. DP: [30] April 1888 (table of contents).

Kraatz, G. 1891: Phyllopertha humeralis n. sp. aus Kleinasien. *Deutsche Entomologische Zeitschrift* **1891** (1): 124. DP: [30] June 1891 (table of contents).

Král, J. 1945: Noví nebo málo známí dřepčíci české fauny. De novis vel minus cognitis Halticinarum speciebus formisque faunae Bohemiae (Coleoptera, Chrysomelidae). *Časopis Československé Společnosti Entomologické* **42** (1/4): 31–33. DP: 1 September 1945 (issue cover).

Král, J. 1947: Příspěvek k poznání dřepčíků fauny Československa (3.). Ad halticidarum faunae Čechoslovakiae cognitionem additamentum III. (Coleoptera, Phytophaga). *Časopis Československé Společnosti Entomologické* **44** (1/2): 21–26. DP: 1 June 1947 (issue cover).

Král, J. 1964a: Drei neue Arten der Familie Alticidae aus Grusien (Coleoptera, Phytophaga). VIII. Beitrag zur Kenntnis der Alticiden. *Entomologische Blätter* **60** (1): 34–39. DP: 15 June 1964 (issue cover).

Král, J. 1964b: Zur Kenntnis der palaearktischen Altica-Arten I (Col. Phytophaga, Alticidae). IX. Beitrag zur Kenntnis der Alticiden. *Entomologische Blätter* **60** (2): 126–133. DP: 25 October 1964 (issue cover).

Král, J. 1965a: Zur Kenntnis der palaearktischen Altica-Arten (II.) (Col., Phytophaga, Alticidae). X. Beitrag zur Kenntnis der Alticiden. *Entomologische Blätter* **60** [1964] (3): 162–166. DP: 31 January 1965 (issue cover).

Král, J. 1965b: Eine neue Aphthona-Art aus Spanien. Coleoptera, Phytophaga, Alticidae. XII. Beitrag zur Kenntnis der Alticiden. *Miscelánea Zoológica* 2 (1): 131–132. DP: [30] November 1965 (issue cover).

Král, J. 1966: Zur Kenntnis der palaearktischen Altica-Arten IV. (Coleoptera, Phytophaga, Alticidae). XVI. Beitrag zur Kenntnis der Alticiden. *Entomologische Blätter* 62 (3): 159–168. DP: 20 December 1966 (issue cover).

Král, J. 1967: 111. Alticidae II. Ergebnisse der zoologischen Forschungen von Dr. Z. Kaszab in der Mongolei (Coleoptera). *Reichenbachia* 9: 181–189. DP: 14 September 1967 (paper title page).

Král, J. 1969a: Zwei neue Arten der Familie Alticidae aus der Sowjetunion. XIX. Beitrag zur Kenntnis der Alticiden. *Entomologische Blätter* 65 (2): 67–71. DP: 1 September 1969 (issue cover).

Král, J. 1969b: Zur Kenntnis der paläarktischen Altica-Arten V. (Coleoptera, Phytophaga, Alticidae). XVIII. Beitrag zur Kenntnis der Alticiden. *Entomologische Blätter* 65 (2): 72–85. DP: 1 September 1969 (issue cover).

Král, J. 1969c: Eine neue Alticiden-Art aus Marokko: Aphthona kocheri n. sp. (Coleoptera, Phytophaga, Chrysomeloidea). XX. Beitrag zur Kenntnis der Alticiden. *Bulletin de la Société des Sciences Naturelles et Physiques du Maroc* 48 [1968] (3/4): 77–79. DP: [31 December] 1969 (issue cover).

Král, J. 1973: Ergebnisse der zoologischen Forschungen von Dr. Z. Kaszab in der Mongolei. 329. Alticidae III. (Coleoptera). *Folia Entomologica Hungarica, Series Nova* 26 (supplement): 111–133. DP: 31 October 1973 (colophon).

Král, J. 1979: Zur Kenntnis der Altica-Arten 7 (Coleoptera, Phytophaga, Alticidae). 24. Beitrag zur Kenntnis der Alticiden. *Entomologische Blätter* 75 (1/2): 98–108. DP: 30 September 1979 (paper title page).

Krátký, J. 2021: New distributional data on beetle species (Coleoptera) from the Canary Islands. *WEEVIL News* 90: 1–6. DP: 1 March 2021 (paper title page).

Krauss, H. 1899: Coleopterologische Beiträge zur Fauna austriaca. *Wiener Entomologische Zeitung* 18 (7): 203–207. DP: 31 July 1899 (footers throughout the issue).

Krauss, H. 1900: Coleopterologische Beiträge zur Fauna Austriaca. *Wiener Entomologische Zeitung* 19 (9): 239–241. DP: 5 November 1900 (footers throughout the issue).

Krausse, A. H. 1911: Cryptocephalus equiseti pallidus m., v. n. *Entomologische Blätter* 7 (7): 145–146. DP: 10 July 1911 (issue title page).

Krynicki, J. 1832: Enumeratio Coleopterorum Rossiae meridionalis et praecipue in Universitatis Caesareae Charkoviensis circulo obvenientium, quae annorum 1827–1831 spatio observavit. *Bulletin de la Société Impériale des Naturalistes de Moscou* 5: 65–179, pls. II–III. DP: 6 December 1832 (Bousquet & Davies 2021).

Krynicki, J. 1834: Addenda et nonnula synonyma Rossiae meridionalis coleopterorum. *Bulletin de la Société Impériale des Naturalistes de Moscou* 7: 166–173, pl. V. DP: [30] September 1834 (Bousquet & Davies 2021).

Kugelann, J. G. 1798: [new taxon]. In: Illiger, J. C. W.: *Verzeichniss der Käfer Preussens. Entworfen von Johann Gottlieb Kugelann Apotheker in Osterode. Ausgearbeitet von Johann Karl Wilhelm*

Illiger. Mit einer Vorrede des Professors und Pagenhofmeisters Hellwig in Braunschweig, und dem angehängten Versuche einer natürlichen Ordnungs- und Gattungs-Folge der Insekten. Halle: Johann Jacob Gebauer, xlii + 510 + [1] pp. DP: 19 September 1798 (Bousquet 2016).

Kühnelt, M. W. 1984: Monographie der Blattkäfergattung Chrysochloa (Coleoptera, Chrysomelidae). *Sitzungsberichte der Österreichischen Akademie der Wissenschaften, Mathematisch-Naturwissenschaftliche Klasse, Abteilung 1* **193** (6/10): 171–287. DP: [31 December] 1984 (issue cover).

Kung, Y.-T. & Chen, S.-H. 1954a: Notes on Chinese Halticinae of the genus Luperomorpha Weise. *Acta Entomologica Sinica* **4** (1): 83–97. DP: [31] March 1954 (paper title page).

Kung, Y.-T. & Chen, S.-H. 1954b: Notes on the Chinese and Tonkinese species of the halticine genus Hespera Weise. *Acta Entomologica Sinica* **4** (2): 149–163. DP: [30] June 1954 (paper title page).

Kung, Y.-T. & Tan, C.-C. [= J.-J.] 1961: [new taxon]. In: Chen, S.-H., Tan, C.-C. [= J.-J.] & Yu, P.-Y.: Results of the zoologico-botanical expedition to Southwest China 1955–57 (Coleoptera, Hispinae I). *Acta Entomologica Sinica* 10 (4/6): 457–481. DP: [31] December 1961 (paper title page).

Kung, Y.-T. & Yu, P.-Y. 1961: [new taxon]. In: Chen, S.-H., Tan, C.-C. [= J.-J.] & Yu, P.-Y.: Results of the zoologico-botanical expedition to Southwest China 1955–57 (Coleoptera, Hispinae I). *Acta Entomologica Sinica* 10 (4/6): 457–481. DP: [31] December 1961 (paper title page).

Kuntzen, H. 1911: Zwei neue spanische Timarchen. (Col.). *Deutsche Entomologische Zeitschrift* **1911** (2): 215–216. DP: 11 March 1911 (table of contents).

Kuntzen, H. 1914: Zur Kenntnis der Sagra-Arten (Coleopt. Chrysomelidae). I. Teil. *Archiv für Naturgeschichte, Abteilung A* 80 (1): 117–135. DP: [30] June 1914 (*Ent. Litteraturbl.* 1914, 14 (6): 132).

Kuntzen, H. 1919: Skizze zur Verbreitung einiger flugunfähiger Blattkäfer (Metallotimarcha). *Sitzungsberichte der Gesellschaft Naturforschender Freunde zu Berlin* **1919** (6): 228–250. DP: 6 October 1919 (issue title page).

Kunze, G. 1818a: Beiträge zur Monographie der Rohrkäfer. *Neue Schriften der Naturforschenden Gesellschaft zu Halle* **2** (4): 1–56. DP: 4 October 1818 (*Allg. Verz. Bücher, Michaelismesse* 1818: 442 and Evenhuis 2014).

Kunze, G. 1818b: Zeugophora (Jochträger), eine neue Käfergattung. *Neue Schriften der Naturforschenden Gesellschaft zu Halle* **2** (4): 71–76. DP: 4 October 1818 (*Allg. Verz. Bücher, Michaelismesse* 1818: 442 and Evenhuis 2014).

Küster, H. C. 1844: *Die Käfer Europa's. Nach der Natur beschrieben. Mit Beiträgen mehrerer Entomologen. Erster Band.* Nürnberg: Bauer & Raspe, [6] + 100 sheets, 2 pls. DP: [31 December] 1844 (Bousquet 2016).

Küster, H. C. 1845a: *Die Käfer Europa's. Nach der Natur beschrieben. Mit Beiträgen mehrerer Entomologen. Zweites Heft.* Nürnberg: Bauer & Raspe, [6] + 100 sheets, 2 pls. DP: 28 August 1845 (Bousquet 2016).

Küster, H. C. 1845b: *Die Käfer Europa's. Nach der Natur beschrieben. Mit Beiträgen mehrerer Entomologen. Drittes Heft*. Nürnberg: Bauer & Raspe, [6] + 100 sheets, 2 pls. DP: 11 December 1845 (Bousquet 2016).

Küster, H. C. 1846a: *Die Käfer Europas. Nach der Natur beschrieben. Mit Beiträgen mehrerer Entomologen. Viertes Heft*. Nürnberg: Bauer und Raspe, [6] + 100 sheets, 3 pls. DP: 9 October 1846 (Bousquet 2016).

Küster, H. C. 1846b: *Die Käfer Europa's. Nach der Natur beschrieben. Mit Beiträgen mehrerer Entomologen. Fünftes Heft*. Nürnberg: Bauer und Raspe, [4] + 100 sheets, 3 pls. DP: 9 October 1846 (Bousquet 2016).

Küster, H. C. 1846c: *Die Käfer Europas. Nach der Natur beschrieben. Mit Beiträgen mehrerer Entomologen. Siebentes Heft*. Nürnberg: Bauer und Raspe, [4] + 100 sheets, 3 pls. DP: [31 December] 1846 (Bousquet 2016).

Küster, H. C. 1847: *Die Käfer Europas. Nach der Natur beschrieben. Mit Beiträgen mehrerer Entomologen. Neuntes Heft*. Nürnberg: Bauer & Raspe, [2] + 100 sheets + 3 pl. DP: 25 April 1847 (Bousquet 2017).

Küster, H. C. 1848a: *Die Käfer Europa's. Nach der Natur beschrieben. Mit Beiträgen mehrerer Entomologen. Dreizehntes Heft*. Nürnberg: Bauer & Raspe, [4] + 100 sheets, 3 pls. DP: 30 November 1848 (Bousquet 2016).

Küster, H. C. 1848b: *Die Käfer Europas. Nach der Natur beschrieben. Mit Beiträgen mehrerer Entomologen. Vierzehntes Heft*. Nürnberg: Bauer & Raspe, [4] + 100 sheets, 2 pls. DP: 30 November 1848 (Bousquet 2016).

Küster, H. C. 1848c: *Die Käfer Europa's. Nach der Natur beschrieben. Mit Beiträgen mehrerer Entomologen. Fünfzehntes Heft*. Nürnberg: Bauer & Raspe, [4] + 100 sheets, 3 pls. DP: [31 December] 1848 (Bousquet 2016).

Küster, H. C. 1849a: *Die Käfer Europas. Nach der Natur beschrieben. Mit Beiträgen mehrerer Entomologen. Sechzehntes Heft*. Nürnberg: Bauer & Raspe, 100 sheets, 3 pls. DP: 29 April 1849 (Bousquet 2017).

Küster, H. C. 1849b: *Die Käfer Europa's. Nach der Natur beschrieben. Mit Beiträgen mehrerer Entomologen. Siebzehntes Heft*. Nürnberg: Bauer & Raspe, [4] + 100 sheets, 2 pls. DP: 15 October 1849 (Bousquet 2016).

Küster, H. C. 1849c: *Die Käfer Europa's. Nach der Natur beschrieben. Mit Beiträgen mehrerer Entomologen. Neunzehntes Heft*. Nürnberg: Bauer & Raspe, [4] + 100 sheets, 2 pls. DP: 15 October 1849 (Bousquet 2016).

Küster, H. C. 1852: *Die Käfer Europa's. Nach der Natur beschrieben. Mit Beiträgen mehrerer Entomologen. Fünfundzwanzigstes Heft*. Nürnberg: Bauer & Raspe, 100 cheets, 2 pls. DP: [31 December] 1849 (Bousquet 2016).

Küster, H. C. 1854: *Die Käfer Europas. Nach der Natur beschrieben. Mit Beiträgen mehrerer Entomologen. Achtundzwanzigstes Heft*. Nürnberg: Bauer & Raspe, [4] + 100 sheets, 2 pls. DP: [31 December] 1854 (Bousquet 2016).

Kutschera, F. 1859: Beiträge zur Kenntnis der europäischen Halticinen. *Wiener Entomologische Monatschrift* 3 (5): 129–141. DP: [31] May 1859 (issue title page).

Kutschera, F. 1860a: Beiträge zur Kenntnis der europäischen Halticinen. *Wiener Entomologische Monatschrift* 4 (1): 1–19. DP: [31] January 1860 (issue title page).

Kutschera, F. 1860b: Beiträge zur Kenntnis der europäischen Halticinen. *Wiener Entomologische Monatschrift* 4 (3): 65–79. DP: [31] March 1860 (issue title page).

Kutschera, F. 1860c: Beiträge zur Kenntnis der europäischen Halticinen. *Wiener Entomologische Monatschrift* 4 (5): 129–143. DP: [31] May 1860 (issue title page).

Kutschera, F. 1860d: Beiträge zur Kenntnis der europäischen Halticinen. *Wiener Entomologische Monatschrift* 4 (7): 193–208. DP: [31] July 1860 (issue title page).

Kutschera, F. 1860e: Beiträge zur Kenntnis der europäischen Halticinen. *Wiener Entomologische Monatschrift* 4 (10): 297–310. DP: [31] October 1860 (issue title page).

Kutschera, F. 1861: Beiträge zur Kenntnis der europäischen Halticinen. *Wiener Entomologische Monatschrift* 5 (8): 233–250. DP: [31] August 1861 (issue title page).

Kutschera, F. 1862a: Beiträge zur Kenntnis der europäischen Halticinen. *Wiener Entomologische Monatschrift* 6 (2): 47–55. DP: [28] February 1862 (issue title page).

Kutschera, F. 1862b: Beiträge zur Kenntnis der europäischen Halticinen. *Wiener Entomologische Monatschrift* 6 (4): 97–109. DP: [30] April 1862 (issue title page).

Kutschera, F. 1862c: Beiträge zur Kenntnis der europäischen Halticinen. *Wiener Entomologische Monatschrift* 6 (7): 215–228. DP: [31] July 1862 (issue title page).

Kutschera, F. 1863a: Beiträge zur Kenntnis der europäischen Halticinen. *Wiener Entomologische Monatschrift* 7 (5): 151–168. DP: [31] May 1863 (issue title page).

Kutschera, F. 1863b: Beiträge zur Kenntnis der europäischen Halticinen. *Wiener Entomologische Monatschrift* 7 (9): 291–320. DP: [30] September 1863 (issue title page).

Kutschera, F. 1864a: Beiträge zur Kenntnis der europäischen Halticinen. *Wiener Entomologische Monatschrift* 8 (2): 33–52. DP: [29] February 1864 (issue title page).

Kutschera, F. 1864b: Beiträge zur Kenntnis der europäischen Halticinen. *Wiener Entomologische Monatschrift* 8 (5): 141–164. DP: [31] May 1864 (issue title page).

Kutschera, F. 1864c: Beiträge zur Kenntnis der europäischen Halticinen. *Wiener Entomologische Monatschrift* 8 (9): 269–288. DP: [30] September 1864 (issue title page).

Kutschera, F. 1864d: Beiträge zur Kenntnis der europäischen Halticinen. *Wiener Entomologische Monatschrift* 8 (10): 303–321. DP: [31] October 1864 (issue title page).

Kutschera, F. 1864e: Beiträge zur Kenntnis der europäischen Halticinen. *Wiener Entomologische Monatschrift* 8 (11): 337–353. DP: [30] November 1864 (issue title page).

Kutschera, F. 1864f: Beiträge zur Kenntnis der europäischen Halticinen. *Wiener Entomologische Monatschrift* 8 (12): 373–472. DP: [31] December 1864 (issue title page).

Kuwayama, S. 1931: Lema oryzae sp. nov. *Insecta Matsumurana* 5 (3): 155. DP: 25 March 1931 (Evenhuis 2022).

Kuwayama, S. 1932: Studies on the morphology and ecology of the rice leaf beetle, Lema oryzae Kuwayama, with special reference to the taxonomic aspects. *Journal of the Faculty of*

Agriculture, Hokkaido Imperial University **33** (1): 1–132, 4 pls. DP: [31] March 1932 (footer of paper title page).

Laboissière, V. 1911: Nouvelle variété de Sermyla halensis Linné (Col. Chrysomelidae). *Bulletin de la Société Entomologique de France* **1911** (15): 303. DP: 25 October 1911 (*dates de publication* at the end of volume).

Laboissière, V. 1912a: Nouvelles variétés de Coléoptères phytophages (Chrysomelidae). *Bulletin de la Société Entomologique de France* **1912** (15): 329–330. DP: 23 October 1912 (*dates de publication* at the end of volume).

Laboissière, V. 1912b: Observations diverses. *Bulletin de la Société Entomologique de France* **1912** (18): 375. DP: 11 December 1912 (*dates de publication* at the end of volume).

Laboissière, V. 1912c: Revision des Galerucini d'Europe et pays limitrophes (suite). *Annales de l'Association des Naturalistes de Levallois-Perret* **18**: 13–55. DP: [31 December] 1912 (volume cover).

Laboissière, V. 1913: Revision des Galerucini d'Europe et pays limitrophes (suite). *Annales de l'Association des Naturalistes de Levallois-Perret* **19**: 14–78. DP: [31 December] 1913 (volume cover).

Laboissière, V. 1914: Descriptions et synonymies de Galerucini paléarctiques (Col. Chrysomelidae). *Bulletin de la Société Entomologique de France* **1914** (2): 82–84. DP: 11 February 1914 (*dates de publication* at the end of volume).

Laboissière, V. 1917: Description d'un Luperus nouveau du Maroc (Col. Chrysomelidae). *Bulletin de la Société Entomologique de France* **1917** (9): 159–160. DP: 6 June 1917 (*dates de publication* at the end of volume).

Laboissière, V. 1919a: Description d'un Luperus nouveau de Sicile (Col. Galerucidae). *Bulletin de la Société Entomologique de France* **1919** (6): 116–117. DP: 16 May 1919 (*dates de publication* at the end of volume).

Laboissière, V. 1919b: Descriptions de deux Oïdes nouveaux, du Tonkin (Col. Galerucini). *Bulletin de la Société Entomologique de France* **1919** (9): 160–161. DP: 28 June 1919 (*dates de publication* at the end of volume).

Laboissière, V. 1920: Descriptions d'espèces nouvelles de Galerucini de l'Afrique occidentale (Col. Chrysomelidae). *Annales de la Société Entomologique de France* **88** [1919] (3/4): 321–327. DP: 30 June 1920 (*dates de publication* on p. 419).

Laboissière, V. 1921: Étude des Galerucini de la collection du Musée du Congo Belge. Première partie. *Revue Zoologique Africaine* **9** (1): 33–86. DP: 15 August 1921 (issue cover).

Laboissière, V. 1922a: Descriptions de Galerucini nouveaux du Yunnan et de l'Indo-Chine (Col. Chrysomelidae). *Bulletin de la Société Entomologique de France* **1922** (8): 101–102. DP: 23 May 1922 (*dates de publication* on p. 348).

Laboissière, V. 1922b: Étude des Galerucini de la collection du Musée du Congo Belge. Première partie (suite). *Revue Zoologique Africaine* **10** (1): 1–44. DP: 15 July 1922 (issue cover).

Laboissière, V. 1922c: Étude des Galerucini de la collection du Musée du Congo Belge. Première partie (fin). *Revue Zoologique Africaine* **10** (3): 219–271. DP: 15 December 1922 (issue cover).

Laboissière, V. 1925: Description de trois espèces nouvelles de Galerucini des Philippines. *Philippine Journal of Sciences* 28 (4): 609–612. DP: 12 December 1925 (volume table of contents).

Laboissière, V. 1926a: Supplément au Catalogus Coleopterorum, Pars 78 (Galerucinae), de J. Weise, précédé de remarques sur la classification des Galerucini. *Encyclopédie Entomologique* (B) 1 [1925–1926] (1): 38–48 & (2): 49–62. DP (pp. 33–48): 15 June 1925, (pp. 49–62): 15 January 1926 (footers on pp. 33 and 49).

Laboissière, V. 1926b: Description d'un nouveau genre et de trois nouvelles espèces de Galerucini africains (Col.). *Bulletin de la Société Entomologique de France* **1926** (18): 190–193. DP: 30 December 1926 (*dates d'apparition* on p. 252).

Laboissière, V. 1927: Contribution a l'étude des Galerucini de l'Indochine et du Yunnan avec descriptions de nouveaux genres et espèces (Col. Chrysomelidae). *Annales de la Société Entomologique de France* **96** (1): 37–62. DP: 31 March 1927 (*dates de publication* on p. 348).

Laboissière, V. 1929a: Observations sur les Galerucini asiatiques principalement du Tonkin et du Yunnan et descriptions de nouveaux genres et espèces. *Annales de la Société Entomologique de France* **98** (3): 251–288, pl. 3. DP: 31 October 1929 (*dates de parution* on p. 557).

Laboissière, V. 1929b: Nouveaux Galerucini de la faune d'Asie (Col. Chrys.). *Revista Chilena de Historia Natural* **33**: 289–294. DP: [31 December] 1929 (volume title page).

Laboissière, V. 1930: Observations sur les Galerucini asiatiques principalement du Tonkin et du Yunnan et descriptions de nouveaux genres et espèces. (2e partie). *Annales de la Société Entomologique de France* **99** (4): 325–368. DP: 31 December 1930 (*dates de parution* on p. 372).

Laboissière, V. 1931a: Galerucini africains nouveaux ou peu connus de la collection du Deutsches Entomologisches Institut. *Wiener Entomologische Zeitung* **48** (1): 42–57. DP: 30 June 1931 (footers throughout the issue).

Laboissière, V. 1931b: Observations sur les Galerucini asiatiques principalement du Tonkin et du Yunnan et descriptions de nouveaux genres et espèces. (3e partie). *Annales de la Société Entomologique de France* 100 (1/2): 133–152. DP: 30 June 1931 (*dates de parution* on p. 472).

Laboissière, V. 1931c: Galerucini (Coleoptera Chrysomelidae) d'Angola. *Revue Suisse de Zoologie* **38** (4): 405–418. DP: [30] November 1931 (paper title page).

Laboissière, V. 1932a: Résultats scientifiques du voyage aux Indes Orientales Néerlandaises de LL. AA. RR. le Prince et la Princesse Léopold de Belgique. Galerucinae. *Mémoires du Musée Royal d'Histoire Naturelle de Belgique, Hors Série* **4** (4): 143–184, 2 pls. DP: 30 September 1932 (issue title page).

Laboissière, V. 1932b: Chrysomelidae nouveaux de l'Indo-Chine et de la presqu'île de Malacca. *Annales de l'Association des Naturalistes de Levallois-Perret* 20 [1914–1931]: 130–142. DP: [31 December] 1932 (see the comments in Dating of publications).

Laboissière, V. 1933a: Galerucini de la collection du Muséum National d'Histoire naturelle recueillis dans l'Himalaya par le Dr. J. Harmand,. 1re note. *Bulletin du Muséum National d'Histoire Naturelle, 2e Série* (Paris) **4** [1932] (8): 960–970. DP: 30 January 1933 [erroneously written 1932] (footer on p. 1096).

Laboissière, V. 1933b: Observations sur les Galerucini asiatiques principalement du Tonkin et du Yunnan et descriptions de nouveaux genres et espèces (4e partie). *Annales de la Société Entomologique de France* 102 (1): 51–72. DP: 30 March 1933 (*dates de publication* on p. 408).

Laboissière, V. 1933c: Descriptions de trois nouveaux Galerucini du Tonkin. *Bulletin du Muséum National d'Histoire Naturelle, 2e Série* (Paris) 5 (3): 203–208. DP: 19 April 1933 (footer on p. 256).

Laboissière, V. 1934a: Schwedisch-chinesische wissenschaftliche Expedition nach den nordwestlichen Provinzen Chinas, unter Leitung von Dr. Sven Hedin und Prof. Sü Ping-chang. Insekten gesammelt vom schwedischen Arzt der Expedition Dr. David Hummel 1927–1930. 20. Coleoptera. 6. Galerucinae. *Arkiv för Zoologi* 27A (6): 1–9. DP: 26 January 1934 (last page of the paper).

Laboissière, V. 1934b: Galerucinae de la faune française (coléoptères). *Annales de la Société Entomologique de France* 103 (1): 1–108. DP: 31 March 1934 (*dates de publication* on p. 400).

Laboissière, V. 1935a: Galerucini (Chrysom. Col.). Pp. 324–325. In: Visser, P. C. & Visser-Hooft, J. (eds.): *Wissenschaftliche Ergebnisse der Niederländischen Expeditionen in den Karakorum und die angrenzende Gebiete 1922, 1925 und 1929/30. Band 1.* Leiden: E. J. Brill, xviii + 499 pp. DP: 15 January 1935 (stamp on the title page of a reprint in NMPC).

Laboissière, V. 1935b: Coléoptères galérucines nouveaux ou peu connus de la faune indo-malaise. *Annales de l'Association des Naturalistes de Levallois-Perret* 21 [1932–1934]: 109–137. DP: [28] February 1935 (footer on table of contents).

Laboissière, V. 1935c: Rectifications au mémoire de E. Allard «Contribution à la faune indo-chinoise. Galérucides et alticides». *Annales de l'Association des Naturalistes de Levallois-Perret* 21 [1932–1934]: 139–145. DP: [28] February 1935 (footer on table of contents).

Laboissière, V. 1936: Observations sur les Galerucini asiatiques principalement du Tonkin et du Yunnan et descriptions de nouveaux genres et espèces (5e partie). *Annales de la Société Entomologique de France* 105 (3): 239–262. DP: 31 October 1936 (*dates de publication* on p. 388).

Laboissière, V. 1937a: Nouveaux Galerucinae paléarctiques. *Bulletin de la Société Entomologique de France* 42 (2): 27–31. DP: 11 March 1937 (*dates de parution* on p. 312).

Laboissière, V. 1937b: Nouvelles espèces de Timarcha des Pyrénées françaises (Col. Chrysomelidae). *Bulletin de la Société Entomologique de France* 42 (11): 164–166. DP: 27 August 1937 (*dates de parution* on p. 312).

Laboissière, V. 1938: Galerucinae (Col.) de la Chine. Du musée de Stockholm. *Arkiv för Zoologi* 30A (11): 1–9. DP: 7 April 1938 (last page of the paper).

Laboissière, V. 1939: Timarcha et Chrysomela nouveaux pour la faune française. *Bulletin de la Société Entomologique de France* 44 (9/10): 127–129. DP: 12 July 1939 (*dates de parution* on p. 240).

Laboissière, V. 1940a: Observations sur les Galerucinae des collections du Musée royal d'histoire naturelle de Belgique et descriptions d'espèces nouvelles. 2e note. *Bulletin du Musée Royal d'Histoire Naturelle de Belgique* 16 (3): 1–31. DP: [31] January 1940 (paper title page).

Laboissière, V. 1940b: Observations sur les Galerucinae des collections du Musée royal d'histoire naturelle de Belgique et descriptions de nouveaux genres et espèces. 4me note. *Bulletin du Musée Royal d'Histoire Naturelle de Belgique* **16** (37): 1–41. DP: [31] May 1940 (paper title page).

Laboissière, V. 1942: Halticinae (Coleoptera Phytophaga) Fam. Chrysomelidae. *Exploration du Parc National Albert, Mission G. F. de Witte (1933–1935)* **39**: 3–131. DP: [31 December] 1942 (cover).

Lacordaire, J. T. 1845: Monographie des coléoptères subpentamères de la famille des phytophages. Tome premier. *Mémoires de la Société Royale des Sciences de Liége* **3** (1): liv + 740 pp. DP (pp. liv + 1–320): 1 May 1845, (pp. 321–740): 15 November 1845 (verso of issue title page).

Lacordaire, J. T. 1848: Monographie des coléoptères subpentamères de la famille des phytophages. Tome second. *Mémoires de la Société Royale des Sciences de Liége* **5**: vi + 890. DP: [31] May 1848 (volume title page).

Laczó, J. 1909: Új levélbogár Trencsén-vármegyéből. (Ein neuer Blattkäfer aus dem Comitat Trencsén). *Rovartani Lapok* **16** (4): 57. DP: 30 April 1909 (issue title page).

Laczó, J. 1912: Új bogarak Trencsén-vármegyéből. (Neue Käfer aus dem Komitat Trencsén). *Rovartani Lapok* **19** (1): 3–5. DP: 30 January 1912 (issue title page).

Laczó, J. 1928a: Coleoptera nova ex Comitatu Trenčín Čecho-Slovakia. Noví chrobáci zo župy Trenčianskej Česko-Slovensko. Pp. 1–8. In: Laczó, J.: *Chrobákoveda. Popis nových chrobákov. Fauna chrobákov Slovenska, župy Trenčianskej, okolia Bratislavy a okolia Bolešov-Piechov. [Beetle-science. Descriptions of new beetles. Fauna of beetles of Slovakia, Trenčín County, surroundings of Bratislava and Bolešov-Piechov]*. Bratislava: Josef Laco, 117 + 8 + 4 + 1 + 10 + 5 + 14 + 35 + 7 + [8] + [5] + [14] pp. DP: [31] December 1928 (introduction dated).

Laczó, J. 1928b: Novae aberrationes coleopterorum ex Slovakia. Pp. 1–4. In: Laczó, J.: *Chrobákoveda. Popis nových chrobákov. Fauna chrobákov Slovenska, župy Trenčianskej, okolia Bratislavy a okolia Bolešov-Piechov. [Beetle-science. Descriptions of new beetles. Fauna of beetles of Slovakia, Trenčín County, surroundings of Bratislava and Bolešov-Piechov]*. Bratislava: Josef Laco, 117 + 8 + 4 + 1 + 10 + 5 + 14 + 35 + 7 + [8] + [5] + [14] pp. DP: [31] December 1928 (introduction dated).

Laicharting, J. N. E. von 1781: *Verzeichniß und Beschreibung der Tyroler-Insekten. I. Theil. Käferartige Insecten. I. Band.* Zürich: Johann Caspar Füessly, [4] + xii + [1] + 248 pp. DP: 6 May 1781 (Bousquet 2017).

Lambelet, J. 2005: Une nouvelle espèce: Pachybrachis canigouensis nova (Coleoptera Chrysomelidae). *Rutilans* 8 (2): 29–34. DP: 10 August 2005 (table of contents).

Laporte [= de Castelnau], F. L. N. de Caumont 1833: Mémoire sur les divisions du genre Colaspis. *Revue Entomologique* (Silbermann) **1**: 18–25. DP: 16 March 1833 (Nagel & Schmidlin 2014).

Laporte [= de Castelnau], F. L. N. de Caumont 1840: *Histoire naturelle des insectes coléoptères, avec une introduction renfermant l'anatomie et la physiologie des animaux articulés, par M. Brullé. Tome deuxième.* Paris: P. Duménil, 563 + [1] pp., 38 pls. DP: 26 December 1840 (Bousquet 2016).

Latreille, P. A. 1802: *Histoire naturelle, générale et particulière, des crustacés et des insectes. Ouvrage faisant suite à l'histoire naturelle générale et particulière, composée par Leclerc de Buffon, et rédigée par C. S. Sonnini, membre de plusieurs sociétés savantes. Tome troisième. Familles naturelles des genres.* Paris: F. Dufart, xii + 13–467 pp. + [1 p. errata]. DP: 6 November 1802 (Bousquet 2016).

Latreille, P. A. 1804a: *Histoire naturelle, générale et particulière, des crustacés et des insectes. Ouvrage faisant suite aux oeuvres de Leclerc de Buffon, et partie du cours complet d'histoire naturelle rédigé par C. S. Sonnini, membre de plusieurs sociétés savantes. Tome onzième.* Paris: F. Dufart, 422 pp., pls. 91–93. DP: 17 September 1804 (Bousquet 2016).

Latreille, P. A. 1804b: *Histoire naturelle, générale et particulière des crustacés et des insectes. Ouvrage faisant suite aux oeuvres de Leclerc de Buffon, et partie du cours complet d'histoire naturelle rédigé par C. S. Sonnini, membre de plusieurs sociétés savantes. Tome douzième.* Paris: F. Dufart, 424 pp., pls. 94–97. DP: 17 September 1804 (Bousquet 2016).

Latreille, P. A. 1805: *Genera crustaceorum et insectorum secundum ordinem naturalem in familias disposita, iconibus exemplisque plurimis explicata. Tomus primus.* Parisii et Argentorati: Amand Koenig, xviii + 302 + [1] pp., 16 pls. DP: 17 September 1805 (Bousquet 2016).

Latreille, P. A. 1807: *Genera crustaceorum et insectorum secundum ordinem naturalem in familias disposita, iconibus exemplisque plurimis explicata. Tomus tertius.* Parisii et Argentorati: Amand Koenig, 258 + [1] pp. DP: 15 April 1807 (Bousquet 2016).

Latreille, P. A. 1825: *Familles naturelles du règne animal, exposées succinctement et dans un ordre analytique, avec l'indication de leurs genres.* Paris: J.-B. Baillière & Baudouin Frères, 570 pp. DP: 16 May 1825 (Bousquet 2016).

Latreille, P. A. 1829: Suite et fin des insectes. In: Cuvier, G.: *Le règne animal distribué d'après son organisation, pour servir de base à l'histoire naturelle des animaux et d'introduction à l'anatomie comparée. Nouvelle édition, revue et augmentée. Tome V.* Paris: Déterville, xxii + 556 pp. DP: 11 April 1829 (Bousquet 2016).

Lea, A. M. 1903: Descriptions of some new species of Australian and Tasmanian Chrysomelidae. *Report of the Australasian Association for the Advancement of Science* **9** [1902]: 384–431. DP: [31 December] 1903 (verso of the volume title page).

Lea, A. M. 1917: Notes on some miscellaneous Coleoptera, with descriptions of new species. Part III. *Transactions and Proceedings of the Royal Society of South Australia* **41**: 121–322, pls. XII–XV. DP: 24 December 1917 (volume title page).

Lea, A. M. 1929: Notes on some miscellaneous Coleoptera, with descriptions on new species. Part VII. *Transactions and Proceedings of the Royal Society of South Australia* **53**: 203–244. DP: 24 December 1929 (volume title page).

Leach, W. E. 1815: Entomology. Pp. 57–172. In: Brewster, D. (ed.): *The Edinburgh Encyclopaedia. Volume IX, part I.* Edinburgh: William Blackwood, 384 pp. DP: [30] April 1815 (Bousquet 2016).

Leach, W. E. & Risso, J. A. 1827: [new taxon]. In: Risso, J. A.: *Histoire naturelle des principales productions de l'Europe méridionale et particulièrement de celles des environs de Nice et des Alpes Maritimes. Tome cinquième.* Paris: F.-G. Levrault, viii + 403 pp. + 10 pls. DP: 22 September 1827 (Bousquet 2016).

Lebedev, A. G. 1931: Zwei neue Coleopteren aus Central-Asien. *Entomologische Nachrichtenblatt* (Troppau) 5 (3): 78–79. DP: [30] September 1931 (footers throughout the issue).

LeConte, J. L. 1850: General remarks upon the Coleoptera of Lake Superior. Pp. 201–242. In: Agassiz, L.: *Lake Superior: its physical character, vegetation, and animals, compared with those of other and similar regions.* (*With narrative of the tour by J. Elliot Cabot*). Boston: Gould, Kendall and Lincoln, x + 428 pp., 17 pls. DP: 9 March 1850 (Bousquet 2016).

LeConte, J. L. 1858: Description of new species of Coleoptera, chiefly collected by the United States and Mexican Boundary Commission, under Major W. H. Emory, U. S. A. *Proceedings of the Academy of Natural Sciences of Philadelphia* 10: 59–89. DP: 19 April 1858 (Nolan 1913).

LeConte, J. L. 1859a: Additions to the coleopterous fauna of northern California and Oregon. *Proceedings of the Academy of Natural Sciences of Philadelphia* 11: 281–292. DP: 31 December 1859 (Bousquet 2012).

LeConte, J. L. 1859b: The Coleoptera of Kansas and eastern New Mexico. *Smithsonian Contributions to Knowledge* 11: 1–58, 2 pls. DP: [31] December 1859 (table of contents).

LeConte, J. L. 1865: On the species of Galeruca and allied genera inhabiting North America. *Proceedings of the Academy of Natural Sciences of Philadelphia* 17 (4): 204–222. DP: 26 December 1865 (Nolan 1913).

LeConte, J. L. 1868: New Coleoptera collected on the survey for the extension of the Union Pacific Railway, E. D. from Kansas to Fort Craig, New Mexico. *Transactions of the American Entomological Society* 2 [1868–1869] (1): 49–59. DP: 13 July 1868 (Brown 1964).

LeConte, J. L. 1878: Additional descriptions of new species. Pp. 373–472. In: Schwarz, E. A.: Coleoptera of Florida. *Proceedings of the American Philosophical Society held at Philadelphia for Promoting Useful Knowledge* 17: 353–472. DP (pp. 417–432): 22 April 1878 (footers on pp. 417 and 425).

LeConte, J. L. 1881: Short studies of North American Coleoptera. *Transactions of the American Entomological Society* 8 (1): 163–168 & [1880] (3/4): 169–218. DP (pp. 163–168): [30] September 1880, (pp. 169–218): 11 February 1881 (Brown 1964).

LeConte, J. L. 1884: Short studies of North American Coleoptera. (No. 2). *Transactions of the American Entomological Society and Proceedings of the Entomological Section of the Academy of Natural Sciences* 12: 1–32. DP: [30] November 1884 (footers throughout the paper).

Lee, C.-F. 2009: A taxonomic revision of Monolepta pallidula species group in Taiwan (Coleoptera: Chrysomelidae: Galerucinae). *Zootaxa* 2170: 15–27. DP: 29 July 2009 (footer on the paper title page).

Lee, C.-F. 2012: The genus Phygasia Chevrolat (Coleoptera: Chrysomelidae: Galerucinae: Alticini) in Taiwan. *Zootaxa* 3256: 27–37. DP: 4 April 2012 (footer on the paper title page).

Lee, C.-F. 2014a: The genus Sphenoraia Clark, 1865 (Coleoptera: Chrysomelidae: Galerucinae) in Taiwan, with description of a new species. *Coleopterists Bulletin* 68 (1): 143–151. DP: 20 March 2014 (last page of the paper).

Lee, C.-F. 2014b: Revision of the genus Sangariola Jacobson (Coleoptera: Chrysomelidae: Galerucinae: Alticini). *Journal of Taiwan Agricultural Research* 63 (3): 188–204. DP: 30 September 2014 (C.-F. Lee, 2019, pers. comm.).

Lee, C.-F. 2014c: Review of the genus Nonarthra Baly (Coleoptera: Chrysomelidae: Galerucinae: Alticini) from Taiwan and Japan, with descriptions of two new species. *Japanese Journal of Systematic Entomology* 20 (2): 251–263. DP: 30 November 2014 (paper title page).

Lee, C.-F. 2015a: The genus Paraplotes Laboissière, 1933 in Taiwan, a speciose group with brachelytrous females (Coleoptera: Chrysomelidae: Galerucinae). *Zootaxa* 3904 (2): 223–248. DP: 6 January 2015 (footer on the paper title page).

Lee, C.-F. 2015b: Agonita yuae n. sp., the fourth species of the genus from Taiwan (Coleoptera: Chrysomelidae: Cassidinae: Gonophorini). *Japanese Journal of Systematic Entomology* 21 (2): 265–267. DP: 30 November 2015 (paper title page).

Lee, C.-F. 2016: Review of the genus Siemssenius Weise, 1922 (Coleoptera: Chrysomelidae: Galerucinae) from Taiwan, with descriptions of five new species. *Zootaxa* 4158 (3): 367–384. DP: 30 August 2016 (footer on the paper title page).

Lee, C.-F. 2017: Revision of the genus Doryscus Jacoby (Coleoptera: Chrysomelidae: Galerucinae). *Zootaxa* 4269 (1): 1–43. DP: 19 May 2017 (footer on the paper title page).

Lee, C.-F. 2018: The genus Clerotalia [sic!] Jacoby of Taiwan (Coleoptera: Chrysomelidae: Galerucinae). *Zootaxa* 4420 (3): 357–378. DP: 17 May 2018 (footer on the paper title page).

Lee, C.-F. 2019: The genus Lochmaea Weise, 1883 in Taiwan: results of taxonomic expeditions by citizen scientists (Coleoptera, Chrysomelidae, Galerucinae). *ZooKeys* 856: 75–100 DP: 17 June 2019 (paper title page).

Lee, C.-F. 2020: Revision of Taiwanese species of Atrachya Chevrolat, 1836 (Coleoptera, Chrysomelidae, Galerucinae): descriptions of three new genera, two new species, and designations of three new synonyms. *ZooKeys* 940: 117–159. DP: 11 June 2020 (paper title page).

Lee, C.-F. 2021: A new species of the genus Dercetina (Coleoptera: Chrysomelidae: Galerucinae) discovered in alpine regions of Taiwan. *Japanese Journal of Systematic Entomology* 27 (2): 316–320. DP: 30 December 2021 (paper title page).

Lee, C.-F. 2022a: Revision of the genus Arthrotus Motschulsky, 1858 (Coleoptera, Chrysomelidae, Galerucinae) of Taiwan, with notes on color polymorphism. *ZooKeys* 1091: 161–208. DP: 1 April 2022 (paper title page).

Lee, C.-F. 2022b: Review of the Zipanginia sakishimana species complex (Coleoptera: Chrysomelidae: Galerucinae: Alticini), with description of a new species from Taiwan and new synonymy of the genus Yoshiakia Takizawa. *Coleopterists Bulletin* 76 (2): 175–185. DP: 21 June 2022 (last page of the paper).

Lee, C.-F. 2022c: The genus Japonitata Strand (Insecta, Coleoptera, Chrysomelidae, Galerucinae) in Taiwan: a redefinition of the genus and descriptions of two new species. *ZooKeys* 1125: 171–192. DP: 26 October 2022 (paper title page).

Lee, C.-F. 2023: Four new species of Sphaeroderma Stephens (Coleoptera, Chrysomelidae, Galerucinae, Alticini) from Taiwan, with discussion on genus boundaries based on S. flavonotatum Chûjô and S. jungchani sp. nov. *ZooKeys* **1185**: 1–19. DP: 27 November 2023 (paper title page).

Lee, C.-F. & Beenen, R. 2010: [new taxon]. In: Beenen, R. & Lee, C.-F. 2010: Two new Erganoides species from P. R. China and Taiwan (Coleoptera: Chrysomelidae: Galerucinae). *Genus* **21** (2): 257–264. DP: 30 July 2010 (paper title page).

Lee, C.-F. & Beenen, R. 2012: Calomicrus jungchangi Lee and Beenen (Coleoptera: Chrysomelidae: Galerucinae), a new species from Taiwan, with redescription of a similar species, Monolepta rufofulva Chûjô, 1938. *Coleopterists Bulletin* **66** (2): 123–130. DP: 20 June 2012 (last page of the paper).

Lee, C.-F. & Beenen, R. 2015: Revision of Taiwanaenidea Kimoto, 1984 (Coleoptera: Chrysomelidae: Galerucinae). *Zootaxa* **4020** (1): 153–168. DP: 21 September 2015 (footer on the paper title page).

Lee, C.-F. & Beenen, R. 2017a: Revision of the genus Shairella Chûjô, 1962 (Coleoptera: Chrysomelidae: Galerucinae) from Taiwan, with descriptions of five new species. *Zootaxa* **4268** (4): 489–507. DP: 18 May 2017 (footer on the paper title page).

Lee, C.-F. & Beenen, R. 2017b: Revision of the Palaearctic and Oriental species of the genus Oides Weber, 1801 (Coleoptera: Chrysomelidae: Galerucinae). *Zootaxa* **4346** (1): 1–125. DP: 9 November 2017 (footer on the paper title page).

Lee, C.-F. & Beenen, R. 2020: Taiwanoshaira Lee & Beenen, a new genus and first record of moss-inhabiting Galerucinae sensu stricto (Coleoptera, Chrysomelidae) from Taiwan. *ZooKeys* **944**: 129–146 DP: 30 June 2020 (paper title page).

Lee, C.-F. & Bezděk, J. 2013: Revision of the genus Dercetina from Taiwan and their similar species, with description of a new species from Myanmar (Insecta, Chrysomelidae, Galerucinae). *ZooKeys* **323**: 1–33. DP: 13 August 2013 (paper title page).

Lee, C.-F. & Bezděk, J. 2014: Revision of the genus Paridea Baly, 1886 from Taiwan (Coleoptera, Chrysomelidae, Galerucinae). *ZooKeys* **405**: 83–125. DP: 28 April 2014 (paper title page).

Lee, C.-F. & Bezděk, J. 2015: Revision of "Phyllobrotica" from Taiwan with description of Jolibrotica gen. n. (Coleoptera, Chrysomelidae, Galerucinae). *ZooKeys* **547**: 75–92. DP: 17 December 2015 (paper title page).

Lee, C.-F. & Bezděk, J. 2016: Revision of the wingless Sikkimia Duvivier (Coleoptera, Chrysomelidae, Galerucinae) from Taiwan, including a new generic synonymy and four new species descriptions. *ZooKeys* **553**: 79–106. DP: 14 January 2016 (paper title page).

Lee, C.-F. & Bezděk, J. 2018: Revision of the genus Theopea Baly (Coleoptera: Chrysomelidae: Galerucinae) of East Asia: species lacking modified clypeus in males and the T. sauteri species group. *Zootaxa* **4508** (3): 334–376. DP: 1 November 2018 (footer on the paper title page).

Lee, C.-F. & Bezděk, J. 2020: Revision of the Theopea genus group (Coleoptera, Chrysomelidae, Galerucinae), part III: Descriptions of two new genera and nine new species. *ZooKeys* **912**: 65–124 DP: 17 February 2020 (paper title page).

Lee, C.-F. & Bezděk, J. 2021a: Revision of the genera Xanthogaleruca Laboissière, 1932 and Pyrrhalta Joannis, 1865 (Coleoptera, Chrysomelidae, Galerucinae) of Taiwan, with type designation of Galerucella lineatipes Takei. *ZooKeys* **1039**: 1–108. DP: 20 May 2021 (paper title page).

Lee, C.-F. & Bezděk, J. 2021b: Revision of the genus Furusawaia Chûjô, 1962 (Coleoptera, Chrysomelidae, Galerucinae). *ZooKeys* 1057: 117–148. DP: 27 August 2021 (paper title page).

Lee, C.-F., Bezděk, J. & Staines, C. L. 2010: A review of the genus Agetocera (Coleoptera: Chrysomelidae: Galerucinae) in Taiwan – are there only two species? *Zootaxa* **2441**: 1–19. DP: 30 April 2010 (footer on the paper title page).

Lee, C.-F., Bezděk, J. & Staines, C. L. 2011: A review of the genus Haplosomoides Duvivier, 1890 in Taiwan and Japan (Coleoptera: Chrysomelidae: Galerucinae). *Zoological Studies* 50 (1): 118–138. DP: [31] January 2011 (issue cover).

Lee, C.-F., Bezděk, J. & Suenaga, H. 2012: Revision of Menippus (Coleoptera: Chrysomelidae: Galerucinae) of Taiwan and Menippus dimidiaticornis species group with a new generic synonymy. *Zootaxa* **3427**: 1–16. DP: 17 August 2012 (footer on the paper title page).

Lee, C.-F. & Chen, J.-C. 2023: Demarchus hsui (Coleoptera, Chrysomelidae, Galerucinae, Alticini), a new species from Taiwan, with notes on immatures and biology. *ZooKeys* **1177**: 3–21. DP: 30 August 2023 (paper title page).

Lee, C.-F. & Hsieh, C.-H. 2022: Integrative taxonomy of the leaf-beetle genus Gonioctena Chevrolat, 1836 in Taiwan (Coleoptera, Chrysomelidae, Chrysomelinae, Gonioctenini) reveals new synonymies and one new species. *ZooKeys* **1120**: 1–46. DP: 5 September 2022 (paper title page).

Lee, C.-F. & Huang, C.-L. 2023: Two new species of the genus Shairella (Coleoptera: Chrysomelidae: Galerucinae) from China and Taiwan, with an updated key to species of the world. *Annales Zoologici* **73** (3): 411–419. DP: 30 September 2023 (paper title page).

Lee, C.-F. & Konstantinov, A. S. 2015: A new species of Burumoseria Csiki, 1939 from Taiwan, with redescription of the genus and new synonymy of Licyllus Jacoby, 1885 with Thrasychroma Jacoby, 1885 (Coleoptera: Chrysomelidae: Galerucinae: Alticini). *Coleopterists Bulletin* **69** (2): 325–331. DP: 18 June 2015 (last page of the paper).

Lee, C.-F. & Sekerka, L. 2018: Taxonomic status of Wallacea dactyliferae Maulik (Coleoptera: Chrysomelidae: Cassidinae: Bothryonopini) from Taiwan. *Japanese Journal of Systematic Entomology* **24** (2): 299–303. DP: 30 November 2018 (paper title page).

Lee, C.-F. & Staines, C. L. 2009: Hemipyxis yui, a new species from Taiwan, with redescription of its allied species H. quadrimaculata (Jacoby, 1892) (Coleoptera: Chrysomelidae: Galerucinae). *Coleopterists Bulletin* **63** (1): 62–70. DP: 22 April 2009 (last page of the paper).

Lee, C.-F. & Staines, C. L. 2010: Monolepta ongi, a new species from Lanyu Island, with redescription of its allied species Monolepta longitarsoides Chûjô, 1938 (Coleoptera: Chrysomelidae: Galerucinae). *Proceedings of the Entomological Society of Washington* **112** (4): 530–540. DP: 16 December 2010 (verso of the issue cover).

Lee, C.-F., Świętojańska, J. & Staines, C. L. 2009: Prionispa houjayi (Coleoptera: Chrysomelidae: Cassidinae: Oncocephalini), a newly recorded genus and new species from Taiwan, with a description of its immature stages and notes on its bionomy. *Zoological Studies* 48 (4): 558–568. DP: [31] July 2009 (issue cover).

Lee, C.-F., Świętojańska, J. & Staines, C. L. 2012: A review of the genus Callispa Baly, 1858 in Taiwan (Coleoptera: Chrysomelidae: Cassidinae: Callispini), with descriptions of two new species and their immature stages, and notes on their bionomy. *Zoological Studies* 51 (6): 832–861. DP: [30] November 2012 (issue cover).

Lee, C.-F., Tian, P.-L. & Staines, C. L. 2010: Monolepta meihuai, a new species from Taiwan, with redescription of its sibling species Monolepta mandibularis Chûjô, 1962 (Coleoptera: Chrysomelidae: Galerucinae). *Coleopterists Bulletin* 64 (4): 303–311. DP: 20 December 2010 (last page of the paper).

Lee, C.-F., Tsai, Ch.-L., Konstantinov, A. & Yeh, W.-B. 2016: Revision of Mandarella Duvivier from Taiwan, with a new species, new synonymies and identities of highly variable species (Insecta, Chrysomelidae, Galerucinae, Alticini). *ZooKeys* 568: 23–49. DP: 23 February 2016 (paper title page).

Leeder, F. 1950: Eine neue Phytodecta aus den Ostalpen (Coleopt., Chrysomelidae). *Zeitschrift der Wiener Entomologischen Gesellschaft* 35 (1/3): 21–22. DP: 20 March 1950 (issue title page).

Lefebvre, A. L. 1827: Description de divers insectes inédits recueillis en Sicile. *Mémoires de la Société Linnéenne de Paris* 6: 94–108, pl. v. DP: [31] March 1827 (volume title page).

Lefèvre, É. 1870: [Note qui suit au sujet de deux espèces de Clythrides]. *Bulletin de la Société Entomologique de France* 1870 (1): xlii–xliii. DP: 13 July 1870 (Lefèvre 1885).

Lefèvre, É. 1872a: Monographie des clytrides d'Europe et du bassin de la Méditerranée. *Annales de la Société Entomologique de France, Cinquième Série* 2 (1): 49–96 & (2) 97–168. DP (pp. 49–96): 10 July 1872, (pp. 97–168): 25 September 1872 (Lefèvre 1885).

Lefèvre, É. 1872b: Monographie des clytrides d'Europe et du bassin de la Méditerranée. *Annales de la Société Entomologique de France, Cinquième Série* 2 (3): 313–352 & (4) 353–396. DP (pp. 313–352): 31 December 1872 (Lefèvre 1885).

Lefèvre, É. 1873a: Monographie des clytrides d'Europe et du bassin de la Méditerranée. *Annales de la Société Entomologique de France, Cinquième Série* 2 (3): 313–352 & (4) 353–396. DP (pp. 353–396): 27 April 1873 (Lefèvre 1885).

Lefèvre, É. 1873b: [new taxon]. *Bulletin de la Société Entomologique de France* 1873 (14): 6–8. DP: 31 October 1873 (Evenhuis 2002).

Lefèvre, É. 1873c: [new taxon]. *Bulletin de la Société Entomologique de France* 1873 (15): 4–5. DP: 20 November 1873 (Evenhuis 2002).

Lefèvre, É. 1875: [new taxa]. *Bulletin de la Société Entomologique de France* 1875 (1): 10–11. DP: 23 January 1875 (Evenhuis 2002).

Lefèvre, É. 1876a: [Quatre nouvelles espèces de clytrides]. *Bulletin de la Société Entomologique de France* 1876 (7): 76–78. DP: 20 April 1876 (Evenhuis 2002).

Lefèvre, É. 1876b: Synopsis des eumolpides d'Europe et confins. *L'Abeille, Journal d'Entomologie* 14 [1874–1876]: 1–20. DP: 14 June 1876 (*Bull. Soc. Ent. Fr.* 1876: 136).

Lefèvre, É. 1876c: Descriptions d'eumolpides nouveaux ou peu connus. *Revue et Magasin de Zoologie Pure et Appliquée* (3e Série) 4 (8): 278–311. DP: 15 November 1876 (*Pet. Nouv. Ent.* 1876, 8 (160): 88).

Lefèvre, É. 1877: Descriptions de coléoptères nouveaux ou peu connus de la famille des eumolpides. 1re partie. *Annales de la Société Entomologique de France, Cinquième Série* 7 (1): 115–128 & (2) 129–166. DP (pp. 115–128): 11 July 1877, (pp. 129–166): 5 September 1877 (Lefèvre 1885).

Lefèvre, É. 1878: Descriptions de coléoptères nouveaux ou peu connus de la famille des eumolpides. 2e partie. *Annales de la Société Entomologique de France, Cinquième Série* 7 [1877] (4): 309–326. DP: 10 April 1878 (Lefèvre 1885).

Lefèvre, É. 1883a: [Descriptions de nouvelles espèces de coléoptères de la famille des Clytrides (1re partie)]. *Bulletin de la Société Entomologique de France* **1883** (17): 149–150. DP: 20 September 1883 (Evenhuis 2002).

Lefèvre, É. 1883b: [Descriptions de nouvelles espèces de coléoptères de la famille des Clytrides (2e partie)]. *Bulletin de la Société Entomologique de France* **1883** (18): 164–165. DP: 6 October 1883 (Evenhuis 2002).

Lefèvre, É. 1883c: [Description suivante d'un nouveau genre de coléoptères de la famille des Eumolpides]. *Bulletin de la Société Entomologique de France* **1883** (19): 169–170. DP: 20 October 1883 (Evenhuis 2002).

Lefèvre, É. 1884a: [Descriptions de trois genres nouveaux de coléoptères, de la famille des Eumolpides]. *Bulletin de la Société Entomologique de France* **1884** (9): 91–93. DP: 20 May 1884 (Evenhuis 2002).

Lefèvre, É. 1884b: [Description suivante d'un nouveau genre et d'une nouvelle espèce de la famille des Eumolpides]. *Bulletin de la Société Entomologique de France* **1884** (11): 106. DP: 20 June 1884 (Evenhuis 2002).

Lefèvre, É. 1884c: Descriptions de quatre genres nouveaux et de plusieurs espèces nouvelles de coléoptères de la famille des eumolpides. *Bulletin ou Comptes-rendus des Séances de la Société Entomologique de Belgique, Série III* **1884** (46): cxciii–ccvi. DP: 9 July 1884 (*Bull. Soc. Ent. Fr.* 1884: 158).

Lefèvre, É. 1885a: [Descriptions suivantes de trois espèces nouvelles de coléoptères de la famille des Eumolpides, qui ont été capturées au Tonkin par M. le lieutenant Martin, et qui lui ont été communiquées par M. l'abbe Umhang]. *Bulletin de la Société Entomologique de France* **1885** (7): lxv–lxvi. DP: 18 April 1885 (Evenhuis 2002).

Lefèvre, É. 1885b: [new taxa]. *Bulletin de la Société Entomologique de France* **1885** (13): cxxiv–cxxvi. DP: 18 July 1885 (Evenhuis 2002).

Lefèvre, É. 1885c: [Description suivante d'un nouveau genre de la famille des Eumolpides]. *Bulletin de la Société Entomologique de France* **1885** (16): cxlvii. DP: 5 September 1885 (Evenhuis 2002).

Lefèvre, É. 1885d: Eumolpidarum hucusque cognitarum catalogus, sectionum conspectu systematico, generum sicut et specierum nonnullarum novarum descriptionibus adjunctis.

Mémoires de la Société Royale des Sciences de Liège, Deuxième Série 11 (16): 1–172. DP (reprint): 23 December 1885 (*Bull. Soc. Ent. Fr.* 1885: ccxxxii).

Lefèvre, É. 1887a: [Quelques espèces de clytrides et d'eumolpides, provenant de la Chine, et qui lui ont été communiquées par M. l'abbé Armand David]. *Bulletin de la Société Entomologique de France* 1887 (6): liv–lvii. DP: 3 April 1887 (Evenhuis 2002).

Lefèvre, É. 1887b: [Quatre nouvelles espèces d'eumolpides, capturées par M. Fabre à Ramnad (Hindoustan), et qui lui ont été communiquées par M. Gounelle]. *Bulletin de la Société Entomologique de France* 1887 (20): clxvi–clxvii. DP: 5 November 1887 (Evenhuis 2002).

Lefèvre, É. 1888: [Diagnoses de quatre nouvelles espèces d'Eumolpides qui lui ont été envoyées par notre collègue M. le Dr Pipitz, de Graz (Styrie)]. *Bulletin de la Société Entomologique de France* 1888 (11): lxxxi–lxxxii. DP: 23 June 1888 (Evenhuis 2002).

Lefèvre, É. 1889a: Contributions à la faune indo-chinoise. 4e Mémorie. Cryptocéphalides, clytrides et eumolpides. *Annales de la Société Entomologique de France, 6e Série* 9 (2): 287–299. DP: 31 October 1889 (Lefèvre 1895).

Lefèvre, É. 1889b: [Description d'une nouvelle espèce de clytrides, decouverte aux environs de Smyrna par M. Delagrange, de Besançon, et qui iui a été commuuiquée par notre collègue M. Ernest Olivier]. *Bulletin de la Société Entomologique de France* 1889 (11): cxix. DP: 22 June 1889 (Evenhuis 2002).

Lefèvre, É. 1890a: [new taxa]. *Bulletin de la Société Entomologique de France* 1890 (8): lvii. DP: 10 May 1890 (Evenhuis 2002).

Lefèvre, É. 1890b: Coléoptères. Clytridae, Eumolpidae. Pp. 189–202. In: Collection d'insectes formée dans l'Indo-Chine par M. Pavie Consul de France au Cambodge. *Nouvelles Archives du Muséum d'Histoire Naturelle, Troisième Série* 2 (2): 177–210. DP: [31 December] 1890 (issue cover).

Lefèvre, É. 1891: Descriptions d'espèces nouvelles de clytrides et d'eumolpides. *Bulletin ou Comptes-Rendus des Séances de la Société Entomologique de Belgique, Série IV* 1891 (20): ccxlviii–cclxxix. DP: 14 October 1891 (DP refers to issue 21 mentioned in *Bull. Soc. Ent. Fr.* 1891: cxlvii).

Lefèvre, É. 1892: [Description d'une nouvelle espèce de clytrides provenant de Chine]. *Bulletin de la Société Entomologique de France* 1892 (7): xciv–xcv. DP: 23 April 1892 (Evenhuis 2002).

Lefèvre, É. 1893: Contribution à la faune indo-chinoise. 12e Mémorie. Clytrides & eumolpides (2e mémorie). *Annales de la Société Entomologique de France* 62 (1): 111–134. DP: 31 July 1893 (issue cover).

Legalov, A. A. 1999: Materialy po faune zhukov-zernovok (Coleoptera, Bruchidae) zapovednika «Daurskiy». (Materials on the seed beetles (Coleoptera, Bruchidae) of the Dahurian State Nature Reserve). Pp. 116–118. In: Dubatolov, V. V. (ed.): *Nasekomye Daurii i sopredelnykh territoriy (Sbornik nauchnykh trudov). Vypusk II. [Insects of Dahuria and adjacent territories. Proceedings of the Dahurskii State Biosphere Nature Reserve. Vol. II]*. Novosibirsk, 264 pp. DP: [31 December] 1999 (book cover).

Legalov, A. A. 2011: K poznaniyu roda Kytorhinus Fisch. (Coleoptera, Chrysomelidae: Bruchinae). (Contribution to the knowledge of the genus Kytorhinus Fisch. (Coleoptera, Chrysomelidae: Bruchinae)). *Amurian Zoological Journal* **3** (3): 262–264. DP: [30] September 2011 (issue cover).

Legalov, A. A. 2021: A new species of the genus Orsodacne Latreille, 1802 (Coleoptera: Orsodacnidae) from China. *Ecologica Montenegrina* **46**: 74–79. DP: 2 November 2021 (paper title page).

Legalov, A. A. 2022: A new species of the genus Orsodacne Latreille, 1802 (Coleoptera: Orsodacnidae) from Iraqi Kurdistan. *Ecologica Montenegrina* **54**: 81–85. DP: 9 August 2022 (paper title page).

LeGrand, G. 1861: Liste des coléoptères du département de l'Aube. *Mémoires de la Société Académique d'Agriculture, des Sciences, Arts et Belles-Lettres du Département de l'Aube, Deuxième Série* **12**: 181–274. DP: 13 July 1861 (*Bibl. France* 1861, 50 (28): 338).

Lei, Q.-L., Yang, X.-K. & Nie, R.-E. 2020: A worldwide species catalogue and key of Sermyloides (Coleoptera, Chrysomelidae, Galerucinae) with description of two new species from China. *Journal of Asia-Pacific Entomology* **23** (2): 411–417. DP: 1 February 2020 (footer on the paper title page).

Lei, Q.-L., Xu, S.-Y., Yang, X.-K. & Nie, R.-E. 2021: Five new species of the leaf-beetle genus Monolepta Chevrolat (Coleoptera, Chrysomelidae, Galerucinae) from China. *ZooKeys* **1056**: 35–57 + supplementary material. DP: 18 August 2021 (paper title page).

Lencina, J. L. & Petitpierre, E. 2011: [new taxon]. In: Petitpierre, E. & Lencina, J. L.: Una nueva especie de Cryptocephalus Geoffroy, 1762, de la Península Ibérica: Cryptocephalus (Heterichnus) gallegoi n. sp. (Coleoptera, Chrysomelidae, Cryptocephalinae). *Nouvelle Revue d'Entomologie, Nouvelle Série* **27** [2010–2011] (3): 283–287. DP: 30 November 2011 (last page of the cover).

Lengerken, H. von 1913: Melasoma lapponicum L. und seine Formen. *Berliner Entomologische Zeitschrift* **57** [1912] (3/4): 123–130. DP: [31] May 1913 (verso of the volume title page).

Leonardi, C. 1971: Considerazioni sulle Psylliodes del gruppo napi e descrizione di una nuova specie (Coleoptera Chrysomelidae). *Atti della Società Italiana di Scienze Naturali e del Museo Civile di Storia Naturale di Milano* **112** (4): 485–533. DP: 15 December 1971 (paper title page).

Leonardi, C. 1973: Longitarsus bertii nome nuovo per il L. ferrugineus (Foudras) sensu Kaszab (Coleoptera Chrysomelidae). *Atti della Società Italiana di Scienze Naturali e del Museo Civico di Storia Naturale di Milano* **114** (4): 465–474. DP: 15 December 1973 (paper title page).

Leonardi, C. 1974: Studi tassonomici su Orestia electra Gredler, con descrizione di due nuove sottospecie (Coleoptera Chrysomelidae). *Atti della Società Italiana di Scienze Naturali e del Museo Civile di Storia Naturale di Milano* **115** (3/4): 385–393. DP: 15 December 1974 (paper title page).

Leonardi, C. 1975a: Dati biogeografici sul popolamento Sardo di Alticini con diagnosi preliminare di una nuova specie di Aphthona (Coleoptera Chrysomelidae). *Atti della Società Italiana di Scienze Naturali e del Museo Civile di Storia Naturale di Milano* **116** (1/2): 3–14. DP: 15 June 1975 (paper title page).

Leonardi, C. 1975b: Note su alcuni Longitarsus con descrizione di due nuove specie e citazione di quattro entitá nuove per la fauna europea (Coleoptera Chrysomelidae). *Atti della Società Italiana di Scienze Naturali e del Museo Civico di Storia Naturale di Milano* **116** (3/4): 199–217. DP: 15 December 1975 (paper title page).

Leonardi, C. 1975c: Le Psylliodes appenniniche del Museo civico di storia naturale di Verona (Coleoptera Chrysomelidae). *Bollettino del Museo Civico di Storia Naturale di Verona* **2**: 51–90. DP: 31 December 1975 (table of contents).

Leonardi, C. 1976: Descrizione di un nuovo alticino europeo: Longitarsus noricus n. sp. (Coleoptera Chrysomelidae). *Atti della Società Italiana di Scienze Naturali e del Museo Civico di Storia Naturale di Milano* **117** (3/4): 239–250. DP: 15 December 1976 (paper title page).

Leonardi, C. 1977: Considerazioni su alcune Orestia con descrizione di una nuova specie dell'Asia Minore (Coleoptera Chrysomelidae). *Natura, Società Italiana di Scienze Naturali, Museo Civico di Storia Naturale e Aquario Civico, Milano* **68** (1/2): 123–131. DP: 15 June 1977 (reprint cover).

Leonardi, C. 1978: Studio critico sulla Psylliodes picina (Marsh.) e sulle forme che le sono state attribuite, con particulare riguardo alla fauna italiana (Coleoptera Chrysomelidae). *Atti della Società Italiana di Scienze Naturali e del Museo Civico di Storia Naturale di Milano* **119** (3/4): 271–299. DP: 15 December 1978 (paper title page).

Leonardi, C. 1979: Descrizione di quattro nuovi Longitarsus della regione Mediterranea (Coleoptera Chrysomelidae). *Atti della Società Italiana di Scienze Naturali e del Museo Civico di Storia Naturale di Milano* **120** (3/4): 291–306. DP: 15 December 1979 (paper title page).

Leonardi, C. 2007: Dati inediti sul genere Psylliodes Latreille, con descrizione di quattro nuove specie mediterranee (Coleoptera Chrysomelidae). *Atti della Società Italiana di Scienze Naturali e del Museo Civico di Storia Naturale in Milano* **148** (2): 161–240. DP: [31] January 2007 (paper title page).

Leonardi, C. 2013: Indagine critica su Psylliodes laevifrons Kutschera con descrizione di due nuove specie (Coleoptera Chrysomelidae). *Atti della Società Italiana di Scienze Naturali e del Museo Civico di Storia Naturale in Milano* **154** (2): 81–114. DP: [31] December 2013 (paper title page).

Leonardi, C. & Arnold, U. 1995: Due nuove specie di Psylliodes della regione mediterranea orientale (Coleoptera Chrysomelidae). *Atti della Società Italiana di Scienze Naturali e del Museo Civico di Storia Naturale in Milano* **134** [1993] (2): 299–311. DP: [31] July 1995 (paper title page).

Leonardi, C. & Gruev, B. 1993: Note sistematiche e geonemiche su alcuni Psylliodes del complesso picinus (Marsh.) con descrizione di una nuova specie (Coleoptera, Chrysomelidae). *Atti della Società Italiana di Scienze Naturali e del Museo Civico di Storia Naturale di Milano* **133** [1992]: 12–32. DP: [31] July 1993 (paper title page).

Leonardi, C. & Mohr, K.-H. 1974: Drei neue mit Longitarsus rubellus verwandte Arten aus den westeuropäischen Berggegenden (Coleoptera Chrysomelidae). *Atti della Società Italiana di Scienze Naturali e del Museo Civico di Storia Naturale di Milano* **115** (2): 191–203. DP: 15 June 1974 (paper title page).

Leonardi, C. & Sassi, D. 1997: I crisomelidi (Coleoptera Chrysomelidae) del Monte Barro (Italia, Lombardia, Lecco). Pp. 189–227. In: Leonardi, C. & Sassi, D. (eds.): Studi geobotanici ed entomofaunistici nel Parco Regionale del Monte Barro. *Memorie della Società Italiana di Scienze Naturali e del Museo Civico di Storia Naturale di Milano* 27 (2): 137–263. DP: 17 December 1997 (issue cover).

Leonardi, C. & Sassi, D. 2001: Studio critico sulle specie di Cryptocephalus del gruppo hypochaeridis (Linné, 1758) e sulle forme ad esse attribuite (Coleoptera Chrysomelidae). *Atti della Società Italiana di Scienze Naturali e del Museo Civico di Storia Naturale in Milano* **142** (1): 3–96. DP: [30] November 2001 (paper title page).

Leoni, G. 1906: I Chilotomini italiani. *Rivista Coleotterologica Italiana* **4** (6/7): 145–148. DP: 15 July 1906 (issue title page).

Lepesme, P. 1941: [new taxon]. In: Vayssière, P. & Lepesme, P.: Sur quelques bruchides nuisibles (Col.). *Revue Français d'Entomologie* 8 (4): 192–202. DP: 30 December 1941 (footer on first page of issue).

Leprieur, C. E. 1883: [new taxon]. *Bulletin de la Sociétè Entomologique de France* **1883** (13): 114–115. DP: 21 July 1883 (Evenhuis 2002).

LeSage, L. 1986: A taxonomic monograph of the Nearctic galerucine genus Ophraella Wilcox (Coleoptera: Chysomelidae). *Memoirs of the Entomological Society of Canada* **133**: 1–75. DP: 14 April 1986 (journal website).

LeSage, L. 1988: Notes on European Longitarsus species introduced in North America (Coleoptera: Chysomelidae: Alticinae). *Canadian Entomologist* **120** (12): 1133–1145. DP: [31] December 1988 (issue cover).

LeSage, L. 1998: Description d'une nouvelle espèce de Clavicornaltica en provenance de Taïwan (Coleoptera, Chrysomelidae). *Nouvelle Revue d'Entomologie, Nouvelle Série* **14** [1997] (3): 239–247. DP: 1 April 1998 (last page of the cover).

Leschen, R. A. B. & Beutel, R. G. (eds.) 2014: *Handbook of Zoology. Coleoptera, beetles. Morphology and systematics. Volume 3*. Berlin/Boston: Walter de Gruyter, 675 pp. DP: [31 December] 2014 (colophon).

Letzner, K. W. 1840: Clythra diversipes, n. sp. *Uebersicht der Arbeiten und Veränderungen der Schlesischen Gesellschaft für Vaterländische Kultur* (Breslau) [**1839**]: 114–116. DP: 18 June 1840 (*Ber. Bekanntmach. Verh. Akad. Wiss. Berlin* 1840: 133).

Letzner, K. W. 1847: Bericht über die Beschäftigungen der entomologischen Section im Jahre 1846. *Uebersicht der Arbeiten und Veränderungen der Schlesischen Gesellschaft für Vaterländische Kultur* (Breslau) **1846**: 73–85. DP: 14 October 1847 (*Ber. Bekanntmach. Verh. Akad. Wiss. Berlin* 1847: 369).

Letzner, K. W. 1849: Käfer. Pp. 95–99. In: Gravenhorst, J. L. C.: Bericht über die Arbeiten der entomologischen Section im Jahre 1848. *Uebersicht der Arbeiten und Veränderungen der Schlesischen Gesellschaft für Vaterländische Kultur* (Breslau) **1848**: 89–111. DP: 25 October 1849 (*Ber. Bekanntmach. Verh. Akad. Wiss. Berlin* 1849: 246).

Letzner, K. W. 1853: Chrysomela fusco-aenea Schum. et Suff. und Chrysomela speciosissima Scop. et Suff. Pp. 87–93. In: Gravenhorst, J. L. C.: Bericht über die Verhandlungen der entomologischen Section im Jahre 1852. Coleoptera. *Dreissigster Jahres-Bericht der Schlesischen Gesellschaft für Vaterländische Kultur* (Breslau) **1852**: 87–104. DP: 24 August 1853 (*Ztschr. Ges. Naturwiss.* 1853, 2: 198).

Lewis, G. 1879: *A catalogue of Coleoptera from the Japanese Archipelago*. London: Taylor & Francis, 31 pp. DP: [31] October 1879 (Bousquet 2016).

Lewis, G. 1896: On new species of Coleoptera from Japan, and notices of others. *Annals and Magazine of Natural History, Sixth Series* **17** (101): 329–343. DP: 1 May 1896 (Evenhuis 2003).

Li, J.-K. 1992: *The Coleoptera fauna of northeast China*. Jilin: Education Publishing House, 2 + 6 + 205 pp. DP: [30] April 1992 (colophon).

Li, J.-K. & Wang, Z.-G. 1993: The soil beetles in Anhui province. Pp. 151–167. In: Li, J.-K. & Chen, P. (eds.): *Studies on fauna and ecogeography of soil animal*. Shengyang: Northeast Normal University, [6] + 265 pp. DP: [31] March 1993 (colophon).

Li, K.-Q. & Liang, H.-B. 2013: [new taxa]. In: Li, K.-Q., Liang, Z.-L. & Liang, H.-B.: Two new species of the genus Temnaspis Lacordaire, 1845, (Coleoptera: Chrysomeloidea: Megalopodidae) from China and Myanmar, with notes on the biology of the genus. *Zootaxa* **3737** (4): 379–398. DP: 21 November 2013 (footer on the paper title page).

Li, K.-Q. & Liang, H.-B. 2018: A check list of the Chinese Zeugophorinae (Coleoptera: Megalopodidae), with new synonym, new record and two new species of subgenus Pedrillia from China. *Zootaxa* **4455** (1): 127–149. DP: 1 August 2018 (footer on the paper title page).

Li, K.-Q. & Liang, H.-B. 2020: Four new species and two new records of genus Zeugophora (Coleoptera, Megalopodidae, Zeugophorinae) from China. *ZooKeys* **975**: 51–78. DP: 12 October 2020 (paper title page).

Li, X.-P. 2013: *A taxonomic study on Eumolpidae (Coleoptera: Chrysomeloidea) from Heilongjiang province*. Master Thesis, Northeast Forestry University Charbin, iv + 74 pp. DP: [31] October 2013 (publisher website).

Li, X.-P., Li, C.-D., Song, L.-W., Chen, Y.-Q. & Wang, Y.-Y. 2008: Faunistic and distribution of Eumolpidae (Coleoptera: Chrysomeloidea) species in Heilongjiang province, China. *Zoological Research* **29** (4): 438–446. DP: [31] August 2008 (paper title page).

Liang, Z.-L., Konstantinov, A. S. & Ge, S.-Q. 2023: [new taxa]. In: Liang, Z.-L., Konstantinov, A. S., Ruan, Y.-Y., Li, Z.-Q., Huang, Z.-Z. & Ge, S.-Q.: Two new species of the Longitarsus violentus group from China (Coleoptera, Chrysomelidae, Galerucinae, Alticini). *ZooKeys* **1181**: 111–123. DP: 3 October 2023 (paper title page).

Lichtenstein, A. A. H. 1796: *Catalogus musei zoologici ditissimi Hamburgi, d III Februar 1796. Auctionis lege distrahendi. Sectio tertia continens Insecta. Verzeichniss von höchstseltenen, aus allen Welttheilen mit vieler Mühe und Kosten zusammen gebrachten, auch aus unterschiedlichen Cabinettern, Sammlungen und Auctionen ausgehobenen Naturalien welche von einem Liebhaber, als Mitglied der batavischen und verschiedener anderer naturforschenden Gesellschaften gesammelt worden. Dritter Abschnitt, bestehend in wohlerhaltenen, mehrentheils ausländischen und höchstseltenen Insecten, die Theils einzeln, Theils mehrere zusammen in Schachteln festgesteckt sind, und welche am Mittewochen, den 3ten Februar 1796 und den folgenden Tagen auf dem Eimbeckschen Hause öffentlich verkauft werden sollen durch dem Mackler Peter Heinrich Packischefsky*. Hamburg: G. F. Schniebes, xii + 222 pp. DP: 3 February 1796 (Bousquet 2016) [surpressed by the Opinion 1820 (ICZN 1995b), but 19 species-group names retained available].

Lindberg Håk. 1926: Die ostfennoskandischen Arten der Gattung Haltica Geoffr. (Col., Chrysom.). *Notulae Entomologicae* 6 (3): 65–76. DP: 25 November 1926 (volume title page).

Lindberg Har. 1938: Nykomlingar till Finlands skalbaggsfauna II. *Notulae Entomologicae* 18 (2): 40–48. DP: 27 May 1938 (verso of the volume title page).

Lindberg Har. 1950: Beitrag zur Kenntnis der Käferfauna der Kanarischen Inseln. *Commentationes Biologicae* 10 (18): 1–20. DP: [30] November 1950 (last page of the paper).

Lindberg Har. 1953: Zweiter Beitrag zur Kenntnis der Käferfauna der Kanarischen Inseln. *Commentationes Biologicae* 13 (12): 1–18. DP: [31] March 1953 (last page of the paper).

Linell, M. L. 1896: List of Coleoptera collected on the Tana River, and on the Jombene Range, East Africa, by Mr. William Astor Chanler, and Lieutenant Ludwig von Höhnel, with descriptions of new genera and species. *Proceedings of the United States National Museum* 18 [1895] (1094): 687–716. DP: 12 August 1896 (table of contents).

Linnaeus, C. 1758: *Systema naturae per regna tria naturae, secundum classes, ordines, genera, species, cum characteribus, differentiis, synonymies, locis. Tomus I. Editio decima, reformata*. Holmiae: Impensis Direct. Laurentii Salvii, iv + 824 + [1] pp. DP: 1 January 1758 (antedated, ICZN 1999: Article 3).

Linnaeus, C. 1761: *Fauna svecica sistens animalia Sveciae Regni: Mammalia, Aves, Amphibia, Pisces, Insecta, Vermes. Distributa per classes & ordines, genera & species, cum differentiis specierum, synonymis auctorum, nominibus incolarum, locis natalium, descriptionibus insectorum. Editio altera, auctior*. Stockholmiae: Laurentii Salvii, [48] + 578 pp., 2 pls. DP: 5 August 1761 (Zilli 2021).

Linnaeus, C. 1763: *D.D. Centuria insectorum rariorum quam consent. experimentiss. fac. med. In Regia Academia Upsaliensi, praeside nobilissimo atque celeberrimo D:o Doct. Carolo von Linné, equite aurat. de Stella Polari. Publico examini submittit Boas Johansson, Calmariensis. In audit Carol. maj. d. XXIII. Junii. anni MDCCLXIII. H.a.m.s.* Upsaliae, [5] + 32 pp. DP: 23 June 1763 (Bousquet 2016).

Linnaeus, C. 1767: *Systema naturae per regna tria naturae, secundum classes, ordines, genera, species, cum characteribus differentiis, synonymis, locis. Editio duodecima, reformata. Tom. I. Pars II*. Holmiae: Laurentii Salvii, [2] + 533–1327 + [37] pp. DP: 14 June 1767 (Bousquet 2016).

Löbl, I. 2010: New acts and comments. Chrysomelidae: Eumolpinae. Pp. 83. In: Löbl, I. & Smetana, A. (eds.): *Catalogue of Palaearctic Coleoptera. Volume 6. Chrysomeloidea*. Stenstrup: Apollo Books, 924 pp. DP: 22 February 2010 (colophon).

Löbl, I. & Löbl, D. (eds.) 2015: *Catalogue of Palaearctic Coleoptera. Volume 2. Revised and updated edition. Hydrophiloidea – Staphylinoidea*. Leiden – Boston: Brill, 1702 pp. DP: 28 August 2015 (colophon).

Löbl, I. & Smetana, A. (eds.) 2011: *Catalogue of Palaearctic Coleoptera. Volume 7. Curculionoidea I*. Stenstrup: Apollo Books, 373 pp. DP: 1 February 2011 (colophon).

Lodewyckx, M. 1995: Une nouvelle espèce de Coptocephala Chevrolat de la Turquie (Coleoptera: Chrysomelidae: Clytrinae). *Genus* 6 (2): 103–106. DP: 30 June 1995 (paper title page).

Lohse, G. A. 1954: Neuheiten der deutschen Käferfauna. Pp. 209–214. In: Titschack, E. (ed.): *Deutscher Entomologentag Hamburg 30. Juli bis 3.* August 1953. Jena: Gustav Fischer, iv + 215 pp. DP: [31 December] 1954 (book title page).

Lohse, G. A. 1956: Neuheiten der deutschen Käferfauna II. *Entomologische Blätter* 51 [1955] (1/2): 84–89. DP: 1 April 1956 (issue cover).

Lohse, G. A. 1980: Eine neue mitteleuropäische Art der Gattung Phyllotreta Stephens (Col. Chrysom.). *Entomologische Blätter* 75 [1979] (3): 155–156. DP: 1 May 1980 (paper title page).

Lohse, G. A. 1989: Hydrogaleruca-Studien (Col. Chrysomelidae, Gattung Galerucella Crotch). *Entomologische Blätter* 85 (1/2): 61–69. DP: 10 July 1989 (paper title page).

Long, J.-G. 1988: Three new species of Criocerinae from South China (Coleoptera: Crioceridae). *Entomotaxonomia* 10 (3/4): 231–234 DP: [31] December 1988 (paper title page).

Long, J.-G. 2000a: New species and new record of Criocerinae to China (Coleoptera: Crioceridae). *Acta Zootaxonomica Sinica* 25 (4): 416–418. DP: [31] October 2000 (paper title page).

Long, J.-G. 2000b: A new species of Criocerinae (Coleoptera: Crioceridae) from Guangdong, China. *Entomotaxonomia* 22 (4): 262–265 DP: [31] December 2000 (paper title page).

Lopatin, I. K. 1952: O novykh vidakh palearkticheskikh zhukov listoedov (Chrysomelidae). [On a new palaearctic species of leaf-beetles (Chrysomelidae)]. *Byulleten' Moskovskogo Obshchestva Ispytateley Prirody, Novaya Seriya, Otdel Biologicheskiy* 57 (4): 27–29. SFP: 13 August 1952, DP: 1 September 1952 (registered at VKP).

Lopatin, I. K. 1953: Novye vidy zhukov listoedov roda Cryptocephalus Geoffr. (Coleoptera, Chrysomelidae). [New species of leaf-beetles of the genus Cryptocephalus Geoffr. (Coleoptera, Chrysomelidae)]. *Entomologicheskoe Obozrenie* 33: 307–310. SFP: 23 December 1953, DP: [31 December] 1953 (paper title page).

Lopatin, I. K. 1954: Novye formy zhukov listoedov (Coleoptera, Chrysomelidae) Palearktiki. [New forms of leaf-beetles (Coleoptera, Chrysomelidae) from the Palaerctic]. *Byulleten' Moskovskogo Obshchestva Ispytateley Prirody, Novaya Seriya, Otdel Biologicheskiy* 59 (5): 96–98. SFP: 30 October 1954, DP: 27 December 1954 (registered at ZIN library).

Lopatin, I. K. 1956a: Novye vidy listoedov (Coleoptera, Chrysomelidae) iz Tadzhikistana. [New species of leaf-beetles (Coleoptera, Chrysomelidae) from Tajikistan]. *Doklady Akademii Nauk Tadzhikskoy SSR* 16: 71–72. SFP: 29 June 1956, DP: 28 August 1956 (registered at VKP).

Lopatin, I. K. 1956b: O novykh i maloizvestnykh vidakh zhukov-listoedov (Coleoptera, Chrysomelidae) SSSR i sopredelnykh stran. [On new and little known leaf-beetles species (Coleoptera, Chrysomelidae) of the USSR and neighboring countries]. *Izvestiya Otdeleniya Yestestvennykh Nauk AN Tadzhikskoy SSR* 16: 157–163. SFP: 21 September 1956, DP: [30] September 1956 (Lopatin 1963c: 122–123).

Lopatin, I. K. 1956c: Materialy k faune listoedov (Coleoptera, Chrysomelidae) Sredney Azii. [Materials on the fauna of leaf-beetles (Coleoptera, Chrysomelidae) of Central Asia]. *Doklady Akademii Nauk Tadzhikskoy* SSR 18: 43–49. SFP: 29 June 1956, DP: 28 December 1956 (registered at VKP).

Lopatin, I. K. 1958a: Novye vidy listoedov (Coleoptera, Chrysomelidae) iz Tadzhikistana. Soobshchenie III. [New species of leaf-beetles (Coleoptera, Chrysomelidae) from Tajikistan. Contribution III]. *Doklady Akademii Nauk Tadzhikskoy* SSR 1 (1): 53–56. SFP: 26 April 1958, DP: 29 July 1958 (registered at VKP).

Lopatin, I. K. 1958b: Materialy k faune listoedov (Coleoptera, Chrysomelidae) Sredney Azii. [Materials on the fauna of leaf-beetles (Coleoptera, Chrysomelidae) of Central Asia]. *Doklady Akademii Nauk Tadzhikskoy* SSR 1 (2): 31–35. SFP: 15 October 1958, DP: [31 December] 1958 (issue cover, paper title page).

Lopatin, I. K. 1960: O rode Thelyterotarsus Wse. (Coleoptera, Chrysomelidae) v Tadzhikistane i v sopredelnykh chastyakh Sredney Azii. [On the genus Thelyterotarsus Wse. (Coleoptera, Chrysomelidae) in Tajikistan and surrounding parts of Central Asia]. *Doklady Akademii Nauk Tadzhikskoy* SSR 3 (2): 37–40. SFP: 28 May 1960, DP: 1 September 1960 (registered at VKP).

Lopatin, I. K. 1961a: O novykh vidakh zhukov-listoedov (Coleoptera, Chrysomelidae) iz Tadzhikistana. (New species of chrysomelid-beetles (Coleoptera) from Tajikistan). *Zoologicheskiy Zhurnal* 40 (2): 201–206. SFP: 15 February 1961, DP: 18 March 1961 (registered at VKP).

Lopatin, I. K. 1961b: Dva novykh vida zemlyanykh bloshek (Coleoptera, Chrysomelidae, Halticinae) iz Tadzhikistana. (Two new species of flea-beetles (Coleoptera, Chrysomelidae, Halticinae) from Tajikistan). *Entomologicheskoe Obozrenie* 40 (1): 144–146. SFP: 14 February 1961, DP: 10 April 1961 (registered at VKP).

Lopatin, I. K. 1961c: Materialy k faune zhukov listoedov (Coleoptera, Chrysomelidae) Sredney Azii. Soobshchenie III. [Materials on the fauna of leaf-beetles (Coleoptera, Chrysomelidae) of Central Asia. Contribution III]. *Trudy Instituta Zoologii i Parazitologii Akademii Nauk Tadzhikskoy* SSR 20: 97–105. SFP: 27 March 1961, DP: 29 June 1961 (registered at VKP).

Lopatin, I. K. 1962a: Novyy rod zhukov listoedov iz Kazahstana (Coleoptera, Chrysomelidae). [New genus of leaf-beetles from Kazakhstan (Coleoptera, Chrysomelidae)]. *Doklady Nauchnykh Uchrezhdeniy Ministerstva Selskogo Khozyaystva Kazakhskoy* SSR 3: 30–31. DP: [31 December] 1962 (volume title page).

Lopatin, I. K. 1962b: Novye sredneaziatskie vidy roda Stylosomus Suffr. (Coleoptera, Chrysomelidae). [New Central Asian species of the genus Stylosomus Suffr. (Coleoptera, Chrysomelidae)]. *Doklady Akademii Nauk Tadzhikskoy* SSR 5 (2): 49–51. SFP: 23 June 1962, DP: [31 December] 1962 (issue cover).

Lopatin, I. K. 1962c: Neue palaearktische Chrysomeliden aus der Sammlung des Ungarischen Naturwissenschaftlichen Museums in Budapest (Coleoptera). *Annales Historico-Naturales Musei Nationalis Hungarici* 54: 319–326. DP: 31 December 1962 (Merkl et al. 2008).

Lopatin, I. K. 1962d: O faune zhukov-listoedov (Coleoptera, Chrysomelidae) Afganistana. (On the fauna of leaf-beetles (Coleoptera, Chrysomelidae) of Afghanistan). *Zoologicheskiy Zhurnal* 41 (12): 1811–1816. SFP: 30 November 1962, DP: [31 December] 1962 (issue cover).

Lopatin, I. K. 1963a: Reviziya roda Chrysochares Mor. (Coleoptera, Chrysomelidae). (Revision du genre Chrysochares Mor. (Coleoptera, Chrysomelidae)). *Časopis Československé Společnosti Entomologické* 60 (4): 296–303. DP: 20 October 1963 (colophon).

Lopatin, I. K. 1963b: Novyy vid zhuka-listoeda roda Jaxartiolus Jacobs. (Coleoptera, Chrysomelidae, Cryptocephalinae) iz Kazakhstana. (A new Jaxartiolus Jacobs. (Coleoptera, Chrysomelidae, Cryptocephalinae) from Kazakhstan). *Entomologicheskoe Obozrenie* 42 (4): 855–857. SFP: 16 December 1963, DP: 30 December 1963 (registered at ZIN library).

Lopatin, I. K. 1963c: Materialy k faune listoedov (Coleoptera, Chrysomelidae) Sredney Azii. [Materials to the leaf-beetles fauna (Coleoptera, Chrysomelidae) of Middle Asia]. *Trudy Instituta Zoologii i Parazitologii Akademii Nauk Tadzhikskoy SSR* 24: 116–123. SFP: 6 September 1963, DP: [31 December] 1963 (volume title page).

Lopatin, I. K. 1963d: Die Chrysomeliden (Coleoptera) Afghanistans auf Grund der Ergebnisse der Forschungsreise des Herrn J. Klapperich in den Jahren 1952/53. *Annales Historico-Naturales Musei Nationalis Hungarici* 55: 349–378. DP: 31 December 1963 (Merkl et al. 2008).

Lopatin, I. K. 1965a: K sistematike roda Cryptocephalus Geoffr. (Coleoptera, Chrysomelidae). (To the systematical position of the genera Cryptocephalus (Coleoptera, Chrysomelidae)). *Acta Entomologica Bohemoslovaca* 62 (6): 451–457. DP: 20 November 1965 (table of contents).

Lopatin, I. K. 1965b: Materialy k faune listoedov (Coleoptera, Chrysomelidae) Kirgizii. [Materials on the leaf-beetles fauna (Coleoptera, Chrysomelidae) of Kyrgyzstan]. Pp. 8–13. In: Paliy, V. F. (ed.): *Sbornik Entomologicheskihk Rabot*. Frunze: Izdatel'stvo "ILIM", 137 pp. SFP: 25 February 1965, DP: 27 December 1965 (registered at ZIN library).

Lopatin, I. K. 1965c: Novye vidy listoedov (Coleoptera, Chrysomelidae) iz Tadzhikistana. [New species of leaf-beetles (Coleoptera, Chrysomelidae) from Tajikistan]. *Doklady Akademii Nauk Tadzhikskoy SSR* 8 (1): 38–41. SFP: 20 January 1965, DP: [31 December] 1965 (issue cover).

Lopatin, I. K. 1966a: Chrysomelidae II. Ergebnisse der zoologischen Forschungen von Dr. Z. Kaszab in der Mongolei (Coleoptera). *Reichenbachia* 7: 229–239. DP: 10 August 1966 (paper title page).

Lopatin, I. K. 1966b: Novye vidy listoedov podsemeystva Halticinae iz Sredney Azii. [New species of leaf-beetles of the subfamily of Halticinae from Central Asia]. *Doklady Akademii Nauk Tadzhikskoy SSR* 9 (8): 45–48. SFP: 3 September 1966, DP: [31 December] 1966 (issue cover).

Lopatin, I. K. 1966c: Die Chrysomeliden Afghanistans. 2. Beitrag zur Kenntnis der afghanischen Chrysomelidenfauna. *Annales Historico-Naturales Musei Nationalis Hungarici* 58: 361–369. DP: 31 December 1966 (Merkl et al. 2008).

Lopatin, I. K. 1967a: Materialy k sistematike zhukov-listoedov podsem. Cryptocephalinae (Coleoptera, Chrysomelidae) Sredney Azii i Kazakhstana. (Materials on the taxonomy of leaf beetles of the subfamily Cryptocephalinae (Coleoptera, Chrysomelidae) from the Middle Asia and Kazakhstan). *Entomologicheskoe Obozrenie* 46 (1): 212–221. SFP: 15 March 1967, DP: 8 April 1967 (registered at ZIN library).

Lopatin, I. K. 1967b: Neue Chrysomeliden-Arten aus Pakistan (Col. Chrysomelidae). *Entomologische Arbeiten aus dem Museum G. Frey* 18: 323–326. DP: 1 July 1967 (table of contents on the cover).

Lopatin, I. K. 1967c: Obzor vidov roda Entomoscelis Chevr. (Coleoptera, Chrysomelidae) fauny SSSR. (The review of species of the genus Entomoscelis Chevr. (Coleoptera, Chrysomelidae) of the fauna of the USSR). *Zoologicheskiy Zhurnal* 46 (6): 951–954. SFP: 30 May 1967, DP: 26 July 1967 (registered at ZIN library).

Lopatin, I. K. 1967d: 106. Chrysomelidae der III. Expedition. Ergebnisse der zoologischen Forschungen von Dr. Z. Kaszab in der Mongolei (Coleoptera). *Reichenbachia* 9: 157–169. DP: 14 September 1967 (paper title page).

Lopatin, I. K. 1967e: Österreichische entomologische Expeditionen nach Persien und Afghanistan. Beiträge zur Coleopterologie. Teil V: Chrysomelidae. *Annalen des Naturhistorischen Museums in Wien* 70 [1966]: 437–441. DP: [31] October 1967 (paper title page).

Lopatin, I. K. 1968a: Novye sredneaziatskie vidy roda Longitarsus Latr. (Coleoptera, Chrysomelidae, Halticinae). [New Central Asian species of the genus Longitarsus Latr. (Coleoptera, Chrysomelidae, Halticinae)]. *Izvestiya Akademii Nauk Tadzhikskoy SSR* (*Otdeleniye Biologicheskikh Nauk*) [1967] (4): 112–116. SFP: 12 January 1968, DP: 24 April 1968 (registered at VKP).

Lopatin, I. K. 1968b: Chrysomelidae IV. Ergebnisse der zoologischen Forschungen von Dr. Z. Kaszab in der Mongolei (Coleoptera). *Reichenbachia* 11: 207–220. DP: 9 September 1968 (paper title page).

Lopatin, I. K. 1968c: Novye vidy zhukov-listoedov (Coleoptera, Chrysomelidae) fauny Sredney Azii i Kazakhstana. (New species of leaf-beetles (Coleoptera, Chrysomelidae) from the fauna of Middle Asia and Kazakhstan). *Entomologicheskoe Obozrenie* 47 (3): 541–552. SFP: 6 September 1968, DP: 9 October 1968 (registered at ZIN library).

Lopatin, I. K. 1970a: 187. Chrysomelidae der V. Expedition. Ergebnisse der zoologischen Forschungen von Dr. Z. Kaszab in der Mongolei (Coleoptera). *Reichenbachia* 12: 249–258. DP: 11 March 1970 (paper title page).

Lopatin, I. K. 1970b: Obzor vidov roda Chrysomela L. podroda Pezocrosita Jcbs. (Coleoptera, Chrysomelidae) fauny Sredney Azii i Kazakhstana. (Review of species of the genus Chrysomela, subgenus Pezocrosita Jcbs. (Coleoptera, Chrysomelidae) from Middle Asia and Kazakhstan). *Entomologicheskoe Obozrenie* 49 (1): 183–196. SFP: 17 March 1970, DP: 9 April 1970 (registered at ZIN library).

Lopatin, I. K. 1970c: Beiträge zur Kenntnis der Fauna Afghanistans (Sammelergebnisse von O. Jakeš 1963–64, D. Povolný 1965, D. Povolný & Fr. Tenora 1966, J. Šimek 1965–66, D. Povolný, J. Gaisler, Z. Šebek & Fr. Tenora 1967). Chrysomelidae, Col. (Nachtrag). 4. Beitrag zur Kenntnis der afghanischen Chrysomeliden. *Časopis Moravského Musea, Vědy přírodní* 54 [1969] (supplementum): 199–204. DP: 15 May 1970 (colophon).

Lopatin, I. K. 1971a: 232. Chrysomelidae VI. Ergebnisse der zoologischen Forschungen von Dr. Z. Kaszab in der Mongolei (Coleoptera). *Faunistische Abhandlungen Staatliches Museum für Tierkunde in Dresden* 3 [1969–1971]: 219–233. DP: 28 December 1971 (paper title page).

Lopatin, I. K. 1971b: [new taxa]. In: Lopatin, I. K. & Kulenova, K. Z.: Obzor vidov roda Cryptocephalus Geoffr. fauny Kazakhstana (Coleoptera, Chrysomelidae). (A review of species of the genus Cryptocephalus Geoffr. in the fauna of Kazakhstan (Coleoptera, Chrysomelidae)). *Trudy Instituta Zoologii Akademii Nauk Kazakhskoy SSR* (Alma Ata) **32**: 67–90. SFP: 5 May 1971, DP: [31 December] 1971 (volume cover).

Lopatin, I. K. 1972: Obzor zhukov listoedov roda Gnathomela Jcbs. (Coleoptera, Chrysomelidae). (Review of the genus Gnathomela Jcbs. (Coleoptera, Chrysomelidae)). *Entomologicheskoe Obozrenie* **51** (2): 346–351. SFP: 30 May 1972, DP: 6 July 1972 (registered at ZIN library).

Lopatin, I. K. 1974a: Obzor listoedov roda Pachybrachys Redt. (Coleoptera, Chrysomelidae) sredney Azii i Kazakhstana. (A review of leaf beetles of the genus Pachybrachys Redt. (Coleoptera, Chrysomelidae) from Middle Asia and Kazakhstan). *Entomologicheskoe Obozrenie* **53** (3): 651–661. SFP: 30 August 1974, DP: 27 September 1974 (registered at ZIN library).

Lopatin, I. K. 1974b: Dva novykh vida zhukov-listoedov (Coleoptera, Chrysomelidae) iz Mongolii. (Two new species of Chrysomelidae (Coleoptera) from Mongolia). Pp. 176–178. In: Kozlov, M. A. (ed.): *Nasekomye Mongolii 2*. (*Insects of Mongolia 2*). St. Petersburg: Academy of Sciences of the USSR, Zoological Institute, 439 pp. DP: 30 September 1974 (Kerzhner 1994).

Lopatin, I. K. 1975a: Ergebnisse der Bhutan-Expedition 1972 des Naturhistorishen Museums in Basel. Coleoptera: Fam. Chrysomelidae, Subfam. Cryptocephalinae. *Entomologica Basiliensia* **1**: 367–369. DP: 1 October 1975 (table of contents).

Lopatin, I. K. 1975b: Novye vidy listoedov iz yuzhnogo Tadzhikistana (Coleoptera: Chrysomelidae). [New species of leaf-beetles from South Tajikistan (Coleoptera: Chrysomelidae)]. *Izvestiya Akademii Nauk Tadzhikskoy SSR* (*Otdelenie Biologicheskikh Nauk*) **1975** (3): 45–51. SFP: 30 October 1975, DP: 19 December 1975 (registered at ZIN library).

Lopatin, I. K. 1975c: Zhuki-listoedy (Coleoptera, Chrysomelidae) Mongol'skoy Narodnoy Respubliki. (The chrysomelid-beetles (Coleoptera, Chrysomelidae) of the Mongolian People's Republic). Pp. 191–233. In: Kerzhner, I. M. (ed.): *Nasekomye Mongolii 3*. (*Insects of Mongolia 3*). St. Petersburg: Academy of Sciences of the USSR, Zoological Institute, 675 pp. DP: 25 December 1975 (Kerzhner 1994).

Lopatin, I. K. 1976a: Novye i maloizvestnye zhuki-listoedy (Coleoptera, Chrysomelidae) fauny SSSR. (New and little known leaf-beetles (Coleoptera, Chrysomelidae) from the USSR). *Entomologicheskoe Obozrenie* **55** (1): 105–116. SFP: 19 February 1976, DP: 29 March 1976 (registered at ZIN library).

Lopatin, I. K. 1976b: Dopolnenie k faune listoedov SSSR. [Supplement to the leaf-beetles fauna of the USSR]. *Vestnik Belorusskogo Gosudarstvennogo Universiteta Imeni V. I. Lenina, Seria II* (Minsk) **1976** (3): 69. DP: [31 December] 1976 (issue cover).

Lopatin, I. K. 1977a: Mongolskiye vidy roda Thelyterotarsus (Coleoptera, Chrysomelidae). (Mongolian species of the genus Thelyterotarsus (Coleoptera, Chrysomelidae)). *Zoologicheskiy Zhurnal* **56** (1): 169–171. SFP: 16 December 1976, DP: 8 February 1977 (registered at ZIN library).

Lopatin, I. K. 1977b: *Zhuki-listoedy (Chrysomelidae) Sredney Azii i Kazakhstana*. [*Leaf-beetles (Chrysomelidae) of Central Asia and Kazakhstan*]. Leningrad: Nauka, 268 pp. SFP: 17 May 1977, DP: 29 June 1977 (registered at ZIN library).

Lopatin, I. K. 1977c: Dva novykh vida zhukov-listoedov podsemeystva Cryptocephalinae (Coleoptera, Chrysomelidae) iz Mongolii. (Two new species of chrysomelid beetles of the subfamily Cryptocephalinae (Coleoptera, Chrysomelidae) from Mongolia). Pp. 304–307. In: Kerzhner, I. M. (ed.): *Nasekomye Mongolii 5*. (*Insects of Mongolia 5*). St. Petersburg: Academy of Sciences of the USSR, Zoological Institute, 757 pp. DP: 2 September 1977 (Kerzhner 1994).

Lopatin, I. K. 1977d: Novye i maloizvestnye vidy listoedov (Coleoptera, Chrysomelidae) evropeyskoy chasti SSSR i Kavkaza. (New and little known species of chrysomelid-beetles (Coleoptera, Chrysomelidae) from the European part of the USSR and Caucasus). Pp. 30–34. In: Skarlato, O. A. (ed.): *Novye i maloizvestnye vidy nasekomykh evropeyskoy chasti SSSR (Sbornik nauchnykh rabot)*. (*New and little known species of insects of the European part of the USSR*). Leningrad: Akademia Nauk SSSR, 116 pp. SFP: 15 April 1977, DP: [31 December] 1977 (book title page).

Lopatin, I. K. 1977e: Novyy podrod listoedov roda Cryptocephalus Geoffr. (Coleoptera, Chrysomelidae) iz Gindukusha. [New leaf-beetles subgenus of the genus Cryptocephalus Geoffr. (Coleoptera, Chrysomelidae) from Hindukush]. *Izvestiya Akademii Nauk Tadzhikskoy SSR (Otdelenie Biologicheskikh Nauk)* **1977** (4): 72–73. SFP: 12 December 1977, DP: [31 December] 1977 (paper title page).

Lopatin, I. K. 1977f: Eine neue Cryptocephalus-Art aus Iran (Coleoptera, Chrysomelidae). *Mitteilungen der Entomologischen Gesellschaft Basel, Neue Folge* 27 (4): 116–117. DP: [31] December 1977 (heading on p. 105).

Lopatin, I. K. 1977g: Weitere Beiträge zur Kenntnis der Chrysomeliden-Fauna der Mongolei (Coleoptera). *Annales Historico-Naturales Musei Nationalis Hungarici* **69**: 153–155. DP: 31 December 1977 (Merkl et al. 2008).

Lopatin, I. K. 1978: Novyy rod i vid zhukov-listoedov podsem. Eumolpinae (Coleoptera, Chrysomelidae) iz Turkmenii. (A new genus and species of leaf-beetles of the subfamily Eumolpinae (Coleoptera, Chrysomelidae) from Turkmenia). *Trudy Zoologicheskogo Instituta Akademii Nauk SSSR* 71: 53–54. SFP: 25 November 1977, DP: 30 June 1978 (Asanovich et al. 2007).

Lopatin, I. K. 1979a: Insects of Saudi Arabia. Coleoptera: Fam. Chrysomelidae, Subfam. Cryptocephalinae. *Fauna of Saudi Arabia* 1: 299–303. DP: 5 March 1979 (table of contents).

Lopatin, I. K. 1979b: Novyy rod i novye vidy zhukov-listoedov (Coleoptera, Chrysomelidae) iz Irana. (A new genus and some new species of the leaf-beetles (Coleoptera, Chrysomelidae) from Iran). *Entomologicheskoe Obozrenie* 58 (3): 586–589. SFP: 29 August 1979, DP: 28 September 1979 (registered at ZIN library).

Lopatin, I. K. 1979c: Neue und wenig bekannte Chrysomeliden (Coleoptera) von Nord-Indien aus der Sammlung des Naturhistorischen Museums Basel. *Entomologica Basiliensia* **4**: 431–441. DP: 10 October 1979 (table of contents).

Lopatin, I. K. 1980: Zhuki-listoedy (Coleoptera, Chrysomelidae) Irana. Rezultaty chekhoslovatsko-iranskoy ekspedicii 1973 g. I. (Leaf-beetles (Coleoptera, Chrysomelidae) of Iran. Results of the Czechoslovak-Iranian expedition in 1973. I). *Entomologicheskoe Obozrenie* 59 (3): 613–623. SFP: 20 August 1980, DP: 29 September 1980 (registered at ZIN library).

Lopatin, I. K. 1981a: Ergebnisse der Tschechoslowakisch-Iranischen entomologischen Expeditionen nach dem Iran (1970). Coleoptera. Chrysomelidae. *Acta Entomologica Musei Nationalis Pragae* 40: 371–376. DP: 25 March 1981 (colophon).

Lopatin, I. K. 1981b: Novye rody i vidy zhukov-listoedov (Coleoptera, Chrysomelidae) iz Irana. Rezultaty Chekhoslovatsko-Iranskoy ekspedicii 1973 g. II. (New genera and species of leaf-beetles (Coleoptera, Chrysomelidae) from Iran. Results of the Czechoslovak-Iranian expedition in 1973. II). *Entomologicheskoe Obozrenie* 60 (3): 623–628. SFP: 4 September 1981, DP: 19 October 1981 (registered at ZIN library).

Lopatin, I. K. 1982a: Insects of Oman. Coleoptera: Fam. Chrysomelidae Subfam. Cryptocephalinae. *Fauna of Saudi Arabia* 4: 465–467. DP: 1 November 1982 (table of contents).

Lopatin, I. K. 1982b: Übersicht der Cryptocephalus-Arten der Gruppe exsulans Suffr. aus dem Himalaya (Coleoptera, Chrysomelidae, Cryptocephalinae). *Entomologica Basiliensia* 7: 413–426. DP: 30 November 1982 (table of contents).

Lopatin, I. K. 1982c: Chrysomeliden Afghanistans aus den Sammlungen des Museo Civico di Storia Naturale, Verona (Coleoptera, Chrysomelidae). *Bollettino del Museo Civico di Storia Naturale di Verona* 8 [1981]: 255–263. DP: 30 November 1982 (table of contents).

Lopatin, I. K. 1983a: Novye vidy zhukov-listoedov (Coleoptera, Chrysomelidae) iz Sredney Azii i yugo-vostochnogo Kazakhstana. (New species of the leaf-beetles (Coleoptera, Chrysomelidae) from Middle Asia and south-eastern Kazakhstan). *Entomologicheskoe Obozrenie* 62 (1): 91–95. SFP: 5 March 1983, DP: 29 March 1983 (registered at ZIN library).

Lopatin, I. K. 1983b: Insects of Saudi Arabia. Coleoptera: Fam. Chrysomelidae, Subfam. Cryptocephalinae and Eumolpinae (Part 2). *Fauna of Saudi Arabia* 5: 205–209. DP: 30 December 1983 (table of contents).

Lopatin, I. K. 1984a: Zhuki-listoedy (Coleoptera, Chrysomelidae) Irana. Rezultaty Chekhoslovatsko-Iranskikh ekspeditsiy 1973–1977 gg. III. (Leaf-beetles (Coleoptera, Chrysomelidae) of Iran. Results of the Czechoslovak-Iranian expeditions in 1973–1977. III). *Entomologicheskoe Obozrenie* 63 (1): 79–93. SFP: 14 March 1984, DP: 9 April 1984 (registered at ZIN library).

Lopatin, I. K. 1984b: Zwei neue Arten der Chrysomeliden (Coleoptera) aus dem Ungarischen Naturwissenschaftlichen Museum. *Annales Historico-Naturales Musei Nationalis Hungarici* 76: 197–199. DP: 31 December 1984 (Merkl et al. 2008).

Lopatin, I. K. 1984c: Chrysomelidae aus dem Himalaja (Coleoptera). *Entomologica Basiliensia* 9: 328–339. DP: 31 December 1984 (table of contents).

Lopatin, I. K. 1985a: [new taxa]. In: Lopatin, I. K. & Kulenova, K. Z.: Tri novykh vida zhukov-listoedov (Coleoptera, Chrysomelidae) iz Kazakhstana. [Three new species of leaf-beetles (Coleoptera, Chrysomelidae) from Kazakhstan]. *Izvestiya Akademii Nauk Kazakhskoy SSR, Seriya Biologicheskaya* 1985 (1): 47–49. SFP: 26 February 1985, DP: 9 April 1985 (registered at ZIN library).

Lopatin, I. K. 1985b: Zhuki-listoedy (Coleoptera, Chrysomelidae) Irana. Rezultaty chekhoslovatsko-iranskikh ekspediciy 1973–1977 gg. IV. (Leaf-beetles (Coleoptera, Chrysomelidae) of Iran.

Results of the Czechoslovak-Iranian expeditions of the 1973–1977. IV). *Entomologicheskoe Obozrenie* **64** (4): 760–772. SFP: 17 December 1985, DP: [31 December] 1985 (issue cover).

Lopatin, I. K. 1986a: Neue Coleoptera aus Irak: Fam. Chrysomelidae, subfam. Clytrinae, Cryptocephalinae, Eumolpinae und Chrysomelidae. *Annales Entomologici Fennici* **51** [1985] (4): 109–110. DP: 28 January 1986 (table of contents).

Lopatin, I. K. 1986b: Novye vidy listoedov podsemeystva Cryptocephalinae (Coleoptera, Chrysomelidae) fauny SSSR. (New species of the subfamily Cryptocephalinae (Coleoptera, Chrysomelidae) of the USSR fauna). *Vestnik Zoologii* **1986** (1): 25–29. SFP: 22 January 1986, DP: 18 March 1986 (registered at ZIN library).

Lopatin, I. K. 1986c: [new name]. In: Lopatin, I. K. & Kulenova, K. Z.: *Zhuki-listoedy (Coleoptera, Chrysomelidae) Kazakhstana. Opredelitel'. [Leaf-beetles (Coleoptera, Chrysomelidae) of Kazakhstan. Identification key]*. Alma-Ata: Nauka, 198 + [1] pp. SFP: 2 September 1986, DP: [31 December] 1986 (book title page).

Lopatin, I. K. 1987a: Novye vidy zhukov-listoedov (Coleoptera, Chrysomelidae) iz gornoy Sredney Azii. (New species of the chrysomelid beetles (Coleoptera) of the montane Middle Asia). *Vestnik Zoologii* **1987** (1): 82–85. SFP: 26 January 1987, DP: 18 March 1987 (registered at ZIN library).

Lopatin, I. K. 1987b: Novye podrod i vid roda Thelyterotarsus (Coleoptera, Chrysomelidae) iz Shri-Lanki. (New subgenus and species of the genus Thelyterotarsus (Coleoptera, Chrysomelidae) from Sri Lanka). *Zoologicheskiy Zhurnal* **66** (5): 778–780. SFP: 21 April 1987, DP: 15 June 1987 (registered at ZIN library).

Lopatin, I. K. 1988a: Asionus Lopatin, nom. n. *Vestnik Zoologii* **1988** (2): 8. SFP: 24 March 1988, DP: 28 April 1988 (registered at ZIN library).

Lopatin, I. K. 1988b: Novye palearkticheskie vidy zhukov-listoedov roda Chrysolina Motsch. (Coleoptera, Chrysomelidae). (New Palaearctic species of leaf-beetles of the genus Chrysolina Motsch. (Coleoptera, Chrysomelidae)). *Entomologicheskoe Obozrenie* **67** (3): 585–592. SFP: 9 September 1988, DP: 26 October 1988 (registered at ZIN library).

Lopatin, I. K. 1988c: Calomicrus ghilarovi sp. n. i blizkie k nemu vidy iz Sredney Azii i Irana (Coleoptera, Chrysomelidae). [Calomicrus ghilarovi sp. n. and species close to it from Central Asia and Iran (Coleoptera, Chrysomelidae)]. *Trudy Vsesoyuznogo Entomologicheskogo Obshchestva* 70: 65–67. SFP: 11 August 1988, DP: 21 December 1988 (registered at ZIN library).

Lopatin, I. K. 1990a: Rod Xenomela Weise (Coleoptera, Chrysomelidae) i ego polozhenie v tribe Entomoscelini. (The genus Xenomela Weise (Coleoptera, Chrysomelidae) and its position in the tribe Entomoscelini). *Trudy Zoologicheskogo Instituta Akademii Nauk SSSR* 208 [1989]: 110–124. SFP: 10 October 1989, DP: 16 May 1990 (Asanovich et al. 2007).

Lopatin, I. K. 1990b: K faune zhukov-listoedov podsem. Alticinae (Coleoptera, Chrysomelidae) Irana. Rezultaty chekhoslovatsko-iranskikh ekspeditsiy 1970–1977 gg. V. (On the fauna of chrysomelid beetles of the subfamily Alticinae (Coleoptera, Chrysomelidae) of Iran. Results of the Czechoslovak-Iranian expeditions in 1970–1977. V). *Entomologicheskoe Obozrenie* **69** (3): 598–608. SFP: 16 October 1990, DP: 26 December 1990 (registered at ZIN library).

Lopatin, I. K. 1990c: Novye i maloizvestnye vidy zhukov-listoedov (Coleoptera, Chrysomelidae) fauny Kazakhstana. [New and little known species of leaf-beetles (Coleoptera, Chrysomelidae) of Kazakhstan fauna]. *Trudy Instituta Zoologii Akademii Nauk Kazakhskoy SSR* **45**: 46–58. SFP: 26 June 1990, DP: 17 December 1990 (registered at IVB CAS library).

Lopatin, I. K. 1990d: Novye i maloizvestnie vidy zhukov-listoedov (Coleoptera, Chrysomelidae) fauny SSSR. (New and little-known chrysomelid beetle species (Coleoptera, Chrysomelidae) of the USSR fauna). Pp. 48–54. In: Akimov, I A. (ed.): *Novosti faunistiki i sistematiki.* Kiev: Naukova Dumka, 184 pp. SFP: 16 November 1990, DP: [31 December] 1990 (book title page).

Lopatin, I. K. 1991a: Novye vidy roda Pachybrachis (Coleoptera, Chrysomelidae) palearktiches-koy fauny s zametkami po sinonimii. (Description of new species of the genus Pachybrachis (Coleoptera, Chrysomelidae) of the Palearctic fauna, with synonymic notes). *Vestnik Zoologii* **1991** (2): 23–27. SFP: 26 March 1991, DP: 7 May 1991 (registered at ZIN library).

Lopatin, I. K. 1991b: Zwei neue Cryptocephalus-Arten aus Südasien (Coleoptera, Chrysomelidae). *Entomologica Basiliensia* **14**: 403–406. DP: 31 December 1991 (table of contents).

Lopatin, I. K. 1992a: Obzor tyanshanskikh vidov zhukov-listoedov roda Chrysolina, blizkikh k Ch. brevilata (Coleoptera, Chrysomelidae). (Review of the chrysomelid beetles of the genus Chrysolina similar to Ch. brevilata from the Tien-Shan (Coleoptera, Chrysomelidae)). *Zoologicheskiy Zhurnal* **71** (5): 66–71. SFP: 10 April 1992, DP: 13 August 1992 (registered at VKP).

Lopatin, I. K. 1992b: Materialy k sistematike listoedov roda Thelyterotarsus Wse. (Coleoptera, Chrysomelidae) fauny Zakavkazya, Sredney Azii i Kazakhstana. (Materials on the systematics of leaf-beetles of the genus Thelyterotarsus Wse. (Coleoptera, Chrysomelidae) from Transcaucasia, Middle Asia and Kazakhstan). *Entomologicheskoe Obozrenie* **71** (2): 395–404. SFP: 25 June 1992, DP: 15 September 1992 (registered at ZIN library).

Lopatin, I. K. 1992c: [new taxa]. In: Konstantinov, A. S. & Lopatin, I. K.: On the taxonomy of Phyllotreta Chevr. of the Palaearctic region (Coleoptera, Chrysomelidae, Alticinae). *Spixiana* **15** (3): 261–267. DP: 1 November 1992 (paper title page).

Lopatin, I. 1992d: Novyy vid listoeda (Coleoptera, Chrysomelidae) iz yugo-vostochnogo Kazakhstana. (A new chrysomelid beetle species (Coleoptera, Chrysomelidae) from South-East Kazakhstan). *Vestnik Zoologii* **1992** (6): 72–74. SFP: 1 December 1992, DP: [31 December] 1992 (footers throughout the paper).

Lopatin, I. K. 1992e: [new taxon]. In: Lopatin, I. K. & Konstantinov, A. S.: Materialy k faune i sistematike zhukov-listoedov Kavkaza (Coleoptera, Chrysomelidae). (Materials to the fauna and systematics of the leaf-cutting beetles of the Caucasus). Pp. 31–50. In: Abdurakhmanov, G. M. (ed.): *Nasekomye Kavkaza (Fauna, sistematika, geograficheskoe rasprostranenie). Tom I. Novye i maloizvestnye nasekomye s Kavkaza. (Insects of the Caucasus (Fauna, systematics, geographical spreading). Volume 1. New and less known insects from the Caucasus).* [1991], Makhachkala: Ministry of Education of Russia, Daghestan State Pedagogical Institute, Daghestan Ecological Centre, 52 pp. DP: [31 December] 1992 (see the comments in Dating of publications).

Lopatin, I. K. 1993a: Eine neue Labidostomis-Art aus Israel (Insecta, Coleoptera, Chrysomelidae). *Spixiana* **16** (2): 169–170. DP: 1 July 1993 (paper title page).

Lopatin, I. K. 1993b: Zwei neue Chrysolina-Arten aus süd-östlichen Kazakhstan (Coleoptera: Chrysomelidae). *Elytron* 6 [1992]: 5–9. DP: [31] October 1993 (last page of the volume).

Lopatin, I. K. 1994: [new taxa]. In: Lopatin, I. K. & Konstantinov, A. S.: New species of Chrysomelidae (Coleoptera) from Palearctic and Oriental regions. *Lambillionea* 94 (4): 524–530. DP: [31] December 1994 (paper title page).

Lopatin, I. K. 1995a: Novye i maloizvestnye zhuki-listoedy (Coleoptera, Chrysomelidae) iz yuzhnoy i vostochnoy Azii. (New and little known leaf-beetles (Coleoptera, Chrysomelidae) from South and East Asia). *Entomologicheskoe Obozrenie* 74 (1): 97–104. SFP: 15 May 1995, DP: 19 September 1995 (registered at ZIN library).

Lopatin, I. K. 1995b: Typenrevision der von Josef Breit beschriebenen Ischyromus-Arten (Insecta, Coleoptera, Chrysomelidae). *Spixiana* 18 (3): 259–262. DP: 1 November 1995 (paper title page).

Lopatin, I. K. 1995c: Neue Cryptocephalinen und Chlamisinen der indischen Region (Coleoptera, Chrysomelidae). *Entomologica Basiliensia* 18: 455–466. DP: 15 December 1995 (table of contents).

Lopatin, I. K. 1995d: Novye vidy roda Pachybrachis (Coleoptera, Chrysomelidae, Cryptocephalinae) iz central'noy Azii. (New species of the genus Pachybrachis (Coleoptera, Chrysomelidae, Cryptocephalinae) from Central Asia). *Vestnik Zoologii* 1995 (1): 18–22. SFP: 10 May 1995, DP: [31 December] 1995 (footers throughout the paper).

Lopatin, I. K. 1995e: Novye vidy listoedov (Coleoptera, Chrysomelidae) iz centralnoy Azii. (New species of leaf-beetles (Coleoptera, Chrysomelidae) from Central Asia). Pp. 174–180. In: Lopatin, I. K. (ed.): *Fauna i sistematika: Trudy Zoologicheskogo muzeya Belorusskogo universiteta. Vyp. 1.* (*Fauna and taxonomy: Proceedings of Zoological Museum of the Byelorussian University. Vol. 1*). Minsk: Navuka i tekhnika, 328 pp. SFP: 7 August 1995, DP: [31 December] 1995 (book title page).

Lopatin, I. K. 1996: Novye i maloizvestnye zhuki-listoedy (Coleoptera, Chrysomelidae) palearktiki. Opisaniya novykh vidov i podvidov i sinonimicheskie zametki. (New and little known leaf-beetles (Coleoptera, Chrysomelidae) of the palaearctic. Description of new species and subspecies and synonymical notes). *Entomologicheskoe Obozrenie* 75 (3): 637–642. SFP: 4 November 1996, DP: [31 December] 1996 (issue cover).

Lopatin, I. K. 1997a: O rode Callipta (Coleoptera, Chrysomelidae). (On the genus Callipta (Coleoptera, Chrysomelidae)). *Zoologicheskiy Zhurnal* 76 (2): 168–172. SFP: 5 March 1997, DP: 4 July 1997 (registered at ZIN library).

Lopatin, I. K. 1997b: Novye vidy zhukov-listoedov (Coleoptera, Chrysomelidae) aziatskoy fauny. (New species of chrysomelid beetles (Coleoptera, Chrysomelidae) of the Asiatic fauna). *Entomologicheskoe Obozrenie* 76 (2): 363–377. SFP: 4 August 1997, DP: 18 November 1997 (registered at ZIN library).

Lopatin, I. K. 1998a: [new taxon]. In: Konstantinov, A. S.: Revision of the Palearctic species of Aphthona Chevrolat and cladistic classification of the Aphthonini (Coleoptera: Chrysomelidae: Alticinae). *Memoirs on Entomology, International* 11: i–viii + 1–429. DP: 1 July 1998 (colophon).

Lopatin, I. K. 1998b: Materialy k sistematike listoedov podsem. Chrysomelinae (Coleoptera, Chrysomelidae). (Contribution to the systematics of leaf beetles of the subfamily Chrysomelinae (Coleoptera, Chrysomelidae)). *Entomologicheskoe Obozrenie* 77 (4): 826–835. SFP: 30 November 1998, DP: [31 December] 1998 (issue cover).

Lopatin, I. K. 1999a: Novye vidy roda Cryptocephalus (Coleoptera, Chrysomelidae). (New species of the genus Cryptocephalus (Coleoptera, Chrysomelidae)). *Vestnik Zoologii* 33 (4/5): 87–90. SFP: 11 October 1999, DP: [31 December] 1999 (paper title page).

Lopatin, I. K. 1999b: Materialy po listoedam podsemeystv Cryptocephalinae i Chrysomelinae (Coleoptera, Chrysomelidae) Kazakhstana i Sredney Azii. (Contribution to the knowledge of leaf beetles of the subfamilies Cryptocephalinae and Chrysomelinae (Coleoptera, Chrysomelidae) from Kazakhstan and Middle Asia). *Entomologicheskoe Obozrenie* 78 (4): 887–894. SFP: 29 November 1999, DP: [31 December] 1999 (issue cover).

Lopatin, I. K. 2000a: [new taxa]. In: Konstantinov, A. S. & Lopatin, I. K.: Review of the Longitarsus asperifoliarum group of species (Coleoptera: Chrysomelidae: Alticinae). *Coleopterists Bulletin* 54 (2): 200–220. DP: 27 July 2000 (verso of the issue cover).

Lopatin, I. K. 2000b: Obzor vidov podroda Colaphoptera Motsch. roda Chrysolina Motsch. (Coleoptera, Chrysomelidae) Kavkaza i severnoy Turcii. (Review of species of the subgenus Colaphoptera, genus Chrysolina Motsch. (Coleoptera, Chrysomelidae) from the Caucasus and northern Turkey). *Entomologicheskoe Obozrenie* 79 (4): 858–864. SFP: 22 November 2000, DP: 22 December 2000 (ZIN proof-print).

Lopatin, I. K. 2000c: Neue Arten der Gattung Chrysolina Motsch. aus China (Coleoptera: Chrysomelidae). *Mitteilungen des Internationalen Entomologischen Vereins e. V.* (Frankfurt am Main) 25 (3/4): 129–136. DP: 29 December 2000 (paper title page).

Lopatin, I. K. 2001a: Review of Iranian species of the genus Tituboea Lacordaire, 1848 (Coleoptera: Chrysomelidae). *Genus* 12 (1): 35–43. DP: 10 April 2001 (paper title page).

Lopatin, I. K. 2001b: Beitrag zur Kenntnis der Chrysomeliden des Jemen (Coleoptera: Chrysomelidae). *Mitteilungen des Internationalen Entomologischen Vereins e. V.* (Frankfurt am Main) 26 (1/2): 13–19. DP: 15 September 2001 (paper title page).

Lopatin, I. K. 2001c: Einige neue und wenig bekannte Cryptocephalinen-Arten (Coleoptera, Chrysomelidae). *Vestnik Zoologii* 35 (1): 91–93. SFP: 30 January 2001, DP: 28 September 2001 (registered at ZIN library).

Lopatin, I. K. 2002a: Novye vidy zhukov-listoedov podsem. Chrysomelinae (Coleoptera, Chrysomelidae) iz Kitaya. I. (New species of the leaf-beetle subfamily Chrysomelinae (Coleoptera, Chrysomelidae) from China. I). *Entomologicheskoe Obozrenie* 81 (1): 111–120. SFP: 26 February 2002, DP: 22 March 2002 (ZIN proof-print).

Lopatin, I. K. 2002b: New data on the taxonomy and distribution of leaf beetles in Israel (Coleoptera: Chrysomelidae). *Zoosystematica Rossica* 10 [2001] (2): 379–380. DP: 22 April 2002 (journal website).

Lopatin, I. K. 2002c: Übersicht der Ost-Mediterranen Calyptorhina-Arten mit zweifarbigen flügeldecken (Coleoptera, Chrysomelidae, Clytrinae). *Vestnik Zoologii* **36** (2): 87–89. SFP: 8 April 2002, DP: 27 June 2002 (registered at ZIN library).

Lopatin, I. K. 2002d: [new taxa]. In: Konstantinov, A. S. & Lingafelter, S.: *Revision of the Oriental species of Aphthona Chevrolat.* Washington: Entomological Society of Washington, 349 pp. DP: 8 July 2002 (colophon).

Lopatin, I. K. 2002e: Novye dannye o listoedakh yuzhnoy i vostochnoy Azii (Coleoptera, Chrysomelidae). Opisania i sinonimicheskie zametki. (New data on the leaf-beetles of the South and East Asia (Coleoptera, Chrysomelidae). Descriptions and synonymic remarks). *Euroasian Entomological Journal* **1** (1): 83–86. DP: 16 August 2002 (journal website).

Lopatin, I. K. 2002f: Novye vidy zhukov-listoedov (Coleoptera, Chrysomelidae) iz Kitaya. II. (New species of leaf-beetles (Coleoptera, Chrysomelidae) from China. II). *Entomologicheskoe Obozrenie* **81** (4): 874–882. SFP: 26 November 2002, DP: 24 December 2002 (ZIN proof-print).

Lopatin, I. K. 2003a: A new genus and its two new species of leaf-beetles from Vietnam (Coleoptera: Chrysomelidae: Galerucinae). *Genus* **14** (1): 103–107. DP: 15 April 2003 (paper title page).

Lopatin, I. K. 2003b: Novye vidy zhukov-listoedov iz Yugo-vostochnoy Azii (Coleoptera, Chrysomelidae). (New species of leaf-beetles from South-East Asia (Coleoptera, Chrysomelidae)). *Euroasian Entomological Journal* **2** (4): 301–304. DP: 28 December 2003 (journal website).

Lopatin, I. K. 2004a: Novye vidy zhukov-listoedov (Coleoptera, Chrysomelidae) iz Kitaya. III. (New species of leaf beetles (Coleoptera, Chrysomelidae) from China. III). *Entomologicheskoe Obozrenie* **83** (3): 614–621. SFP: 26 August 2004, DP: 5 November 2004 (registered at ZIN library).

Lopatin, I. K. 2004b: [new name]. In: Lopatin, I. K., Aleksandrovich, O. R. & Konstantinov, A. S.: *Checklist of leaf-beetles (Coleoptera, Chrysomelidae) of the Eastern Europe and Northern Asia.* Olsztyn: Mantis, 343 pp. DP: 10 November 2004 (colophon).

Lopatin, I. K. 2004c: Novye vidy zhukov-listoedov (Coleoptera, Chrysomelidae) iz Kitaya i V'etnama. (New species of leaf-beetles (Coleoptera, Chrysomelidae) from China and Vietnam). *Euroasian Entomological Journal* **3** (3): 187–191. DP: 27 December 2004 (journal website).

Lopatin, I. K. 2004d: Three new species of the genus Smaragdina from China (Coleoptera: Chrysomelidae). *Zoosystematica Rossica* **13** (1): 29–31. DP: 28 December 2004 (journal website).

Lopatin, I. K. 2005a: Oreomela dolini sp. n. (Coleoptera, Chrysomelidae) – novyy vid central'noaziatskogo roda zhukov-listoedov. (Oreomela dolini sp. n. (Coleoptera, Chrysomelidae), a new species of a Central Asian genus of leaf beetles). *Vestnik Zoologii* **39** (1): 63–64. SFP: 23 February 2005, DP: 5 May 2005 (registered at ZIN library).

Lopatin, I. K. 2005b: Novye vidy zhukov-listoedov (Coleoptera, Chrysomelidae) iz Kitaya. IV. (New species of leaf beetles (Coleoptera, Chrysomelidae) from China. IV). *Entomologicheskoe Obozrenie* 84 (3): 569–575. SFP: 31 August 2005, DP: 8 November 2005 (registered at ZIN library).

Lopatin, I. K. 2005c: Tri novykh vida zhukov-listoedov (Coleoptera, Chrysomelidae) iz Nepala: opisaniya i sinonimicheskie zametki. (Three new species of leaf beetles (Coleoptera, Chrysomelidae) from Nepal, descriptions and synonymies). *Euroasian Entomological Journal* 4 (3): 231–232. DP: 30 November 2005 (journal website).

Lopatin, I. K. 2005d: Novye vidy zhukov-listoedov (Coleoptera, Chrysomelidae) iz Kitaya. V. (New species of leaf-beetles (Coleoptera, Chrysomelidae) from China. V). *Entomologicheskoe Obozrenie* 84 (4): 873–880. SFP: 30 November 2005, DP: [31 December] 2005 (issue cover).

Lopatin, I. K. 2006a: A new species of Calomicrus from the United Arab Emirates (Coleoptera: Chrysomelidae: Galerucinae). *Zoosystematica Rossica* 14 [2005] (2): 261–262. DP: 27 January 2006 (journal website).

Lopatin, I. K. 2006b: Hyphaenia medvedevi sp. n. (Coleoptera: Chrysomelidae: Galerucinae) – novyy vid zhukov-listoedov iz Kitaya. (Hyphaenia medvedevi sp. n. (Coleoptera: Chrysomelidae: Galerucinae), a new leaf-beetle species from China). *Proceedings of the Russian Entomological Society* (St. Petersburg) 77: 211–212. SFP: 6 March 2006, DP: 3 April 2006 (registered at ZIN library).

Lopatin, I. K. 2006c: Novye vidy zhukov-listoedov (Coleoptera, Chrysomelidae) iz Kitaya. VI. (New species of leaf beetles (Coleoptera, Chrysomelidae) from China. VI). *Entomologicheskoe Obozrenie* 85 (3): 593–601. SFP: 31 August 2006, DP: 29 September 2006 (based on invoice deposited in Russian Entomological Society, A. G. Moseyko 2022, pers. comm.).

Lopatin, I. K. 2007: Novye vidy zhukov-listoedov (Coleoptera, Chrysomelidae) iz Kitaya. VII. (New species of leaf beetles (Coleoptera, Chrysomelidae) from China. VII). *Entomologicheskoe Obozrenie* 86 (1): 176–184. SFP: 1 March 2007, DP: 11 May 2007 (registered at ZIN library).

Lopatin, I. K. 2008a: Order Coleoptera, family Chrysomelidae. Pp. 312–324. In: Harten, A. van (ed.): *Arthropod fauna of the United Arab Emirates. Volume 1.* Abu Dhabi: Dar Al Ummah Printing, Publishing, Distribution & Advertising, 754 pp. DP: 20 January 2008 (title page of pdf version) [note: the date 30 September 2007 in printed version is wrong].

Lopatin, I. K. 2008b: A new replacement name for junior homonym Sichuania Lopatin, 2002 (Coleoptera: Chrysomelidae: Galerucinae. *Zoosystematica Rossica* 17 (1): 138. DP: 23 November 2008 (journal website).

Lopatin, I. K. 2008c: Acolastus mitjaevi sp. n. – novyy vid zhukov-listoedov iz Sredney Azii i ego polozhenie v sisteme roda (Coleoptera, Chrysomelidae, Cryptocephalinae). (Acolastus mitjaevi sp. n. – new species of chrysomelids from Middle Asia and its status in genus system (Coleoptera, Chrysomelidae, Cryptocephalinae)). *Tethys Entomological Research* 16: 37–38. DP: 25 November 2008 (paper title page).

Lopatin, I. K. 2008d: Novye vidy zhukov-listoedov (Coleoptera, Chrysomelidae) iz Kitaya. IX. (New species of the leaf beetles (Coleoptera, Chrysomelidae) from China. IX). *Entomologicheskoe*

Obozrenie 87 (4): 831–841. SFP: 21 November 2008, DP: 31 December 2008 (based on invoice deposited in Russian Entomological Society, A. G. Moseyko 2022, pers. comm.).

Lopatin, I. K. 2009a: [new taxa]. In: Lopatin, I. K. & Konstantinov, A. S.: New genera and species of leaf beetles (Coleoptera: Chrysomelidae) from China and South Korea. *Zootaxa* 2083: 1–18. DP: 27 April 2009 (footer on the paper title page).

Lopatin, I. K. 2009b: Novye vidy zhukov-listoedov (Coleoptera, Chrysomelidae) iz Kitaya. VIII. (New species of the leaf beetles (Coleoptera, Chrysomelidae) from China. VIII). *Entomologicheskoe Obozrenie* 88 (2): 430–437. SFP: 1 June 2009, DP: 30 June 2009 (based on invoice deposited in Russian Entomological Society, A. G. Moseyko 2022, pers. comm.).

Lopatin, I. K. 2009c: Obzor roda Acolastus Gerstaecker (Coleoptera, Chrysomelidae, Cryptocephalinae) aziatskoy chasti areala. (A review of the genus Acolastus Gerstaecker (Coleoptera, Chrysomelidae, Cryptocephalinae) from Asian part of areals). *Euroasian Entomological Journal* 8 (2): 145–176. DP: 17 August 2009 (journal website).

Lopatin, I. K. 2010: *Zhuki-listoedy (Insecta, Coleoptera, Chrysomelidae) centralnoy Azii.* (*The leaf beetles (Insecta, Coleoptera, Chrysomelidae) of Central Asia.* Minsk: BGU, 512 pp. SFP: 24 September 2010, DP: [31 December] 2010 (book title page).

Lopatin, I. K. 2011a: Novyy rod i vid zhukov-listoedov (Coleoptera, Chrysomelidae, Galerucinae) iz Kitaya. (New genus and species of leaf-beetles (Coleoptera, Chrysomelidae, Galerucinae) from China). *Euroasian Entomological Journal* 10 (2): 143–144. DP: 16 June 2011 (journal website).

Lopatin, I. K. 2011b: Novye vidy zhukov-listoedov (Coleoptera, Chrysomelidae) iz Kitaya. X. (New species of leaf-beetles (Coleoptera, Chrysomelidae) from China. X). *Entomologicheskoe Obozrenie* 90 (2): 375–387. SFP: 25 May 2011, DP: 25 July 2011 (registered at ZIN library).

Lopatin, I. K. 2012: [new taxon]. In: Ge, S.-Q., Daccordi, M., Lopatin, I. K., Cui, J.-Z., Li, W.-Z. & Yang, X.-K.: A new genus and two new species of Chrysomelinae (Coleoptera: Chrysomelidae) from China. *Entomotaxonomia* 34 (4): 633–640. DP: 25 December 2012 (last page of the cover).

Lopatin, I. K. 2013: Novye vidy zhukov-listoedov (Coleoptera, Chrysomelidae) iz Kitaya. V. (New species of leaf-beetles (Coleoptera, Chrysomelidae) from China. XI). *Entomologicheskoe Obozrenie* 92 (4): 765–776. SFP: 28 November 2013, DP: 31 December 2013 (based on invoice deposited in Russian Entomological Society, A. G. Moseyko 2022, pers. comm.).

Lopatin, I. K., Aleksandrovich, O. R. & Konstantinov, A. S. 2004: *Checklist of leaf-beetles (Coleoptera, Chrysomelidae) of the Eastern Europe and Northern Asia.* Olsztyn: Mantis, 343 pp. DP: 10 November 2004 (colophon).

Lopatin, I. & Chikatunov, V. 1997: The Cryptocephalinae (Coleoptera: Chrysomelidae) of Israel, Jordan and the Sinai part of Egypt. *Israel Journal of Entomology* 31: 97–119. DP: 31 December 1997 (verso of the issue cover).

Lopatin, I. K. & Chikatunov, V. I. 2000a: Two new species of Cryptocephalus from Israel (Coleoptera: Chrysomelidae). *Zoosystematica Rossica* 8 [1999] (2): 329–330. DP: 16 March 2000 (journal website).

Lopatin, I. & Chikatunov, V. 2000b: Rhaebus amnoni n. sp. – the first representative of the Central-Asian genus Rhaebus in Israel (Coleoptera: Bruchidae). *Mitteilungen des Internationalen Entomologischen Vereins e. V.* (Frankfurt am Main) **25** (1/2): 31–34. DP: 31 July 2000 (paper title page).

Lopatin, I. K. & Chikatunov, V. I. 2001: Two new species of the genus Antipa (subgenus Tituboea) from Israel (Coleoptera: Chrysomelidae). *Zoosystematica Rossica* **9** [2000] (2): 437–438. DP: 15 June 2001 (journal website).

Lopatin, I. K. & Friedmann, A. L. L. 2003: Gonioctena israelita sp. n., a new species of leaf-beetles from Israel (Coleoptera: Chrysomelidae). *Zoosystematica Rossica* **11** [2002] (2): 343–345. DP: 25 June 2003 (journal website).

Lopatin, I. K. & Kulenova, K. Z. 1982: Dva novykh vida zhukov-listoedov (Coleoptera, Chrysomelidae) iz Kazakhstana. [Two new species of leaf-beetles (Coleoptera, Chrysomelidae) from Kazakhstan]. *Izvestiya Akademii Nauk Kazakhskoy SSR, Seriya Biologicheskaya* **1982** (6): 43–46. SFP: 21 December 1982, DP: [31 December] 1982 (issue cover).

Lopatin, I. K. & Kulenova, K. Z. 1986: *Zhuki-listoedy (Coleoptera, Chrysomelidae) Kazakhstana. Opredelitel'.* [*Leaf-beetles (Coleoptera, Chrysomelidae) of Kazakhstan. Identification key*]. Alma-Ata: Nauka, 198 + [1] pp. SFP: 2 September 1986, DP: [31 December] 1986 (book title page).

Lopatin, I. K. & Kulenova, K. Z. 1987: Novyy vid zhukov-listoedov roda Chrysolina Motsch. (Coleoptera, Chrysomelidae) iz vysokogoriy yugo-vostochnogo Kazakhstana. [A new species of leaf-beetles from the genus Chrysolina Motsch. (Coleoptera, Chrysomelidae) from the high mountains of southeastern Kazakhstan]. *Izvestiya Akademii Nauk Kazakhskoy SSR, Seriya Biologicheskaya* **1987** (4): 39–40. SFP: 7 August 1987, DP: 17 October 1987 (registered at ZIN library).

Lopatin, I. K. & Nesterova, O. L. 2006: A new species of the genus Calomicrus Stephens, 1832 from United Arab Emirates (Coleoptera: Chrysomelidae: Galerucinae). *Genus* **17** (4): 537–539. DP: 28 December 2006 (paper title page).

Lopatin, I. K. & Nesterova, O. L. 2007a: A new species of Acolastus from United Arab Emirates (Coleoptera: Chrysomelidae: Cryptocephalinae). *Zoosystematica Rossica* **15** [2006] (2): 308. DP: 2 March 2007 (paper title page).

Lopatin, I. K. & Nesterova, O. L. 2007b: K sistematike zhukov-listoedov roda Labidostomis Germ. (Coleoptera, Chrysomelidae, Clytrinae). (A contribution on the taxonomy of the genus Labidostomis Germ. (Coleoptera, Chrysomelidae, Clytrinae)). *Euroasian Entomological Journal* 6 (4): 429–432. DP: 28 December 2007 (journal website).

Lopatin, I. K. & Nesterova, O. L. 2010: Two new species and identification note on two poorly known species of the genus Pachybrachis Chevr. (Coleoptera: Chrysomelidae: Cryptocephalinae). *Genus* **21** (1): 83–88. DP: 30 March 2010 (paper title page).

Lopatin, I. K. & Nesterova, O. L. 2013: Dva novykh vida zhukov-listoedov roda Calomicrus Dillwyn, 1829 (Coleoptera: Chrysomelidae: Galerucinae) iz Turcii. (Two new species of the genus Calomicrus Dillwyn, 1829 (Coleoptera: Chrysomelidae: Galerucinae) from Turkey). *Caucasian Entomological Bulletin* 9 (1): 95–96. DP: 4 August 2013 (journal website).

Lopatin, I. K., Smetana, A. & Schöller, M. 2010: Subfamily Cryptocephalinae Gyllenhal, 1813. Tribe Cryptocephalini Gyllenhal, 1813, genus Cryptocephalus Geoffroy, 1762. Pp. 580–606. In: Löbl, I. & Smetana, A. (eds.): *Catalogue of Palaearctic Coleoptera. Volume 6. Chrysomeloidea.* Stenstrup: Apollo Books, 924 pp. DP: 22 February 2010 (colophon).

López-Colón, J. I. 2004: Cryptocephalus (Cryptocephalus) bahilloi sp. n., nueva especie de crisomélido Ibérico (Coleoptera: Chrysomelidae). *Biocosme Mésogéen* 20 [2003] (2): 93–102. DP: 15 March 2004 (mailed to J. Bezděk, J. I. López-Colón 2022, pers. comm.).

López-Pérez, S., Zaragoza-Caballero, S., Ochoterena, H. & Morrone, J. J. 2017: A phylogenetic study of the worldwide tribe Cassidini Gyllenhal, 1813 (Coleoptera: Chrysomelidae: Cassidinae) based on morphological data. *Systematic Entomology* 43 [2018] (2): 372–386. DP: 8 December 2017 (journal website).

Lou, Q.-Z. & Liang, H.-B. 2011: [new taxon]. In: Lou, Q.-Z., Yu, P.-Y. & Liang, H.-B.: Two new species of Macroplea Samouelle (Coleoptera: Chrysomelidae: Donaciinae) from China, with a key to all known species. *Zootaxa* 3003: 1–21. DP: 25 August 2011 (footer on the paper title page).

Lou, Q.-Z. & Yu, P.-Y. 2011: [new taxon]. In: Lou, Q.-Z., Yu, P.-Y. & Liang, H.-B.: Two new species of Macroplea Samouelle (Coleoptera: Chrysomelidae: Donaciinae) from China, with a key to all known species. *Zootaxa* 3003: 1–21. DP: 25 August 2011 (footer on the paper title page).

Lou, Q.-Z., Yu, P.-Y. & Liang, H.-B. 2011: Two new species of Macroplea Samouelle (Coleoptera: Chrysomelidae: Donaciinae) from China, with a key to all known species. *Zootaxa* 3003: 1–21. DP: 25 August 2011 (footer on the paper title page).

Luca, Y. de 1960: Au sujet de Bruchidius algiricus All. et Bruchidius incarnatus Boh. (Col. Bruchidae). *Annales de l'École Nationale d'Agriculture d'Alger* 2 (2): 3–12. DP: [31] May 1960 (issue cover).

Luca, Y. de 1961: Contribution aux bruchides (Coléoptères) d'Algérie. Leurs hôtes – leurs parasites – leurs stations. *Mémoires de la Société d'Histoire Naturelle de l'Afrique du Nord, Nouvelle Série* 7: 1–115. DP: [31 December] 1961 (paper title page).

Lucas, P. H. 1845: Note sur quelques nouvelles espèces d'insectes qui habitent les possessions françaises du nord de l'Afrique. *Revue Zoologique, par la Société Cuvierienne* 8 (4): 120–127. DP: [31] July 1845 (Dickinson et al. 2011).

Lucas, P. H. 1846a: Pl. 22. In: *Exploration scientifique de l'Algérie pendant les années 1840, 1841, 1842 publiée par ordre du Gouvernement et avec le concours d'une Commission Académique. Sciences Physiques. Zoologie. II. Histoire naturelle des animaux articulés. Deuxième partie. Insectes.* [1846–1849], Paris: A. Bertrand, 590 pp. + 47 pls. DP (pl. 22): 15 August 1846 (Evenhuis 2012, Bousquet 2016).

Lucas, P. H. 1846b: Pls. 35, 46. In: *Exploration scientifique de l'Algérie pendant les années 1840, 1841, 1842 publiée par ordre du Gouvernement et avec le concours d'une Commission Académique. Sciences Physiques. Zoologie. II. Histoire naturelle des animaux articulés. Deuxième partie.*

Insectes. [1846–1849], Paris: A. Bertrand, 590 pp. + 47 pls. DP (pls. 35, 46): [30] November 1846 (Evenhuis 2012, Bousquet 2016).

Lucas, P. H. 1847a: Pls. 44–45. In: *Exploration scientifique de l'Algérie pendant les années 1840, 1841, 1842 publiée par ordre du Gouvernement et avec le concours d'une Commission Académique. Sciences Physiques. Zoologie. II. Histoire naturelle des animaux articulés. Deuxième partie. Insectes*. [1846–1849], Paris: A. Bertrand, 590 pp. + 47 pls. DP (pls. 44–45): 15 May 1847 (Evenhuis 2012, Bousquet 2016).

Lucas, P. H. 1847b: *Exploration scientifique de l'Algérie pendant les années 1840, 1841, 1842 publiée par ordre du Gouvernement et avec le concours d'une Commission Académique. Sciences Physiques. Zoologie. II. Histoire naturelle des animaux articulés. Deuxième partie. Insectes*. [1846–1849], Paris: A. Bertrand, 590 pp. + 47 pls. DP (pp. 201–240): 15 May 1847 (Evenhuis 2012, Bousquet 2016).

Lucas, P. H. 1847c: *Exploration scientifique de l'Algérie pendant les années 1840, 1841, 1842 publiée par ordre du Gouvernement et avec le concours d'une Commission Académique. Sciences Physiques. Zoologie. II. Histoire naturelle des animaux articulés. Deuxième partie. Insectes*. [1846–1849], Paris: A. Bertrand, 590 pp. + 47 pls. DP (pp. 361–536): 20 November 1847 (Evenhuis 2012, Bousquet 2016).

Lucas, P. H. 1849: *Exploration scientifique de l'Algérie pendant les années 1840, 1841, 1842 publiée par ordre du Gouvernement et avec le concours d'une Commission Académique. Sciences Physiques. Zoologie. II. Histoire naturelle des animaux articulés. Deuxième partie. Insectes*. [1846–1849], Paris: A. Bertrand, 590 pp. + 47 pls. DP (pp. 537–590): 13 June 1849 (Evenhuis 2012, Bousquet 2016).

Lucas, R. 1920: Catalogus alphabeticus generum et subgenerum Coleopterorum orbis terrarum totius (famil., trib., subtrib., sect. incl.). Pars I. *Archiv für Naturgeschichte, Abteilung A* **84** [1918] (3): 289–448. DP: [31] January 1920 (issue cover).

Luigioni, P. 1920: Contributo allo studio della fauna coleotterologica del Lazio. *Atti della Pontificia Accademia Romana dei Nuovi Lincei* **73**: 1–29. DP: [31 December] 1920 (reprint cover).

Luigioni, P. 1930: Secondo contributo alla conscenza della fauna entomologica del Parco Nazionale d'Abruzzo. Una nuova specie di Chrysochloa dell'Italia centrale. Coleoptera (Fam. Chrysomelidae). *Atti della Pontificia Academia delle Scienze Nuovi Lincei* **83** (5): 206–211. DP: [31 December] 1930 (reprint cover).

Luigioni, P. 1933: Quarto contributo alla conscenza della fauna entomologica del Parco Nazionale d'Abruzzo. Un'altra specie nuova del gen. Chrysochloa Hope dell'Italia centrale. (Coleoptera Fam. Chrysomelidae). *Atti della Pontificia Academia delle Scienze Nuovi Lincei* **86** (3): 121–126. DP: [31 December] 1933 (reprint cover).

Lukjanovitch, F. K. 1939: Zhuki roda Rhaebus Fisch.-W. (Coleoptera, Bruchidae) i ikh svyaz s Nitraria (Zygophyllaceae). [Beetles of the genus Rhaebus Fisch.-W. (Coleoptera, Bruchidae) and their association with Nitraria (Zygophyllaceae)]. Pp. 546–566. In: Galperin, V. M. (ed.): *Prezidentu Akademii nauk SSSR akademiku Vladimiru Leontevichu Komarovu k semidesyatiletiyu so dnya rozhdeniya i sorokapyatiletiyu nauchnoy deyatelnosti.* Leningrad: Akademia nauk SSSR, 808 pp. SFP: 28 December 1939, DP: [31 December] 1939 (book title page).

Lukjanovitch, F. K. & Ter-Minassian, M. E. 1954: [new taxa]. In: Ter-Minassian, M. E.: Novye vidy zernovok (Coleoptera, Bruchidae) fauny SSSR. [New species of seed-beetles (Coleoptera, Bruchidae) of the USSR fauna]. *Trudy Zoologicheskogo Instituta Akademii Nauk SSSR* 15: 67–73. SFP: 1 March 1954, DP: 17 April 1954 (Asanovich et al. 2007).

Lukjanovitch, F. K. & Ter-Minassian, M. E. 1957: Zhuki-zernovki (Bruchidae). *Fauna SSSR. Zhestkokrylye. 24(1), n. s. 67.* Moskva: Zoolcgicheskiy Inststitut Akademii Nauk SSSR, 209 pp. SFP: 3 June 1957, DP: 12 September 1957 (registered at ZIN library).

Lyubishchev, A. A. 1963: Dva novykh palearkticheskikh vida roda Chaetocnema gruppy Ch. concinna Marsh. (Coleoptera, Chrysomelidae, Halticinae). (Two new palearctic species of the genus Chaetocnema of the group Ch. concinna Marsh. (Coleoptera, Chrysomelidae, Halticinae)). *Entomologicheskoe Obozrenie* 42 (4): 858–863. SFP: 16 December 1963, DP: 30 December 1963 (registered at ZIN library).

Machado, A. & Oromí, P. 2000: *Elenco de los Coleópteros de las Islas Canarias.* La Laguna (Tenerife): Instituto de Estudios Canarios, 306 pp. DP: 25 February 2000 (last page of the book).

Machatschke, J. W. 1954: Zur Kenntnis des Genus Colaphellus Weise (Coleoptera: Chrysomelidae). *Beiträge zur Entomologie* 4 (2): 208–218. DP: [30] June 1954 (issue title page).

Machatschke, J. W. 1959: Über die verwandtschaftliche Stellung des Cryptocephalus angaricus Franz (Coleoptera: Chrysomelidae). *Beiträge zur Entomologie* 9 (7/8): 746–752. DP: [31] December 1959 (issue title page).

Madar, Jind. 1959: Ueber die geographische Verbreitung und das interessante Vorkommen der gelbstreifigen Phyllotreten in den ostpalaearktischen Grenzgebieten. *Niponius, Acta Coleopterologica* 1 (2): 1–7. DP: 30 November 1959 (paper title page).

Madar, Jind. 1960: Zur Frage der zoogeographischen Verbreitung der Chaetocnema concinnicollis Baly mit Beschreibung zweier neuen Halticinen-Formen (Halticinae, Col.). *Mushi* 33 (7): 47–49. DP: 18 February 1960 (paper title page).

Madar, Jind. & Madar, Jiř. 1963: Beitrag zur Zoogeographie der Halticiden von Madeira (Mit einer Neubeschreibung). Pp. 122–125. In: Lindberg, Håk. (ed.): A contribution to the study of beetles in the Madeira Islands. Results of expeditions in 1957 and 1959. *Commentationes Biologicae* 25 (2): 1–156. DP: [31] May 1963 (last page of the paper).

Madar, Jind. & Madar, Jiř. 1964: Zoogeographische Resultate einer Sammelreise von Lindberg und Meinander in Marokko (Halticinae, Col.) (Mit einer Neubeschreibung). *Notulae Entomologicae* 44 (4): 113–120. DP: 29 December 1964 (verso of the volume title page).

Madar, Jind. & Madar, Jiř. 1965: Beitrag zur Systematik und Zoogeographie einiger paläarktischen Vertreter der Subfamilie Halticinae (Col., Chrysomelidae). *Acta Entomologica Bohemoslovaca* 62 (1): 8–13, 2 pls. DP: 20 January 1965 (table of contents)

Madar, Jind. & Madar, Jiř. 1968: Zur Systematik und Zoogeographie der Halticinen (Col.) der Kanarischen Inseln. *Acta Entomologica Bohemoslovaca* 65 (2): 138–151. DP: 20 March 1968 (table of contents).

Madar, Jind. & Madar, Jiř. 1969: Beiträge zur Kenntnis der Fauna Afghanistans (Sammelergebnisse von O. Jakeš 1963–64, D. Povolný 1965, D. Povolný & Fr. Tenora 1966, J. Šimek 1965–66, D. Povolný, J. Gaisler, Z. Šebek & Fr. Tenora 1967). Halticinae, Chrysomelidae, Col. (1. Teil). *Časopis Moravského Musea, Vědy přírodní* 53 [1968] (supplementum): 139–146. DP: 1 March 1969 (colophon).

Madar, Jind. & Madar, Jiř. 1970: Beiträge zur Kenntnis der Fauna Afghanistans (Sammelergebnisse von O. Jakeš 1963–64, D. Povolný 1965, D. Povolný & Fr. Tenora 1966, J. Šimek 1965–66, D. Povolný, J. Gaisler, Z. Šebek & Fr. Tenora 1967). Halticinae, Chrysomelidae, Col. (2. Teil). *Časopis Moravského Musea, Vědy přírodní* 54 [1969] (supplementum): 205–214. DP: 15 May 1970 (colophon).

Mader, L. 1927: Coleopterologische Notizen. *Entomologischer Anzeiger* 7 (17): 193–197. DP: 20 September 1927 (issue title page).

Mader, L. 1931: Über paläarktische Coleopteren, I. *Entomologischer Anzeiger* 11 (1): 19–20. DP: 5 January 1931 (issue cover).

Mader, L. 1937: Coleopterologische Notizen. III. *Entomologische Zeitschrift* 51 (30): 284–285. DP: 8 November 1937 (issue title page).

Mader, L. 1938: Neue Coleopteren aus China und Japan nebst Notizen. *Entomologisches Nachrichtenblatt* (Troppau) 12 (1): 40–61. DP: [31] March 1938 (footers throughout the paper).

Mader, L. 1943: Beitrag zur Kenntnis der Coleopteren. *Entomologische Blätter* 39 (1/2): 41–46. DP: 30 April 1943 (issue cover).

Maican, S. & Serafim, R. 2012: Overview on the Chrysomeloidea superfamily (Coleoptera: Cerambycidae, Orsodacnidae, Chrysomelidae) in Dobrogea (Romania). *Travaux du Muséum National d'Histoire Naturelle "Grigore Antipa"* 55 (1): 65–123. DP: 31 August 2012 (paper title page).

Maican, S. & Serafim, R. 2017: Catalogue of Cassidinae (Coleoptera: Chrysomelidae) from the new leaf beetles collection from "Grigore Antipa" National Museum of Natural History (Bucharest) (Part II). *Travaux du Muséum National d'Histoire Naturelle "Grigore Antipa"* 60 (2): 477–494. DP: 31 December 2017 (paper title page).

Makhan, D. 2012: Galerucella rishwani sp. nov. (Coleoptera: Chrysomelidae), a new leaf beetle from Chal Godarzi, Borujerd Lorestan Province, Iran. *Calodema* 230: 1–4. DP: 30 September 2012 (last page of the paper).

Makhan, D., Hawkeswood, T. J. & Kabir, A. 2020: Galerucella rishwani Makhan 2012, is distinct Iranian species (Coleoptera: Chrysomelidae: Galerucinae). *Calodema* 757: 1–3. DP: 17 March 2020 (last page of the paper).

Mäklin, F. W. 1878: Diagnoser öfver några nya sibirska insektarter. *Öfversigt af Finska Vetenskaps-Societetens Förhandlingar* 19 [1876–1877]: 15–32. DP: 15 June 1878 (*Ent. Nachricht.* 1878, 12: 167).

Mäklin, F. W. 1881: Coleoptera insamlade under den Nordenskiöld'ska Expeditionen 1875 på några öar vid Norges Nordvestkust, på Novaja Semlja och ön Waigatsch samt vid Jenissej i

Sibirien. *Kongliga Svenska Vetenskaps-Akademiens Handlingar, Ny Följd* **18** [1880] (4): 1–48. DP: 24 October 1881 (*Öfv. Fin. Vetens.-Soc. Förhand.* 1881, 24: ii).

Mallet, P.-M. 1924: Note sur les variétés de la Chrysomela graminis L. *Miscellanea Entomologica* **27** (10/11): 76–77. DP: 31 May 1924 (footer on p. 73).

Mallet, P.-M. 1925: Note sur une aberration de Chrysochloa coerulea Ol. *Miscellanea Entomologica* **29** (2/3): 27–28. DP: 15 November 1925 (footer on p. 17).

Mallet, P.-M. 1933: Le mélanisme chez Chrysomela graminis L. (Col. Chrysomeidae). *Bulletin de l'Association des Naturalistes de la Vallée du Loing* **16** (1/2): 74–75. DP: [31 December] 1933 (issue cover).

Mallet, P.-M. 1936: "Chrysomela varians" Schall. ab. "nigricollis" nov. (Coléopt.). *Bulletin Mensuel de la Société Linnéenne de Lyon* **5** (9): 141–142. DP: [30] November 1936 (issue title page).

Mandl, K. 1965: Ergebnisse der österreichischen Iran-Expeditionen 1949/50 und der österreichischen Karakorum-Expedition 1958. *Koleopterologische Rundschau* **42** [1964]: 18–26. DP: 1 June 1965 (volume cover).

Mandl, K. 1968: Revision der Gattung Theone Gistl (Coleoptera – Chrysomelidae – Galerucinae). *Entomologische Arbeiten aus dem Museum G. Frey* **19**: 5–31. DP: 1 June 1968 (table of contents).

Mandl, K. 1970: Revision der Gattung Theone Gistl (Coleoptera – Chrysomelidae – Galerucinae). 2. Teil. *Entomologische Arbeiten aus dem Museum G. Frey* **21**: 154–169. DP: 1 November 1970 (table of contents).

Mandl, K. 1981: Neue Coleopteren-Taxa vom Nahen bis zum Fernen Osten (Col. Cicindelidae, Carabidae und Chrysomelidae). *Entomologica Basiliensia* **6**: 167–182. DP: 30 November 1981 (table of contents).

Mandl, K. 1986: Pseudadimonia holzschuhi, eine neue Galerucinae-Art aus Nepal, und Bemerkungen zu einer vergessenen Art: Rhabdotilla rosti Jakobson 1911 (Chrysomelidae Col.). *Mitteilungen der Entomologischen Gesellschaft Basel, Neue Folge* **36** (2): 71–77. DP: [30] June 1986 (paper title page).

Mannerheim, C. G. von 1825: Novae coleopterorum species Imperii Rossici incolae. Pp. 19–41. In: Hummel, A. D.: *Essais Entomologiques No. IV. Insectes de 1824. Novae species*. St. Petersbourg: Chancellerie privée du Ministère de l'Intérieur, 71 pp. + index. DP: 1 May 1825 (Bousquet 2016).

Mannerheim, C. G. von 1827: [new taxon]. In: Hummel, A. D.: Insectes de 1826. Pp. 1–20. In: Hummel, A. D.: *Essais Entomologiques No. VI*. St. Petersbourg: Chancellerie privée du Ministère de l'Intérieur, 49 pp. DP: [31 December] 1827 (Bousquet 2016).

Mannerheim, C. G. von 1843: Beitrag zur Kaefer-Fauna der Aleutischen Inseln, der Insel Sitkha und Neu-Californiens. *Bulletin de la Société Impériale des Naturalistes de Moscou* **16** (2): 175–314. DP: [31] July 1843 (Bousquet & Davies 2021).

Mannerheim, C. G. von 1844: Description de quelques nouvelles espèces de coléoptères de Finlande. *Bulletin de la Société Impériale des Naturalistes de Moscou* **17** (1): 189–202. DP: 28 March 1844 (Bousquet & Davies 2021).

Mannerheim, C. G. von 1845: Lettre a S. E. Mr. Fischer de Waldheim ou relation d'un voyage fait en 1844, en Suède, en Danemarck et dans le nord de l'Allemagne. *Bulletin de la Société*

Impériale des Naturalistes de Moscou **17** [1844] (4): 844–872. DP: 27 January 1845 (Bousquet & Davies 2021).

Mannerheim, C. G. von 1849: Insectes coléopteres de la Sibérie orientale, nouveaux ou peu connus. *Bulletin de la Société Impériale des Naturalistes de Moscou* **22** (1): 220–249. DP: 8 March 1849 (Bousquet & Davies 2021).

Mannerheim, C. G. von 1852: Zweiter Nachtrag zur Kaefer-Fauna der nord-amerikanischen Laender des Russischen Reiches. *Bulletin de la Société Impériale des Naturalistes de Moscou* **25** (2): 283–387. DP: 30 October 1852 (Bousquet & Davies 2021).

Mannerheim, C. G. von 1853: Dritter Nachtrag zur Kaefer-Fauna der nord-amerikanischen Laender des Russischen Reiches. *Bulletin de la Société Impériale des Naturalistes de Moscou* **26** (3): 95–273. DP: 10 November 18 53 (Bousquet & Davies 2021).

Marchand, H. 1933: Zwei neue Aberrationen von Chrysochloa gloriosa Fabr. *Mitteilungen der Schweizer Entomologischen Gesellschaft* **15** [1930–1933] (10): 455–456. DP: 15 March 1933 (issue title page).

Marchand, H. 1938: Ein interessanter Fund aus der Gattung Chrysochloa Hope (Coleoptera). *Mitteilungen der Schweizerischen Entomologischen Gesellschaft* **15** (5): 205–208. DP: 15 March 1938 (issue title page).

Marchand, H. 1939: Zwei neue Rassen der Gattung Chrysochloa Hope. *Acta Entomologica Musei Nationalis Pragae* **17**: 156–157. DP: 1 December 1939 (volume title page).

Marcu, O. 1938: Einige neue Käferformen aus Rumänien. *Comptes Rendus des Séances de l'Académie des Sciences de Roumanie* **2** (3): 336. DP: [31 December] 1938 (paper title page).

Marcu, O. 1957: Contribuţii la cunoaşterea faunei coleopterelor Transilvaniei. (Contribution à la connaissance de la faune des coléoptères de Transylvanie). *Buletinul Universităţilor "V. Babeş" şi "Bolyai", Seria Ştiinţele naturii* (Cluj) **1** (1/2): 527–544. DP: [31 December] 1957 (issue cover).

Märkel, J. C. F. 1841: Beiträge zur Kenntniss der unter Ameisen lebenden Insekten. *Zeitschrift für die Entomologie* (E. F. Germar) **3** (1/2): 203–225. DP: 13 August 1841 (*Allg. Bibl. Deutsch.* 1841 (33): 359).

Märkel, J. C. F. 1847: [new taxon]. In: Kiesenwetter, E. A. H. & Märkel, J. C. F.: Eine entomologische Excursion im Riesengebirge im Juli 1846. *Entomologische Zeitung* (Stettin) **8** (3): 73–87. DP: [31] March 1847 (issue title page).

Marquet, M. 1864: Espéces nouvelles de Coléoptères de France. *L'Abeille, Mémoires d'Entomologie* **1**: 372–373. DP: [31 December] 1864 (volume title page).

Marseul, S. A. de 1866: Descriptions d'espèces nouvelles. *L'Abeille, Mémoires d'Entomologie* **4** [1865–1867]: lxviii. DP: [31 December] 1866 (*Avertissement* of the volume).

Marseul, S. A. de 1868: Descriptions des espèces nouvelles. *L'Abeille, Journal d'Entomologie* **5** [1868–1869]: 171–218. DP: [31 December] 1868 (footers on pp. 169, 181, 217).

Marseul, S. A. de 1871: Descriptions de coléoptères nouveaux. *L'Abeille, Mémoires d'Entomologie* **6** [1869–1871] (2): 368–389. DP (pp. 368–376): [31] May 1870, (pp. 377–384): [30] June 1870, (pp. 385–389): [30] June 1871 (footers on pp. 369, 377, 385).

Marseul, S. A. de 1874: Monographie des cryptocéphales du nord de l'ancien monde. *L'Abeille, Journal d'Entomologie* **13** [1873–1875]: 1–326. DP: [31 December] 1874 (footers on pp. 57, 229, 269, 293).

Marseul, S. A. de 1875: Nouvelles et faits divers de l'Abeille. No. 10. *L'Abeille, Journal d'Entomologie* **14** [1874–1876]: xxxvii–xl. DP: [31 December] 1875 (paper title page).

Marseul, S. A. de 1876: Nouvelles et faits divers de l'Abeille. Deuxième série, No 9. *L'Abeille, Journal d'Entomologie* **16** [1876–1878]: 35–36. DP: [31 December] 1876 (noticed on p. 33).

Marseul, S. A. de 1878: Nouvelles et faits divers de l'Abeille. Deuxième série, No 21. *L'Abeille, Journal d'Entomologie* **17** [1878–1879]: 81–84. DP: [31 December] 1878 (paper title page).

Marseul, S. A. de 1883: Monographie des chrysomélides de l'ancien-monde. *L'Abeille, Journal d'Entomologie* **21**: 1–108. DP: [31 December] 1883 (footers on pp. 1, 25, 49, 73, 97).

Marseul, S. A. de 1884: Nouvelles et faits divers de l'Abeille. Deuxième série, No. 47. *L'Abeille, Journal d'Entomologie* **21** [1883]: 188. DP: [31 December] 1884 (paper title page).

Marseul, S. A. de 1886: Monographie des chrysomélides. *L'Abeille, Journal d'Entomologie* **24** [1886–1887]: 1–48 [= 109–156]. DP (pp. 1–24, No. 309, Livr. 9): 24 September 1886, (pp. 25–48, No. 312, Livr. 12): 30 December 1886 (issue covers in the BHL copy).

Marseul, S. A. de 1887: Monographie des chrysomélides. *L'Abeille, Journal d'Entomologie* **24** [1886–1887]: 49–190 [= 157–298]. DP (pp. 49–72, No. 314, Livr. 2): 28 February 1887, (pp. 73–96, No. 316, Livr. 4): 20 April 1887, (pp. 97–120, No. 317, Livr. 5): 28 May 1887, (pp. 121–144, No. 319, Livr. 7): 9 July 1887, (pp. 145–168, No. 320, Livr. 8): 31 August 1887, (pp. 169–190, Nos. 322–323, Livr. 10–11): 30 November 1887 (issue covers in the BHL copy).

Marseul, S. A. de 1888a: Monographie des chrysomélides. *L'Abeille, Journal d'Entomologie* **25** [1887–1888]: 1–96 [= 299–394]. DP (pp. 299–322, No. 330, Livr. 6): 28 June 1888, (pp. 323–346, Nos. 331–332, Livr. 7–8) 28 October 1888, (pp. 347–370, No. 333, Livr. 9): 28 November 1888, (pp. 371–394, No. 334–335, Livr. 10–11): [31] December 1888 (issue covers in the BHL copy).

Marseul, S. A. de. 1888b: Catalogue synonymique et geographique des coleopteres de l'Ancien Monde, Europe et contrees limitrophes en Afrique et en Asie. *L'Abeille, Journal d'Entomologie* **25** [1887–1888]: 361–480. DP (pp. 457–480, Nos. 331–332, Livr. 7–8): 28 October 1888 (issue cover in the BHL copy).

Marseul, S. A. de. 1889: [Monographie des chrysomélides (Suite)]. *L'Abeille, Journal d'Entomologie* **26** [1888–1889]: 97–148 [= 395–446]. DP (pp. 97–132, No. 337, Livr. 1): [31] January 1889, (pp. 133–148, No. 339, Livr. 3): [31] March 1889 (issue covers in the BHL copy).

Marshall, T. A. 1864: Corynodinorum recensio. *Journal of the Linnean Society of London, Zoology* 8 [1865] (29): 24–50. DP: 3 September 1864 (Kappel 1896).

Marsham, T. 1802: *Entomologia Britannica, sistens insecta Britanniae indigena, secundum methodum Linnaeanam disposita. Tomus I. Coleoptera*. Londini: Wilks & Taylor, xxxi + 547 + [1] pp. DP: [31] August 1802 (Bousquet 2016).

Matcha, J. 1917: Chrysochloa vittigera ab. nova Matheyi m. *Koleopterologische Rundschau* 6 (7/9): 54. DP: 30 September 1917 (issue cover).

Matis, E. G. 1970: Opisanie lichinok i kukolok nekotorykh maloizvestnykh vidov i imago odnogo novogo vida (Coleoptera, Cassidinae). [Descriptions of larvae and pupae of some little known species and imago of one new species (Coleoptera, Cassidinae)]. Pp. 23–33. In: Protsenko, A. I. (ed.): *Entomologicheskie issledovaniya v Kirgizii*. [*Entomological research in*

Kyrgyzstan]. Frunze: Izdatel'stvo "ILIM", 87 pp. SFP: 23 July 1970, DP: 28 October 1970 (registered at VKP).

Matis, E. G. 1974: Novyy vid roda Cassida (Coleoptera, Chrysomelidae) iz Primorya. (A new species of the genus Cassida (Coleoptera, Chrysomelidae) from the Marine territory). *Zoologicheskiy Zhurnal* **53** (7): 1100–1102. SFP: 18 June 1974, DP: 29 July 1974 (registered at ZIN library).

Matsumura, S. 1911a: [new taxon]. In: Otsuka, T.: [New insects collected by myself]. *Insect World* (Gifu) **15** (2): 59–63 (in Japanese) DP: 15 January 1911 (issue cover).

Matsumura, S. 1911b: Erster Beitrag zur Insekten-Fauna von Sachalin. *Journal of the College of Agriculture, Tohoku Imperial University, Sapporo, Japan* **4** (1): 1–145, 2 pls. DP: [31] March 1911 (issue cover).

Matsumura, S. 1911c: Beschreibungen von am Zuckerrohr Formosas schädlichen oder nützlichen Insecten. *Mémoires de la Société Entomologique de Belgique* **18**: 129–150. DP: 9 June 1911 (stamp on the BHL copy).

Matsumura, S. 1934: Insects collected at the foot of Mt. Yatsugadake and its environment. *Insecta Matsumurana* **9** (1/2): 60–80. DP: 30 November 1934 (table of contents).

Matsumura, Y., Yao, I., Beutel, R. G. & Yoshizawa, K. 2014: Molecular phylogeny of the leaf beetle subfamily Criocerinae (Coleoptera: Chrysomelidae) and the correlated evolution of reproductive organs. *Arthropod Systematics & Phylogeny* **72** (2): 95–110. DP: 18 July 2014 (paper title page).

Matzek, C. A. E. 1843: Schlesische Chrysomelen. *Uebersicht der Arbeiten und Veränderungen der Schlesischen Gesellschaft für Vaterländische Kultur* (Breslau) [1842]: 154–157. DP: 17 August 1843 (*Ber. Bekanntmach. Verh. Akad. Wiss. Berlin* 1843: 224).

Maulik, S. 1913: Cryptostomes from the Indian Museum. *Records of the Indian Museum* **9** (2): 105–118. DP: [30] June 1913 (table of contents).

Maulik, S. 1915: Cryptostomes of the Indian Museum. Part V. *Records of the Indian Museum* **11** (5): 367–381. DP: 27 September 1915 (table of contents).

Maulik, S. 1916: On cryptostome beetles in the Cambridge University Museum of Zoology. *Proceedings of the Zoological Society of London* **1916** (4): 567–589. DP: 15 December 1916 (Duncan 1937).

Maulik, S. 1918: Two new Indian Cassidinae beetles. *Annals and Magazine of Natural History, Ninth Series* **1** (4): 322–325. DP: 28 March 1918 (Evenhuis 2003).

Maulik, S. 1919a: *Coleoptera. Chrysomelidae (Hispinae and Cassidinae)*. In: Shipley, A. E. (ed.): *The Fauna of British India including Ceylon and Burma*. London: Taylor and Francis, xi + 439 pp. DP: [31] July 1919 (colophon).

Maulik, S. 1919b: New Hispinae. *Annals and Magazine of Natural History, Ninth Series* **4** (24): 407–411. DP: 29 November 1919 (Evenhuis 2003).

Maulik, S. 1919c: New species of Anisoderopsis collected by Mr. R. Vitalis de Salvaza. *Annali del Museo Civico di Storia Naturale Giacomo Doria* **48**: 382–384. DP: 10 December 1919 (Poggi 2010).

Maulik, S. 1920: A new Chlamys from Darjiling. *Records of the Indian Museum* **19** (1): 19–21. DP: 22 March 1920 (table of contents).

Maulik, S. 1921: A new hispid beetle injurious to nipa palm. *Annals and Magazine of Natural History, Ninth Series* 7 (41): 451–452. DP: 1 May 1921 (Evenhuis 2003).

Maulik, S. 1923: New cryptostome beetles. *Proceedings of the Zoological Society of London* **1923** (3): 599–608. DP: 3 September 1923 (Duncan 1937).

Maulik, S. 1926: *Coleoptera. Chrysomelidae (Chrysomelinae and Halticinae)*. In: Shipley, A. E. (ed.): *The Fauna of British India, including Ceylon and Burma*. London: Taylor and Francis, xiv + 442 pp., 1 map. DP: [30] April 1926 (title page).

Maulik, S. 1927: Two new Hispinae (Col.) from Fiji and New Britain. *Annals and Magazine of Natural History, Ninth Series* **20** (115): 105–108. DP: 1 July 1927 (Evenhuis 2003).

Maulik, S. 1928: New chrysomelid beetles from India. With a note on the scales of Coleoptera. *Proceedings of the Zoological Society of London* **1928** (1): 151–161. DP: 16 April 1928 (reprint cover).

Maulik, S. 1929: Chrysomelidae; with a note on the comparative anatomy of some halticine tibiae. Pp. 177–215. In: *Insects of Samoa and other Samoan terrestrial Arthropoda. Part IV. Coleoptera. Fasc. 3*. London: British Museum (Natural History), pp. 175–215. DP: 23 February 1929 (cover).

Maulik, S. 1931: Reports of the Percy Sladen Trust Expedition to the Indian Ocean in 1905, under the leadership of Mr. J. Stanley Gardiner. Coleoptera, Chrysomelidae: Eumolpinae, Galerucinae and Halticinae. *Transactions of the Linnean Society of London, Second Series, Zoology* **19** (2): 241–260. DP: [31] August 1931 (table of contents).

Maulik, S. 1933: A new galerucine beetle from China. *Annals and Magazine of Natural History, Tenth Series* **12** (72): 563–564. DP: 1 December 1933 (Evenhuis 2003).

Maulik, S. 1934: [new taxon]. In: Hering, M.: Minen von Shanghai, gesammelt von H. Hoene. I. mit Beschreibung einer neuen Argopistes-Art (Col. Haltic.). *Mitteilungen der Deutschen Entomologischen Gesellschaft* **5** (3/4): 23–28. DP: 25 May 1934 (footer on p. 48).

Maulik, S. 1936: *Coleoptera. Chrysomelidae (Galerucinae)*. In: Sewell, R. B. S. (ed.): *The Fauna of British India, including Ceylon and Burma*. London: Taylor & Francis, xiv + 648 pp., 1 pl. DP: 30 January 1936 (book title page).

Mayr, E. & Ashlock, P. D. 1991: *Principles of Systematic Zoology. Second edition*. McGraw-Hill, New York, xx + 475 pp. DP: [31 December] 1991 (colophon).

Medvedev, L. N. 1955: K faune listoedov (Coleoptera, Chrysomelidae) Sredney Azii. [To the leaf-beetles fauna (Coleoptera, Chrysomelidae) of Central Asia]. *Zoologicheskiy Zhurnal* **34** (5): 1177–1178. SFP: 19 September 1955, DP: 17 October 1955 (registered at VKP).

Medvedev, L. N. 1956a: Materialy k faune listoedov (Coleoptera, Chrysomelidae) Dal'nego Vostoka. [Materials on the leaf-beetles fauna (Coleoptera, Chrysomelidae) of the Far East]. *Zoologicheskiy Zhurnal* **35** (1): 156–157. SFP: 27 December 1955, DP: 6 February 1956 (registered at ZIN library).

Medvedev, L. N. 1956b: Novyy rod i vid listoeda Allecumolpus gen. nov. iz yugo-vostochnoy Turkmenii. [New genus and species of leaf-beetle Allecumolpus gen. nov. from south-eastern Turkmenistan]. *Trudy Instituta Biologii Akademii Nauk Turkmenskoy SSR, Seriya Zoologicheskaya* **4**: 242. SFP: 8 December 1956, DP: [31 December] 1956 (volume cover).

Medvedev, L. N. 1956c: Obzor turkmenskikh vidov Cryptocephalus Geoffr. [Review of Turkmenian species of Cryptocephalus Geoffr.]. *Trudy Instituta Biologii Akademii Nauk Turkmenskoy SSR, Seriya Zoologicheskaya* 4: 243–250. SFP: 8 December 1956, DP: [31 December] 1956 (volume cover).

Medvedev, L. N. 1956d: [new taxa]. In: Ogloblin, D. A. & Medvedev, L. N.: Novye formy zhukov-listoedov (Coleoptera, Chrysomelidae) palearkticheskoy fauny. (New palearctic chrysomelid-beetles (Coleoptera, Chrysomelidae)). *Entomologicheskoe Obozrenie* 35 (4): 895–898. SFP: 18 December 1956, DP: [31] December 1956 (Lopatin 1963c: 122).

Medvedev, L. N. 1957a: Listoedy-shiponoski (Coleoptera, Chrysomelidae, Hispinae) fauny SSSR. (Hispid beetles (Coleoptera, Chrysomelidae, Hispinae) of the fauna of the USSR). *Zoologicheskiy Zhurnal* 36 (2): 293–297. SFP: 27 March 1957, DP: 7 May 1957 (registered at ZIN library).

Medvedev, L. N. 1957b: Neue palaearktische Blattkäferarten (Coleoptera: Chrysomelidae). *Beiträge zur Entomologie* 7 (3/4): 326–333. DP: [31] August 1957 (issue title page).

Medvedev, L. N. 1957c: Obzor Eumolpinae (Coleoptera, Chrysomelidae) Sredney Azii. (A survey of Eumolpinae (Coleoptera, Chrysomelidae) of Central Asia). *Zoologicheskiy Zhurnal* 36 (9): 1323–1337. SFP: 20 September 1957, DP: 29 October 1957 (registered at ZIN library).

Medvedev, L. N. 1957d: Neue Cassidinen aus China (Coleoptera: Chrysomelidae). *Beiträge zur Entomologie* 7 (5/6): 553–556. DP: [30] November 1957 (issue title page).

Medvedev, L. N. 1958: Chinesische und japanische Criocerinen aus der Kollektion des Museums G. Frey (Col. Chrysom.). *Entomologische Arbeiten aus dem Museum G. Frey* 9 (1): 106–113. DP: 15 February 1958 (table of contents on the cover).

Medvedev, L. N. 1961: Obzor palearkticheskikh vidov roda Clytra Laich. (Coleoptera, Chrysomelidae). (Review of Palearctic species species of the genus Clytra Laich. (Coleoptera, Chrysomelidae)). *Entomologicheskoe Obozrenie* 40 (3): 636–651. SFP: 28 September 1961, DP: 25 November 1961 (registered at VKP).

Medvedev, L. N. 1962a: Fauna zhukov-listoedov severo-zapadnogo Kavkaza (Coleoptera, Chrysomelidae). Soobshchenie 1. Opisanie novykh form. (Leaf-beetle fauna of north-west Caucasus (Coleoptera, Chrysomelidae). Part 1. Description of new species). *Zoologicheskiy Zhurnal* 41 (3): 384–390. SFP: 7 March 1962, DP: 28 July 1962 (registered at ZIN library).

Medvedev, L. N. 1962b: Dva novykh vida Labidostomis (Coleoptera, Chrysomelidae) iz SSSR. (Two new species of Labidostomis (Coleoptera, Chrysomelidae) from the USSR). *Trudy Zoologicheskogo Instituta Akademii Nauk SSSR* (Leningrad) 30: 254–257. SFP: 28 July 1962, DP: 21 September 1962 (Asanovich et al. 2007).

Medvedev, L. N. 1962c: Obzor zhukov-listoedov roda Antipa (Coleoptera, Chrysomelidae) fauny SSSR i sopredel'nykh stran. (Review of leaf beetles Antipa (Coleoptera, Chrysomelidae) of the fauna of the USSR and contiguous countries). *Entomologicheskoe Obozrenie* 41 (3): 613–624. SFP: 24 August 1962, DP: 1 October 1962 (registered at ZIN library).

Medvedev, L. N. 1962d: New and interesting species of Paleartic and Oriental Clytrinae (Coleoptera, Chrysomelidae). *Annales Historico-Naturales Musei Nationalis Hungarici* 54: 333–337. DP: 31 December 1962 (Merkl et al. 2008).

Medvedev, L. N. 1963a: Obzor vidov Cryptocephalus gruppy C. macrodactylus Gebl. [Review of Cryptocephalus species of the group C. macrodactylus Gebl.]. Pp. 38–47. In: Paliy, V. F. (ed.): *Sbornik Entomologicheskihk Rabot 2.* [*Collection of entomological papers 2*]. Frunze: Izdatel'stvo akademii nauk Kirgizskoy SSR, 106 pp. SFP: 10 November 1963, DP: [31 December] 1963 (book title page).

Medvedev, L. N. 1963b: K faune listoedov (Coleoptera, Chrysomelidae) Kamchatskoy oblasti. (On the fauna of Chrysomelidae (Coleoptera) from Kamchatka). Pp. 113–117. In: *Fauna Kamchatskoy oblasti.* [*Fauna of Kamchatka Region*]. Moskva, Leningrad: Izdatelstvo Akademii nauk SSSR, 119 pp. SFP: 29 October 1963, DP: [31 December] 1962 (reprint title page).

Medvedev, L. N. 1964: Novyy vid zhuka listoeda iz roda Phytodecta (Coleoptera, Chrysomelidae) s Dal'nego Vostoka. (A new species of Phytodecta (Coleoptera, Chrysomelidae) from Far-East). *Entomologicheskoe Obozrenie* 43 (1): 180–181. SFP: 20 February 1964, DP: 28 March 1964 (registered at ZIN library).

Medvedev, L. N. 1965a: Sredneaziatskie formy Coptocephala unifasciata L. (Coleoptera, Chrysomelidae). [Central Asian forms of Coptocephala unifasciata L. (Coleoptera, Chrysomelidae)]. Pp. 9–15. In: Protsenko, A. I. (ed.): *Entomologicheskie issledovaniya v Kirgizii.* [*Entomological research in Kyrgyzstan*] Frunze: Izdatel'stvo "ILIM", 121 pp. SFP: 30 June 1965, DP: [31 December] 1965 (book title page).

Medvedev, L. N. 1965b: [new taxon]. In: Medvedev, L. N. & Shapiro, D. S.: Sem. Chrysomelidae – listoedy. Pp. 419–474. In: Bey-Bienko, G. A. (ed.): *Opredelitel nasekomykh evropeyskoy chasti SSSR v pyati tomakh. 2. Zhestkokrylye i veerokrylye.* [*Identification key to insects of European part of USSR in five volumes. 2. Coleoptera and Strepsiptera*]. Moskva – Leningrad: Nauka, 668 pp. SFP: 26 June 1965, DP: [31 December] 1965 (book title page).

Medvedev, L. N. 1966: Novye formy zhukov-listoedov (Coleoptera, Chrysomelidae) s Kurilskikh ostrovov. [New forms of leaf-beetles (Coleoptera, Chrysomelidae) from Kuril Islands]. Pp. 39–44. In: Soboleva, R. G. & Ivliev, L. A. (eds.): *Entomofauna lesov Kurilskikh ostrovov, poluostrova Kamchatka i Magadanskoy oblasti.* [*Forest entomofauna of Kuril Islands, Kamchatka peninsula and Madagan Region*]. Moskva – Leningrad: Nauka, 137 pp. SFP: 16 December 1966, DP: [31 December] 1966 (reprint cover).

Medvedev, L. N. 1968: [new taxa]. In: Ivliev, L. A., Kononov, D. G. & Medvedev, L. N.: Fauna listoedov Magadanskoy oblasti i severa Khabarovskogo kraya. (The fauna of the leaf beetles of Madagan region and the North of Khabarovsk territory). Pp. 62–87. In: Kurentsov, A. I. & Konovalova, Z. A. (eds.): *Fauna i ekologiya nasekomykh Dal'nego Vostoka.* (*Fauna and ecology of the insects of Far East*). Vladivostok: Akademiya Nauk SSSR, Sibirskoe otdelenie, 171 pp. SFP: 4 December 1968, DP: [31 December] 1968 (book title page).

Medvedev, L. N. 1969: [new taxon]. In: Medvedev, L. N. & Chernov, Y. I.: Novyy vid listoeda roda Chrysomela (Coleoptera, Chrysomelidae) – vazhnyy komponent biocenozov Taymyra. (A new species of Chrysomela (Coleoptera, Chrysomelidae) – an important component of the Taimyr Peninsula biocoenoses). *Zoologicheskiy Zhurnal* 48 (4): 532–537. SFP: 1 April 1969, DP: 29 April 1969 (registered at ZIN library).

Medvedev, L. N. 1970a: A list of Chrysomelidae collected by Dr. W. Wittmer in Turkey (Coleoptera). *Revue Suisse de Zoologie* 77 (2): 309–319. DP: [30] June 1970 (paper title page).

Medvedev, L. N. 1970b: A contribution to the knowledge of African and Oriental Clytrinae (Coleoptera – Chrysomelidae). *Entomologischen Arbeiten aus dem Museum G. Frey* **21**: 184–195. DP: 1 November 1970 (table of contents).

Medvedev, L. N. 1970c: K poznaniyu roda Chrysolina Motsch. (Chrysomelidae) Dal'nego Vostoka i prilegayushchikh rayonov. [To the knowledge of the genus Chrysolina Motsch. (Chrysomelidae) of the Far East and surrounding areas]. *Trudy Biologo-Pochvennogo Instituta* **2** (1): 159–162. DP: [31 December] 1970 (paper title page).

Medvedev, L. N. 1970d: Contribution à la faune de l'Iran. 20. Coléoptères Chrysomelidae Clytrinae. *Annales de la Société Entomologique de France, Nouvelle Série* **6** (4): 999–1002. DP: [31 December] 1970 (issue cover).

Medvedev, L. N. 1971a: Novye formy listoedov podsemeystva Clytrinae (Coleoptera, Chrysomelidae) SSSR i prilegayushchikh stran. (New forms of the subfamily Clytrinae (Coleoptera, Chrysomelidae) in the USSR and adjacent countries). *Zoologicheskiy Zhurnal* 50 (5): 686–695. SFP: 28 April 1971, DP: 20 June 1971 (registered at ZIN library).

Medvedev, L. N. 1971b: African Clytrinae from the collection of the Zoological Museum Berlin. *Mitteilungen aus dem Zoologischen Museum in Berlin* 47 (1): 21–32. DP: 15 July 1971 (verso of issue title page).

Medvedev, L. N. 1971c: Neue und wenig bekannte palaearktische Arten der Gattung Cryptocephalus Geoffr. (Coleoptera, Chrysomelidae). *Deutsche Entomologische Zeitschrift, Neue Folge* **18** (1/3): 129–132. DP: 15 August 1971 (issue title page).

Medvedev, L. N. 1971d: Contribution à l'étude biologique du Sénégal septentrional. XII. Coléoptères Chrysomelidae Clytrinae. *Bulletin de l'Institut Français d'Afrique Noire, Série A* **33** (3): 655–658. DP: 1 September 1971 (last page of the issue).

Medvedev, L. N. 1973a: Obzor roda Cryptocephalus Geoffr. (Chrysomelidae) Sibiri i Dal'nego Vostoka. [Review of the genus Cryptocephalus Geoffr. (Chrysomelidae) of Siberia and Far East]. *Trudy Biologo-pochvennogo Instituta, Novaya Seriya* **9**: 104–127. SFP: 3 August 1973, DP: [31 December] 1973 (paper title page).

Medvedev, L. N. 1973b: Novye zhuki-listoedy (Coleoptera, Chrysomelidae) palearktiki. (New leaf-beetles (Coleoptera, Chrysomelidae) from Palaearctic). *Entomologicheskoe Obozrenie* **52** (4): 876–885. SFP: 4 December 1973, DP: [31 December] 1973 (paper title page).

Medvedev, L. N. 1974: Novye vidy listoedov (Coleoptera, Chrysomelidae) iz Mongolii i sopredel'nykh rayonov. (New species of Chrysomelidae (Coleoptera) from Mongolia and adjacent areas). Pp. 179–182. In: Kozlov, M. A. (ed.): *Nasekomye Mongolii 2*. (*Insects of Mongolia 2*). St. Petersburg: Academy of Sciences of the USSR, Zoological Institute, 439 pp. DP: 30 September 1974 (Kerzhner 1994).

Medvedev, L. N. 1975a: Chrysomelidae collected by Dr. W. Wittmer in Turkey and Iran. *Mitteilungen der Entomologischen Gesellschaft Basel, Neue Folge* **25** (1): 12–19. DP: [31] March 1975 (heading on p. 1).

Medvedev, L. N. 1975b: [new taxa]. In: Medvedev, L. N. & Korotyaev, B. A.: K faune zhukov-listoedov (Coleoptera, Chrysomelidae) Tuvinskoy Avtonomnoy Respubliki i severo-zapadnoy Mongolii, 1. (On the fauna of chrysomelid-beetles (Coleoptera, Chrysomelidae) of the Tuvinian Autonomous Republic and of the north-western Mongolia, 1). Pp. 177–190. In: Kerzhner, I. M. (ed.): *Nasekomye Mongolii 3*. (*Insects of Mongolia 3*). St. Petersburg: Academy of Sciences of the USSR, Zoological Institute, 675 pp. DP: 25 December 1975 (Kerzhner 1994).

Medvedev, L. N. 1976a: [new taxa]. In: Medvedev, L. N. & Voronova, N. V. 1976: K faune listoedov (Coleoptera, Chrysomelidae) Mongolii. (On the fauna of chrysomelid beetles (Coleoptera, Chrysomelidae) of Mongolia). Pp. 222–236. In: Kerzhner, I. M. & Kozlov, M. A. (eds.): *Nasekomye Mongolii 4*. (*Insects of Mongolia 4*). St. Petersburg: Academy of Sciences of the USSR, Zoological Institute, 639 pp. DP: 9 July 1976 (Kerzhner 1994).

Medvedev, L. N. 1976b: [new taxon]. In: Medvedev, L. N. & Korotyaev, B. A.: K faune zhukov-listoedov (Coleoptera, Chrysomelidae) Tuvinskoy Avtonomnoy Respubliki i severo-zapadnoy Mongolii, II. [On the fauna of chrysomelid beetles (Coleoptera, Chrysomelidae) of the Tuvinian Autonomous Republic and of the north-western Mongolia, II]. Pp. 241–244. In: Kerzhner, I. M. & Kozlov, M. A. (eds.): *Nasekomye Mongolii 4*. (*Insects of Mongolia 4*). St. Petersburg: Academy of Sciences of the USSR, Zoological Institute, 639 pp. DP: 9 July 1976 (Kerzhner 1994).

Medvedev, L. N. 1976c: Reviziya zhukov-listoedov roda Crosita Motsch. (Coleoptera, Chrysomelidae). (Revision of the genus Crosita Motsch. (Coleoptera, Chrysomelidae)). *Entomologicheskoe Obozrenie* 55 (4): 903–911. SFP: 3 December 1976, DP: 28 December 1976 (registered at ZIN library).

Medvedev, L. N. 1977a: [new taxa]. In: Medvedev, L. N. & Voronova, N. V. 1977: K faune listoedov (Coleoptera, Chrysomelidae) Mongolii, 2. (On the chrysomelid-beetle fauna of Mongolia, 2 (Coleoptera, Chrysomelidae)). Pp. 319–352. In: Kerzhner, I. M. (ed.): *Nasekomye Mongolii 5*. (*Insects of Mongolia 5*). St. Petersburg: Academy of Sciences of the USSR, Zoological Institute, 757 pp. DP: 2 September 1977 (Kerzhner 1994).

Medvedev, L. N. 1977b: Nakhozhdenie predstavitelya amerikanskogo roda Neohaemonia Szek. (Coleoptera, Chrysomelidae) v Mongolii. *Doklady Akademii Nauk SSSR* **236** (2): 488–490. SFP: 31 August 1977, DP: 11 September 1977 (heading on p. 273).

Medvedev, L. N. 1977c: Dva novykh podvida zhukov-listoedov (Coleoptera, Chrysomelidae) iz evropeyskoy chasti SSSR. (Two new subspecies of chrysomelid-beetles (Coleoptera, Chrysomelidae) from the European part of USSR). Pp. 34–36. In: Skarlato, O. A. (ed.): *Novye i maloizvestnye vidy nasekomykh evropeyskoy chasti SSSR* (*Sbornik nauchnykh rabot*). (*New and little known species of insects of the European part of the USSR*). Leningrad: Akademia Nauk SSSR, 116 pp. SFP: 15 April 1977, DP: [31 December] 1977 (book title page).

Medvedev, L. N. 1978a: Materialy k izucheniyu fauny listoedov (Coleoptera Chrysomelidae) MNR. [Material to the knowledge of leaf-beetles fauna (Coleoptera Chrysomelidae) of Mongolia]. Pp. 189–196. In: Bannikova, I. A. & Medvedev, L. N. (eds.): *Geografia i dinamika rastitelnogo i zhivotnego mira MNR*. [*Geography and dynamics of flora and fauna of Mongolia*]. Moskva: Nauka, 224 pp. SFP: 13 June 1978, DP: 15 September 1978 (registered at ZIN library).

Medvedev, L. N. 1978b: Clytrinae from Africa (Coleoptera, Chrysomelidae). *Faunistische Abhandlungen Staatliches Museum für Tierkunde in Dresden* 7 [1978–1979] (1): 55–59. DP: 28 December 1978 (paper title page).

Medvedev, L. N. 1978c: K faune listoedov (Coleoptera, Chrysomelidae) Afganistana. (On the fauna of leaf-beetles (Coleoptera, Chrysomelidae) of Afghanistan). *Entomologicheskoe Obozrenie* 57 (4): 859–869. SFP: 30 November 1978, DP: 29 December 1978 (registered at ZIN library).

Medvedev, L. N. 1978d: Taksonomicheskie zametki o zhukakh-listoedakh (Coleoptera, Chrysomelidae) Sakhalina i Kuril'skih ostrovov. [Taxonomic notes on leaf-beetles (Coleoptera, Chrysomelidae) of Sakhalin and the Kuril Islands]. *Trudy Biologo-pochvennogo Instituta, Novaya Seriya* 50: 82–86. SFP: 4 April 1978, DP: [31 December] 1978 (paper title page).

Medvedev, L. N. 1978e: [new taxa]. In: Medvedev, L. N. & Ammosov, Y. N.: Fauna listoedov (Coleoptera, Chrysomelidae) Yakutii i ee glavneyshie zoogeograficheskie osobennosti. [Fauna of leaf-beetles (Coleoptera, Chrysomelidae) of Yakutia and its main zoogeographical characteristics]. Pp. 113–128. In: Ammosov, Y. N. (ed.): *Ekologo-faunisticheskie issledovaniya nasekomykh Yakutii. [Ecological-faunistic research on insects of Yakutia]*. Yakutsk: Izdanie Yakutskogo filiala SO AN SSSR, 148 pp. SFP: 21 June 1978, DP: [31 December] 1978 (book title page).

Medvedev, L. N. 1979a: Insects of Saudi Arabia. Coleoptera: fam. Chrysomelidae subfam. Clytrinae. *Fauna of Saudi Arabia* 1: 295–298. DP: 5 March 1979 (table of contents).

Medvedev, L. N. 1979b: Novyy vid roda Chrysolina (Coleoptera, Chrysomelidae) iz Sibiri. [New species of the genus Chrysolina (Coleoptera, Chrysomelidae) from Siberia]. *Trudy Vsesoyuznogo Entomologicheskogo Obshchestva* 61: 83–85. SFP: 7 August 1979, DP: 16 October 1979 (registered at ZIN library).

Medvedev, L. N. 1979c: Further contribution to the knowledge of Chrysomelidae from Mongolia (Coleoptera). *Folia Entomologica Hungarica, Series Nova* 32 (2): 161–167. DP: 15 December 1979 (verso of issue title page).

Medvedev, L. N. 1980a: Novye listoedy (Coleoptera, Chrysomelidae) iz Mongolii. (New leaf-beetles (Coleoptera, Chrysomelidae) from Mongolia). *Entomologicheskoe Obozrenie* 59 (2): 311–321. SFP: 29 May 1980, DP: 2 July 1980 (registered at ZIN library).

Medvedev, L. N. 1980b: [new taxa]. In: Medvedev, L. N. & Korotyaev, B. A.: Ocherki po faune listoedov (Coleoptera, Chrysomelidae) arkticheskoy Azii i Kamchatki. [Outlines of leaf-beetles fauna (Coleoptera, Chrysomelidae) of arctic Asia and Kamchatka]. Pp. 77–95. In: Medvedev, G. S. & Matis, E. G. (eds.): *Issledovaniya po entomofaune severo-vostoka SSR. [Research on entomofauna of North-East USSR]*. Vladivostok: DVNC AN SSSR, 171 pp. SFP: 20 May 1980, DP: [31 December] 1980 (book title page and colophon).

Medvedev, L. N. 1982: *Listoedy MNR. Opredelitel. [Leaf-beetles of Mongolia. Key to identification]*. Moskva: Nauka, 304 pp. SFP: 3 November 1982, DP: [31 December] 1982 (book title page and colophon).

Medvedev, L. N. 1983: Chrysomelidae from Iran (Insecta: Coleoptera). *Senckenbergiana Biologica* 64 (1/3): 133–140. DP: 30 November 1983 (paper title page).

Medvedev, L. N. 1984: Chrysomelidae from the Nepal Himalayas. I. Alticinae (Insecta: Coleoptera). *Senckenbergiana Biologica* **65** (1/2): 47–63. DP: 15 November 1984 (issue cover).

Medvedev, L. N. 1985a: K faune listoedov (Coleoptera, Chrysomelidae) Afganistana. II. (On the fauna of leaf-beetles (Coleoptera, Chrysomelidae) of Afghanistan. II.). *Entomologicheskoe Obozrenie* **64** (2): 370–377. SFP: 19 June 1985, DP: 17 July 1985 (registered at ZIN library).

Medvedev, L. N. 1985b: Fauna Criocerinae (Coleoptera, Chrysomelidae) V'etnama. [Criocerinae fauna (Coleoptera, Chrysomelidae) of Vietnam]. Pp. 64–79. In: Medvedev, L. N. (ed.): *Nasekomye V'etnama. [Insects of Vietnam]*. Moskva: Nauka, 181 pp. SFP: 17 June 1985, DP: [31 December] 1985 (book title page and colophon).

Medvedev, L. N. 1987: Fauna Chrysomelinae (Coleoptera, Chrysomelidae) V'etnama. [Chrysomelinae fauna (Coleoptera, Chrysomelidae) of Vietnam]. Pp. 69–80. In: Medvedev, L. N. (ed.): *Entomofauna V'etnama. [Entomofauna of Vietnam]*. Moskva: Nauka, 192 pp. SFP: 20 November 1986, DP: 7 September 1987 (registered at ZIN library).

Medvedev, L. N. 1988a: Clytrinae (Coleoptera, Chrysomelidae) of the Himalayas from Basel Museum of Natural History. *Entomologica Basiliensia* **12**: 469–480. DP: 29 April 1988 (table of contents).

Medvedev, L. N. 1988b: Zhuki-listoedy podesemeystva Clytrinae (Coleoptera, Chrysomelidae) fauny Vietnama. [Leaf-beetles of the subfamily Clytrinae (Coleoptera, Chrysomelidae) of Vietnam fauna]. Pp. 21–45. In: Medvedev, L. N. & Striganova, B. R. (eds.): *Fauna i ekologiya nasekomykh V'etnama. [Fauna and ecology of insects of Vietnam]*. Moskva: Nauka, 197 + [3] pp. SFP: 20 April 1988, DP: 26 September 1988 (registered at ZIN library).

Medvedev, L. N. 1988c: Novye i maloizvestnye vidy listoedov (Coleoptera, Chrysomelidae) iz yuzhnoy Azii. [New and little known species of leaf-beetles (Coleoptera, Chrysomelidae) from South Asia]. Pp. 46–50. In: Medvedev, L. N. & Striganova, B. R. (eds.): *Fauna i ekologiya nasekomykh V'etnama. [Fauna and ecology of insects of Vietnam]*. Moskva: Nauka, 197 + [3] pp. SFP: 20 April 1988, DP: 26 September 1988 (registered at ZIN library).

Medvedev, L. N. 1989: Chrysomelidae (Coleoptera) of Sri Lanka. The subfamilies Cryptocephalinae and Chlamisinae. *Annotationes Zoologicae et Botanicae* (Bratislava) **193**: 1–8. DP: 27 February 1989 (paper title page).

Medvedev, L. N. 1990a: Zhuki-listoedy podsemeystva Clytrinae (Coleoptera, Chrysomelidae) Kavkaza. (The leaf-beetle subfamily Clytrinae (Coleoptera, Chrysomelidae) in the Caucasus). Pp. 118–136. In: Striganova, B. R. (ed.): *Fauna nazemnykh bespozvonochnykh Kavkaza. (Fauna of terrestrial unvertebrates of the Caucasus)*. Moskva: Nauka, 234 pp. SFP: 6 February 1990, DP: 13 August 1990 (registered at ZIN library).

Medvedev, L. N. 1990b: Chrysomelidae from the Nepal Himalayas, II (Insecta: Coleoptera). *Stuttgarter Beiträge zur Naturkunde, Serie A (Biologie)* **453**: 1–46. DP: 31 December 1990 (paper title page).

Medvedev, L. N. 1992a: 105. Sem. Chrysomelidae – Listoedy. [Family Chrysomelidae – leaf beetles]. Pp. 533–602. In: Ler, P. A. (ed.): *Opredelitel' nasekomykh Dal'nego Vostoka SSSR v shesti tomakh. Tom III Zhestkokrylye, ili zhuki. Chast' 2. [Identification key to insects of Far East of*

USSR in six volumes. Volume III Coleoptera. Part 2]. Sankt Petersburg: Nauka, 704 pp. SFP: 28 January 1992, DP: 18 September 1992 (registered at ZIN library).

Medvedev, L. N. 1992b: Novye i maloizvestnye vidy listoedov (Coleoptera, Chrysomelidae) s ostrovov Vietnamskogo poberezhya. (New and little-known chrysomelid species (Coleoptera, Chrysomelidae) from Vietnamese coastal islands). Pp. 71–77. In: Medvedev, L. N. (ed.): *Sistematika i ekologia nasekomykh Vietnama.* (*Systematization and ecology of insects of Vietnam*). Moskva, 264 pp. SFP: 12 February 1992, DP: 7 October 1992 (registered at ZIN library).

Medvedev, L. N. 1992c: Reviziya zhukov-listoedov podsemeystva Hispinae (Coleoptera, Chrysomelidae) fauny Vietnama. Chast' I. (Revision of the chrysomelid subfamily Hispinae (Coleoptera, Chrysomelidae) in the fauna of Vietnam. Part. 1). Pp. 127–156. In: Medvedev, L. N. (ed.): *Sistematika i ekologia nasekomykh Vietnama.* (*Systematization and ecology of insects of Vietnam*). Moskva, 264 pp. SFP: 12 February 1992, DP: 7 October 1992 (registered at ZIN library).

Medvedev, L. N. 1992d: Chrysomelidae from the Nepal Himalayas, III (Insecta: Coleoptera). *Stuttgarter Beiträge zur Naturkunde, Serie A* (*Biologie*) **485**: 1–36. DP: 1 December 1992 (paper title page).

Medvedev, L. N. 1992e: The Clytrinae (Coleoptera Chrysomelidae) of Israel. *Russian Entomological Journal* **1** (1): 51–55. DP: 25 December 1992 (journal website).

Medvedev, L. N. 1992f: New and interesting species of Chrysomelidae (Coleoptera) from South Asia. *Entomologica Basiliensia* **15**: 379–388. DP: 31 December 1992 (table of contents).

Medvedev, L. N. 1992g: [new taxa]. In: Medvedev, L. N. & Dubeshko, L. N.: *Opredelitel listoedov Sibiri.* [*Key to identification of leaf-beetles of Siberia*]. Irkutsk: Izdatel'stvo Irkutskogo Universiteta, 220 + [4] pp. SFP: 12 May 1992, DP: [31 December] 1992 (book title page).

Medvedev, L. N. 1993a: Further records of Clytrinae (Coleoptera: Chrysomelidae) from Arabia. *Fauna of Saudi Arabia* **13**: 130–136. DP: 30 April 1993 (table of contents).

Medvedev, L. N. 1993b: Alticinae of the Philippine Islands (Coleoptera, Chrysomelidae). Part 1. *Russian Entomological Journal* **2** (3/4): 41–58. DP: 13 December 1993 (journal website).

Medvedev, L. N. 1993c: New species of Chrysomelidae from South Asia from the Natural History Museum in Basel. *Entomologica Basiliensia* **16**: 359–376. DP: 29 December 1993 (table of contents).

Medvedev, L. N. 1994: Alticinae of the Philippine Islands (Coleoptera, Chrysomelidae). Part 2. *Russian Entomological Journal* **2** [1993] (5/6): 11–32. DP: 9 January 1994 (journal website).

Medvedev, L. N. 1995a: New species of Clytrinae (Coleoptera, Chrysomelidae) from China and Indochina. *Russian Entomological Journal* **3** [1994] (3/4): 87–91. DP: [30] June 1995 (journal website).

Medvedev, L. N. 1995b: New species of Alticinae (Coleoptera, Chrysomelidae) from South Asia. *Entomologica Basiliensia* **18**: 479–488. DP: 15 December 1995 (table of contents).

Medvedev, L. N. 1996: The Chrysomelidae of Arabia. *Fauna of Saudi Arabia* **15**: 211–263. DP: 15 April 1996 (table of contents).

Medvedev, L. N. 1997a: New species of the Alticinae (Coleoptera, Chrysomelidae) from Nepal and adjacent regions. *Elytra* 25 (1): 13–22. DP: 15 May 1997 (paper title page).

Medvedev, L. N. 1997b: New and interesting Chrysomelidae (Coleoptera) from the Philippines collected by Dr. M. Satô. *Japanese Journal of Systematic Entomology* 3 (1): 99–104. DP: 15 May 1997 (paper title page).

Medvedev, L. N. 1997c: New data on Alticinae (Coleoptera, Chrysomelidae) from the Philippines. *Russian Entomological Journal* 5 (1/4): 65–83. DP: [31] May 1997 (journal website).

Medvedev, L. N. 1997d: A new subgenus and species of Xuthea Baly from Burma (Coleoptera Chrysomelidae, Alticinae). *Atti della Società Italiana di Scienze Naturali e del Museo Civico di Storia Naturale in Milano* 137 [1996] (1/2): 75–78. DP: [30] September 1997 (paper title page).

Medvedev, L. N. 1997e: New data on the Chrysomelidae from Nepal. *Elytra* 25 (2): 255–266. DP: 15 November 1997 (paper title page).

Medvedev, L. N. 1997f: New records and new species of Chrysomelidae from Arabia. *Fauna of Saudi Arabia* 16: 319–326. DP: 30 November 1997 (table of contents).

Medvedev, L. N. 1997g: Novyy aberrantnyy rod Alticinae (Coleoptera, Chrysomelidae) – predstavitel' ekologicheskoy gruppy detritobiontov. (A new aberrant genus of Alticinae (Coleoptera, Chrysomelidae) as a representative of humicole ecological group). *Zoologicheskiy Zhurnal* 76 (10): 1218–1220. SFP: 19 September 1997, DP: 11 December 1997 (registered at IVB CAS library).

Medvedev, L. N. 1998a: A revision of the genus Miochira Lacordaire (Coleoptera: Chrysomelidae: Clytrinae). *Serangga* 3 (1): 23–38. DP: [31] January 1998 (issue cover).

Medvedev, L. N. 1998b: New Alticinae and Galerucinae (Coleoptera, Chrysomelidae) from South Asia. *Acta Zoologica Academiae Scientiarum Hungaricae* 43 [1997] (3): 207–211. DP: 15 February 1998 (last page of the paper).

Medvedev, L. N. 1998c: [new taxon]. In: Medvedev, L. N. & Sprecher-Uebersax, E.: New data on Chrysomelidae of Nepal (Insecta, Coleoptera). *Spixiana* 21 (1): 25–42. DP: 1 March 1998 (paper title page).

Medvedev, L. N. 1998d: Novyy vid roda Luperus (Coleoptera, Chrysomelidae) s Dal'nego Vostoka. (A new species of the genus Luperus (Coleoptera, Chrysomelidae) from the Far East). *Zoologicheskiy Zhurnal* 77 (5): 613–614. SFP: 22 April 1998, DP: 17 July 1998 (registered at ZIN library).

Medvedev, L. N. 1998e: To the knowledge of Zeugophorinae (Chrysomelidae) of the Old World. *Russian Entomological Journal* 6 [1997] (3/4): 65–69. DP: 21 October 1998 (journal website).

Medvedev, L. N. 1998f: New Chrysomelidae (Coleoptera) from Southeast Asia in the Hungarian Natural History Museum. *Annales Historico-Naturales Musei Nationalis Hungarici* 90: 163–174. DP: 31 December 1998 (Merkl et al. 2008).

Medvedev, L. N. 1999a: To the knowledge of Oriental Alticinae (Coleoptera: Chrysomelidae). Genera Euphitrea Baly, 1875, Chilocoristes Weise, 1895, Pentamesa Harold, 1876, and Maulika Basu & Sengupta, 1980. *Russian Entomological Journal* 7 [1998] (3/4): 147–156. DP: 2 July 1999 (journal website).

Medvedev, L. N. 1999b: Novyy aberrantnyy vid Chrysolina (Coleoptera, Chrysomelidae) iz Sibiri. (A new aberrant species Chrysolina (Coleoptera, Chrysomelidae) from Siberia).

Zoologicheskiy Zhurnal 78 (8): 1014–1015. SFP: 14 July 1999, DP: 21 October 1999 (registered at ZIN library).

Medvedev, L. N. 1999c: To the knowledge of Chrysomelidae (Coleoptera) from Nepal and adjacent regions. *Veröffentlichungen des Naturkundemuseums Erfurt* **18**: 181–187. DP: 20 December 1999 (editor of journal M. Hartmann, 2019, pers. comm.).

Medvedev, L. N. 2000a: Jacoby's types of Chrysomelidae (Coleoptera) from Burma in the Museo Civico di Storia Naturale "Giacomo Doria", Genoa. Part 1. *Annali del Museo Civico di Storia Naturale "Giacomo Doria"* **93** [1999–2000]: 167–184. DP: 25 January 2000 (table of contents).

Medvedev, L. N. 2000b: A revision of the genus Hoplasoma Jacoby, 1884 (Coleoptera: Chrysomelidae). *Russian Entomological Journal* 8 [1999] (2): 123–128. DP: 31 January 2000 (journal website).

Medvedev, L. N. 2000c: A revision of the group Blepharidini (Chrysomelidae: Alticinae) from the Oriental region. *Russian Entomological Journal* 8 [1999] (3): 175–184. DP: 13 April 2000 (journal website).

Medvedev, L. N. 2000d: To the knowledge of Oriental Chrysomelidae. *Russian Entomological Journal* 8 [1999] (4): 259–264. DP: 14 June 2000 (journal website).

Medvedev, L. N. 2000e: Chrysomelidae from the Nepal Himalayas, with revision of the genus Haplosomoides (Insecta: Coleoptera). *Stuttgarter Beiträge zur Naturkunde, Serie A* (*Biologie*) **616**: 1–32. DP: 15 November 2000 (paper title page).

Medvedev, L. N. 2000f: Chrysomelidae (Coleoptera) of Laos from the collection of the Hungarian Natural History Museum. *Annales Historico-Naturales Musei Nationalis Hungarici* **92**: 161–182. DP: 31 December 2000 (Merkl et al. 2008).

Medvedev, L. N. 2001: Chrysomelidae of southern Asia (Coleoptera). *Entomologica Basiliensia* **23**: 159–191. DP: 1 September 2001 (table of contents).

Medvedev, L. N. 2002a: New and poorly known Chrysomelidae (Coleoptera) from northern India. *Entomologica Basiliensia* **24**: 245–253. DP: 24 September 2002 (table of contents).

Medvedev, L. N. 2002b: Revision of the genus Podagricomela Heikertinger, 1924 (Coleoptera: Chrysomelidae: Alticinae). *Russian Entomological Journal* 11 (2): 209–214. DP: 18 October 2002 (journal website).

Medvedev, L. N. 2003a: New species of leaf beetles (Insecta: Coleoptera: Chrysomelidae) from Nepal. Pp. 317–322. In: Hartmann, M. & Baumbach, H. (eds.): *Biodiversität und Naturausstattung im Himalaya*. Erfurt: Verein der Freunde und Förderer des Naturkundemuseums Erfurt e. V., 389 pp., 16 pls. DP: 5 October 2003 (editor of the book M. Hartmann 2019, pers. comm.).

Medvedev, L. N. 2003b: Towards knowledge of the Indian Clytrinae (Coleoptera, Chrysomelidae). *Entomologica Basiliensia* **25**: 275–291. DP: 24 October 2003 (table of contents).

Medvedev, L. N. 2004a: Revision of the genus Colaspoides Laporte, 1833 (Chrysomelidae, Eumolpinae) from continental Asia. *Russian Entomological Journal* **12** [2003] (3): 257–297. DP: 27 March 2004 (journal website).

Medvedev, L. N. 2004b: Revision of the genus Cheilotoma Chevrolat, 1837 (Coleoptera: Chrysomelidae: Clytrinae). *Russian Entomological Journal* **13** (1/2): 35–39. DP: 27 August 2004 (journal website).

Medvedev, L. N. 2004c: New genera and species of Oriental Chrysomelidae (Coleoptera). *Entomologica Basiliensia* **26**: 325–338. DP: 24 November 2004 (table of contents).

Medvedev, L. N. 2004d: New species of Chrysomelidae (Coleoptera) from Nepal. *Veröffentlichungen des Naturkundemuseums Erfurt* **23**: 203–206. DP: 18 December 2004 (editor of journal M. Hartmann, 2019, pers. comm.).

Medvedev, L. N. 2005a: A new genus and new species of Eumolpinae (Coleoptera, Chrysomelidae) from Turkey. *Entomologische Blätter* 100 [2004] (3): 191–196. DP: 30 May 2005 (paper title page).

Medvedev, L. N. 2005b: A revision of the continental Asian species of the genus Atysa Baly, 1864 (Chrysomelidae, Galerucinae). *Entomologica Basiliensia et Collectionis Frey* **27**: 227–237. DP: 15 December 2005 (last page of the cover).

Medvedev, L. N. 2005c: New and poorly-known genera and species of Oriental Chrysomelidae (Coleoptera). *Entomologica Basiliensia et Collectionis Frey* **27**: 279–295. DP: 15 December 2005 (last page of the cover).

Medvedev, L. N. 2006: K poznaniyu listoedov (Coleoptera, Chrysomelidae) Rossii. (Contribution to the knowledge of the leaf-beetles (Coleoptera, Chrysomelidae) of Russia). *Entomologicheskoe Obozrenie* 85 (2): 385–388. SFP: 2 June 2006, DP: 7 September 2006 (registered at ZIN library).

Medvedev, L. N. 2007a: To the knowledge of Chrysomelidae (Coleoptera) described by V. Motschulsky,. *Russian Entomological Journal* 15 [2006] (4): 409–417. DP: 9 February 2007 (journal website).

Medvedev, L. N. 2007b: Revision of the genus Eudolia Jacoby, 1885 (Chrysomelidae: Alticinae). *Russian Entomological Journal* 16 (1): 75–78. DP: 31 March 2007 (journal website).

Medvedev, L. N. 2007c: New species of Oriental Chrysomelidae (Coleoptera). *Entomologica Basiliensia et Collectionis Frey* **29**: 289–305. DP: 20 December 2007 (table of contents on the cover).

Medvedev, L. N. 2007d: New taxa of Oriental Chrysomelidae (Coleoptera). *Euroasian Entomological Journal* 6 (4): 433–438. DP: 28 December 2007 (journal website).

Medvedev, L. N. 2008a: To the knowledge of the genus Adiscus Gistl, 1857 (Coleoptera: Chrysomelidae: Cryptocephalinae). *Russian Entomological Journal* 17 (2): 193–205. DP: 12 November 2008 (journal website).

Medvedev, L. N. 2008b: New and poorly known Clytrinae and Cryptocephalinae (Coleoptera, Chrysomelidae) from the Institut Royal des Sciences Naturalles de Belgique. *Bulletin de l'Institut Royal des Sciences Naturelles de Belgique, Entomologie* 78: 279–283. DP: 18 December 2008 (verso of the volume title page).

Medvedev, L. N. 2009a: Taksonomicheskie zametky po listoedam (Coleoptera, Chrysomelidae). (Taxonomical notes on leaf beetles (Coleoptera, Chrysomelidae)). *Euroasian Entomological Journal* 8 (1): 57–58. DP: 17 April 2009 (journal website).

Medvedev, L. N. 2009b: New species of leaf beetles from the collection of the Naturkundemuseum Erfurt (Insecta: Coleoptera: Chrysomelidae). Pp. 407–410. In: Hartmann, M. & Weipert, J. (eds.): *Biodiversität und Naturausstattung im Himalaya III*. Erfurt: Verein der

Freunde & Förderer des Naturkundemuseums Erfurt e. V., 477 pp. DP: 2 July 2009 (editor of the book M. Hartmann 2019, pers. comm.).

Medvedev, L. N. 2009c: A revision of the genus Cassena Weise, 1892 (Coleoptera, Chrysomelidae). *Entomologica Basiliensia et Collectionis Frey* **31**: 219–238. DP: 28 December 2009 (table of contents on the cover).

Medvedev, L. N. 2009d: New species of Chrysomelidae from Nepal in the collection of the Naturkundemuseum Erfurt (Insecta: Coleoptera). *Vernate* **28**: 409–412. DP: 29 December 2009 (editor of journal M. Hartmann, 2019, pers. comm.).

Medvedev, L. N. 2009e: *Alticinae of Indochina*. Moskva: KMK Scientific Press, 224 pp. SFP: 30 November 2009, DP: [31 December] 2009 (book title page).

Medvedev, L. N. 2010a: Chrysomelidae (Coleoptera) of high mountain regions of North-West Vietnam. *Russian Entomological Journal* **18** [2009] (3): 201–208. DP: 5 April 2010 (journal website).

Medvedev, L. N. 2010b: Revision of the genus Ryparida Baly, 1863 (Chrysomelidae: Eumolpinae) from continental Asia. *Russian Entomological Journal* **18** [2009] (4): 287–292. DP: 26 April 2010 (journal website).

Medvedev, L. N. 2010c: New and poorly known Alticinae (Coleoptera: Chrysomelidae) from New Guinea and islands of Southeast Asia. *Stuttgarter Beiträge zur Naturkunde A, Neue Serie* **3**: 291–304. DP: 30 April 2010 (paper title page).

Medvedev, L. N. 2010d: K faune zhukov listoedov (Coleoptera, Chrysomelidae) Lazovskogo zapovednika. (Leaf beetle (Coleoptera, Chrysomelidae) fauna of Lazovsky Natural Reserve). *Euroasian Entomological Journal* **9** (3): 485–488. DP: 22 November 2010 (journal website).

Medvedev, L. N. 2010e: A key to Oriental continental species of Smaragdina Chevrolat, 1936 (Chrysomelidae, Clytrinae). *Entomologica Basiliensia et Collectionis Frey* **32**: 259–287. DP: 20 December 2010 (table of contents on the cover).

Medvedev, L. N. 2010f: New and interesting Chrysomelidae (Insecta: Coleoptera) from the collection of the Naturkundemuseum Erfurt. *Vernate* **29**: 195–199. DP: 21 December 2010 (editor of journal M. Hartmann, 2019, pers. comm.).

Medvedev, L. N. 2011a: New and poorly known Oriental species of leaf beetles (Coleoptera: Chrysomelidae). *Russian Entomological Journal* **20** (2): 189–192. DP: 15 July 2011 (journal website).

Medvedev, L. N. 2011b: [new taxon]. In: Medvedev, L. N. & Khruleva, O. A.: K poznaniyu arkticheskikh form roda Chrysomela L. (Coleoptera, Chrysomelidae). (Contribution to the knowledge of the Arctic forms of the genus Chrysomela L. (Coleoptera, Chrysomelidae)). *Entomologicheskoe Obozrenie* **90** (3): 570–591. SFP: 6 September 2011, DP: 30 September 2011 (registered at ZIN library).

Medvedev, L. N. 2011c: New taxa of Chrysomelidae (Coleoptera) from Nepal. *Vernate* **30**: 249–254. DP: 18 December 2011 (editor of journal M. Hartmann, 2019, pers. comm.).

Medvedev, L. N. 2011d: New and poorly known species of Cryptocephalus Geoffroy, 1762 from the Palearctic and Oriental regions (Chrysomelidae, Cryptocephalinae). *Entomologica Basiliensia et Collectionis Frey* **33**: 329–349. DP: 22 December 2011 (table of contents on the cover).

Medvedev, L. N. 2011e: A contribution to the knowledge of Oriental species of Cneorane Baly, 1865 (Chrysomelidae, Galerucinae). *Entomologica Basiliensia et Collectionis Frey* **33**: 351–368. DP: 22 December 2011 (last page of the cover).

Medvedev, L. N. 2012a: Revision of the genus Aoria Baly, 1863 (Chrysomelidae: Eumolpinae) from China and Indochina. *Russian Entomological Journal* **21** (1): 45–52. DP: 23 May 2012 (journal website).

Medvedev, L. N. 2012b: To the knowledge of the genera Mandarella Duvivier and Stenoluperus Ogloblin (Insecta: Chrysomelidae: Alticinae) from the Himalayas. Pp. 423–427. In: Hartmann, M. & Weipert, J. (eds.): *Biodiversität und Naturausstattung im Himalaya IV*. Erfurt: Verein der Freunde und Förderer des Naturkundemuseums Erfurt e. V., 492 pp. DP: 15 August 2012 (editor of the book M. Hartmann 2019, pers. comm.).

Medvedev, L. N. 2012c: New species of Luperus Geoffroi, 1762 (Coleoptera: Chrysomelidae: Galerucinae) from China and Vietnam. *Caucasian Entomological Bulletin* **8** (2): 252–253. DP: 12 December 2012 (journal website).

Medvedev, L. N. 2012d: New and interesting Chrysomelidae (Insecta: Coleoptera) from the collection of the Naturkundemuseum Erfurt. *Vernate* **31**: 501–515. DP: 20 December 2012 (editor of journal M. Hartmann, 2019, pers. comm.) [note: widely distributed pdf has different pagination 489–503]

Medvedev, L. N. 2012e: New species of leaf-beetles (Chrysomelidae) from Arabia. *Russian Entomological Journal* **21** (4): 415–418. DP: 29 December 2012 (journal website).

Medvedev, L. N. 2013a: New species of Calomicrus Stephens, 1834 (Chrysomelidae: Galerucinae) from China and Indochina. *Russian Entomological Journal* **22** (1): 37–42. DP: 6 May 2013 (journal website).

Medvedev, L. N. 2013b: New taxa of Oriental leaf beetles (Coleoptera: Chrysomelidae). *Russian Entomological Journal* **22** (4): 293–296. DP: 6 December 2013 (journal website).

Medvedev, L. N. 2013c: New and poorly known Oriental Chrysomelidae (Coleoptera). *Acta Biologica Universitatis Daugavpiliensis* **13** (2): 85–87. DP: 16 December 2013 (verso of the issue cover).

Medvedev, L. N. 2013d: A key for identification with description of new species of the genus Pyrrhalta Joannis, 1865 (Coleoptera: Chrysomelidae: Galerucinae) from Indochina. *Caucasian Entomological Bulletin* **9** (2): 267–272. DP: 25 December 2013 (journal website).

Medvedev, L. N. 2013e: To the knowledge of the species of Chlamisus pilifrons-group (Insecta: Coleoptera: Chrysomelidae: Chlamisinae). *Vernate* **32**: 421–424. DP: 27 December 2013 (colophon).

Medvedev, L. N. 2013f: Novye vidy listoedov roda Trichomimastra Weise, 1922 (Coleoptera, Chrysomelidae) iz Pakistana. (New species of the chrysomelid genus Trichomimastra Weise, 1922 (Coleoptera, Chrysomelidae) from Pakistan). *Entomologicheskoe Obozrenie* **92** (4): 777–779. SFP: 28 November 2013, DP: 31 December 2013 (based on invoice deposited in Russian Entomological Society, A. G. Moseyko, 2022, pers. comm.).

Medvedev, L. N. 2014: New and poorly known leaf beetles (Insecta: Coleoptera: Chrysomelidae) from China and Indochina. *Vernate* **33**: 237–241. DP: 29 December 2014 (colophon).

Medvedev, L. N. 2015a: New and poorly known Chrysomelidae (Insecta: Coleoptera) from the Himalayas. Pp. 475–478. In: Hartmann, M. & Weipert, J. (eds.): *Biodiversität und Naturausstattung im Himalaya V.* Erfurt: Verein der Freunde und Förderer des Naturkundemuseums Erfurt e. v., 580 pp. DP: 15 August 2015 (colophon).

Medvedev, L. N. 2015b: To the knowledge of leaf beetles (Coleoptera: Chrysomelidae) from Turkey. *Caucasian Entomological Bulletin* 11 (2): 391–394. DP: 28 November 2015 (journal website).

Medvedev, L. N. 2015c: New and poorly known Oriental Chrysomelidae (Insecta: Coleoptera) in the collection of the Naturkundemuseum Erfurt. *Vernate* **34**: 319–335. DP: 30 November 2015 (editor of journal M. Hartmann, 2019, pers. comm.).

Medvedev, L. N. 2016: New and poorly known Oriental Alticinae (Coleoptera: Chrysomelidae). *Russian Entomological Journal* 25 (3): 259–264. DP: 27 September 2016 (journal website).

Medvedev, L. N. 2019a: New and poorly known Oriental Chrysomelidae (Coleoptera). *Russian Entomological Journal* 28 (2): 165–168 DP: 29 June 2019 (journal website).

Medvedev, L. N. 2019b: To the knowledge of Oriental leaf beetles (Coleoptera: Chrysomelidae). *Russian Entomological Journal* 28 (3): 282–285. DP: 27 September 2019 (journal website).

Medvedev, L. N. & Chernov, Y. I. 1969: Novyy vid listoeda roda Chrysomela (Coleoptera, Chrysomelidae) – vazhnyy komponent biocenozov Taymyra. (A new species of Chrysomela (Coleoptera, Chrysomelidae) – an important component of the Taimyr Peninsula biocoenoses). *Zoologicheskiy Zhurnal* 48 (4): 532–537. SFP: 1 April 1969, DP: 29 April 1969 (registered at ZIN library).

Medvedev, L. N. & Dang, D. T. 1981: Novye rody i vidy zhukov-listoedov podsem. Galerucinae (Coleoptera, Chrysomelidae) iz V'etnama. (New genera and species of leaf-beetles of the subfamily Galerucinae (Coleoptera, Chrysomelidae) from Vietnam). *Entomologicheskoe Obozrenie* 60 (3): 629–635. SFP: 4 September 1981, DP: 19 October 1981 (registered at ZIN library).

Medvedev, L. N. & Dubeshko, L. N. 1972: [new taxon]. In: Dubeshko, L. N. & Medvedev, L. N.: Novyy vid listoeda (Coleoptera Chrysomelidae) – vreditel' ol'khi v Pribaykal'e. [New species of leaf-beetle (Coleoptera Chrysomelidae) – pest of alder in Pribaykalsky District]. Pp. 79–87. In: *Ekologiya vrednykh i poleznykh nasekomykh.* [*Ecology of harmful and useful insects*]. Voronezh: Central'no-chernozemnoe knizhnoe izdatel'stvo, 160 pp. SFP: 17 March 1972, DP: 29 May 1972 (registered at VKP).

Medvedev, L. N. & Dubeshko, L. N. 1992: *Opredelitel listoedov Sibiri.* [*Key to identification of leaf-beetles of Siberia*]. Irkutsk: Izdatel'stvo Irkutskogo Universiteta, 220 + [4] pp. SFP: 12 May 1992, DP: [31 December] 1992 (book title page).

Medvedev, L. N., El Torkey, A. M. & Al Dhafer, H. M. 2014: A new genus of Clytrinae from Saudi Arabia (Coleoptera: Chrysomelidae). *Entomologische Zeitschrift* **124** (1): 7–9. DP: 15 March 2014 (table of contents).

Medvedev, L. N. & Eroshkina, G. A. 1982: Fauna i ekologiya Cassidinae (Coleoptera, Chrysomelidae) plato Tay-Nguyen. [Fauna and ecology of Cassidinae (Coleoptera, Chrysomelidae) of

Tay-Nguyen plateau]. Pp. 103–115. In: Medvedev, L. N. (ed.): *Zhivotnyy mir V'etnama.* [*Animal world of Vietnam*]. Moskva: Nauka, 168 pp. SFP: 20 July 1982, DP: [31 December] 1982 (book title page).

Medvedev, L. N. & Eroshkina, G. A. 1988: Reviziya listoedov-shchitonosok (Chrysomelidae, Cassidinae) fauny V'etnama. [Revision of hispid-beetles (Chrysomelidae, Cassidinae) of Vietnam fauna]. Pp. 105–142. In: Medvedev, L. N. & Striganova, B. R. (eds.): *Fauna i ekologiya nasekomykh V'etnama.* [*Fauna and ecology of insects of Vietnam*]. Moskva: Nauka, 197 + [3] pp. SFP: 20 April 1988, DP: 26 September 1988 (registered at ZIN library).

Medvedev, L. N. & Kantner, F. 2002: Some new and poorly known Clytrinae (Coleoptera, Chrysomelidae) of the Old World. *Entomologica Basiliensia* **24**: 259–269. DP: 24 September 2002 (table of contents).

Medvedev, L. N. & Kantner, F. 2003: Description of a new species of Cheilotoma Chevrolat, 1837 (Coleoptera: Chrysomelidae) from Turkey. *Entomologische Zeitschrift* **113** (9): 268–269. DP: 18 September 2003 (table of contents).

Medvedev, L. N. & Katbeh-Bader, A. 2002: New species of Smaragdina Chevrolat, 1837 from Jordan (Chrysomelidae, Clytrinae). *Entomologica Basiliensia* **24**: 255–258. DP: 24 September 2002 (table of contents).

Medvedev, L. N. & Kulenova, K. Z. 1980: Novyy vid Cryptocephalus (Coleoptera, Chrysomelidae) iz Kazakhstana. [New species of Cryptocephalus (Coleoptera, Chrysomelidae) from Kazakhstan]. *Trudy Instituta Zoologii Akademii Nauk Kazakhskoy SSR* **39**: 119–122. SFP: 27 May 1980, DP: 13 August 1980 (registered at VKP).

Medvedev, L. N. & Khruleva, O. A. 2011: K poznaniyu arkticheskikh form roda Chrysomela L. (Coleoptera, Chrysomelidae). (Contribution to the knowledge of the Arctic forms of the genus Chrysomela L. (Coleoptera, Chrysomelidae)). *Entomologicheskoe Obozrenie* **90** (3): 570–591. SFP: 6 September 2011, DP: 30 September 2011 (registered at ZIN library).

Medvedev, L. N. & Mirzoeva, N. B. 1969: Novyy vid iz roda Galeruca Geoffr. (Chrysomelidae, Coleoptera) iz Azerbaydzhana. *Doklady Akademii Nauk Azerbaydzhanskoy SSR* **25** (9): 56–58. SFP: 29 December 1969, DP: [31 December] 1969 (volume and paper title pages).

Medvedev, L. N. & Okhrimenko, N. V. 1992: K poznaniyu zhukov-listoedov roda Chrysolina Motsch. (Coleoptera, Chrysomelidae) Kavkaza. (To the knowledge of leaf-beetles of the genus Chrysolina Motsch. (Coleoptera, Chrysomelidae) of the Caucasus). *Entomologicheskoe Obozrenie* 70 [1991] (4): 866–874. SFP: 14 January 1992, DP: 18 May 1992 (registered at ZIN library).

Medvedev, L. N. & Regalin, R. 1998: Nuove o interessanti specie di Clytrinae afrotropicali e orientali (Coleoptera Chrysomelidae). *Bollettino di Zoologia Agraria e di Bachicoltura, Serie II* 30 (1): 1–32. DP: 30 June 1998 (paper title page).

Medvedev, L. N. & Roginskaya, E. Ya. 1988: *Katalog kormovykh rasteniy listoedov SSSR.* [*Catalogue of host-plants of the USSR leaf-beetles*]. Moskva: Akademia Nauk SSR, 193 pp. SFP: 25 May 1988, DP: [31 December] 1988 (book title page).

Medvedev, L. N. & Romantsov, P. 2012: New and poorly known Chrysomelidae from the Oriental Region (Coleoptera). *Entomologische Zeitschrift* **122** (2): 75–78. DP: 16 April 2012 (table of contents).

Medvedev, L. N. & Romantsov, P. V. 2013: New and poorly known Chrysomelidae (Coleoptera) from South-East Asia. *Caucasian Entomological Bulletin* **9** (1): 137–140. DP: 4 August 2013 (journal website).

Medvedev, L. N. & Rybakova, N. A. 1980: New species and subspecies of Pachybrachys Redt. from Mongolia and China (Coleoptera, Chrysomelidae). *Reichenbachia* **18**: 43–48. DP: 10 December 1980 (paper title page).

Medvedev, L. N. & Samoderzhenkov, E. V. 1987: Fauna Cryptocephalinae (Coleoptera, Chrysomelidae) V'etnama. [Cryptocephalinae fauna (Coleoptera, Chrysomelidae) of Vietnam]. Pp. 20–40. In: Medvedev, L. N. (ed.): *Entomofauna V'etnama. [Entomofauna of Vietnam]*. Moskva: Nauka, 192 pp. SFP: 20 November 1986, DP: 7 September 1987 (registered at ZIN library).

Medvedev, L. N. & Samoderzhenkov, E. V. 1989: New Galerucinae from Vietnam (Coleoptera, Chrysomelidae). *Entomofauna* 10 (29): 453–462. DP: 5 December 1989 (paper title page).

Medvedev, L. N. & Samoderzhenkov, E. V. 1998: Revision of Paridea Baly, 1886 (Chrysomelidae, Galerucinae) from the Himalaya and adjacent regions. *Russian Entomological Journal* **6** [1997] (1/2): 57–65. DP: 20 October 1998 (journal website).

Medvedev, L. N. & Seperteladze, M. P. 1962: Novyy podvid Cryptocephalus ergenensis Mor. iz Gruzii. [A new subspecies of Cryptocephalus ergenensis Mor. from Georgia]. *Soobshcheniya Akademii Nauk Gruzinskoy SSR* **28** (6): 699–700. DP: 15 September 1962 (registered at VKP).

Medvedev, L. N. & Sprecher-Uebersax, E. 1997: Chrysomelidae of Nepal and neighbouring regions. (Coleoptera: Chrysomelidae). *Coleoptera* **1**: 203–247. DP: 20 December 1997 (reprint cover).

Medvedev, L. N. & Sprecher-Uebersax, E. 1998: New data on Chrysomelidae of Nepal (Insecta, Coleoptera). *Spixiana* **21** (1): 25–42. DP: 1 March 1998 (paper title page).

Medvedev, L. N. & Sprecher-Uebersax, E. 1999: Taxonomical study of Chrysomelidae (Coleoptera) from Nepal. *Entomologica Basiliensia* **21**: 355–370. DP: 20 April 1999 (table of contents).

Medvedev, L. N. & Voronova, N. V. 1977: Obzor roda Thelyterotarsus Wse. (Coleoptera, Chrysomelidae) fauny Mongolii. (Review of the genus Thelyterotarsus Wse. (Coleoptera, Chrysomelidae) from Mongolia). Pp. 308–318. In: Kerzhner, I. M. (ed.): *Nasekomye Mongolii 5*. (*Insects of Mongolia 5*). St. Petersburg: Academy of Sciences of the USSR, Zoological Institute, 757 pp. DP: 2 September 1977 (Kerzhner 1994).

Medvedev, L. N. & Voronova, N. V. 1979: K faune listoedov (Coleoptera, Chrysomelidae) Mongolii, 3. (On the chrysomelid-beetle fauna of Mongolia, 3 (Coleoptera, Chrysomelidae)). Pp. 97–134. In: Kerzhner, I. M. (ed.): *Nasekomye Mongolii 6*. (*Insects of Mongolia 6*). St. Petersburg: Academy of Sciences of the USSR, Zoological Institute, 512 pp. DP: 26 June 1979 (Kerzhner 1994).

Medvedev, L. N. & Yavorskaya, M. I. 2008: Monolepta septentrionalis – novyy vid s Dal'nego Vostoka Rosii (Coleoptera, Chrysomelidae, Galerucinae). (Monolepta septentrionalis, a new species from the Russian Far East (Coleoptera, Chrysomelidae, Galerucinae)). *Euroasian Entomological Journal* 7 (4): 284–286. DP: 31 December 2008 (journal website).

Medvedev, L. N. & Zoia, S. 1996: New and interesting species of Eumolpinae from Pakistan and Thailand (Coleoptera Chrysomelidae). *Bollettino di Zoologia Agraria e di Bachicoltura, Serie II* 28 (2): 111–117. DP: 30 December 1996 (last page of the paper).

Medvedev, V. L. 1999: New Oriental forms of the genus Gonioctena Chevrolat (Coleoptera: Chrysomelidae: Chrysomelinae). *Serangga* 4 (1): 11–16. DP: [31] January 1999 (issue cover).

Medvedev, V. L. 2004: New species of the genus Gonioctena Chevr. 1837 (Coleoptera: Chrysomelidae) from China. *Russian Entomological Journal* 13 (1/2): 41–42. DP: 27 August 2004 (journal website).

Megerle von Mühlfeld, J. K. 1815: Bemerkungen, Berichtigungen und Zusaetze zu Illigers Zusaetzen, Berichtigungen und Bemerkungen zu Fabricii Systema Eleutheratorum, Linz 1812. 8v. 44 S. *Magazin der Entomologie* 1 (2): 135–179. DP: [31 December] 1815 (issue title page).

Meier, W. 1900a: Timarcha v. Frascassii – Chrysomela sirentensis – Hippodamia v. equiseti. *Entomologische Nachrichten* 26 (5): 78. DP: [31] March 1900 (issue title page).

Meier, W. 1900b: Timarcha Schenklingi n. sp. *Entomologische Nachrichten* 26 (11): 161–162. DP: [30] June 1900 (issue title page).

Melichar L. 1912a: Neue palaearktische Coleopteren. *Coleopterologische Rundschau* 1 (3): 33–36. DP: [31] March 1912 (*Ent. Litteraturbl.* 1912, 12 (5): 89).

Melichar L. 1912b: Eine neue Kytorrhinus-Art (Col. Lariidae) aus dem nordwestlichen Kaukasus. *Coleopterologische Rundschau* 1 (12): 170–172. DP: [31] December 1912 (*Ent. Litteraturbl.* 1913, 13 (2): 33).

Melsheimer F. E. 1847: Descriptions of new species of Coleoptera of the United States. *Proceedings of the Academy of Natural Sciences of Philadelphia* 3 (7): 158–181. DP: 7 May 1847 (Nolan 1913).

Ménétriés, E. 1832: *Catalogue raisonné des objets de zoologie recueillis dans un voyage au Caucase et jusqu'aux frontières actuelles de la Perse entrepris par l'ordre de S. M. l'Empereur.* St.-Pétersbourg: Académie Impérial des Sciences, xxxiii + 272 + iv + [1] pp., 5 pls. DP: [30] September 1832 (Bousquet 2016).

Ménétriés, E. 1836: Insectes nouveaux de la Turquie. *Bulletin Scientifique de l'Académie Impériale des Sciences de Saint-Pétersbourg* 1 (19): 149–151. DP: 15 November 1836 (last page of the issue).

Ménétriés, E. 1837: Sur quelques insectes de la Russie. *Bulletin Scientifique de l'Académie Impériale des Sciences de Saint-Pétersbourg* 1 (23): 180–181. DP: 31 January 1837 (last page of the issue).

Ménétriés, E. 1848: *Descriptions des insectes recueillis par feu M. Lehmann.* St.-Pétersbourg: Académie impériale des sciences, 112 + [6] pp., pls. I–VI. DP: [31 December] 1848 (preprint title page) [preprint of Ménétriés E. 1849: Catalogue des insectes recueillis par feu M. Lehmann avec les descriptions des nouvelles espèces. Seconde et dernière partie. *Mémoires de l'Académie Impériale des Sciences de St. Pétersbourg, Sixième Série* 6: 217–328, pls. III–V]

Ménétriés, E. 1851: Insecten. Pp. 43–76, pl. 3. In: *Reise in den äussersten Norden und Osten Sibiriens während der Jahre 1843 und 1844 mit allerhöchster Genehmigung auf Veranstaltung der Kaiserlichen Akademie der Wissenschaften zu St. Petersburg ausgeführt und in Verbindung mit vielen Gelehrten heraugegeben von Dr. A. Th. v. Middendorff. Zweiter Band. Teil 1.* St. Petersburg: Eggers & Comp., v + 516 pp. + 32 pls. DP: 12 October 1851 (verso of second title page, Bousquet 2016).

Ménétriés, E. 1854: Coléoptères recueillis dans la Mongolie Chinoise et aux environs de Pékin. *Etudes Entomologiques* 3: 26–41. DP: [31] March 1854 (Bousquet 2016).

Méquignon A. 1945: Notes diverses sur les coléoptères de France (3e note). *Bulletin de la Société Entomologique de France* 50 (2): 29–31. DP: 16 June 1945 (*dates de publication* on p. 144).

Méquignon A. 1948: Notes diverses sur des coléoptères de France (Sixième série). *Bulletin de la Société Entomologique de France* 53 (5/6): 75–80. DP: 13 August 1948 (*dates de publication* on p. 160).

Merkl E. 1897: Adalék Biharmegye bogár-faunájához. II. (Beitrag zur Käferfauna von Bihar. II). *Rovartani Lapok* 4 (10): 207–210. DP: [31] December 1897 (issue cover).

Meschnigg J. 1934: Beitrag zur Kenntnis der Coleopteren vom Balkan. *Entomologischer Anzeiger* 14 (6): 65–68. DP: 10 June 1934 (issue title page).

Mikhailov, Y. E. 2000a: A new species of the genus Cryptocephalus Geoffr., 1762 (Coleoptera: Chrysomelidae) from the Russian Far East. *Russian Entomological Journal* 8 [1999] (2): 129–130. DP: 31 January 2000 (journal website).

Mikhailov, Y. E. 2000b: New and little known leaf beetles of the genus Chrysolina Motschulsky, 1860 from Altai and Sayany mountains in South Siberia (Coleoptera: Chrysomelidae). *Genus* 11 (2): 129–146. DP: 30 June 2000 (paper title page).

Mikhailov, Y. E. 2001: Ecology, preimaginal stages of Chrysolina (Altailina) dudkoi Mikhailov, 2000, and keys to similar species from the Altai Mountains (Coleoptera: Chrysomelidae). *Genus* 12 (3): 325–334. DP: 30 September 2001 (paper title page).

Mikhailov, Y. E. 2002: Contribution to the knowledge of the subgenus Chrysolina (Pezocrosita Jacobson, 1901): three new species from Tuva and the Urals (Coleoptera, Chrysomelidae). *Nouvelle Revue d'Entomologie, Nouvelle Série* 19 (1): 57–71. DP: 26 August 2002 (last page of the cover).

Mikhailov, Y. E. 2003: Vzglyad na podrody Timarchoptera Motschulsky, 1860 i Paraheliostola L. Medvedev, 1992 roda Chrysolina Motschulsky posle opisaniya dvukh novykh form iz Khakasii (Coleoptera, Chrysomelidae). (A view on the subgenera Timarchoptera Motschulsky, 1860 and Paraheliostola L. Medvedev, 1992 of the genus Chrysolina Motschulsky after the description of two new forms from Khakassia (Coleoptera, Chrysomelidae)). *Euroasian Entomological Journal* 1 [2002] (2): 219–228. DP: 15 February 2003 (journal website).

Mikhailov, Y. E. 2006a: Novyy vid iz podroda Anopachys Motschulsky, 1860 roda Chrysolina Motschulsky, 1860 (Coleoptera, Chrysomelidae) iz yuzhnogo Primor'ya. (A new species of the subgenus Anopachys Motschulsky, 1860 of the genus Chrysolina Motschulsky,

1860 (Coleoptera, Chrysomelidae) from South Maritime Province of Russia). *Euroasian Entomological Journal* 5 (2): 135–136. DP: 30 June 2006 (journal website).

Mikhailov, Y. E. 2006b: O tipakh i statuse Chrysolina sylvatica (Gebler, 1823) i Chrysolina subcostata (Gebler, 1848) iz podroda Pleurosticha Motschulsky, 1860 (Coleoptera, Chrysomelidae). (On the types and status of Chrysolina sylvatica (Gebler, 1823) and Chrysolina subcostata (Gebler, 1848) from the subgenus Pleurosticha Motschulsky, 1860 (Coleoptera, Chrysomelidae)). *Euroasian Entomological Journal* 5 (4): 292–302. DP: 29 December 2006 (journal website).

Mikhailov, Y. E. 2007a: O severnoy granitse rasprostraneniya zhukov-listoedov roda Oreomela (Coleoptera: Chrysomelidae): novye vidy iz gor yuzhnoy Sibiri. (On the northernmost distribution of leaf beetles of the genus Oreomela (Coleoptera, Chrysomelidae): new species from mountains of southern Siberia). *Zoologicheskiy Zhurnal* 86 (4): 434–443. SFP: 5 March 2007, DP: 15 June 2007 (registered at ZIN library).

Mikhailov, Y. E. 2007b: Novyy vid roda Chrysolina Motsch. (Coleoptera, Chrysomelidae) iz gor Tuvy i Buryatii. (A new species of the genus Chrysolina Motsch. (Coleoptera, Chrysomelidae) from the mountains of Tuva and Buryatia). *Entomologicheskoe Obozrenie* 86 (3): 683–686. SFP: 28 August 2007, DP: 29 September 2007 (based on invoice deposited in Russian Entomological Society, A. G. Moseyko 2022, pers. comm.).

Mikhailov, Y. E. 2007c: K poznaniyu podroda Pleurosticha Motschulsky, 1860 (Coleoptera, Chrysomelidae, Chrysolina). 2. Vidy s yuzhnogo kraya areala. (Contribution to our knowledge of the subgenus Pleurosticha Motschulsky, 1860 (Coleoptera, Chrysomelidae, Chrysolina). 2. Species from the southern edge of their distribution). *Euroasian Entomological Journal* 6 (3): 255–263, pl. VIII. DP: 23 November 2007 (journal website).

Mikhailov, Y. E. 2007d: Zhuki-listoedy iz arkto-al'piyskikh podrodov Arctolina Kontkanen i Pleurosticha Motschulsky roda Chrysolina Motschulsky na Urale (Coleoptera, Chrysomelidae). [Leaf-beetles from arcto-alpine subgenera Arctolina Kontkanen and Pleurosticha Motschulsky of the genus Chrysolina Motschulsky in Ural mountains (Coleoptera, Chrysomelidae)]. *Izvestiya Chelyabinskogo Nauchnogo Centra* 34 [2006] (4): 110–114. SFP: 21 March 2007, DP: [31 December] 2007 (based on the SFP date).

Mikhailov, Y. E. 2009a: Vysokogorno-predgornye dizyunkcii i vysotno-poyasnye ekologicheskie rasy u zhukov-listoedov (Coleoptera, Chrysomelidae). [Alpine-foothill disjunctions and altitudinal ecological races in leaf beetles (Coleoptera Chrysomelidae)]. *Vestnik Orenburgskogo Gosudarstvennogo Universiteta* 2009 (5): 141–145. DP: [31] May 2009 (footers throughout the paper).

Mikhailov, Y. E. 2009b: Novye dannye po sistematike i ekologii zhukov-listoedov (Coleoptera, Chrysomelidae) iz gor Yuzhnoy Sibiri. (New data on systematics and ecology of leaf beetles (Coleoptera, Chrysomelidae) from South Siberian mountains). *Euroasian Entomological Journal* 8 (2): 177–188. DP: 17 August 2009 (journal website).

Mikhailov, Y. E. 2010: [new taxon]. In: Lopatin, I. K. & Mikhailov, Y. E.: K poznaniyu fauny zhukov-listoedov (Coleoptera, Chrysomelidae) vostochnogo Kazakhstana. (Contribution to the knowledge of the leaf-beetle fauna (Coleoptera, Chrysomelidae) of East Kazakhstan).

Entomologicheskoe Obozrenie 89 (3): 603–611. SFP: 26 August 2010, DP: 30 September 2010 (based on invoice deposited in Russian Entomological Society, A. G. Moseyko 2022, pers. comm.).

Mikhailov, Y. E. 2013: K poznaniyu zhukov-listoedov roda Cystocnemis (Coleoptera, Chrysomelidae). (To the knowledge of leaf-beetles of the genus Cystocnemis (Coleoptera, Chrysomelidae)). *Zoologicheskiy Zhurnal* 92 (3): 300–312. SFP: 11 February 2013, DP: 25 June 2013 (registered at ZIN library).

Mikhailov, Y. E. 2018: O dvukh maloizvestnykh vidakh zhukov-listoedov (Coleoptera, Chrysomelidae) s yuzhnogo Urala. (On two little-known species of leaf beetles (Coleoptera, Chrysomelidae) from the southern Urals). *Zoologicheskiy Zhurnal* 97 (3): 286–298. DP: 26 March 2018 (colophon).

Mikhailov, Y. E. 2019: Review of the genus Chrysolina Motschulsky, 1860 (Coleoptera: Chrysomelidae) from Nepal Himalaya with description of a new species. *Russian Entomological Journal* 28 (1): 58–63. DP: 30 March 2019 (journal website).

Mikhailov, Y. E. 2020: [new taxon]. In: Kippenberg, H. & Mikhailov, Y. E.: Contribution to the knowledge of Entomoscelis adonidis (Pallas, 1771) and allied species (Coleoptera: Chrysomelidae: Chrysomelinae). *Koleopterologische Rundschau* 90: 257–290. DP: [30] September 2020 (paper title page).

Mikhailov, Y. E. 2021a: [new taxon]. In: Mikhailov, Y. E., Simanchuk, E., Yunussova, G. & Kazkenova, G.: A new subspecies of Cryptocephalus ergenensis Morawitz, 1863 (Coleoptera, Chrysomelidae) from Kazakhstan and significance of colour pattern polymorphism examination. *Ecologica Montenegrina* 49: 79–87. DP: 23 December 2021 (paper title page).

Mikhailov, Y. E. 2021b: New and little known representatives of the genus Chrysolina Motschulsky, 1860 (Coleoptera: Chrysomelidae) from China. *Far Eastern Entomologist* 443: 6–16. DP: [31] December 2021 (paper title page).

Mikhailov, Y. E. 2022: Review of Chrysolina immarginata species group from the subgenus Chalcoidea (Coleoptera: Chrysomelidae) with description of a new species. *Russian Entomological Journal* 31 (3): 278–282. DP: 29 September 2022 (journal website).

Mikhailov, Y. E. 2023: Genus Apterocuris Jacobson, 1901 (Coleoptera: Chrysomelidae) in the subfamily Chrysomelinae with description of a new species and remarks on significance of preimaginal characters. *Diversity* 15 (228): 1–18. DP: 6 February 2023 (paper title page).

Mikhailov, Y. E. & Guskova, E. V. 2013a: Novyy vid Chrysolina Motschulsky, 1860 (Coleoptera: Chrysomelidae) – pervyy predstavitel' podroda Pleurosticha iz Mongolskogo Altaya. (New species of Chrysolina Motschulsky, 1860 (Coleoptera: Chrysomelidae) – the first representative of the subgenus Pleurosticha from Mongolian Altai). *Caucasian Entomological Bulletin* 9 (1): 125–129. DP: 4 August 2013 (journal website).

Mikhailov, Y. E. & Guskova, E. V. 2013b: Review of the genus Cystocnemis (Coleoptera: Chrysomelidae: Chrysomelinae) with descriptions of two new species. *Zootaxa* 3736 (5): 536–548. DP: 15 November 2013 (footer on the paper title page).

Mikhailov, Y. E. & Reshetnikov, S. V. 2021: Reviziya maloizuchennoy vidovoy gruppy zhukov-listoedov Chrysolina (sahlbergiana) podroda Pezocrosita (Coleoptera, Chrysomelidae).

(Revision of leaf beetles from the poorly studied Chrysolina (sahlbergiana) species group in the subgenus Pezocrosita (Coleoptera, Chrysomelidae). *Zoologicheskiy Zhurnal* 100 (9): 1007–1018. SFP: 24 July 2021, DP: [30] September 2021 (issue cover).

Mikšić, R. 1954: Jedan nov kromatski varijetet Phytodecta (Spartophila) fornicatus Brügg. (Eine neue chromatische Varietät der Phytodecta (Spartophila) fornicatus Brügg. *Zbornik Matice Srpske, Serija Prirodnih Nauka* 6: 81–84. DP: [31 December] 1954 (volume title page).

Milani, S. & Ruffo, S. 1963: Osservazioni sulla distribuzione geografica in Italia delle specie di Phytodecta del sottogenere Goniomena (Coleoptera Chrysomelidae). *Memorie del Museo Civico di Storia Naturale di Verona* 11: 105–123. DP: 31 December 1963 (reprint cover).

Miller, L. 1862: Ergebnisse einer entomologischen Reise nach Cephalonia. *Wiener Entomologische Monatschrift* 6 (9): 269–282. DP: [30] September 1862 (issue title page).

Miller, L. 1867: Timarcha Lomnickii n. sp. *Verhandlungen der Kaiserlich-Königlichen Zoologisch-Botanischen Gesellschaft in Wien* 17 (2/3): 503–504. DP: 2 October 1867 (Evenhuis 2023).

Miller, L. 1868: Eine entomologische Reise in die ostgalizischen Karpathen. *Verhandlungen der Kaiserlich-Königlichen Zoologisch-Botanischen Gesellschaft in Wien* 18 (1/2): 3–34. DP: [31 December] 1868 (Evenhuis 2023).

Miller, L. 1880: Bericht über eine im Frühling 1879 nach Dalmatien unternommene coleopterologische Reise. *Verhandlungen der Kaiserlich-Königlichen Zoologisch-Botanischen Gesellschaft in Wien* 30 (1): 1–8. DP: [31 December] 1880 (Evenhuis 2023).

Miller, L. 1881: [new taxon]. In: Reitter, E.: Neue und seltene Coleopteren, im Jahre 1880 in Süddalmatien und Montenegro gesammelt und beschrieben. *Deutsche Entomologische Zeitschrift* 25 (1): 177–230, pls. VI–VII. DP: [31] May 1881 (table of contents).

Minelli A. & Vittorelli S. 1976: Il rassenkreis Timarcha tenebricosa (F.) (Coleoptera Chrysomelidae). *Bollettino del Museo Civico di Storia Naturale di Verona* 3: 1–33. DP: 31 December 1976 (table of contents).

Mjöberg E. 1917: Results of Dr. E. Mjöberg's Swedish scientific expeditions to Australia 1910–1913. 14. Cetonidae, Rutelidae, Passalidae, Chrysomelidae: Subfam. Sagrinae, Cassidinae, Hispinae. *Arkiv för Zoologi* 11 [1917–1918] (3): 1–19. DP: 10 March 1917 (last page of the paper).

Mohamedsaid M. S. 1997: Kinabalua, a new genus of chrysomelid beetle from Sabah, Malaysia (Coleoptera: Chrysomelidae: Galerucinae). *Serangga* 2 (2): 131–135. DP: [30] June 1997 (issue cover).

Mohr, K.-H. 1958: Bemerkenswerte deutsche Halticinenfunde. *Mitteilungen der Deutschen Entomologischen Gesellschaft* 17 (3): 50–53. DP: 12 July 1958 (issue title page).

Mohr, K.-H. 1962: Neue palaearktische Halticinen (Col. Chrys. Halticinae). *Deutsche Entomologische Zeitschrift, Neue Folge* 9 (3/4): 316–324. DP: 30 August 1962 (table of contents).

Mohr, K.-H. 1965: Ergebnisse der Albanien-Expedition 1961 des Deutschen Entomologischen Institutes. 39. Beitrag. Coleoptera: Chrysomelidae I (Halticinae). *Beiträge zur Entomologie* 15 (5/6): 701–729. DP: [31] December 1965 (issue title page).

Mohr, K.-H. 1966: Bestimmungstabelle der Untergattung Pezocrosita Jacobson 1902 (Allohypericia Bechyné 1950) (Coleoptera, Chrysomelidae). *Reichenbachia* 8: 89–104. DP: 22 December 1966 (paper title page).

Mohr, K.-H. 1968a: Eine neue Phyllodecta Kirby 1837 (Phratora Chevr. 1837) aus dem Kaukasus (Col., Chrysomelidae). *Nachrichtenblatt der Bayerischen Entomologen* **17** (3): 47–50. DP: 15 June 1968 (issue title page).

Mohr, K.-H. 1968b: Beitrag zur Kenntnis der Gattung Epithrix Foudr. (Col. Chrys. Halt.). *Entomologische Blätter* **64** (1): 58–60. DP: 15 July 1968 (issue cover).

Mohr, K.-H. 1968c: Eine übersehene Exosoma-Art aus Kleinasien (Coleoptera, Chrysomelidae, Gal.). *Reichenbachia* 10: 239–243. DP: 5 September 1968 (paper title page).

Mohr, K.-H. 1969: Beitrag zur Chrysomelidenfauna Bulgariens (Coleoptera: Chrysomelidae). *Beiträge zur Entomologie* **19** (3/6): 389–394. DP: [31 December] 1969 (issue title page).

Mohr, K.-H. 1977: Beiträge zur Insektenfauna der DDR: Coleoptera – Chrysomelidae: Cryptocephalinae. *Beiträge zur Entomologie* **27** (2): 197–231. DP: [31 December] 1977 (issue title page).

Mohr, K.-H. 1981: Revision der paläarktischen Arten der Gattung Dibolia Latreille, 1829 (Coleoptera, Chrysomelidae, Halticinae). *Polskie Pismo Entomologiczne* **51** (3): 393–469. DP: 30 August 1981 (paper title page).

Mohr, K.-H. 1984: Zwei neue Longitarsus-Arten aus Asien (Coleoptera, Chrysomelidae, Halticinae). *Reichenbachia* **22** (2): 171–174. DP: 10 December 1984 (paper title page).

Moll C. E. von 1784a: Versuch eines Verzeichnisses der Salzburgischen Insecten. Fortsetzung. *Neues Magazin für die Liebhaber der Entomologie* (Fuessly) **2** (1): 27–44. DP: 26 April 1784 (*Allg. Verz. Bücher, Ostermesse* 1784: 840 and Evenhuis 2014).

Moll, C. E. von 1784b: Verzeichniss der Salzburger-Insekten. IIIte Lieferung. *Neues Magazin für die Liebhaber der Entomologie* (Fuessly) **2** (2): 169–198. DP: [31 December] 1784 (issue cover).

Monnot, E. 1913: Notes pour servir à l'étude des espèces françaises du genre Phyllotreta à élytres unicolores et description d'une espèce nouvelle. *Insecta, Revue Illustrée d'Entomologie* **3** (30): 223–235. DP: [30] June 1913 (issue title page).

Monnot, E. 1914: Description d'une nouvelle espèce appartenant au genre Phyllotreta (Col. Chrysomelidae). *Insecta, Revue Illustrée d'Entomologie* **4** (41): 170–172. DP: [31] May 1914 (issue title page).

Monrós, F. 1949: El genero "Mylassa" Stal (Col., Chrysomelidae, Cryptocephalinae). *Acta Zoologica Lilloana* **7**: 489–525. DP: 31 December 1949 (last page of the volume).

Monrós, F. 1951: Notas sobre Criocerinae (Col., Chrysomelidae). *Acta Zoologica Lilloana* **11**: 467–482. DP: 13 November 1951 (last page of the volume).

Monrós, F. 1953: Some corrections in the nomenclature of Clytrinae (Chrysomelidae). *Coleopterists Bulletin* **7** (6): 45–50. DP: 10 December 1953 (last page of the cover).

Monrós, F. 1956: Revision generica de Lamprosominae con descripcion de algunos generos y especies nuevas (Col., Chrysomelidae). *Revista Agronómica del Noroeste Argentino* (Tucumán) **2** (1): 25–77. DP: [31] July 1956 (title page of paper).

Monrós, F. 1960: Los generos de Chrysomelidae (Coleoptera). *Opera Lilloana* (Tucumán) **3** [1959]: 1–337, 3 pls. DP: 11 February 1960 (last page of the volume).

Monrós, F. & Bechyné J. 1956: Über einige verkannte Chrysomeliden-Namen. *Entomologischen Arbeiten aus dem Museum G. Frey* 7 (3): 1118–1137. DP: 15 December 1956 (issue table of contents).

Montagna, M. 2011: Pachybrachis sassii, a new species from the Mediterranean Giglio Island (Italy) (Coleoptera, Chrysomelidae, Cryptocephalinae). *ZooKeys* 155: 51–60. DP: 15 December 2011 (paper title page).

Montagna, M. & Sassi D. 2013: [new taxon]. In: Montagna, M., Sassi, D. & Giorgi, A.: Pachybrachis holerorum (Coleoptera: Chrysomelidae: Cryptocephalinae), a new species from the Apennines, Italy, identified by integration of morphological and molecular data. *Zootaxa* 3741 (2): 243–253. DP: 27 November 2013 (footer on the paper title page).

Montrouzier, X. 1856: Essai sur la faune de l'île de Woodlark ou Moiou. *Annales des Sciences Physiques et Naturelles, d'Agriculture et d'Industrie, Deuxième Série* 7 [1855] (1): 1–114. DP: 21 June 1856 (*Bibl. France* 1856, 45 (25): 663).

Montrouzier, X. 1865: [new taxon]. In: Perroud, B.-P. & Montrouzier, X.: Essai sur la faune entomologique de Kanala (Nouvelle-Calédonie) et description de quelques espèces nouvelles ou peu connues. *Annales de la Société Linnéenne de Lyon, Nouvelle Série* 11 [1864]: 46–257. DP: [29] February 1865 (volume title page).

Morawitz, F. 1860: Beschreibung einiger neuen Chrysomelidae aus der Umgegend von Sarepta. *Bulletin de la Société Impériale des Naturalistes de Moscou* 33 (1): 298–302. DP: [30] June 1860 (*Ztschr. Ges. Naturwiss.* 1860, 15 (6): 523).

Morawitz, F. 1861: Zur Kenntniss der russischen Eumolpiden. *Horae Societatis Entomologie Rossicae* 1: 159–164. DP: 16 September 1861 (Kerzhner 1984, Standfuss & Kerzhner 2004).

Morawitz, F. 1863: Ein paar kleinere Mittheilungen über russische Coleopteren. *Horae Societatis Entomologicae Rossicae* 2: 167–175. DP: [31 December] 1863 (Kerzhner 1984, Standfuss & Kerzhner 2004).

Morimoto, K. 1990: Bruchidius serricollis, a new species of Bruchidae from Japan (Coleoptera). *Esakia, Special Issue* 1: 93–95. DP: 20 April 1990 (paper title page).

Moseyko, A. G. 2012: Zamechaniya o rode Aphilenia Weise in Reitter, 1889 (Coleoptera, Chrysomelidae: Eumolpinae) s opisaniem novogo vida iz Rossii. (Comments on the genus Aphilenia Weise in Reitter, 1889 (Coleoptera, Chrysomelidae: Eumolpinae), with a description of a new species from Russia). *Entomologicheskoe Obozrenie* 91 (1): 99–105. SFP: 1 March 2012, DP: 28 March 2012 (based on invoice deposited in Russian Entomological Society, A. G. Moseyko 2022, pers. comm.).

Moseyko, A. G. 2013a: New data on the leaf beetle genus Aphilenia Weise in Reitter, 1889, with descriptions of two new species from Kazakhstan (Coleoptera: Chrysomelidae: Eumolpinae). *Caucasian Entomological Bulletin* 9 (1): 108–111. DP: 4 August 2013 (journal website).

Moseyko, A. G. 2013b: [new name]. In: Errata for Volume 6. Pp. 41–46. In: Löbl, I., Smetana, A. (eds.): *Catalogue of Palaearctic Coleoptera. Volume 8. Curculionoidea II*. Leiden-Boston: Brill, 700 pp. DP: 5 August 2013 (publisher website).

Moseyko, A. G. 2017: O sisteme trib zhukov-listoedov podsemeystva Eumolpinae (Coleoptera: Chrysomelidae). (On the tribe system of leaf beetles of the subfamily Eumolpinae (Coleoptera: Chrysomelidae)). Pp. 337–338. In: Barkalov, A. V. (ed.): *XV Congress of the Russian Entomological Society. Russia, Novosibirsk, July 31 – August 7, 2017. Materials of the congress.* Novosibirsk: Garamond, 576 pp. DP: [31 December] 2017 (book title page).

Moseyko, A. G. 2020: A revision of the genus Atomyria Jacobson, 1894 (Coleoptera: Chrysomelidae: Eumolpinae). *Caucasian Entomological Bulletin* **16** (2): 341–346. DP: 22 December 2020 (last page of the paper).

Moseyko, A. G. 2021: Notes on Asiatic Eumolpinae (Coleoptera, Chrysomelidae). *Entomological Review* 100 [2020] (6): 843–862. DP: 11 February 2021 (based on invoice received by editor-in-chief, A. G. Moseyko 2022, pers. comm.).

Moseyko, A. G., Guskova,, E. V., Kolov, S. V. & Orlova-Benkovskaya, M. Ya. 2018: New data on the distribution of Colasposoma dauricum (Mannerheim, 1849) (Coleoptera, Chrysomelidae) in Russia and Kazakhstan. *Entomological Review* **98** (3): 324–328. DP: 26 June 2018 (journal website).

Moseyko, A. G. & Kirejtshuk, A. G. 2013: On the taxonomic position of new taxa of the subfamily Eumolpinae (Coleoptera, Chrysomelidae) from Baltic amber. *Terrestrial Arthropod Reviews* **6** (1/2): 163–172. DP: 1 January 2013 (journal website).

Moseyko, A. G. & Medvedev, L. N. 2005: K sistematike roda Pagria Lefevre, 1884, s zamechaniyami po rodu Rhyparida Baly, 1861 (Coleoptera, Chrysomelidae, Eumolpinae). (On the systematics of the genus Pagria Lefevre, 1884, with comments on the genus Rhyparida Baly, 1861 (Coleoptera, Chrysomelidae, Eumolpinae)). *Entomologicheskoe Obozrenie* **84** (1): 108–124. SFP: 1 March 2005, DP: 22 April 2005 (registered at ZIN library).

Moseyko, A. G., Ponomarev, K. B., Teploukhov, V. Yu. & Knyazev, S. A. 2019: A review of the leaf-beetle fauna (Coleoptera, Chrysomelidae sensu lato) of Omsk Province. *Entomological Review* **98** [2018] (8): 1064–1087. DP: 22 February 2019 (journal website).

Moseyko, A. G. & Romantsov, P. V. 2022: O nekotorykh vidakh listoedov iz rodov Colaspoides Laporte, 1833 i Colaspedusa L. Medvedev, 1998 (Coleoptera, Chrysomelidae: Eumolpinae). (On some species of the leaf-beetle genera Colaspoides Laporte, 1833 and Colaspedusa L. Medvedev, 1998 (Coleoptera, Chrysomelidae: Eumolpinae)). *Entomologicheskoe Obozrenie* 101 (1): 148–162. DP: 1 April 2022 (based on invoice deposited in Russian Entomological Society, A. G. Moseyko 2022, pers. comm.).

Moseyko, A. G. & Sprecher-Uebersax,, E. 2010: New acts and comments. Chrysomelinae [sic!]: Eumolpinae. Pp. 80–83. In: Löbl, I. & Smetana, A. (eds.): *Catalogue of Palaearctic Coleoptera. Volume 6. Chrysomeloidea.* Stenstrup: Apollo Books, 924 pp. DP: 22 February 2010 (colophon).

Motschulsky, V. de [= Victor T.] 1838: Coléoptères du Caucase et des provinces Transcaucasiennes. *Bulletin de la Société Impériale des Naturalistes de Moscou* 11 (2): 175–188, pl. III. DP: [31] July 1838 (Bousquet & Davies 2021).

Motschulsky, V. de [= Victor T.] 1839: Insectes du Caucase et des provinces Transcaucasiennes. *Bulletin de la Société Impériale des Naturalistes de Moscou* 12 (1): 44–67, pls. I–II. DP: [31 December] 1839 (Bousquet & Davies 2021).

Motschulsky, V. de 1845a: Remarques sur la collection de coléoptères russes. Article I. *Bulletin de la Société Impériale des Naturalistes de Moscou* **18** (1): 1–127, pls. 5–7. [Corrections in **18** (2) Additum, post p. 549 (unnumbered)]. DP: 8 May 1845 (Bousquet & Davies 2021).

Motschulsky, V. de 1845b: Observations sur le Musée entomologique de l'Université Impériale de Moscou. *Bulletin de la Société Impériale des Naturalistes de Moscou* **18** (4): 332–388, pls. 5–7. DP: 25 December 1845 (Bousquet & Davies 2021).

Motschulsky, V. de 1846: Remarques sur la collection de coléoptères russes. 3me article. *Bulletin de la Société Impériale des Naturalistes de Moscou* **19** (2): 372–418. DP: 30 April 1846 (Bousquet & Davies 2021).

Motschulsky, V. de 1849: Coléoptères reçus d'un voyage de M. Handschuh dans le midi de l'Espagne, énumérés et suivis de notes. *Bulletin de la Société Impériale des Naturalistes de Moscou* **22** (3): 52–163. DP: 1 November 1849 (Bousquet & Davies 2021).

Motschulsky, V. de 1850: Lettre adressée au second secrétaire. *Bulletin de la Société Impériale des Naturalistes de Moscou* **23** (3): 354–368. DP: 31 October 1850 (Bousquet & Davies 2021).

Motschulsky, V. de 1852: Correspondance. Lettres adressées a Monsieur le premier secrétaire de la Société Imperiale des Naturalistes de Moscou. *Bulletin de la Société Impériale des Naturalistes de Moscou* **24** [1851] (4): 592–616. DP: 5 March 1852 (Bousquet & Davies 2021).

Motschulsky, V. de 1854a: Diagnoses de coléoptères nouveaux, trouvés par M. M. Tatarinoff et Gaschkéwitsch aux environs de Pékin. *Etudes Entomologiques* **2**: 44–51. DP: [31] March 1854 (Bousquet 2016).

Motschulsky, V. de 1854b: O gorokhovikakh (Bruchus, Linné), portyashchikh struchkovyya rasteniya. [On the pea beetles harmful to legume plants]. *Trudy Imperatorskago Vol'nago Ekonomicheskago Obshchestva* **1854** (1): 1–19. DP: [31 December] 1854 (issue title page).

Motschulsky, V. de 1857: Insectes du Japon. *Etudes Entomologiques* **6**: 25–41. DP: 26 November 1857 (Bousquet 2016).

Motschulsky, V. de 1858: Insectes des Indes orientales 1:ière serie. *Etudes Entomologiques* **7**: 20–122. DP: 30 December 1858 (Bousquet 2016).

Motschulsky, V. de 1859: Coléoptères du gouvernement de Iakoutsk, recueillis par M. Pavlofski. *Bulletin de la Classe Physico-Mathématique de l'Académie Impériale des Sciences de St.-Pétersbourg* **17** [1858–1859] (35/36): 567–574. DP: 27 June 1859 (last page of the issue).

Motschulsky, V. de 1860a: Catalogue des insectes rapportés des environs du fl. Amour, depuis la Schilka jusqu'à Nikolaëvsk. *Bulletin de la Société Impériales des Naturalistes de Moscou* **32** [1859] (4): 487–507. DP: 9 March 1860 (Bousquet & Davies 2021).

Motschulsky, V. de 1860b: Coléoptères rapportés de la Sibérie orientale et notamment des pays situés sur le bords du fleuve Amour par MM. Schrenck, Maack, Ditmar, Voznessenski etc. In: *Reisen und Forschungen im Amur-Lande in den Jahren 1854–1856 im Auftrage der Kaiserl. Akademie der Wissenschaften zu St. Petersburg ausgeführt und in Verbindung mit mehreren Gelehrten herausgegeben von Dr. Leopold v. Schrenck. Band II. Zweite Lieferung. Coleopteren.* St. Petersburg: Eggers & Comp., pp. 77–257 + [1], pls. VI–XI, 1 map. DP: [31] December 1860 (Bousquet 2016).

Motschulsky, V. de 1861a: Insectes du Japon. *Etudes Entomologiques* 9 [1860]: 4–39. DP: 24 October 1861 (Bousquet 2016).

Motschulsky, V. de 1861b: Diagnoses d'insectes nouveaux des rives du fl. Amur et de la Daourie méridionale. *Etudes Entomologiques* 9 [1860]: 39–41. DP: 24 October 1861 (Bousquet 2016).

Motschulsky, V. de 1862: Insectes du Japon. (Continuation). *Etudes Entomologiques* 10 [1861]: 3–24. DP: 9 April 1862 (*Öfv. Kongl. Vetens.-Akad. Forhand. 1862, 19: 270*).

Motschulsky, V. de 1866a: Catalogue des insectes reçus du Japon. *Bulletin de la Société Impériales des Naturalistes de Moscou* 39 (1): 163–200. DP: 2 August 1866 (*Monatsber. Königl. Preuss. Akad. Wiss. Berlin* 1866: 576).

Motschulsky, V. de 1866b: Essai d'un catalogue des insectes de l'ile de Ceylan. Supplément. *Bulletin de la Société Impériale des Naturalistes de Moscou* 39 (2): 393–446. DP: 1 November 1866 (Bousquet & Davies 2021).

Motschulsky, V. de 1868: Genres et espèces d'insectes, publiés dans différents ouvrages par Victor Motschoulsky. *Horae Societatis Entomologicae Rossicae* 6 (Supplément): 1–118. DP: [31 December] 1868 (supplement title page).

Motschulsky, V. de 1874: Enumération des nouvelles espèces de coléoptères rapportés de ses voyages. *Bulletin de la Société Impériale des Naturalistes de Moscou* 46 [1873] (4): 203–252. DP: 30 April 1874 (Bousquet & Davies 2021).

Mroczkowski M. 1990: [new name]. In: Burakowski, B., Mroczkowski, M. & Stefańska, J.: Chrząszcze Coleoptera, Stonkowate – Chrysomelidae, część 1. *Katalog Fauny Polski. Catalogus faunae Poloniae. Część XXIII, tom 16.* Warszawa: Polska Akademia Nauk, Instytut Zoologii, 279 pp. DP: [30] September 1990 (colophon).

Müller, J. 1916: Coleopterologische Beiträge zur Fauna der österreichischen Karstprovinzen und ihrer Grenzgebiete. *Entomologische Blätter* 12 (4/6): 73–109. DP: 6 July 1916 (issue title page).

Müller, J. 1921: Zur Systematik einiger phytophager Koleopteren, vorwiegend aus dem julischen und dalmatinischen Faunengebiet. *Koleopterologische Rundschau* 9 (4/9): 65–87. DP: 30 September 1921 (issue cover).

Müller, J. [= G.] 1924a: Nuovi coleotteri paleartici. *Bollettino della Società Entomologica Italiana* 56 (5): 68–80. DP: 15 May 1924 (Poggi 2008).

Müller, J. [= G.] 1924b: Una nuova Chrysomela. *Bollettino della Società Entomologica Italiana* 56 (8): 118. DP: 29 October 1924 (Poggi 2008).

Müller, J. [= G.] 1925: Nuovi coleotteri della regione adriatica. *Studi Entomologici* 1 [1925–1926] (1): 25–31. DP: [31 December] 1925 (table of contents).

Müller, J. [= G.] 1948: Contributo alla conoscenza dei coleotteri fitofagi (Cerambycidae e Chrysomelidae). *Atti del Museo Civico di Storia Naturale di Trieste* 17 [1948–1949] (2): 61–98. DP: 31 December 1948 (footer on the paper title page).

Müller, J. [= G.] 1951: *I coleotteri della Venezia Giulia. Catalogo ragionato con tabelle dicotomiche per la classificazione delle specie della regione Adriatica orientale, del Veneto e della Pianura Padana. Volume II: Coleoptera Phytophaga (Cerambycidae, Chrysomelidae, Bruchidae).*

[1949–1953], Trieste: Centro sperimentale agrario e forestale, 685 + [1] pp. DP (pp. 225–368): [30] November 1951 (last page of the book).

Müller, J. [= G.] 1952: *I coleotteri della Venezia Giulia. Catalogo ragionato con tabelle dicotomiche per la classificazione delle specie della regione Adriatica orientale, del Veneto e della Pianura Padana. Volume II: Coleoptera Phytophaga (Cerambycidae, Chrysomelidae, Bruchidae).* [1949–1953], Trieste: Centro sperimentale agrario e forestale, 685 + [1] pp. DP (pp. 369–480): [30] September 1952 (last page of the book).

Müller, J. [= G.] 1955: I Pachybrachis del grupo haliciensis Mill. ed hippophaës Suffr. (Col. Chrysomelidae). *Bollettino della Società Entomologica Italiana* 85 (5/6): 67–70. DP: 20 July 1955 (Poggi 2008).

Müller, O. F. 1764: *Fauna insectorum Fridrichsdalina, sive methodica descriptio insectorum agri fridrichsdalensis, cum characteribus genericis et specificis, nominibus trivialibus, iconibus allegatis, novisque pluribus speciebus additis.* Hafniae et Lipsiae: Gleditsch, xxiv + 96 pp. DP: 26 April 1764 (Bousquet 2016).

Müller, O. F. 1766: [new taxa]. In: Allioni, C.: Manipulus Insectorum Taurinensium. *Mélanges de Philosophie et de Mathématique de la Société Royale de Turin* 3 [1762–1765]: 185–198. DP: [31 December] 1766 (volume title page).

Müller, O. F. 1776: *Zoologiae Danicae prodromus, seu animalium Daniae et Norvegiae indigenarum characteres, nomina et synonyma imprimis popularium.* Hafniae: Hallageriis, xxxii + 282 pp. DP: 28 April 1776 (Bousquet 2017).

Mulsant, E. 1854: Description d'une nouvelle espèce du genre Chrysomela. *Mémoires de l'Académie Impériale des Sciences, Belles-Lettres et Arts de Lyon, Classe des Sciences, Nouvelle Série* 4: 18–20. DP: [31 December] 1854 (Bousquet 2016: 378).

Mulsant, E. 1856: Description d'une nouvelle espèce du genre Chrysomela. *Annales de la Société Linnéenne de Lyon, Nouvelle Série* 3: 12–14, 1 pl. DP: [31 December] 1856 (volume title page).

Mulsant, E. & Rey, C. 1852a: Description de deux espèces nouvelles du genre Cryptocephalus. *Opuscules Entomologiques* 1: 5–10, 1 pl. DP: 22 July 1852 (Bousquet 2016).

Mulsant, E. & Rey, C. 1852b: Description de deux espèces nouvelles du genre Cryptocephalus. *Mémoires de l'Académie Impériale des Sciences, Belles-Lettres et Arts de Lyon, Classe des Sciences, Nouvelle Série* 1 [1851]: 145–150, pl. I. DP: [31 December] 1852 (paper on p. 212 was presented to the Society on 12 January 1852).

Mulsant, E. & Rey, C. 1858: Étude sur les coléoptères du genre Bruchus qui se trouvent en France. *Opuscules Entomologiques* 8: 1–44. DP: [31 December] 1858 (Bousquet 2016).

Mulsant, E. & Rey, C. 1859: Description de deux cryptocéphalides nouveaux ou peu connus. *Opuscules Entomologiques* 9: 45–51. DP: 19 November 1859 (Bousquet 2016).

Mulsant, E. & Rey, C. 1861a: Description de quelques coléoptères nouveaux ou peu connus. *Annales de la Société Linnéenne de Lyon, Nouvelle Série* 7 [1860]: 300–345. DP: [28] February 1861 (volume title page).

Mulsant, E. & Rey, C. 1861b: Description de quelques coléoptères nouveaux ou peu connus. *Opuscules Entomologiques* 12: 47–92. DP: [31 December] 1861 (Bousquet 2016).

Mulsant, E. & Rey, C. 1874: Supplément aux altisides de feu M. Foudras. *Annales de la Société Linnéenne de Lyon, Nouvelle Série* 20 [1873]: 215–258. DP: 9 September 1874 (*Bull. Soc. Ent. Fr.* 1874: 205).

Mulsant, E. & Rey, C. 1876: Description d'une espèce d'altiside nouvelle ou peu connue. *Annales de la Société Linnéenne de Lyon, Nouvelle Série* 22 [1875]: 253–254. DP: 9 August 1876 (*Bull. Soc. Ent. Fr.* 1876: 167).

Mulsant, E. & Wachanru, A. 1850: Notes pour servir a l'historie du Cyrtonus rotundatus, suivies de la description de cet insecte et d'une espèce voisine. *Mémoires de l'Académie Royale des Sciences, Belles-Lettres et Arts de Lyon, Section des Sciences* 2 [1847]: 401–415, 1 pl. DP: 8 August 1850 (*Ber. Bekanntmach. Verh. Akad. Wiss. Berlin* 1850: 347).

Mulsant, E. & Wachanru, A. 1852: Première série de coléoptères nouveaux, ou peu connus. *Opuscules Entomologiques* 1: 161–177. DP: 22 July 1852 (Bousquet 2016).

Mulsant, E. & Wachanru, A. 1853a: Première série de coléoptères nouveaux, ou peu connus. *Mémoires de l'Académie Impériale des Sciences, Belles-Lettres et Arts de Lyon. Classe des Sciences, Nouvelle Série* 2 [1852]: 1–17. DP: [31 December] 1853 (Bousquet 2016: 378).

Mulsant, E. & Wachanru, A. 1853b: Description d'un coléoptère nouveau. *Opuscules Entomologiques* 2: 127–128. DP: [31 December] 1853 (Bousquet 2016).

Mulsant, E. & Wachanru, A. 1853c: Description d'un coléoptère nouveau. *Mémoires de l'Académie Impériale des Sciences, Belles-Lettres et Arts de Lyon. Classe des Sciences, Nouvelle Série* 3: 159–160. DP: [31 December] 1853 (Bousquet 2016: 378).

Münzner, J. 1934: Eine Aberration von Lilioceris lilii. *Entomologische Rundschau* 51 (3): 31. DP: 1 February 1934 (issue title page).

Nadein, K. 2005: Obzor zhukov-listoedov gruppy Psylliodes saulcyi All. (Coleoptera, Chrysomelidae, Alticinae). (A review of leaf beetles of the Psylliodes saulcyi species group (Coleoptera, Chrysomelidae, Alticinae)). *Entomologicheskoe Obozrenie* 84 (1): 125–139. SFP: 1 March 2005, DP: 22 April 2005 (registered at ZIN library).

Nadein, K. S. 2006: Novye vidy roda Psylliodes Latr. (Coleoptera, Chrysomelidae) iz Palearktiki. (New species of the genus Psylliodes Latr. (Coleoptera, Chrysomelidae) from the Palaearctic). *Entomologicheskoe Obozrenie* 85 (4): 862–870, 4 pls. SFP: 24 November 2006, DP: 28 December 2006 (registered at ZIN library).

Nadein, K. S. 2007a: On the taxonomy and classification of the genus Psylliodes Latreille, 1825 (Coleoptera, Chrysomelidae, Galerucinae). *Entomologica Basiliensia et Collectionis Frey* 29: 307–332. DP: 20 December 2007 (table of contents on the cover).

Nadein, K. S. 2007b: Review of the cucullatus species group of the genus Psylliodes Latreille (Coleoptera: Chrysomelidae: Galerucinae). *Genus* 18 (4): 637–660. DP: 28 December 2007 (paper title page).

Nadein, K. S. 2008: Review of the Psylliodes gibbosus species group, with descriptions of two new species (Coleoptera: Chrysomelidae: Galerucinae). *Koleopterologische Rundschau* 78: 333–366. DP: [31] July 2008 (paper title page).

Nadein, K. S. 2009: Revision of the genus Mniophila Stephens, 1831 (Coleoptera: Chrysomelidae). *Beiträge zur Entomologie* 59 (1): 103–131. DP: 15 July 2009 (paper title page).

Nadein, K. S. 2011: Revision of the genus Aeschrocnemis Weise, 1888 (Coleoptera, Chrysomelidae). *Zoosystematics and Evolution* 87 (2): 243–289. DP: 23 September 2011 (paper title page).

Nadein, K. 2012: [new taxon]. In: Nadein, K. & Lee, C.-F.: New data about some Alticini from Taiwan with descriptions of two new species (Coleoptera: Chrysomelidae). *Bonn Zoological Bulletin* 61: 41–48. DP: [31] July 2012 (paper title page).

Nadein, K. S. 2013a: Systematics of Manobiina with cladistic analysis based on morphological data (Coleoptera: Chrysomelidae: Galerucinae: Alticini). *Insect Systematics & Evolution* 44 (1): 45–92. DP: 15 April 2013 (paper title page).

Nadein, K. S. 2013b: Ivalia Jacoby – a flea beetle genus new to Australia (Coleoptera: Chrysomelidae: Galerucinae). *Zootaxa* 3669 (3): 384–400. DP: 7 June 2013 (footer on the paper title page).

Nadein, K. S. 2013c: Febraina: a new subtribe of Alticini with cladistics analysis based on morphology (Coleoptera: Chrysomelidae: Galerucinae). *Systematic Entomology* 38 (3): 491–506. DP: 18 June 2013 (volume contents).

Nadein, K. 2015: Phylogeny of Diboliina inferred from a morphologically based cladistic analysis (Coleoptera: Chrysomelidae: Galerucinae). *Arthropod Systematics and Phylogeny* 73 (1): 65–83. DP: 17 April 2015 (paper title page).

Nadein, K. S. & Gök, A. 2009: A new species of the genus Aeschrocnemis Weise, 1888 from Southwest Turkey (Coleoptera: Chrysomelidae). *Annales Zoologici* (Warszawa) 59 (2): 193–196. DP: 30 June 2009 (verso of the issue cover).

Nadein, K. & Lee, C.-F. 2012: New data about some Alticini from Taiwan with descriptions of two new species (Coleoptera: Chrysomelidae). *Bonn Zoological Bulletin* 61 (1): 41–48. DP: [31] July 2012 (paper title page).

Nakane, T. 1950: [A short report of Coleoptera from Hyuga-takanabe (5)]. *Entomological Review of Japan* 5 (1): 53–55 (in Japanese). DP: [31] July 1950 (issue cover).

Nakane, T. 1954: A list of Coleoptera (Polyphaga) from Oza with descriptions of some new species. Pp. 727–740. In: Ogura, Y. (ed.): *Scientific Researches of the Ozegahara Moor*. Tokyo, 841 pp. DP: 28 February 1954 (colophon).

Nakane, T. 1955: New or little-known Coleoptera from Japan and its adjacent regions, XII. *Scientific Reports of the Saikyo University, Natural Science and Living Science, A Series* 2 (1): 24–42. DP: [31] March 1955 (issue cover).

Nakane, T. 1956: New or little-known Coleoptera from Japan and its adjacent regions, XIII. *Scientific Reports of the Saikyo University, Natural Science and Living Science, A Series* 2 (3): 159–172. DP: [31] December 1956 (reprint cover).

Nakane, T. 1958: The Coleoptera of Yakushima Island, Chrysomelidae. *Scientific Reports of the Saikyo University, Natural Science and Living Science, A Series* 2 (5): 303–314. DP: [31] October 1958 (issue cover).

Nakane, T. 1963a: New or little-known Coleoptera from Japan and its adjacent regions. XVI. *Fragmenta Coleopterologica* **1963** (4): 18 & (5): 19–20. DP: [31] May 1963 (issues title pages).

Nakane, T. 1963b: New or little-known Coleoptera from Japan and its adjacent regions. XXV. *Scientific Reports of the Kyoto Prefectural University, Natural Science and Living Science, A Series* **3** (5): 221–226. DP: [31] December 1963 (issue cover).

Nakane, T. 1984: On the Coleoptera occurring in Yaku-shima Island. Pp. 587–631. In: *Conservation Reports of the Yaku-shima Wilderness Area, Kyushu, Japan.* Tokyo: Nature Conservation Bureau, Environment Agency of Japan, 714 pp. DP: [31] March 1984 (reprint cover).

Nakane, T. 1985: Notes on some leaf beetles from Japan (Coleoptera, Chrysomelidae). *Kita-Kyûshû no Konchû* **32** (1): 1–4, 1 pl. DP: [28] February 1985 (paper title page).

Nakane, T. 1999: Notes on the specific name of "Plateumaris sericea" in Japan (Coleoptera, Chrysomelidae). *Nature and Insects* (Tokyo) **34** (8): 45–46. DP: 30 July 1999 (based on the record in library of National Museum of Nature and Science, Tokyo).

Nakane, T. & Kimoto, S. 1959: A list of chrysomelid-beetles from Amami-Ôshima collected by Prof. M. Sasakawa and Mr. R. Ishikawa, with description of a new species. *Scientific Reports of the Kyoto Prefectural University, Natural Science and Living Science, A Series* **3** (1): 67–68. DP: [31] December 1959 (issue cover).

Nakane, T. & Kimoto, S. 1961a: A list of chrysomelid-beetles collected by Dr. T. Shiraki from the Loochoo Islands, with descriptions of new species I (Coleoptera). *Kontyû* **29** (1): 14–21. DP: 31 March 1961 (heading on p. 1).

Nakane, T. & Kimoto, S. 1961b: Entomological results from the scientific survey of the Tokara Islands. Coleoptera: Chrysomelidae. *Bulletin of the Osaka Museum of Natural History* **13**: 71–79. DP: [31] March 1961 (reprint title page).

Nakane, T. & Kimoto, S. 1961c: A list of chrysomelid-beetles collected by Dr. T. Shiraki, from the Loochoo Islands, with descriptions of some new species II (Coleoptera). *Kontyû* **29** (2): 106–110. DP: 30 June 1961 (heading on p. 85).

Nash, R. 1983: A brief summary of the development of entomology in Ireland during the years 1790–1870. *Irish Naturalists' Journal* **21** (4): 145–150. DP: [31] October 1983 (issue cover).

Navajas, E. 1946: Os genótipos da fam. Fulcidacidae (Col. Chrysomeloidea). *Papéis Avulsos do Departamento de Zoologia* (Sao Paulo) 7 [1947]: 243–246. DP: 27 October 1946 (paper title page).

Nesterova, O. L. & Lopatin, I. K. 2009: A new species of the leaf-beetle genus Acolastus Gerstaecker, 1855 (Coleoptera: Chrysomelidae) from Iran. *Zoosystematica Rossica* **18** (2): 273–274. DP: 25 December 2009 (paper title page).

Newman, E. 1834: Attempted division of British insects into natural orders. *Entomological Magazine* **2** [1834–1835] (4): 379–431. DP: 1 October 1834 (p. 4 of the volume 1 and p. 329 of the volume 2).

Nie, R.-E., Andújar, C., Gómez-Rodríguez, C., Bai, M., Xue, H.-J., Tang, M., Yang, C.-T., Tang, P., Yang, X.-K. & Vogler, A. P. 2020: The phylogeny of leaf beetles (Chrysomelidae) inferred from mitochondrial genomes. *Systematic Entomology* **45** (1): 188–204. DP: [31] January 2020 (journal website).

Nie, R.-E., Breeschoten, T., Timmermans, M. J. T. N., Nadein, K., Xue, H.-J., Bai, M., Huang, Y., Yang, X.-K. & Vogler, A. P. 2018: The phylogeny of Galerucinae (Coleoptera: Chrysomelidae) and the performance of mitochondrial genomes in phylogenetic inference compared to nuclear rRNA genes. *Cladistics* **34** (2): 113–130. DP: [30] April 2018 (journal website).

Nie, R.-E. & Yang, X.-K. 2013: [new taxon]. In: Nie, R.-E., Zhou, D.-K., Xue, H.-J. & Yang, X.-K.: Notes on black elytron species of Pyrrhalta Joannis and the description of a new species from China (Coleoptera, Chrysomelidae, Galerucinae). *ZooKeys* **289**: 41–56. DP: 12 April 2013 (paper title page).

Nie, R.-E. & Yang, X.-K. 2017: Galerucinae. Pp. 81–176, 266. In: Yang, X.-K. & Zhang, R.-Z. (eds.): Insect fauna of the Qinling Mountains. Volume VII (Coleoptera III). Xi'an: World Publishing Corporation, [16] + 10 + 485 + [1] pp. DP: [31] December 2017 (colophon).

Nilsson, J. A. & Johnson, C. D. 1993: A taxonomic revision of the palm bruchids (Pachymerini) and a description of the world genera of Pachymerinae. *Memoirs of the American Entomological Society* **41**: 1–104. DP: 1 December 1993 (verso of issue title page).

Normand, H. 1937: Contribution au catalogue des coléoptères de la Tunisie (11e fascicule). *Bulletin de la Société d'Histoire Naturelle de l'Afrique du Nord* **28** (2): 116–143. DP: 8 April 1937 (last page of the issue).

Normand, H. 1938: Contribution au catalogue des coléoptères de la Tunisie (13e fascicule). *Bulletin de la Société d'Histoire Naturelle de l'Afrique du Nord* **29** (5): 337–371. DP: 30 June 1938 (last page of the issue).

Normand, H. 1947a: Remarques sur les Cryptocephalus voisins du C. blandulus Har. *Revue Française d'Entomologie* **14** (2): 141–145. DP 30 June 1947 (footer on p. 121).

Normand, H. 1947b: Nouvelles espèces de chrysomèlides du Nord de l'Afrique. *Bulletin de la Société Entomologique de France* **52** (6): 84–87. DP: 15 September 1947 (*dates de publication* on p. 176).

Normand, H. 1950a: Contribution au catalogue des coléoptères de la Tunisie. Troisième supplement (fascicule 4). *Bulletin de la Société de Sciences Naturelles de Tunisie* **2** [1949] (2): 79–104, pls. VI–VII. DP: 5 April 1950 (registered at ZIN library, 1950 also mentioned on the last page of the issue cover).

Normand, H. 1950b: Nouveaux coléoptères du nord de l'Afrique. *Revue Française d'Entomologie* **17** (2): 105–108. DP: 30 June 1950 (footer on p. 93).

Nowicki, M. 1872: *Beschreibung einer neuen Käferart nebst Ausweis der Literatur über die Käferfauna Galiziens*. Krakau: Universitäts-Buchdruckerei, 7 pp. DP: [31 December] 1872 (paper title page).

Obenberger, J. 1917: II. Beitrag zur Kenntnis der palaearktischen Käferfauna. *Archiv für Naturgeschichte, Abteilung A* **82** (4) [1916]: 9–45, 2 pls. DP: [31] August 1917 (issue cover).

Ochs, G. 1925: Eine neue Varietät von Pyrrhalta viburni Payk. nebst biologischen Bemerkungen. *Senckenbergiana* **7** (3/4): 79–81. DP: 20 April 1925 (issue title page).

Ogloblin, D. A. 1921: Novye vidy triby Halticina palearkticheskoy oblasti (Coleoptera, Chrysomelidae). (Espèces nouvelles de la tribu Halticina de la région paléarctique

(Coleoptera, Chrysomelidae)). *Russkoe Entomologicheskoe Obozrenie* 17 [1917]: 20–39. DP: [31 December] 1921 (volume cover).

Ogloblin, D. A. 1925: Einige neue Haltica-Formen aus der paläarktischen Region. *Russkoe Entomologicheskoe Obozrenie* 19 (2): 91–96. DP: [31 December] 1925 (issue cover).

Ogloblin, D. A. 1927: Novye vidy r. Aphthona Chevr. (Coleoptera, Halticini) v kollekcii Zoologicheskogo Muzeya Akademii Nauk SSSR. (Espèces nouvelles du genre Aphthona Chevr. (Coleoptera, Halticini) des collections du Musée Zoologique de l'Académie des Sciences de l'URSS). *Annuaire du Musée Zoologique de l'Académie des Sciences de l'URSS* 27 [1926] (4): 283–303. DP: [30] April 1927 (Asanovich et al. 2007).

Ogloblin, D. A. 1928: [new taxon]. In: Gavrilenko, M.: Zbirki prirodnichi. (Collections of natural science). Pp. 17–28. In: Benderovskiy, V., Rizhenko, Y. & Gavrilenko, M. (eds.): *Almanach. Vol. 1. Dedicated to the 35-th anniversary of the museum.* Poltava: Poltava State Museum of the name V. Korolenko, xii + 296 pp., 8 pls. DP: 26 June 1928 (registered at ZIN library).

Ogloblin, D. A. 1930: De quelques espèces de Halticinae (Col. Chrysomelidae) de la collection de V. Motschoulsky. *Eos, Revista Española de Entomología* 6 (1): 83–112, pl. I. DP: 10 April 1930 (issue cover).

Ogloblin, D. A. 1936: *Listoedy, Galerucinae. Fauna SSSR. Nasekomye Zhestkokrylye. n. s. 8, 26 (1). [Chrysomelidae, Galerucinae. Faune de l'URSS, Insectes Coléoptères]*. Moskva – Leningrad, Izdatel'stvo Akademii Nauk SSSR, xiv + 455 pp. SFP: 19 December 1936, DP: [31 December] 1936 (book title page).

Ogloblin, D. A. 1941: [new taxa]. In: Semenov, A. E. & Ogloblin, D. A.: Dva novykh vida zhukov listoedov. [Two new species of leaf-beetles]. *Doklady Vsesoyuznoy Akademii Sel'skokhozyaystvennykh Nauk im. V. I. Lenina. (Proceedings of the Lenin Academy of Agricultural Sciences of U.S.S.R.)* 6 (1): 20–23. SFP: 28 January 1941, DP: [31 December] 1941 (volume title page).

Ogloblin, D. A. 1948: Chrysomelidae – Listoedy. Pp. 472–507. In: Tarbinskiy, C. P. & Plavilshchikov, N. N. (ed.): *Opredelitel nasekomykh evropeyskoy chasti SSSR. [Key to identification of insect of European part of USSR]*. Moskva – Leningrad: Ogiz – Selkhozgiz, 1128 pp. SFP: 23 March 1946, DP: [31 December] 1948 (book title page).

Ogloblin, D. A. 1956: Novye vidy roda Cryptocephalus Latr. (Coleoptera, Chrysomelidae) palearkticheskoy fauny. (New palearctic species of the genus Cryptocephalus Latr. (Coleoptera, Chrysomelidae)). *Entomologicheskoe Obozrenie* 35 (4): 890–894. SFP: 18 December 1956, DP: [31 December] 1956 (paper title page).

Ogloblin, D. A. & Medvedev, L. N. 1956: Novye formy zhukov-listoedov (Coleoptera, Chrysomelidae) palearkticheskoy fauny. (New palearctic chrysomelid-beetles (Coleoptera, Chrysomelidae)). *Entomologicheskoe Obozrenie* 35 (4): 895–898. SFP: 18 December 1956, DP: [31] December 1956 (paper title page, see also Lopatin 1963c: 122).

Ogloblin, D. A. & Medvedev, L. N. 1971: [new taxon]. In: Medvedev, L. N.: Novye formy listoedov podsemeystva Clytrinae (Coleoptera, Chrysomelidae) SSSR i prilegayushchikh stran. (New forms of the subfamily Clytrinae (Coleoptera, Chrysomelidae) in the USSR and adjacent countries). *Zoologicheskiy Zhurnal* 50 (5): 686–695. SFP: 28 April 1971, DP: 20 June 1971 (registered at ZIN library).

Ohno, M. 1958a: On the genus Paropsides Motschulsky of Japan. (Studies on the Chrysomelinae of Japan (2)). *Kontyû* **26** (1): 33–40, pls. 2–3. DP: 28 February 1958 (heading on p. 1).

Ohno, M. 1958b: On the genus Lypesthes Baly from Japan (Coleoptera: Chrysomelidae, Alticinae). *Journal of the Toyo University* **12**: 173–181, pl. VI. DP: [28] February 1958 (reprint cover).

Ohno, M. 1959: On the genus Liprus Motschulsky from Japan. (Studies on the flea-beetles of Japan (Coleoptera: Chrysomelidae, Alticinae) (2)). *Bulletin of the Biogeographical Society of Japan* **21** (4): 33–44, 1 pl. DP: [30] June 1959 (issue title page).

Ohno, M. 1960a: Description of a new Demotina-species from Japan (Coleoptera: Chrysomelidae – Eumolpinae). *Transactions of the Shikoku Entomological Society* **6** [1958–1960] (5): 65–66. DP: 30 January 1960 (verso of volume cover).

Ohno, M. 1960b: The chrysomelid-beetles from Hachijo Island including Kojima and Aogashima Chrysomelidae of the Izu Islands, Japan (1) (Coleoptera). *Mushi* **33** (9): 63–71. DP: 28 February 1960 (paper title page).

Ohno, M. 1960c: On the species of the genus Altica occuring in Japan. Studies on the flea-beetles of Japan (1) (Coleoptera, Chrysomelidae). *Bulletin of the Department of Liberal Arts, Toyo University, Tokyo* **1**: 77–95, 2 pls. DP: [29] February 1960 (reprint cover).

Ohno, M. 1960d: Notes on a leaf-beetle, Adoxus obscurus (Linné) from Japan (Coleoptera, Chrysomelidae, Eumolpinae). *Dobutsugaku Zasshi (Zoological Magazine)* **69** (7): 239–242. DP: [31] July 1960 (issue cover).

Ohno, M. 1960e: Notes on the genus Lipromela Chen, with the description of a new species from Formosa (Coleoptera: Chrysomelidae, Alticinae). *Annotationes Zoologicae Japonenses* **33** (3): 197–199. DP: [30] September 1960 (paper title page).

Ohno, M. 1961a: On the species of the genus Trachyaphthona Heikertinger and the new genus Sphaeraltica (Coleoptera, Chrysomelidae). Studies on the flea-beetles of Japan (3). *Bulletin of the Department of Liberal Arts, Toyo University, Tokyo* **2**: 73–91. DP: [28] February 1961 (reprint cover).

Ohno, M. 1961b: Description of a new Zeugophora-species from Japan, with some notes on the other known-species (Coleoptera, Chrysomelidae, Zeugophorinae). *Kontyû* **29** (1): 36–38. DP: 31 March 1961 (heading on p. 1).

Ohno, M. 1961c: Description of a new Chlamisus-species from Yakushima, Japan (Coleoptera, Chrysomelidae, Chlamisinae). *Entomological Review of Japan* **13** (1): 1–3. DP: 30 August 1961 (General index 1960–1963).

Ohno, M. 1961d: Notes on the species of Lamprosominae occuring in Japan and the Loo-Choos (Col., Chrysomelidae). *Entomological Review of Japan* **13** (2): 47–52. DP: 28 December 1961 (General index 1960–1963).

Ohno, M. 1962a: Description of a new Oulema-species from Japan (Coleoptera, Chrysomelidae, Criocerinae). *Mushi* **36** (9): 99–101. DP: 15 March 1962 (paper title page).

Ohno, M. 1962b: A new genus Zipanginia, with the descriptions of a new species and a new subspecies (Coleoptera, Chrysomelidae, Alticinae). *Annotationes Zoologicae Japonenses* **35** (1): 23–28. DP: [31] March 1962 (paper title page).

Ohno, M. 1962c: Description of a new Pyrrhalta-species from the Loochoo Islands (Coleoptera, Chrysomelidae, Galerucinae). *Kontyû* 30 (1): 27–29. DP: 31 March 1962 (heading on p. 1).

Ohno, M. 1962d: On the species of the genus Aphthona Chevrolat occuring in Japan and the Loo-Choo Islands (Coleoptera, Chrysomelidae, Alticinae). *Bulletin of the Department of Liberal Arts, Toyo University* (Tokyo) 3: 61–84. DP: [31] July 1962 (reprint cover).

Ohno, M. 1962e: Revision of the species belonging to the genus Oulema Des Gozis occurring in Japan (Coleoptera, Chrysomelidae, Criocerinae). *Entomological Review of Japan* 14 (2): 42–47. DP: 30 August 1962 (General index 1960–1963).

Ohno, M. 1963: Notes on Aulacophora nigripennis Motshulsky, with the description of a new subspecies (Coleoptera, Chrysomelidae, Galerucinae). *Bulletin of the Biogeographical Society of Japan* 22 (8): 85–88. DP: [31] January 1963 (paper title page).

Ohno, M. 1964a: Revision of Sphaeroderma-species occuring in Japan and the Loo-Choos (Coleoptera, Chrysomelidae, Alticinae). *Journal of the Toyo University, General Education, Natural Science* 5: 29–60. DP: [31] May 1964 (reprint cover).

Ohno, M. 1964b: Descriptions of two new Zipanginia-species from Japan (Coleoptera: Chrysomelidae – Alticinae). *Transactions of the Shikoku Entomological Society* 8 [1963–1965] (2): 35–38. DP: 30 June 1964 (issue cover).

Ohno, M. 1964c: Description of a new Sphaeroderma-species from Formosa (Col., Chrysomelidae, Alticinae). *Entomological Review of Japan* 17 (1): 15–16. DP: 30 August 1964 (General index 1963–1968).

Ohno, M. 1965a: Description of a new Trachyaphthona-species from Japan (Coleoptera, Chrysomelidae, Alticinae). *New Entomologist* 14 (3): 25–28. DP: [31] May 1965 (paper title page).

Ohno, M. 1965b: Description of a new genus Parazipangia from Japan (Coleoptera, Chrysomelidae, Alticinae). *Annotationes Zoologicae Japonenses* 38 (2): 101–103. DP: [30] June 1965 (paper title page).

Ohno, M. 1966: Supplementary notes on the genus Pseudoliprus of Japan (Coleoptera: Chrysomelidae – Alticinae). *Transactions of the Shikoku Entomological Society* 9 [1966–1968] (2): 44–46. DP: 30 September 1966 (issue cover).

Ohno, M. 1967: A revision of Argopus-species occurring in Japan. *Journal of the Toyo University, General Education, Natural Science* 7: 35–51. DP: [31] March 1967 (reprint cover).

Ohno, M. 1968a: A revision of Longitarsus-species occurring in Japan (Coleoptera, Chrysomelidae, Alticinae). *Journal of the Toyo University, General Education, Natural Science* 9: 1–56. DP: [31] March 1968 (reprint cover).

Ohno, M. 1968b: The chrysomelid-beetles from the Island Miyakejima, the Izu Isls., Japan (II). *Entomological Review of Japan* 21 (1): 33–38. DP: 30 November 1968 (General index 1968–1973).

Oke, C. 1932: Notes on Australian Coleoptera, with descriptions of new species. II. *Proceedings of the Linnean Society of New South Wales* 57 (3/4): 148–172. DP: 15 September 1932 (table of contents).

Okhrimenko, N. V. 1990: Novyy vid roda Chrysolina (Coleoptera, Chrysomelidae) s zapadnogo Kavkaza. (A new species of the genus Chrysolina (Coleoptera, Chrysomelidae) from the West

Caucasus). *Vestnik Zoologii* 1990 (5): 64–67. SFP: 20 September 1990, DP: 29 October 1990 (registered at ZIN library).

Olivier, A. G. 1789: *Encyclopédie méthodique ou par ordre de matières; par une société de gens de lettres, de savans et d'artistes; précédée d'un vocabulaire universel, servant de table pour tout l'ouvrage, ornée des portraits de Mm. Diderot et d'Alembert, premiers éditeurs de l'Encyclopédie. Histoire Naturelle. Tome quatrième. Insectes.* Paris: C.-J. Panckoucke et Liége: Plomteux, i – ccclxxiii + 331 pp. DP: 27 April 1789 (Evenhuis 2003).

Olivier, A. G. 1790: *Encyclopédie méthodique, ou par ordre de matières; par une société de gens de lettres, de savans et d'artistes; précédée d'un vocabulaire universel, servant de table pour tout l'ouvrage, ornée des portraits de Mm. Diderot et d'Alembert, premiers éditeurs de l'Encyclopédie. Histoire naturelle. Insectes. Tome cinquième.* Paris: Panckoucke, 793 pp. DP (Pars I, pp. 1–368): 17 November 1790 (Evenhuis 2003, Bousquet 2016).

Olivier, A. G. 1791a: *Encyclopédie méthodique, ou par ordre de matières; par une société de gens de lettres, de savans et d'artistes; précédée d'un vocabulaire universel, servant de table pour tout l'ouvrage, ornée des portraits de Mm. Diderot et d'Alembert, premiers éditeurs de l'Encyclopédie. Histoire naturelle. Insectes. Tome cinquième.* Paris: Panckoucke, 793 pp. DP (Pars II, pp. 369–793): 3 June 1791 (Evenhuis 2003, Bousquet 2016).

Olivier, A. G. 1791b: *Encyclopédie métodique, ou par ordre de matières; par une société de gens de lettres, de savans et d'artistes; précédée d'un vocabulaire universel, servant de table pour tout l'ouvrage, ornée des portraits de Mm. Diderot et d'Alembert, premiers éditeurs de l'Encyclopédie. Histoire Naturelle. Insectes. Tome sixième.* Paris: Panckoucke, 704 pp. DP (Pars I, pp. 1–368): 21 November 1791 (Evenhuis 2003, Bousquet 2016).

Olivier, A. G. 1792: *Encyclopédie métodique, ou par ordre de matières; par une société de gens de lettres, de savans et d'artistes; précédée d'un vocabulaire universel, servant de table pour tout l'ouvrage, ornée des portraits de Mm. Diderot et d'Alembert, premiers éditeurs de l'Encyclopédie. Histoire Naturelle. Insectes. Tome sixième.* Paris: Panckoucke, 704 pp. DP (Pars II, pp. 369–704): 1 October 1792 (Evenhuis 2003, Bousquet 2016).

Olivier, A. G. 1793: *Encyclopédie métodique, ou par ordre de matières; par une société de gens de lettres, de savans et d'artistes; précédée d'un vocabulaire universel, servant de table pour tout l'ouvrage, ornée des portraits de Mm. Diderot et d'Alembert, premiers éditeurs de l'Encyclopédie. Histoire Naturelle. Insectes. Tome septième.* Paris: Panckoucke, 827 pp. DP (Pars I, pp. 1–368): 13 May 1793 (Evenhuis 2003, Bousquet 2016).

Olivier, A. G. 1800: *Entomologie, ou histoire naturelle des insectes. Avec leur caractéres génériques et spécifiques, leur description, leur synonymie, et leur figure enluminée. Coléoptères. Tome quatrième.* Paris: de Lanneau, 519 pp. + 72 pls. DP: 22 September 1800 [each genus paginated separately, the genera Nos. 66–67 (to page 80) were published on 31 December 1795, Nos. 67 (from page 81)–81 on 22 September 1800 (Bousquet 2016, 2018)]

Olivier, A. G. 1807: *Entomologie, ou histoire naturelle des insectes, avec leurs caractères génériques et spécifiques, leur description, leur synonymie et leur figure enluminée. Coléoptères. Tome cinquième.* Paris: Desray, 612 pp. DP: [31 December] 1807 (Bousquet 2016).

Olivier, A. G. 1808: *Entomologie, ou histoire naturelle des insectes, avec leurs caractères génériques et spécifiques, leur description, leur synonymie et leur figure enluminée. Coléoptères. Tome sixième.* Paris: Desray, [4] + 613–1104. DP: [31 December] 1808 (Bousquet 2016).

Olsoufieff [= Olsufiew], G. 1903: Notes sur quelques Donacia de Russie (Col.). *Bulletin de la Société Entomologique de France* **1903** (18): 302–303. DP: 19 December 1903 (*dates d'apparition* at the end on volume).

Orlova-Bienkowskaya, M. J. 2007: Diagnostical characters and distribution of leaf-beetles Oreina coerulea (Olivier, 1790) and O. auricollis Stierlin, 1887 (Coleoptera: Chrysomelidae). *Russian Entomological Journal* **16** (3): 323–328. DP: 4 December 2007 (journal website).

Orlova-Bienkowskaya, M. J. 2015: Epitrix papa sp. n. (Coleoptera: Chrysomelidae: Galerucinae: Alticini), previously misidentified as Epitrix similaris, is a threat to potato production in Europe. *European Journal of Entomology* **112** (4): 824–830. DP: 20 November 2015 (last page of the paper).

Oromí, P. & García, R. 1995: Contribución al conocimiento de la fauna de coleópteros de Canarias y su distribución. *Vieraea* **24**: 175–186. DP: [31] December 1995 (paper title page).

Osculati, G. 1844: Coleopterorum species novae quae in ista enumeratione commemoratae sunt diagnosibus atque observationibus illustratae. P. 72. In: *Note d'un viaggio nella Persia e nelle Indie Orientali negli anni 1841, 1842.* Monza: Luca Corbetta, 72 + [1] pp., 1 pl. DP: 19 September 1844 (Bousquet 2016).

Özdikmen, H. 2006: Nomenclatural changes in Chrysomelidae (Coleoptera). *Munis Entomology & Zoology* **1** (1): 155–156. DP: 1 January 2006 (headings throughtout the paper, see also the comments in the Dating of publications chapter).

Özdikmen, H. 2008: Substitute names for some preoccupied leaf beetles genus group names described by L. N. Medvedev, (Coleoptera: Chrysomelidae). *Munis Entomology & Zoology* **3** (2): 643–647. DP: 1 June 2008 (headings throughtout the paper, see also the comments in the Dating of publications chapter).

Özdikmen, H. 2011: A comprehensive contribution for leaf beetles of Turkey with a zoogeographical evaluation for all Turkish fauna (Coleoptera: Chrysomelidae). *Munis Entomology & Zoology* **6** (2): 540–638. DP: 1 June 2011 (headings throughtout the paper, see also the comments in the Dating of publications chapter).

Özdikmen, H. 2021: A review: a new subgeneric arrangement of the genus Chaetocnema Stephens (Chrysomelidae: Galerucinae: Alticini) with new subgenera based on spermathecal structures. *Munis Entomology & Zoology* **16** (1): 41–105. DP: 1 January 2021 (headings throughtout the paper, see also the comments in the Dating of publications chapter).

Özdikmen, H. & Bal, N. 2016a: A new species of Cheilotoma Chevrolat from Turkey with an updated list (Coleoptera: Chrysomelidae: Clytrinae). *Munis Entomology & Zoology* **11** (2): 303–311. DP: 1 June 2016 (headings throughtout the paper, see also the comments in the Dating of publications chapter).

Özdikmen, H. & Bal, N. 2016b: [new taxon]. In: Özdikmen, H., Bal, N. & Kıyak, S.: The genus Labidostomis Germar of Turkey with a new species and a new record (Coleoptera: Chrysomelidae: Clytrinae). *Munis Entomology & Zoology* **11** (2): 515–538. DP: 1 June 2016 (headings throughtout the paper, see also the comments in the Dating of publications chapter).

Özdikmen, H. & Bal, N. 2021: The structure of spermatheca in the subgenus Cassida (Cassida) Linnaeus, 1758 (Coleoptera: Chrysomelidae: Cassidinae) and its taxonomic significance. *Fresenius Environmental Bulletin* 30 (5): 5270–5291. DP: [31] May 2021 (journal website).

Özdikmen, H., Bal, N. & Bolu, H. 2021: A new species of Aphthona Chevrolat from Turkey (Chrysomelidae: Galerucinae: Alticini). *Munis Entomology & Zoology* **16** (1): 27–32. DP: 1 January 2021 (headings throughtout the paper, see also the comments in the Dating of publications chapter).

Özdikmen, H., Bal, N., Amutkan Mutlu, D. & Suludere, Z. 2023: Morphology and ultrasturucture of aedeagus and spermatheca of the monotypic palaearctic genus Pilemostoma Desbrochers Des Loges, 1891 (Chrysomelidae: Cassidinae: Cassidini) from Turkey and their taxonomic significance. *Transactions of the American Entomological Society* **149** (2): 247–260. DP: 23 October 2023 (paper title page).

Özdikmen, H. & Coral Şahin, D. 2017: A new species, Phyllotreta bilgeae sp. nov., from Turkey (Chrysomelidae: Galerucinae: Alticini). *Munis Entomology & Zoology* **12** (1): 175–179. DP: 1 January 2017 (headings throughtout the paper, see also the comments in the Dating of publications chapter).

Özdikmen, H., Coral Şahin, D. & Bal, N. 2017: A new species of Phyllotreta Chevrolat from Turkey (Chrysomelidae: Galerucinae: Alticini). *Munis Entomology & Zoology* **12** (1): 147–150. DP: 1 January 2017 (headings throughtout the paper, see also the comments in the Dating of publications chapter).

Özdikmen, H., Coral Şahin, D. & Bal, N. 2020a: A new subspecies of Chrysolina sanguineocincta (Coleoptera: Chrysomelidae: Chrysomelinae) from Turkey. *Munis Entomology & Zoology* **15** (1): 85–90 DP: 1 January 2020 (headings throughtout the paper, see also the comments in the Dating of publications chapter).

Özdikmen, H., Coral Şahin, D. & Bal, N. 2020b: A new species of Cassida Linnaeus, 1758, from Turkey (Chrysomelidae: Cassidinae). *Microscopy Research & Technique* **83** (10): 1171–1177. DP: 28 September 2020 (table of contents).

Özdikmen, H., Kılıç, T., Coral Şahin, D. & Bal, N. 2019: A new species of flea beetle genus Argopus Fisher von Waldheim, 1824 from Turkey (Chrysomelidae: Galerucinae: Alticini). *Munis Entomology & Zoology* **14** (2): 448–453. DP: 1 June 2019 (headings throughtout the paper, see also the comments in the Dating of publications chapter).

Özdikmen, H., Mercan, N., Cihan, N., Kaya, G., Topcu, N. N. & Kavak, M. 2014: The importance of superfamily Chrysomeloidea for Turkish biodiversity (Coleoptera). *Munis Entomology & Zoology* **9** (1): 17–45. DP: 1 January 2014 (headings throughtout the paper, see also the comments in the Dating of publications chapter).

Ødegaard, F. & Hanssen, O. 2020: Galerucella (Neogalerucella) anserina Ødegaard & Hanssen, sp. nov., a new species of Chrysomelidae (Coleoptera, Chrysomelidae, Galerucinae) from Norway. *Zootaxa* **4755** (2): 341–352. DP: 24 March 2020 (footer on the paper title page).

Paiva do Castello, B. de 1861: Descriptions of two new species of Coleoptera from the Canary Islands. *Annals and Magazine of Natural History, Third Series* 8 (45): 210–212. DP: 1 September 1861 (Evenhuis 2003).

Pajni, H. R. & Singh, T. 1977: Caryedon punjabensis sp. nov. on Cassia tora (Coleoptera: Bruchidae). *Oriental Insects* 11 (2): 225–228 DP: [31] August 1977 (footer on the paper title page).

Paliy, V. F. 1961: Novyy vid zemlyanoy bloshki Psylliodes agropyri, sp. nova (Coleoptera, Chrysomelidae), iz SSSR. (A new species of Psylliodes (Coleoptera, Chrysomelidae) from the USSR). *Entomologicheskoe Obozrenie* 40 (2): 390–391. SFP: 30 May 1961, DP: [31 December] 1961 (issue cover).

Paliy, V. F. 1963: Dva novykh vida zemlyanykh bloshek (Coleoptera, Chrysomelidae). [Two new species of flea-beetles (Coleoptera, Chrysomelidae)]. Pp. 33–37. In: Paliy, V. F. (ed.): *Sbornik Entomologicheskihk Rabot 2*. [*Collection of entomological papers 2*] Frunze: Izdatel'stvo akademii nauk Kirgizskoy SSR, 106 pp. SFP: 10 November 1963, DP: [31 December] 1963 (book title page).

Paliy, V. F. 1968a: Novye vidy zemlyanykh bloshek (Coleoptera, Chrysomelidae, Halticinae) iz sredney Azii. Soobshchenie, I. (New species of Halticinae (Coleoptera, Chrysomelidae) from the Central Asia. Communication I). *Vestnik Zoologii* **1968** (4): 15–21. SFP: 22 July 1968, DP: 3 October 1968 (registered at ZIN library).

Paliy, V. F. 1968b: Novye vidy zemlyanykh bloshek (Coleoptera, Chrysomelidae, Halticinae) iz sredney Azii. Soobshchenie, II. (New species of Coleoptera, Chrysomelidae, Halticinae from the Central Asia subregion of the Palearctic. Communication II). *Vestnik Zoologii* **1968** (6): 26–31. SFP: 21 November 1968, DP: [31 December] 1968 (paper title page).

Paliy, V. F. 1970a: Novyy vid zemlyanoy bloshki (Coleoptera, Chrysomelidae) iz vostochnoy Sibiri i Primorya. (A new species of flea-beetles (Coleoptera, Chrysomelidae) from East Siberia). *Entomologicheskoe Obozrenie* 49 (1): 197–198. SFP: 17 March 1970, DP: 9 April 1970 (registered at ZIN library).

Paliy, V. F. 1970b: Novye podrody i vidy zemlyanykh bloshek (Coleoptera, Chrysomelidae, Halticinae) iz sredney Azii. [New subgenera and species of flea-beetles (Coleoptera, Chrysomelidae, Halticinae) from Central Asia]. Pp. 3–15. In: Protsenko, A. I. (ed.): *Entomologicheskie issledovaniya v Kirgizii.* [*Entomological research in Kyrgyzstan*]. Frunze: Izdatel'stvo "ILIM", 87 pp. SFP: 23 July 1970, DP: 28 October 1970 (registered at VKP).

Paliy, V. F. 1970c: Novyy yuzhnopalearkticheskiy vid zemlyanoy bloshki (Coleoptera, Chrysomelidae, Halticinae) Longitarsus cynoglossi Paliy sp. n. (A new Southpalaearctic species – Longitarsus cynoglossi sp. n. – Col., Chrysom., Halticinae). *Natura* (Plovdiv) **3** (1): 135–138. DP: [31 December] 1970 (issue cover).

Pallas, P. S. 1771: *Reise durch verschiedene Provinzen des Russischen Reichs. Erster Theil.* St. Petersburg: Kayserliche Akademie der Wissenschaften, [10] + 3–504 pp., 11 pls. DP: 6 September 1771 (Bousquet 2017).

Pallas, P. S. 1773: *Reise durch verschiedene Provinzen des Russischen Reichs. Zweyter Theil. Zweytes Buch vom Jahr 1771.* St. Petersburg: Kayserliche Akademie der Wissenschaften, pp. 371–744. DP: [31 December] 1773 (Bousquet 2016).

Pallister, J. C. 1953: The leaf beetles of North Central Mexico collected on the David Rockefeller Mexican expedition (Coleoptera, Chrysomelidae). *American Museum Novitates* **1623**: 1–95. DP: 5 May 1953 (paper title page).

Palm, T. 1976: Zur Kenntnis der Käferfauna der kanarischen Inseln 17–18. *Entomologica Scandinavica* 7: 96–102. DP: 15 June 1976 (paper title page).

Palm, T. 1977: Zur Kenntnis der Käferfauna der kanarischen Inseln 17–18. Berichtigung. *Entomologica Scandinavica* 8: 240. DP: 15 September 1977 (paper title page).

Palmén, E. 1946: Zur Systematik finnischer Chrysomeliden. 3. Untergattung Phytodecta s. str. *Annales Entomologici Fennici* 11 [1945] (4): 227–234. DP: 26 March 1946 (reprint cover).

Panin, S. 1943: Supplément à la connaissance des coléoptères roumains (2-ème supplément). *Bulletin de la Section Scientifique de l'Académie Roumaine* **24** [1941–1942]: 178–182. DP: [31 December] 1943 (volume title page).

Pankow, W. & Schmitt, M. 2017: Lilioceris schneideri (Weise, 1900) in Central Europe, with lectotype designation and the description of a new subspecies, L. s. abnobae (Coleoptera: Chrysomelidae: Criocerinae). *Entomologische Blätter und Coleoptera* **113** (1): 191–200. DP: [30] November 2017 (issue cover).

Panzer, G. W. F. 1789: Einige seltene Insecten. *Der Naturforscher* **24**: 1–35. DP: 3 May 1789 (*Allg. Verz. Bücher, Ostermesse* 1789: 90 and Evenhuis 2014).

Panzer, G. W. F. 1794a: *Faunae insectorum Germanicae initia oder Deutschlands Insecten. Heft 21.* Norimbergae: Felsecker, 24 pp. + 24 pls. DP 3 May 1794 (Alonso-Zarazaga & Evenhuis 2017).

Panzer, G. W. F. 1794b: *Faunae Insectorum Germanicae initia oder Deutschlands Insecten. Heft 24.* Norimbergae: Felsecker, 24 pp. + 24 pls. DP: 21 August 1794 (Alonso-Zarazaga & Evenhuis 2017).

Panzer, G. W. F. 1795a: *Entomologia Germanica exhibens insecta per Germaniam indigena secundum classes, ordines, genera, species adiectis synonymis, locis, observationibus. I. Eleuterata.* Norimbergae: Felsecker, [8] + 12 + 370 + 2 pp. 12 pls. DP: 14 February 1795 (Bousquet 2016).

Panzer, G. W. F. 1795b: *Faunae Insectorum Germanicae initia oder Deutschlands Insecten. Heft 29.* Norimbergae: Felsecker, 24 pp. + 24 pls. DP: 26 April 1795 (Alonso-Zarazaga & Evenhuis 2017, Bousquet 2017).

Panzer, G. W. F. 1796a: *Faunae Insectorum Germanicae initia oder Deutschlands Insecten. Heft 38.* Norimbergae: Felsecker, 24 pp. + 24 pls. DP: 17 April 1796 (Alonso-Zarazaga & Evenhuis 2017, Bousquet 2017).

Panzer, G. W. F. 1796b: *Faunae Insectorum Germanicae initia oder Deutschlands Insecten. Heft 39.* Norimbergae: Felsecker, 24 pp. + 24 pls. DP: 17 April 1796 (Alonso-Zarazaga & Evenhuis 2017, Bousquet 2017).

Panzer, G. W. F. 1799a: *Faunae Insectorum Germanicae initia oder Deutschlands Insecten. Heft 61.* Norimbergae: Felsecker, 24 pp., 24 pls. DP: 14 April 1799 (Alonso-Zarazaga & Evenhuis 2017, Bousquet 2017).

Panzer, G. W. F. 1799b: *Fauna Insectorum Germanicae initia; oder Deutschlands Insecten. Heft 67.* Norimbergae: Felsecker, 24 pp. + 24 pls. DP: 14 April 1799 (Alonso-Zarazaga & Evenhuis 2017, Bousquet 2017).

Panzer, G. W. F. 1799c: *Fauna Insectorum Germanicae initia; oder Deutschlands Insecten. Heft 68.* Norimbergae: Felsecker, 24 pp. + 24 pls. DP: 14 April 1799 (Alonso-Zarazaga & Evenhuis 2017, Bousquet 2017).

Panzer, G. W. F. 1802: *Fauna Insectorum Germanicae initia; oder Deutschlands Insecten. Heft 87.* Norimbergae: Felsecker, 24 pp. + 24 pls. DP: 9 May 1802 (Alonso-Zarazaga & Evenhuis 2017).

Panzer, G. W. F. 1806: *Faunae Insectorum Germanicae initia oder Deutschlands Insecten. Heft 99.* Norimbergae: Felsecker, 24 pp., 24 pls. DP: 27 April 1806 (Alonso-Zarazaga & Evenhuis 2017).

Papp, K. [= C. S.] 1945: *A magyar bogárfauna határozója. Kézikönyv a történelmi Magyarország területén elöforduló jellegzetes bogarak meghatározására.* [*An identification key of beetles of Hungary. A handbook for identification of characteristic beetles found in the territory of the former Kingdom of Hungary*]. Budapest: Ruszkabányai Könyvkiadó Vállalat, 477 pp. DP: [31 December] 1945 (title page dated 1943 but the book includes an announcement that will be published during the year 1945).

Papp, C. S. 1946a: Colour-variations of the Colasposoma oberthuri Jac. (Chrysomelidae) (Nr. 21.). *Additamenta Faunistica Coleopterorum* 4 (1): 6–7. DP: [31 December] 1946 (issue title page).

Papp, C. S. 1946b: Synonymischen Bemerkungen. *Additamenta Faunistica Coleopterorum* 4 (1): 7. DP: [31 December] 1946 (issue title page).

Papp, C. S. 1946c: Zweiten Beitrag zur Kenntnis der Chrysomeliden des Karpathen-Beckens. (22. Beitrag zur Kenntnis der Chrysomeliden). *Additamenta Faunistica Coleopterorum* 4 (2): 20–24. DP: [31 December] 1946 (issue cover).

Papp, C. S. 1946d: Zur Revision des Genus Pedrillia (Chrysom.). (23. Beitrag zur Kenntnis der Chrysomeliden). *Additamenta Faunistica Coleopterorum* 4 (2): 24–27. DP: [31 December] 1946 (issue cover).

Papp, C. S. 1951a: Some new Phytophaga from South-America and India (42d Contribution to the knowledge of the Chrysomelidae). *Tijdschrift voor Entomologie* **93** [1950]: 131–133. DP: 20 March 1951 (volume title page).

Papp, C. S. 1951b: Deux nouveaux chrysomélides (Col.) nord-africains (Contribution à l'étude des chrysomélides, no 36). *Bulletin de la Société des Sciences Naturelles du Maroc* 30 [1950] (1): 81–84. DP: 30 August 1951 (last page of the issue).

Park, J. 2013a: [new taxon]. In: Park, J., Lee, J. E. & Park, J. K.: A new species of the genus Demotina Baly (Coleoptera: Chrysomelidae: Eumolpinae) from Korea. *Entomological Research Bulletin* **29** (2): 116–118. DP: 30 December 2013 (HeeWook Cho, 2020, pers. comm.).

Park, J. 2013b: [new taxon]. In: Park, J., Lee, J. E. & Park, J. K.: A new species of the genus Phyllotreta Chevrolat (Coleoptera: Chrysomelidae: Alticinae) from Korea. *Entomological Research Bulletin* **29** (2): 119–123. DP: 30 December 2013 (HeeWook Cho, 2020, pers. comm.).

Parry, F. J. S. 1844a: A decade of new Coleoptera from Assam. *Annals and Magazine of Natural History* **14** (93): 454. DP: 1 December 1844 (Evenhuis 2003).

Parry, F. J. S. 1844b: A decade of new Coleoptera from Assam. *Journal of Proceedings of the Entomological Society of London* [**1843**]: 86–87. DP: 1 December 1844 (Wheeler 1912).

Pasquet, C. O. 1923: Coléoptères de la Manche. *Mémoires de la Société National des Sciences Naturelles et Mathématiques de Cherbourg* **39** [1913–1923] (9): 1–332, 1 map. DP: [31 December] 1923 (volume title page).

Paykull, G. 1799a: *Fauna Svecica. Insecta. Tomus II*. Upsaliae: Joh. F. Edman, 234 pp. DP: 20 March 1799 (Bousquet 2016).

Paykull, G. 1799b: Beskrifning öfver nya Svanska Insecter. *Kongl. Vetenskaps Academiens nya Handlingar* 20 (2): 115–116. DP: [31 December] 1799 (volume title page).

Paykull, G. 1800: *Fauna Svecica. Insecta. Tomus III*. Upsaliae: Joh. F. Edman, 459 pp. DP: 5 October 1800 (Bousquet 2017).

Penecke, K. A. B. J. 1898: Coleopterologische Miscellen. *Wiener Entomologische Zeitung* **17** (9): 251–255. DP: 25 November 1898 (footers throughout the issue).

Penecke, K. A. B. J. 1901: Coleopterologische Miscellen. II. *Wiener Entomologische Zeitung* 20 (1/2): 11–21. DP: 31 January 1901 (footers throughout the issue).

Penecke, K. A. B. J. 1907: [new taxon]. In: Penecke, K. A. B. J. & Müller, J.: Koleopterologische Ergebnisse einer Sammelreise nach Dalmatien im Sommer 1905. *Verhandlungen der Kaiserlich-Königlichgen Zoologisch-Botanischen Gesellschaft in Wien* 57 (1): 1–19. DP: 8 March 1907 (volume title page).

Pérez Arcas, L. 1865a: Insectos nuevos ó poco conocidos de la fauna española. Primera parte. *Revista de los Progresos de las Ciencias Exactas, Fisicas y Naturales* **15** (3): 166–186. DP: [31 December] 1865 (see the comments in Dating of publications).

Pérez Arcas, L. 1865b: Insectos nuevos ó poco conocidos de la fauna española. Segunda parte. *Revista de los Progresos de las Ciencias Exactas, Fisicas y Naturales* **15** (7): 413–444. DP: [31 December] 1865 (see the comments in Dating of publications).

Pérez Arcas, L. 1872: Especies nuevoas ó críticas de la fauna española. *Anales de la Sociedad Española de Historia Natural* **1** (1): 89–137, pls. I–III. DP: 15 August 1872 (*Pet. Nouv. Ent.* 1872, 4 (58): 233).

Pérez Arcas, L. 1874: Especies nuevas ó críticas de la fauna española. Tercera parte. *Anales de la Sociedad Española de Historia Natural* **3** (2): 111–155, pls. I–III. DP: 7 October 1874 (issue cover).

Perris, E. 1852: Seconde excursion dans les Grandes-Landes. Lettre adressée à M. Mulsant,. *Annales de la Société Linnéenne de Lyon* [1850/1852]: 145–216. DP: [31 December] 1852 (volume title page).

Perris, E. 1856: [Signalement de sept coléoptères pris dans le département des Landes]. *Bulletin de la Société Entomologique de France* [**1855**] (3): lxxvii–lxxx. DP: 8 January 1856 (Evenhuis 2022).

Perris, E. 1864: Description de quelques espèces nouvelles de coléoptères et notes diverses. *Annales de la Société Entomologique de France, Quatrième Série* **4** (2): 275–303. DP: 12 October 1864 (Evenhuis 2022).

Perris, E. 1866: Descriptions de quelques nouvelles espèces de coléoptères, rectifications et notes. *Annales de la Société Entomologique de France, Quatrième Série* 5 [1865] (4): 505–512. DP: 26 May 1866 (Evenhuis 2022).

Perris, E. 1869: Descriptions de quelques coléoptères nouveaux, rectifications et notes. *L'Abeille, Mémoires d'Entomologie* 7 [1869–1870]: 3–33. DP: [30] November 1869 (*L'Abeille* 7, *Nouvelles et faits divers* No. 7: xxv and *Bull. Soc. Ent. Fr.* 1870: xv–xvi).

Perroud, B.-P. 1864: [new taxon]. In: Perroud, B.-P. & Montrouzier, X.: Essai sur la faune entomologique de Kanala (Nouvelle-Calédonie) et description de quelques espèces nouvelles ou peu connues. *Annales de la Société Linnéenne de Lyon, Nouvelle Série* 11: 46–257. DP: [29] February 1864 (volume title page).

Petagna, L. 1819: Memoria su di alcuni insetti del Regno di Napoli. *Atti della Reale Accademia delle Scienze, Classe de Fisica di Storia Naturale* (Napoli) 1: 19–40, 4 pls. DP: [31 December] 1819 (volume title page).

Petitpierre, E. 1993: Dos nuevas especies de Cryptocephalus Geoffroy ibero-baleares (Coleoptera, Chrysomelidae). *Nouvelle Revue d'Entomologie, Nouvelle Série* 10 (3): 297–302. DP: 31 December 1993 (last page of the cover).

Petitpierre, E. 2007: Cryptocephalus (Burlinius) chafarinensis n. sp. from the Mediterranean northwest Africa (Coleoptera: Chrysomelidae: Cryptocephalinae). *Genus* 18 (4): 571–574. DP: 28 December 2007 (paper title page).

Petitpierre, E. 2013: [new taxon]. In: Petitpierre, E. & Daccordi, M.: Chrysomelidae (Coleoptera) de las sierras del Altiplano de Granada (Granada, Andalucía). *Zoologica Baetica* 24: 53–78. DP: [31 December] 2013 (paper title page).

Petitpierre, E. 2019: *Coleoptera, Chrysomelidae II. Fauna Iberica, vol. 46.* Madrid: Museo Nacional de Ciencias Naturales, CSIS, 596 pp. DP: 17 June 2019 (colophon).

Petitpierre, E. 2021: The genetic and cytogenetic relationships among subgenera of Chrysolina Motschulsky, 1860 and Oreina Chevrolat, 1837 (Coleoptera: Chrysomelidae: Chrysomelinae). *International Journal of Zoology and Animal Biology* 4 (3): 000305. DP: 28 May 2021 (paper title page).

Petitpierre, E. 2023: Insights into the subgenera Chrysolina s. str. and Rhyssoloma Woll. of Chrysolina Mots. (Coleoptera: Chrysomelidae: Chrysomelinae). *Natural History Sciences, Atti della Società Italiana di Scienze Naturali e del Museo Civile di Storia Naturale di Milano* 10 (**Supplement** 1): 31–32. DP: 13 December 2023 (journal website).

Petitpierre, E. & Alonso-Zarazaga, M. A. 2000: [new names]. In: Petitpierre, E.: *Coleoptera, Chrysomelidae I. Fauna Ibérica, vol. 13.* Madrid: Museo Nacional de Ciencias Naturales, CSIS, 521 pp. DP: 29 December 2000 (colophon).

Petitpierre, E. & Alonso-Zarazaga, M. A. 2019a: [new name]. In: Petitpierre, E.: *Coleoptera, Chrysomelidae II. Fauna Iberica, vol. 46.* Madrid: Museo Nacional de Ciencias Naturales, CSIS, 596 pp. DP: 17 June 2019 (colophon).

Petitpierre, E. & Alonso-Zarazaga, M. A. 2019b: Apéndice 1. Nomenclatura: lista de sinónimos y combinaciones. Pp. 497–550. In: Petitpierre, E.: *Coleoptera, Chrysomelidae II. Fauna*

Iberica, vol. 46. Madrid: Museo Nacional de Ciencias Naturales, CSIS, 596 pp. DP: 17 June 2019 (colophon).

Petitpierre, E. & Anichtchenko, A. 2018: Endophallus structure: a promising tool for cryptic species identification in Timarcha Samouelle, 1819 (Coleoptera: Chrysomelidae: Chrysomelinae). *Zootaxa* **4446** (3): 361–383. DP: 18 July 2018 (footer on the paper title page).

Petitpierre, E. & Cuadras, C. M. 1977: The canonical analysis applied to the taxonomy and evolution of the genus Timarcha Latr. (Coleoptera, Chrysomelidae). *Mediterranea* **2**: 13–28. DP: [31] December 1977 (volume cover).

Petri, K. 1907: Vier neue Rüssler aus Turkestan und China und eine neue Crepidodera aus Siebenbürgen. *Wiener Entomologische Zeitung* **16** (2): 57–61. DP: 15 February 1907 (footers throughout the issue).

Petri, K. 1912: *Siebenbürgens Käferfauna auf Grund ihrer Erforschung bis zum Jahre 1911*. Hermannstadt: Jos. Drotleff, ix + [1] pp., 374 columns. DP: 30 November 1912 (*Deutsche Ent. Zeitschr.* 1912 (6): 750).

Petri, K. 1926: Ergänzungen und Berichtigungen zur Käferfauna Siebenbürgens 1912. *Verhandlungen und Mitteilungen des Siebenbürgischen Vereins für naturwissenschaften zu Hermannstadt* **75/76** [1925–1926]: 165–206. DP: [31 December] 1926 (volume title page).

Peyerimhoff, P. M. de 1908: Nouveaux coléoptères du Nord-Africain (sixième note: faune du Djurjura). *Bulletin de la Société Entomologique de France* **1908** (8): 117–125. DP: 3 June 1908 (*dates d'apparition* at the end of volume).

Peyerimhoff, P. M. de 1911: Description et discussion de deux nouvelles races de Thyamis (Col. Chrysomelidae). *Bulletin de la Société Entomologique de France* **1911** (10): 211–212. DP: 14 June 1911 (*dates de publication* at the end of volume).

Peyerimhoff, P. M. de 1912: Notes sur la biologie de quelques coléoptères phytophages du Nord Africain (première série) (avec les descriptions de deux espèces nouvelles). *Annales de la Société Entomologique de France* 80 [1911–1912] (3): 283–314. DP: 29 February 1912 (*Ann. Soc. Ent. Fr.* 1912, 81: 527).

Peyerimhoff, P. M. de 1915a: Notes sur la biologie de quelques coléoptères phytophages du Nord-Africain (deuxième série) (avec les descriptions de cinq espèces nouvelles et de cinq sous-espèces). *Annales de la Société Entomologique de France* **84** [1915–1916] (1): 19–61. DP: 23 June 1915 (*dates d'apparition* on p. 617).

Peyerimhoff, P. M. de 1915b: Nouveaux coléoptères du Nord-Africain (Vingt et unième note: faune du Djurdjura). *Bulletin de la Société Entomologique de France* **1915** (14): 219–223. DP: 2 September 1915 (*dates de publication* at the end of volume).

Peyerimhoff, P. M. de 1919: Notes sur la biologie de quelques coléoptères phytophages du Nord-Africain (troisième série) avec les descriptions de cinq espèces nouvelles et de sept sous-espèces ou variétés. *Annales de la Société Entomologique de France* 88 (1/2): 169–258. DP: 24 November 1919 (*dates de publication* on p. 419).

Peyerimhoff, P. M. de 1920: Nouveaux coléoptères du Nord Africain. Trente-quatrième note: récoltes de M. le capitaine Surcouf dans la région d'El-Goléa. *Bulletin de la Société Entomologique de France* **1920** (3): 42–48. DP: 16 March 1920 (*dates de publication* at the end of volume).

Peyerimhoff, P. M. de 1922: Nouveaux coléoptères du Nord Africain. Quarantième note: récoltes de M. le Dr R. Maire vers le sommet du Tachdirt (Grand Atlas marocain). *Bulletin de la Société Entomologique de France* **1922** (5): 60–66. DP: 7 April 1922 (*dates de publication* on p. 348).

Peyerimhoff, P. M. de 1923: Nouveaux coléoptères du Nord-Africain. Quarante-cinquième note: autres récoltes de M. le Dr R. Maire dans le Grand-Atlas marocain. *Bulletin de la Société Entomologique de France* **1923** (5): 78–81. DP: 22 April 1923 (*dates de publication* at the end of volume).

Peyerimhoff, P. M. de 1924: Nouveaux coléoptères du Nord-Africain. Quarante sixième note: récoltes de M. C. Dumont dans la région d'El-Goléa. *Bulletin de la Société Entomologique de France* **1924** (4): 62–67. DP: 15 April 1924 (*dates de publication* on p. 278).

Peyerimhoff, P. M. de 1925a: Nouveaux coléoptères du Nord-Africain. Cinquantième note (avec table). *Annales de la Société Entomologique de France* **94** (1): 1–29. DP: 31 March 1925 (*dates de publication* on p. 451).

Peyerimhoff, P. M. de 1925b: [new taxon]. In: Dodero, A.: Missione zoologica del Dr. E. Festa in Cirenaica. XIV. *Bollettino del Musei di Zoologia ed Anatomia Comparata della R. Università di Torino* **39**: 1–31. DP: 31 March 1925 (last page of the paper).

Peyerimhoff, P. de 1925c: Nouveaux coléoptères du Nord-Africain. Cinquante-troisième note: faune du Grand-Atlas marocain (suite). *Bulletin de la Société Entomologique de France* **1925** (16): 267–269. DP: 20 December 1925 (*dates de publication* on p. 356).

Peyerimhoff, P. M. de 1926: [new taxa]. In: Sainte-Claire Deville, J. & Peyerimhoff, P. M. de.: Nouveaux coléoptères du Nord-Africain. Cinquante-cinquième note. Faune du Grand-Atlas marocain (suite). *Bulletin de la Société Entomologique de France* **1926** (8): 93–96. DP: 29 May 1926 (*dates d'apparition* on p. 252).

Peyerimhoff, P. M. de 1928: Nouveaux coléoptères du Nord-Africain. Soixante et unième note: récoltes de M. le Dr R. Maire au Moussa-ou-Salah (Moyen-Atlas marocain). *Bulletin de la Société Entomologique de France* **1928** (3): 39–42. DP: 6 March 1928 (*dates d'apparition* on p. 332).

Peyerimhoff, P. de 1929: Nouveaux coléoptères du Nord-Africain. Soixante-huitième note. Faune du Hoggar et des massifs voisins (suite). *Bulletin de la Société Entomologique de France* **1929** (8): 142–146. DP: 5 June 1929 (*dates d'apparition* on p. 336).

Peyerimhoff, P. M. de 1931: Coléoptères. Mission scientifique du Hoggar envoyée de Février à Mai 1928 par M. Pierre Bordes Gouverneur Général de l'Algérie. *Mémoires de la Société d'Histoire Naturelle de l'Afrique du Nord* **2**: 1–172, 3 pls, 2 maps. DP: 15 September 1931 (reprint cover).

Peyerimhoff, P. M. de 1938: Coléoptères nouveaux ou mal connus de Berbérie. IX. – Les Threnosoma Motsch. des Chrysomela L. (Chrysomelidae). *Bulletin de la Société Entomologique de France* **43** (3/4): 52–56. DP: 6 April 1938 (*dates de parution* on p. 260).

Peyerimhoff, P. M. de 1939: Études et descriptions de coléoptères marocains. I. – Chrysomelidae. *Bulletin de la Société des Sciences Naturelles du Maroc* **19** (1): 33–39. DP: 15 November 1939 (last page of the issue).

Peyerimhoff, P. M. de 1941: Matériaux pour un catalogue des coléoptères sahariens. I. Observations et rectifications à la liste de 1931. *Bulletin de la Société d'Histoire Naturelle de l'Afrique du Nord* **32** (1): 7–17. DP: 4 August 1941 (publication date of issue 4, see p. 152).

Peyerimhoff, P. M. de 1949: Etudes et descriptions de coleopteres marocains II. *Bulletin de la Société des Sciences Naturelles du Maroc* **25/27** [1945–1947]: 248–308. DP: 8 March 1949 (*Bull. Soc. Sci. Nat. Maroc* 1949, 29: 24).

Philippi, R. A. & Philippi, F. H. E. 1864: Beschreibung einiger neuen chilenischer Käfer. *Entomologische Zeitung* (Stettin) **25** (10/12): 313–406. DP: [31] December 1864 (issue title page).

Pic, M. 1891: Sur Crioceris tibialis Villa et Allecula morio Fabr. *L'Échange, Revue Linnéenne* 7 (79): 51. DP: 15 July 1891 (issue title page).

Pic, M. 1894a: Habitats de quelques coléoptères algériens et descriptions d'espèces nouvelles. *Annales de la Société Entomologique de France* **63** (1): 101–106. DP: 10 April 1894 (issue cover).

Pic, M. 1894b: Descriptions de deux coléoptères. *L'Échange, Revue Linnéenne* 10 (113): 65–66. DP: [31] May 1894 (issue title page).

Pic, M. 1894c: Quelques notes en passant sur plusieurs coléoptères de la collection Henri Tournier. *L'Échange, Revue Linnéenne* 10 (113): 71–72. DP: [31] May 1894 (issue title page).

Pic, M. 1894d: Contribution à la faune coléoptèrologique d'Algérie. Descriptions d'espèces nouvelles. *Le Naturaliste, 2e Série* **16** (181): 210–211. DP: 15 September 1894 (issue title page).

Pic, M. 1894e: Description d'un coléoptère nouveau. *Le Naturaliste, 2e Série* **16** (182): 221. DP: 1 October 1894 (issue title page).

Pic, M. 1894f: Descriptions d'espèces et variétés de coléoptères asiatiques. *L'Échange, Revue Linnéenne* 10 (120): 142–144. DP: [31] December 1894 (issue title page).

Pic, M. 1894g: [Diagnoses de plusieurs coléoptères rapportés de Syrie par M. C. Delagrange]. *Bulletin de la Société Entomologique de France* **1894** (20): cclxxxiv–cclxxxv. DP: 31 December 1894 (Evenhuis 2002).

Pic, M. 1895a: Excursion entomologique dans la Province d'Oran (Algérie). *Revue Scientifique du Bourbonnais et du Centre de la France* 8 (85): 10–13. DP: 15 January 1895 (issue cover).

Pic, M. 1895b: Notes sur des coléoptères rares ou nouveaux d'Algérie. *Bulletin de la Société Entomologique de France* **1895** (4): cxxvi–cxxx. DP: 13 March 1895 (*Bull. Soc. Ent. Fr.* 1895 (5): cliii – errata mentioned in the issue 4).

Pic, M. 1895c: Descriptions de coléoptères d'Algérie. *L'Échange, Revue Linnéenne* 11 (127): 78–82. DP: [31] July 1895 (issue title page).

Pic, M. 1895d: A propos des variétés. *L'Échange, Revue Linnéenne* 11 (128): 87–89. DP: [31] August 1895 (issue title page).

Pic, M. 1895e: Description de coléoptères. *Le Naturaliste, 2e Série* 17 (207): 243. DP: 15 October 1895 (issue title page).

Pic, M. 1896a: Coléoptères d'Asie Mineure et de Syrie. *Miscellanea Entomologica* 4 (3): 35–36. DP: 15 March 1896 (issue cover).

Pic, M. 1896b: Remarques synonymiques et descriptives sul le genre Cercomorphus, sur divers phytophages, sur des Anthicides, etc. (Col.). *Bulletin de la Société Entomologique de France* **1896** (2): 29–31. DP: 18 May 1896 (*Bibl. Zool.* 1896, 1: 269).

Pic, M. 1896c: Descriptions et notes diverses. *L'Échange, Revue Linnéenne* **12** (138): 61–62. DP: [30] June 1896 (issue title page).

Pic, M. 1896d: Coléoptères du nord de l'Afrique. *Miscellanea Entomologica* 4 (7): 93–94. DP: 15 July 1896 (issue cover).

Pic, M. 1896e: Sur les Pachybrachis voisins de P. vermicularis Suff. *Bulletin de la Société Entomologique de France* **1896** (8): 203–205. DP: 10 August 1896 (*Bibl. Zool.* 1896, 1: 427).

Pic, M. 1896f: Notes et descriptions (Coléoptères). *La Feuille des Jeunes Naturalistes, IIIe Série* **26** [1895–1896] (310): 201–202. DP: [31] August 1896 (issue cover).

Pic, M. 1896g: Notes et diagnoses. *L'Échange, Revue Linnéenne* **12** (140): 87–88. DP: [31] August 1896 (issue title page).

Pic, M. 1896h: Descriptions de coléoptères d'Algérie et de Syrie. *Miscellanea Entomologica* 4 (11/12): 140–142. DP: 15 December 1896 (issue cover).

Pic, M. 1897a: Études sur les coléoptères phytophages (Clytridae). *Bulletin de la Société Zoologique de France* **22** (2): 82–88. DP: [28] February 1897 (issue cover).

Pic, M. 1897b: Note sur Otiocephala opaca Rosh. et ses variétés (Col.). *Bulletin de la Société Entomologique de France* **1897** (7): 135. DP: 14 June 1897 (*Bibl. Zool.* 1897, 2: 357).

Pic, M. 1897c: Complément a mes études sur les coléoptères phytophages (Clytridae). *Bulletin de la Société Zoologique de France* **22** (7): 164–165. DP: [31] July 1897 (issue cover).

Pic, M. 1897d: *Variétés (2me Article)*. Lyon: L. Jacquet, 4 pp. DP: [31] August 1897 (footer on p. 4).

Pic, M. 1897e: Notes et descriptions diverses (Coléoptères). *La Feuille des Jeunes Naturalistes, IIIe Série* **27** [1896–1897] (323): 202–204. DP: 1 September 1897 (issue cover).

Pic, M. 1897f: Descriptions de coléoptères. *Bulletin de la Societé d'Histoire Naturelle d'Autun* 10 (2): 194–198. DP: [30] September 1897 (*Bull. Soc. Zool. Fr.* 1897, 22 (8): 203).

Pic, M. 1897g: Complément a mes études sur les coléoptères phytophages (Clytridae). *Bulletin de la Société Zoologique de France* **22** (8): 202–206. DP: [30] November 1897 (issue cover).

Pic, M. 1897h: Descriptions de coléoptères. *Bulletin de la Société d'Histoire Naturelle d'Autun* 10 (2): 295–300. DP: 8 December 1897 (*Bull. Soc. Ent. Fr.* 1897: 304).

Pic, M. 1898a: Notes descriptives sur plusieurs coléoptères et sur un Ichneumon (Hyménoptère). *Miscellanea Entomologica* **6** (6): 73–75. DP: 15 June 1898 (issue cover).

Pic, M. 1898b: Descriptions de coléoptères d'Anatolie & de Syrie. *Revue Scientifique du Bourbonnais et du Centre France* **11** (125/126): 91–94. DP: 22 June 1898 (*Bull. Soc. Ent. Fr.* 1898: 252).

Pic, M. 1898c: Diagnoses de coléoptères malacodermes et phytophages. *Miscellanea Entomologica* **6** (11): 137–140. DP: 15 November 1898 (issue cover).

Pic, M. 1898d: Description d'un genre nouveaux et de sept coléoptères exotiques. *Bulletin de la Société Zoologique de France* **23** (9/10): 169–175. DP: [30] November 1898 (issue cover).

Pic, M. 1898e: Diagnoses de coléoptères malacodermes et phytophages (2me article). *Miscellanea Entomologica* **6** (12): 153–154. DP: 15 December 1898 (issue cover).

Pic, M. 1898f: Description de deux Luperus nouveaux (Col.). *Bulletin de la Société Entomologique de France* **1898** (16): 311–313. DP: [31 December] 1898 (issue title page).

Pic, M. 1899a: Diagnoses de coléoptères d'Orient. *Bulletin de la Société d'Histoire Naturelle d'Autun* **12**: 204–210. DP: 22 November 1899 (*Bull. Soc. Ent. Fr.* 1899: 363).

Pic, M. 1899b: Diagnoses de coléoptères d'Orient. Deuxième article. *Bulletin de la Société d'Histoire Naturelle d'Autun* **12**: 252–261. DP: 22 November 1899 (*Bull. Soc. Ent. Fr.* 1899: 363).

Pic, M. 1900a: Notes diverses sur les coléoptères. *Le Frelon* **8** [1899–1900] (4): 1–8. DP: 3 February 1900 (Gaedike & Behne 2008).

Pic, M. 1900b: Diagnoses de coléoptères d'Orient recoltés en 1899. *Bulletin de la Société d'Histoire Naturelle d'Autun* **13**: 25–32. DP: 13 June 1900 (*Bull. Soc. Ent. Fr.* 1900: 244).

Pic, M. 1900c: Notes et diagnoses. *L'Échange, Revue Linnéenne* **16** (188): 61–63. DP: [31] August 1900 (issue cover).

Pic, M. 1900d: Notes descriptives et biologiques. *L'Échange, Revue Linnéenne* **16** (189): 65–66. DP: [30] September 1900 (issue cover).

Pic, M. 1900e: Sur le genre « Cryptocephalus » Geof. *L'Échange, Revue Linnéenne* **16** (189): 66–70. DP: [30] September 1900 (issue cover).

Pic, M. 1900f: Notes et diagnoses. 2e article. *L'Échange, Revue Linnéenne* **16** (190): 77–80. DP: [31] October 1900 (issue cover).

Pic, M. 1900g: Diagnoses de malacodermes et d'un Cryptocephalus. *L'Échange, Revue Linnéenne* **16** (191): 85–88. DP: [30] November 1900 (issue cover).

Pic, M. 1900h: Diagnoses de coléoptères du globe. *L'Échange, Revue Linnéenne* **16** (192): 89–91. DP: [31] December 1900 (issue cover).

Pic, M. 1900i: Sur divers « Cryptocephalus » du nord de l'Afrique. *L'Échange, Revue Linnéenne* **16** (192): 95–96. DP: [31] December 1900 (issue cover).

Pic, M. 1901a: Notes diverses et diagnoses (3e article). *L'Échange, Revue Linnéenne* **17** (195): 17–20. DP: [31] March 1901 (issue cover).

Pic, M. 1901b: Notes diverses et diagnoses (4e article). *L'Échange, Revue Linnéenne* **17** (196): 25–27. DP: [30] April 1901 (issue cover).

Pic, M. 1901c: Notes diverses et diagnoses (5e article). *L'Échange, Revue Linnéenne* **17** (197): 33–35. DP: [31] May 1901 (issue cover).

Pic, M. 1901d: « Cryptocephalus crassus » Ol. et ses variétés. *L'Échange, Revue Linnéenne* **17** (197): 37–39. DP: [31] May 1901 (issue cover).

Pic, M. 1901e: Notes diverses et diagnoses (6e article). *L'Échange, Revue Linnéenne* **17** (200): 57–59. DP: [31] August 1901 (issue cover).

Pic, M. 1901f: Notes diverses et diagnoses (7e article). *L'Échange, Revue Linnéenne* **17** (201): 65–68. DP: [30] September 1901 (issue cover).

Pic, M. 1901g: Espéces et variétés nouvelles de coléoptères paléarctiques. *L'Échange, Revue Linnéenne* **17** (202): 79–80. DP: [31] October 1901 (issue cover).

Pic, M. 1901h: Notes diverses et diagnoses (8e article). *L'Échange, Revue Linnéenne* **17** (203): 81–83. DP: [30] November 1901 (issue cover).

Pic, M. 1901i: Coléoptères nouveaux d'Orient, Arménie et Tunisie. *L'Échange, Revue Linnéenne* **17** (204): 89–91. DP: [31] December 1901 (issue cover).

Pic, M. 1901j: Coléoptères d'Orient. *Bulletin de la Société d'Histoire Naturelle d'Autun* **14**: 33–39. DP: [31 December] 1901 (volume title page).

Pic, M. 1902a: Coléoptères asiatiques nouveaux. *L'Échange, Revue Linnéenne* **17** [= **18**] (205): 1–3. DP: [31] January 1902 (issue cover).

Pic, M. 1902b: Coléoptères paléarctiques présumés nouveaux. *L'Échange, Revue Linnéenne* **17** [= **18**] (208): 25–27 DP: [30] April 1902 (issue cover).

Pic, M. 1902c: Diagnoses de coléoptères de l'Ancien et du Nouveau monde. *L'Échange, Revue Linnéenne* **17** [= **18**] (209): 23–26 [note: p. 23 following p. 32]. DP: [31] May 1902 (issue cover).

Pic,, M. 1902d: Description de bruchides nouveaux. *Le Naturaliste, 2e Série* **24** (367): 146. DP: 15 June 1902 (issue title page).

Pic, M. 1902e: Coléoptères présumés nouveaux de la Rhodésie. *Revue d'Entomologie* **21** (6/7): 4–7. DP: [31] July 1902 (footer on p. 1).

Pic, M. 1902f: Descriptions et notes diverses (3e article). *L'Échange, Revue Linnéenne* **17** [= **18**] (212): 47–49. DP: [31] August 1902 (issue cover).

Pic, M. 1902g: Descriptions et notes diverses (5e article). *L'Échange, Revue Linnéenne* **17** [= **18**] (215): 71–73. DP: [30] November 1902 (issue cover).

Pic, M. 1902h: Nouvelles espèces et variétés de coléoptères paléarctiques (2e article). *L'Échange, Revue Linnéenne* **17** [= **18**] (216): 79–80. DP: [31] December 1902 (issue cover).

Pic, M. 1903a: Descriptions et notes complémentaires sur divers coléoptères paléarctiques. *L'Échange, Revue Linnéenne* **19** (218): 97–98. DP: [28] February 1903 (issue cover).

Pic, M. 1903b: Descriptions et habitats de divers coléoptères du Nord de l'Afrique. *L'Échange, Revue Linnéenne* **19** (220): 113–115. DP: [30] April 1903 (issue cover).

Pic, M. 1903c: Nouveaux coléoptères d'Europe. *L'Échange, Revue Linnéenne* **19** (220): 115. DP: [30] April 1903 (issue cover).

Pic, M. 1903d: Notes entomologiques et descriptions. *L'Échange, Revue Linnéenne* **19** (221): 121–125. DP: [31] May 1903 (issue cover).

Pic, M. 1903e: Captures, notes diverses et diagnoses. *L'Échange, Revue Linnéenne* **19** (222): 129–131. DP: [30] June 1903 (issue cover).

Pic, M. 1903f: Coléoptères de l'Afrique australe. *Revue d'Entomologie* **22** (6): 165–171. DP: [30] June 1903 (footer on p. 161).

Pic, M. 1903g: Coléoptères français nouveaux ou nouveaux pour la France. *L'Échange, Revue Linnéenne* **19** (223): 139–140. DP: [31] July 1903 (issue cover).

Pic, M. 1903h: Deux captures intéressantes. Diagnoses de divers coléoptères. *L'Échange, Revue Linnéenne* **19** (224): 145–147. DP: [31] August 1903 (issue cover).

Pic, M. 1903i: Quelques coléoptères rares ou nouveaux pour la faune gallo-rhénane. *L'Échange, Revue Linnéenne* **19** (225): 153–155. DP: [30] September 1903 (issue cover).

Pic, M. 1903j: Espèces et variétés nouvelles de coléoptères. *L'Échange, Revue Linnéenne* **19** (226): 161–163. DP: [31] October 1903 (issue cover).

Pic, M. 1904a: Diagnoses de coléoptères asiatiques provenant surtout de Sibérie. *L'Échange, Revue Linnéenne* **20** (232): 25–27. DP: [30] April 1904 (issue cover).

Pic, M. 1904b: Diagnoses de coléoptères paléarctiques et exotiques. *L'Échange, Revue Linnéenne* **20** (233): 33–36. DP: [31] May 1904 (issue cover).

Pic, M. 1904c: Sur divers Lariidae ou Bruchidae et Urodon anciens et nouveaux. *L'Échange, Revue Linnéenne* **20** (233): 39–40. DP: [31] May 1904 (issue cover).

Pic, M. 1904d: Sur divers Lariidae ou Bruchidae et Urodon anciens et nouveaux. *L'Échange, Revue Linnéenne* **20** (234): 42. DP: [30] June 1904 (issue cover).

Pic, M. 1904e: Diagnoses de divers coléoptères d'Europe et Turquie d'Asie. *L'Échange, Revue Linnéenne* **20** (236): 57–58. DP: [31] August 1904 (issue cover).

Pic, M. 1904f: Diagnoses ou descriptions abrégées de coléoptères paléarctiques. *L'Échange, Revue Linnéenne* **20** (240): 89–94. DP: [31] December 1904 (issue cover).

Pic, M. 1905a: Espèces et variétés nouvelles de coléoptères paléarctiques. *L'Èchange, Revue Linnéenne* **21** (241): 97–100. DP: [31] January 1905 (issue cover).

Pic, M. 1905b: Descriptions abrégées et notes diverses. *L'Èchange, Revue Linnéenne* **21** (242): 105–107. DP: [28] February 1905 (issue cover).

Pic, M. 1905c: Descriptions abrégées et notes diverses (3e article). *L'Èchange, Revue Linnéenne* **21** (244): 121–122. DP: [30] April 1905 (issue cover).

Pic, M. 1905d: Diagnoses de coléoptères algériens. *L'Èchange, Revue Linnéenne* **21** (247): 145–148. DP: [31] July 1905 (issue cover).

Pic, M. 1905e: Sur Cryptocephalus oranensis Weise et ses variétés. *Miscellanea Entomologica* **13** (7): 95–96. DP: [31] July 1905 (issue cover).

Pic, M. 1905f: Nouveaux coléoptères d'Algérie & d'Asie occidentale. *L'Échange, Revue Linnéenne* **21** (248): 153–156. DP: [31] August 1905 (issue cover).

Pic, M. 1905g: Coléoptères nouveaux provenant de France, Grèce, Algérie et Turqui d'Asie. *L'Èchange, Revue Linnéenne* **21** (249): 161–163. DP: [30] September 1905 (issue cover).

Pic, M. 1905h: Espèces et variétés nouvelles de coléoptères de France et Turquie d'Asie. *L'Èchange, Revue Linnéenne* **21** (251): 177–179. DP: [30] November 1905 (issue cover).

Pic, M. 1906a: Coléoptères européens, algériens et asiatiques nouveaux. *L'Échange, Revue Linnénne* **22** (255): 17–20. DP: [31] March 1906 (issue cover).

Pic, M. 1906b: Habitats et descriptions de divers coléoptères paléarctiques. *L'Échange, Revue Linnéenne* **22** (256): 25–27. DP: [30] April 1906 (issue cover).

Pic, M. 1906c: Habitats et descriptions de divers coléoptères paléarctiques. *L'Échange, Revue Linnéenne* **22** (257): 33–35. DP: [31] May 1906 (issue cover).

Pic, M. 1906d: Sur Crioceris asparagi L. et ses variétés (Col.). *Bulletin de la Société Entomologique de France* **1906** (8): 119–123. DP: 10 June 1906 (*Bibl. Zool.* 1906, 13: 370).

Pic, M. 1906e: Habitats et descriptions de divers coléoptères paléarctiques. *L'Échange, Revue Linnéenne* **22** (258): 41–42. DP: [30] June 1906 (issue cover).

Pic, M. 1906f: Coléoptères en partie nouveaux provenant de la Perse et des régions avoisinantes. *L'Échange, Revue Linnéenne* **22** (259): 49–51. DP: [31] July 1906 (issue cover).

Pic, M. 1906g: Sur divers coléoptères paléarctiques nouveaux, rares ou peu connus. *L'Échange, Revue Linnéenne* **22** (261): 65–67. DP: [30] September 1906 (issue cover).

Pic, M. 1906h: Notes entomologiques diverses (Suite). *L'Échange, Revue Linnénne* **22** (264): 96. DP: [31] December 1906 (issue cover).

Pic, M. 1907a: Coléoptères paléarctiques nouveaux. *L'Échange, Revue Linnéenne* 23 (265): 97–100. DP: [31] January 1907 (issue cover).

Pic, M. 1907b: Notes entomologiques diverses (Suite). *L'Échange, Revue Linnéenne* **23** (266): 111–112. DP: [28] February 1907 (issue cover).

Pic, M. 1907c: Sur divers « Cryptocephalus » et « Pachybrachis » peu connus ou présumés nouveaux. *L'Échange, Revue Linnéenne* **23** (266) [hors-texte]: 1–4. DP: [28] February 1907 (issue cover).

Pic, M. 1907d: Notes entomologiques diverses (Suite). *L'Échange, Revue Linnéenne* **23** (267): 113–115. DP: [31] March 1907 (issue cover).

Pic, M. 1907e: Coléoptères exotiques nouveaux ou peu connus (Suite). *L'Échange, Revue Linnéenne* **23** (269): 134–135. DP: [31] May 1907 (issue cover).

Pic, M. 1907f: Notes entomologiques diverses (Suite). *L'Échange, Revue Linnéenne* **23** (270): 137–139. DP: [30] June 1907 (issue cover).

Pic, M. 1907g: Notes entomologiques diverses (Suite). *L'Échange, Revue Linnéenne* **23** (271): 145–146. DP: [31] July 1907 (issue cover).

Pic, M. 1907h: Notes entomologiques diverses (Suite). *L'Échange, Revue Linnéenne* **23** (272): 153–154. DP: [31] August 1907 (issue cover).

Pic, M. 1907i: Notes entomologiques diverses (Suite). *L'Échange, Revue Linnéenne* **23** (274): 169–170. DP: [31] October 1907 (issue cover).

Pic, M. 1907j: Notes entomologiques diverses (Suite). *L'Échange, Revue Linnéenne* **23** (275): 177–179. DP: [30] November 1907 (issue cover).

Pic, M. 1907k: Notes entomologiques diverses (Suite). *L'Échange, Revue Linnéenne* **23** (276): 185–187. DP: [31] December 1907 (issue cover).

Pic, M. 1908a: Observations, renseignement divers et diagnoses sur le genre « Cryptocephalus » Geof. *L'Échange, Revue Linnéenne* **24** (278): 12–16. DP: [29] February 1908 (issue cover).

Pic, M. 1908b: Descriptions ou diagnoses et notes diverses. Suite. *L'Échange, Revue Linnéenne* **24** (282): 41–46. DP: [30] June 1908 (issue cover).

Pic, M. 1908c: Sur divers coléoptères français rares ou nouveaux. *L'Échange, Revue Linnéenne* **24** (282): 46–47. DP: [30] June 1908 (issue cover).

Pic, M. 1908d: Descriptions ou diagnoses et notes diverses. Suite. *L'Échange, Revue Linnéenne* **24** (283): 49–52. DP: [31] July 1908 (issue cover).

Pic, M. 1908e: Descriptions ou diagnoses et notes diverses. Suite. *L'Échange, Revue Linnéenne* **24** (284): 57–58. DP: [31] August 1908 (issue cover).

Pic, M. 1908f: Descriptions ou diagnoses et notes diverses. Suite. *L'Échange, Revue Linnéenne* **24** (285): 65–66. DP: [30] September 1908 (issue cover).

Pic, M. 1908g: Descriptions ou diagnoses et notes diverses. Suite. *L'Échange, Revue Linnéenne* **24** (287): 81–83. DP: [30] November 1908 (issue cover).

Pic, M. 1908h: Coléoptères exotiques nouveaux ou peu connus. *L'Échange, Revue Linnéenne* **24** (287): 85–86. DP: [30] November 1908 (issue cover).

Pic, M. 1908i: Descriptions ou diagnoses et notes diverses. Suite. *L'Échange, Revue Linnéenne* **24** (288): 89–91. DP: [31] December 1908 (issue cover).

Pic, M. 1908j: Sur le genre « Cryptocephalus ». *L'Échange, Revue Linnéenne* **24** (288): 91–94. DP: [31] December 1908 (issue cover).

Pic, M. 1909a: Descriptions ou diagnoses et notes diverses (Suite). *L'Échange, Revue Linnéenne* **25** (289): 97–100. DP: [31] January 1909 (issue cover).

Pic, M. 1909b: Descriptions de coléoptères exotiques nouveaux. *Le Naturaliste, 2e Série* **31** (526): 34. DP: 1 February 1909 (issue cover).

Pic, M. 1909c: Descriptions ou diagnoses et notes diverses (Suite). *L'Échange, Revue Linnéenne* **25** (290): 105–106. DP: [28] February 1909 (issue cover).

Pic, M. 1909d: Descriptions ou diagnoses et notes diverses (Suite). *L'Échange, Revue Linnéenne* **25** (293): 129–131. DP: [31] May 1909 (issue cover).

Pic, M. 1909e: Descriptions ou diagnoses et notes diverses (Suite). *L'Échange, Revue Linnéenne* **25** (294): 137–139. DP: [30] June 1909 (issue cover).

Pic, M. 1909f: Descriptions ou diagnoses et notes diverses. Suite. *L'Échange, Revue Linnéenne* **25** (295): 145–146. DP: [31] July 1909 (issue cover).

Pic, M. 1909g: Sur divers altisides de la Turquie d'Asie principalement et de l'Europe méridionale (Col., Phytophaga). *Bulletin de la Société Entomologique de France* **1909** (13): 226–228. DP: 9 August 1909 (*dates d'apparition* at the end of volume).

Pic, M. 1909h: Descriptions ou diagnoses et notes diverses. Suite. *L'Échange, Revue Linnéenne* **25** (296): 153–156. DP: [31] August 1909 (issue cover).

Pic, M. 1909i: Sur quelques coléoptères recueillis dans les Hautes-Alpes en 1908. *L'Échange, Revue Linnéenne* **25** (296): 156–157. DP: [31] August 1909 (issue cover).

Pic, M. 1909j: Descriptions ou diagnoses et notes diverses. Suite. *L'Échange, Revue Linnéenne* **25** (297): 161–162. DP: [30] September 1909 (issue cover).

Pic, M. 1909k: Descriptions ou diagnoses et notes diverses. Suite. *L'Échange, Revue Linnéenne* **25** (299): 177–179. DP: [30] November 1909 (issue cover).

Pic, M. 1909l: Descriptions ou diagnoses et notes diverses. Suite. *L'Échange, Revue Linnéenne* **25** (300): 185–186. DP: [31] December 1909 (issue cover).

Pic, M. 1910a: Descriptions ou diagnoses et notes diverses. Suite. *L'Échange, Revue Linnéenne* **26** (301): 1–2. DP: [31] January 1910 (issue cover).

Pic, M. 1910b: Descriptions ou diagnoses et notes diverses. Suite. *L'Échange, Revue Linnéenne* **26** (303): 17–19. DP: [31] March 1910 (issue cover).

Pic, M. 1910c: Descriptions ou diagnoses et notes diverses. Suite. *L'Échange, Revue Linnéenne* **26** (304): 25–27. DP: [30] April 1910 (issue cover).

Pic, M. 1910d: Descriptions ou diagnoses et notes diverses. Suite. *L'Échange, Revue Linnéenne* **26** (308): 57–58. DP: [31] August 1910 (issue cover).

Pic, M. 1910e: Descriptions ou diagnoses et notes diverses. Suite. *L'Échange, Revue Linnéenne* **26** (309): 65–66. DP: [30] September 1910 (issue cover).

Pic, M. 1910f: Sur divers phytophages paléarctiques et exotiques (Col.). *Bulletin de la Société Entomologique de France* **1910** (17): 305–307. DP: 23 November 1910 (*dates de publication* at the end of volume).

Pic, M. 1910g: Descriptions ou diagnoses et notes diverses (Suite). *L'Échange, Revue Linnéenne* **26** (311): 81–83. DP: [30] November 1910 (issue cover).

Pic, M. 1910h: Descriptions ou diagnoses et notes diverses (Suite). *L'Échange, Revue Linnéenne* **26** (312): 89–91. DP: [31] December 1910 (issue cover).

Pic, M. 1911a: Sur divers altisides du Nord de l'Afrique avec description de plusieurs formes nouvelles (Col. Chrysomelidae). *Bulletin de la Société Entomologique de France* **1911** (1): 9–11. DP: 25 January 1911 (*dates de publication* at the end of volume).

Pic, M. 1911b: Descriptions ou diagnoses et notes diverses (Suite). *L'Échange, Revue Linnéenne* **27** (313): 97–98. DP: [31] January 1911 (issue cover).

Pic, M. 1911c: Descriptions ou diagnoses et notes diverses (Suite). *L'Échange, Revue Linnéenne* **27** (314): 105–107. DP: [28] February 1911 (issue cover).

Pic, M. 1911d: Contribution à l'étude du genre "Derocrepis" Weise. *L'Échange, Revue Linnéenne* **27** (314): 108–110. DP: [28] February 1911 (issue cover).

Pic, M. 1911e: Descriptions ou diagnoses et notes diverses (Suite). *L'Échange, Revue Linnéenne* **27** (315): 113–114. DP: [31] March 1911 (issue cover).

Pic, M. 1911f: Descriptions ou diagnoses et notes diverses (Suite). *L'Échange, Revue Linnéenne* **27** (317): 129–130. DP: [31] May 1911 (issue cover).

Pic, M. 1911g: Descriptions ou diagnoses et notes diverses (Suite). *L'Échange, Revue Linnéenne* **27** (318): 137–139. DP: [30] June 1911 (issue cover).

Pic, M. 1911h: Coléoptères exotiques nouveaux ou peu connus. *L'Échange, Revue Linnéenne* **27** (318): 142–144. DP: [30] June 1911 (issue cover).

Pic, M. 1911i: Descriptions ou diagnoses et notes diverses (Suite). *L'Échange, Revue Linnéenne* **27** (323): 177–178. DP: [30] November 1911 (issue cover).

Pic, M. 1911j: Descriptions ou diagnoses et notes diverses (Suite). *L'Échange, Revue Linnéenne* **27** (324): 185–186. DP: [31] December 1911 (issue cover).

Pic, M. 1912a: Descriptions ou diagnoses et notes diverses (Suite). *L'Échange, Revue Linnénne* **28** (325): 1–3. DP: [31] January 1912 (issue cover).

Pic, M. 1912b: Descriptions ou diagnoses et notes diverses (Suite). *L'Échange, Revue Linnénne* **28** (327): 17. DP: [31] March 1912 (issue cover).

Pic, M. 1912c: Notes et observations sur divers alticides (Col. Phytophaga). *Bulletin de la Société Entomologique de France* **1912** (6): 135–136. DP: 10 April 1912 (*dates de publication* at the end of volume).

Pic, M. 1912d: Descriptions ou diagnoses et notes diverses (Suite). *L'Échange, Revue Linnénne* **28** (329): 33–35. DP: [31] May 1912 (issue cover).

Pic, M. 1912e: Descriptions ou diagnoses et notes diverses (Suite). *L'Échange, Revue Linnénne* **28** (330): 41–43. DP: [30] June 1912 (issue cover).

Pic, M. 1912f: Descriptions ou diagnoses et notes diverses (Suite). *L'Échange, Revue Linnénne* **28** (331): 49–51. DP: [31] July 1912 (issue cover).

Pic, M. 1912g: Coléoptères égyptiens nouveaux ou plus ou moins rares. *Bulletin de la Société Entomologique d'Égypte* 4 [1911] (2): 71–76. DP: [31] July 1912 (*Natur. Novit.* 1912, 34 (13): 375).

Pic, M. 1912h: Descriptions ou diagnoses et notes diverses (Suite). *L'Échange, Revue Linnénne* 28 (334): 73–74. DP: [31] October 1912 (issue cover).

Pic, M. 1912i: Anomalies, captures et nouveautés de coléoptères. *L'Échange, Revue Linnéenne* 28 (335): 81–83. DP: [30] November 1912 (issue cover).

Pic, M. 1912j: Coléoptères du Maroc. *Bulletin du Muséum National d'Histoire Naturelle* (Paris) 18 (4): 219–221. DP: [31] December 1912 (*Bibl. Zool.* 23: 449, 480).

Pic, M. 1912k: Nouveaux coléoptères paléarctiques. *L'Échange, Revue Linnéenne* 28 (336): 89–90. DP: [31] December 1912 (issue cover).

Pic, M. 1912l: Renseignements généraux sur les Bruchidae. *L'Échange, Revue Linnéenne* 28 (336): 91–93. DP: [31] December 1912 (issue cover).

Pic, M. 1912m: Synopsis pour aider a distinguer "Cryptocephalus carinthiacus" Suff. des sinuatus Harold et voisins. *Annales de la Société Linnéenne de Lyon, Nouvelle Série* 59: 99–102. DP: [31 December] 1912 (volume cover).

Pic, M. 1913a: Sur quelques coléoptères de Sibérie. *L'Échange, Revue Linnéenne* 29 (337): 101–102. DP: [31] January 1913 (issue cover).

Pic, M. 1913b: Coléoptères exotiques en partie nouveaux (Suite). *L'Échange, Revue Linnéenne* 29 (338): 106–110. DP: [28] February 1913 (issue cover).

Pic, M. 1913c: Descriptions de 29 espèces et de plusieurs variétés. *Mélanges Exotico-Entomologiques* 5: 7–20. DP: 25 March 1913 (issue cover).

Pic, M. 1913d: Notes diverses, descriptions et diagnoses (Suite.). *L'Échange, Revue Linnéenne* 29 (339): 113–114. DP: [31] March 1913 (issue cover).

Pic, M. 1913e: Notes diverses, descriptions et diagnoses (Suite.). *L'Échange, Revue Linnéenne* 29 (341): 129–130. DP: [31] May 1913 (issue cover).

Pic, M. 1913f: Notes diverses, descriptions et diagnoses (Suite.). *L'Échange, Revue Linnéenne* 29 (342): 137–139. DP: [30] June 1913 (issue cover).

Pic, M. 1913g: Coléoptères exotiques en partie nouveaux (Suite). *L'Échange, Revue Linnéenne* 29 (342): 142–144. DP: [30] June 1913 (issue cover).

Pic, M. 1913h: Espèces et variétés nouvelles appartenant à diverses familles. *Mélanges Exotico-Entomologiques* 6: 8–16. DP: 12 July 1913 (issue cover).

Pic, M. 1913i: Notes diverses, descriptions et diagnoses (Suite.). *L'Échange, Revue Linnéenne* 29 (344): 153–154. DP: [31] August 1913 (issue cover).

Pic, M. 1913j: Notes diverses, descriptions et diagnoses (Suite.). *L'Échange, Revue Linnéenne* 29 (346): 169–171. DP: [31] October 1913 (issue cover).

Pic, M. 1913k: Notes diverses, descriptions et diagnoses (Suite.). *L'Échange, Revue Linnéenne* 29 (347): 177–180. DP: [30] November 1913 (issue cover).

Pic, M. 1913l: Notes diverses, descriptions et diagnoses (Suite.). *L'Échange, Revue Linnéenne* 29 (348): 185–187. DP: [31] December 1913 (issue cover).

Pic, M. 1913m: « Cryptocephalus 4-punctatus » Oliv. et ses variétés. *L'Échange, Revue Linnéenne* 29 (348): 190–191. DP: [31] December 1913 (issue cover).

Pic, M. 1914a: Notes diverses, descriptions et diagnoses. (Suite.). *L'Échange, Revue Linnéenne* 30 (349): 1–2. DP: [31] January 1914 (issue cover).

Pic, M. 1914b: Notes diverses, descriptions et diagnoses. (Suite.). *L'Échange, Revue Linnéenne* 30 (350): 10–13. DP: [28] February 1914 (issue cover).

Pic, M. 1914c: Note sur Cyaniris thoracica Fisch. et ses variétés (Col. Phytophaga). *Bulletin de la Société Entomologique de France* **1914** (4): 139. DP: 11 March 1914 (*dates de publication* at the end of volume).

Pic, M. 1914d: Notes diverses, descriptions et diagnoses. (Suite.). *L'Échange, Revue Linnéenne* 30 (351): 17–18. DP: [31] March 1914 (issue cover).

Pic, M. 1914e: Observations et corrections concernant divers Cryptocephalus Geof. paléarctiques. *Bulletin de la Société Zoologique de France* **39** (2): 106–108. DP: 16 April 1914 (issue cover).

Pic, M. 1914f: Nouveaux coléoptères de diverses familles. *Mélanges Exotico-Entomologiques* 10: 7–20. DP: 30 October 1914 (issue cover).

Pic, M. 1914g: Notes diverses, descriptions et diagnoses. (Suite.). *L'Échange, Revue Linnéenne* 30 (359): 73–75. DP: 10 November 1914 (issue cover).

Pic, M. 1914h: Nouveau genre, espèces et variétés nouvelles de diverses familles. *Mélanges Exotico-Entomologiques* **11**: 2–8. DP: 25 November 1914 (issue cover).

Pic, M. 1914i: Notes diverses, descriptions et diagnoses. (Suite.). *L'Échange, Revue Linnéenne* 30 (360): 77–79. DP: 10 December 1914 (issue cover).

Pic, M. 1915a: Notes diverses, descriptions et diagnoses (Suite). *L'Èchange, Revue Linnéenne* **31** (361): 1–3. DP: 10 January 1915 (issue cover).

Pic, M. 1915b: Notes diverses, descriptions et diagnoses (Suite). *L'Èchange, Revue Linnéenne* **31** (362): 5–6. DP: 12 February 1915 (issue cover).

Pic, M. 1915c: Nouveautés de diverses familles. *Mélanges Exotico-Entomologiques* **13**: 2–13. DP: 1 March 1915 (issue cover).

Pic, M. 1915d: Notes diverses, descriptions et diagnoses (Suite). *L'Èchange, Revue Linnéenne* **31** (364): 13–14. DP: 10 April 1915 (issue cover).

Pic, M. 1915e: Notes diverses, descriptions et diagnoses (Suite). *L'Èchange, Revue Linnéenne* **31** (365): 17–18. DP: 10 May 1915 (issue cover).

Pic, M. 1915f: Notes diverses, descriptions et diagnoses (Suite). *L'Èchange, Revue Linnéenne* **31** (369): 33–34. DP: 10 September 1915 (issue cover).

Pic, M. 1915g: Genres nouveaux, espèces et variétés nouvelles. *Mélanges Exotico-Entomologiques* **16**: 2–13. DP: 20 October 1915 (issue cover).

Pic, M. 1915h: Notes diverses, descriptions et diagnoses (Suite). *L'Èchange, Revue Linnéenne* **31** (371): 41–42. DP: 12 November 1915 (issue cover).

Pic, M. 1915i: Coléoptères exotiques en partie nouveaux (Suite). *L'Échange, Revue Linnéenne* **31** (372): 48. DP: 12 December 1915 (issue cover).

Pic, M. 1916a: Notes diverses, descriptions et diagnoses (Suite). *L'Échange, Revue Linnéenne* **32** (374): 5–7. DP: 3 April 1916 (issue cover).

Pic, M. 1916b: Notes diverses, descriptions et diagnoses (Suite). *L'Échange, Revue Linnéenne* **32** (375): 9–11. DP: 3 June 1916 (issue cover).

Pic, M. 1916c: Espèces et variétés nouvelles. *Mélanges Exotico-Entomologiques* 19: 6–20. DP: 28 June 1916 (issue cover).

Pic, M. 1916d: Descriptions abrégées diverses. *Mélanges Exotico-Entomologiques* 20: 1–20. DP: 24 July 1916 (issue cover).

Pic, M. 1916e: Notes diverses, descriptions et diagnoses (Suite). *L'Échange, Revue Linnéenne* 32 (376): 13–14. DP: 2 August 1916 (issue cover).

Pic, M. 1916f: Descriptions abrégées diverses. *Mélanges Exotico-Entomologiques* 21: 2–20. DP: 4 October 1916 (issue cover).

Pic, M. 1916g: Notes diverses, descriptions et diagnoses (Suite). *L'Échange, Revue Linnéenne* 32 (378): 21–23. DP: 2 December 1916 (issue cover).

Pic, M. 1917a: Descriptions abrégées diverses. *Mélanges Exotico-Entomologiques* 24: 2–24. DP: 20 June 1917 (issue cover).

Pic, M. 1917b: Notes diverses, descriptions et diagnoses. *L'Échange, Revue Linnéenne* 33 (382): 13–15. DP: 2 August 1917 (issue cover).

Pic, M. 1917c: Notes diverses, descriptions et diagnoses. *L'Échange, Revue Linnéenne* 33 (384): 21–23. DP: 2 December 1917 (issue cover).

Pic, M. 1917d: Descriptions abrégées diverses. *Mélanges Exotico-Entomologiques* 26: 2–24. DP: 12 December 1917 (issue cover).

Pic, M. 1918a: Courtes descriptions diverses. *Mélanges Exotico-Entomologiques* 27: 1–24. DP: 15 March 1918 (issue cover).

Pic, M. 1918b: Notes diverses, descriptions et diagnoses (Suite). *L'Échange, Revue Linnéenne* 34 (387): 9–11. DP: 4 June 1918 (issue cover).

Pic, M. 1918c: Notes diverses, descriptions et diagnoses (Suite). *L'Échange, Revue Linnéenne* 34 (388): 13–15. DP: 2 August 1918 (issue cover).

Pic, M. 1918d: Notes diverses, descriptions et diagnoses (Suite). *L'Échange, Revue Linnéenne* 34 (389): 17–19. DP: 2 October 1918 (issue cover).

Pic, M. 1918e: Notes diverses, descriptions et diagnoses (Suite). *L'Échange, Revue Linnéenne* 34 (390): 21–24. DP: 1 December 1918 (issue cover).

Pic, M. 1919a: Notes diverses, descriptions et diagnoses (Suite). *L'Échange, Revue Linnéenne* 35 (391): 1–2. DP: 3 February 1919 (issue cover).

Pic, M. 1919b: Notes diverses, descriptions et diagnoses (Suite). *L'Échange, Revue Linnéenne* 35 (394): 13–14. DP: 2 August 1919 (issue cover).

Pic, M. 1919c: Notes diverses, descriptions et diagnoses (Suite). *L'Échange, Revue Linnéenne* 35 (395): 17–20. DP: 2 October 1919 (issue cover).

Pic, M. 1919d: Notes diverses, descriptions et diagnoses (Suite). *L'Échange, Revue Linnéenne* 35 (396): 21–23. DP: 2 December 1919 (issue cover).

Pic, M. 1919e: Étude détaillée de divers insectes. *L'Échange, Revue Linnéenne* 35 (396) [hors-texte]: 17–20. DP: 2 December 1919 (issue cover).

Pic, M. 1920a: Notes diverses, descriptions et diagnoses (Suite). *L'Échange, Revue Linnéenne* 36 (398): 5–8. DP: 8 April 1920 (issue cover).

Pic, M. 1920b: Notes diverses, descriptions et diagnoses (Suite). *L'Échange, Revue Linnéenne* **36** (399): 9–12. DP: 2 June 1920 (issue cover).

Pic, M. 1920c: Nouveautés diverses. *Mélanges Exotico-Entomologiques* **32**: 1–28. DP: 10 June 1920 (issue cover).

Pic, M. 1920d: Diagnoses de coléoptères exotiques (Suite). *L'Échange, Revue Linnéenne* **36** (401): 18–20. DP: 2 October 1920 (issue cover).

Pic, M. 1920e: Notes diverses, descriptions et diagnoses (Suite). *L'Échange, Revue Linnéenne* **36** (402): 21–22. DP: 2 December 1920 (issue cover).

Pic, M. 1921a: Notes diverses, descriptions et diagnoses (Suite). *L'Échange, Revue Linnéenne* **37** (403): 1–4. DP: 10 February 1921 (issue cover).

Pic, M. 1921b: Nouveautés diverses. *Mélanges Exotico-Entomologiques* **33**: 1–32. DP: 12 May 1921 (issue cover).

Pic, M. 1921c: Deux nouveaux Crioceris Geoffr. asiatiques (Col. Chrysomelidae). *Bulletin de la Société Entomologique de France* **1921** (9): 136–137. DP: 20 June 1921 (*dates de publication* at the end of volume).

Pic, M. 1921d: Nouveautés diverses. *Mélanges Exotico-Entomologiques* **34**: 1–33. DP: 10 October 1921 (issue cover).

Pic, M. 1921e: Notes diverses, descriptions et diagnoses (Suite). *L'Échange, Revue Linnéenne* **37** (406): 13–15. DP: 31 October 1921 (issue cover).

Pic, M. 1922a: Notes diverses, descriptions et diagnoses (Suite). *L'Échange, Revue Linnéenne* **38** (407): 17–19. DP: 15 January 1922 (issue cover).

Pic, M. 1922b: Coléoptères exotiques en partie nouveaux (Suite). *L'Échange, Revue Linnéenne* **38** (407): 19–20. DP: 15 January 1922 (issue cover).

Pic, M. 1922c: Nouveautés diverses. *Mélanges Exotico-Entomologiques* **35**: 1–32. DP: 10 February 1922 (issue cover).

Pic, M. 1922d: Notes diverses, descriptions et diagnoses (Suite). *L'Échange, Revue Linnéenne* **38** (409): 25–28. DP: 2 August 1922 (issue cover).

Pic, M. 1922e: Nouveautés diverses. *Mélanges Exotico-Entomologiques* **36**: 1–32. DP: 1 September 1922 (issue cover).

Pic, M. 1922f: Nouveautés diverses. *Mélanges Exotico-Entomologiques* **37**: 1–32. DP: 28 November 1922 (issue cover).

Pic, M. 1923a: Un nouveau coléoptère phytophage d'Italie. *Bollettino della Società Entomologica Italiana* **55** (2): 28–29. DP: 19 February 1923 (issue cover and Poggi 2008).

Pic, M. 1923b: Nouveautés diverses. *Mélanges Exotico-Entomologiques* **38**: 1–32. DP: 20 March 1923 (issue cover).

Pic, M. 1923c: Notes diverses, descriptions et diagnoses (Suite). *L'Échange, Revue Linnéenne* **39** (412): 5–7. DP: 15 April 1923 (issue cover).

Pic, M. 1923d: Notes diverses, descriptions et diagnoses (Suite). *L'Échange, Revue Linnéenne* **39** (413): 9–11. DP: 12 July 1923 (issue cover).

Pic, M. 1923e: Nouveautés diverses. *Mélanges Exotico-Entomologiques* 39: 3–32. DP: 20 August 1923 (issue cover).

Pic, M. 1923f: Sur diverses coléoptères intéressants ou nouveaux d'Egypte. *Bulletin de la Société Royale Entomologique d'Égypte* 7 [1922–1923]: 95–104. DP: 13 October 1923 (stamp on the BHL copy).

Pic, M. 1923g: Nouveautés diverses. *Mélanges Exotico-Entomologiques* 40: 3–32. DP: 18 November 1923 (issue cover).

Pic, M. 1924a: Notes diverses, descriptions et diagnoses (Suite). *L'Échange, Revue Linnéenne* 39 [= 40] (415): 17–18. DP: 5 February 1924 (issue cover).

Pic, M. 1924b: Nouveautés diverses. *Mélanges Exotico-Entomologiques* 41: 1–32. DP: 15 May 1924 (issue cover).

Pic, M. 1924c: Notes diverses, descriptions et diagnoses (Suite). *L'Échange, Revue Linnéenne* 39 [= 40] (416): 21–23. DP: 23 May 1924 (issue cover).

Pic, M. 1924d: Hispides nouveaux (Col.). *Bulletin de la Société Entomologique de France* **1924** (8/9): 99–100. DP: 24 June 1924 (*dates de publication* on p. 278).

Pic, M. 1924e: Notes diverses, descriptions et diagnoses (Suite). *L'Échange, Revue Linnéenne* 39 [= 40] (417): 25–27. DP: 20 August 1924 (issue cover).

Pic, M. 1924f: Nouveaux Bruchidae du Congo Belge. *Revue de Zoologie Africaine* **12** (4): 455–460. DP: 1 November 1924 (paper title page).

Pic, M. 1924g: Nouveautés diverses. *Mélanges Exotico-Entomologiques* 42: 1–32. DP: 10 November 1924 (issue cover).

Pic, M. 1924h: Notes diverses, descriptions et diagnoses (Suite). *L'Échange, Revue Linnéenne* 39 [= 40] (418): 29–30. DP: 20 November 1924 (issue cover).

Pic, M. 1924i: Coléoptères exotiques en partie nouveaux (Suite). *L'Échange, Revue Linnéenne* 39 [= 40] (418): 30–32. DP: 20 November 1924 (issue cover).

Pic, M. 1925a: Nouveautés diverses. *Mélanges Exotico-Entomologiques* 43: 1–32. DP: 18 April 1925 (issue cover).

Pic, M. 1925b: Notes diverses, descriptions et diagnoses (Suite). *L'Échange, Revue Linnéenne* 41 (421): 9–11. DP: 20 August 1925 (issue cover).

Pic, M. 1925c: Coléoptères d'Egypte et du Sinaï. *Bulletin de la Société Royale Entomologique d'Egypte* 8 [1924] (2/4): 392–397. DP: [31 December] 1925 (issue title page).

Pic, M. 1926a: Nouveaux coléoptères exotiques. *Bulletin de la Société Entomologique de France* [**1925**] (19): 322–324. DP: 8 February 1926 (*dates de publication* on p. 356).

Pic, M. 1926b: Nouveautés diverses. *Mélanges Exotico-Entomologiques* 45: 1–32. DP: 18 March 1926 (issue cover).

Pic, M. 1926c: Nouveaux coléoptères du Tonkin. III. *Bulletin de la Société Zoologique de France* 51 (2): 143–145. DP: 30 April 1926 (issue cover).

Pic, M. 1926d: Nouveautés diverses. *Mélanges Exotico-Entomologiques* 46: 1–32. DP: 16 July 1926 (issue cover).

Pic, M. 1926e: Notes diverses, descriptions et diagnoses (Suite). *L'Échange, Revue Linnéenne* **42** (425): 9–10. DP: 20 August 1926 (issue cover).

Pic, M. 1926f: Nouveautés diverses. *Mélanges Exotico-Entomologiques* **47**: 1–32. DP: 20 November 1926 (issue cover).

Pic, M. 1927a: Diagnoses de coléoptères exotiques. *Bulletin Bi-Mensuel de la Société Linnéenne de Lyon* **6** (1): 5–6. DP: 7 January 1927 (issue cover).

Pic, M. 1927b: Nouveaux coléoptères du globe (2e article). *Bulletin de la Société Zoologique de France* **51** [1926] (5): 392–394. DP: 30 January 1927 (issue cover).

Pic, M. 1927c: Nouveautés diverses. *Mélanges Exotico-Entomologiques* **48**: 1–32. DP: 9 April 1927 (issue cover).

Pic, M. 1927d: Nouveaux coléoptères du Globe II. *Bulletin de la Société Zoologique de France* **52** (2): 106–110. DP: 30 April 1927 (issue cover).

Pic, M. 1927e: Notes diverses, descriptions et diagnoses (Suite). *L'Échange, Revue Linnéenne* **43** (428): 5–7. DP: 12 May 1927 (issue cover).

Pic, M. 1927f: Coléoptères exotiques en partie nouveaux (Suite). *L'Échange, Revue Linnéenne* **43** (428): 7–8. DP: 12 May 1927 (issue cover).

Pic, M. 1927g: Coléoptères nouveaux de Chine et du Japon. *Bulletin de la Société Entomologique de France* **1927** (9): 152–154. DP: 7 June 1927 (issue cover).

Pic, M. 1927h: Nouveaux coléoptères du Globe (IV). *Bulletin de la Société Zoologique de France* **52** (3): 185–189. DP: 30 June 1927 (issue cover).

Pic, M. 1927i: Nouveaux coleoptères exotiques. *Entomologische Mitteilungen* **16** (4): 246–250. DP: 1 July 1927 (issue cover).

Pic, M. 1927j: Nouveaux coléoptères exotiques. *Bulletin de la Société Entomologique de France* **1927** (11): 183–184. DP: 5 July 1927 (issue cover).

Pic, M. 1927k: Coléoptères de l'Indochine. *Mélanges Exotico-Entomologiques* **49**: 1–36. DP: 8 July 1927 (issue cover).

Pic, M. 1927l: Notes diverses, descriptions et diagnoses (Suite). *L'Échange, Revue Linnéenne* **43** (429): 9–11. DP: 16 August 1927 (issue cover).

Pic, M. 1927m: Coléoptères du globe. *Mélanges Exotico-Entomologiques* **50**: 1–36. DP: 14 October 1927 (issue cover).

Pic, M. 1927n: Coléoptères asiatiques nouveaux. *Bulletin Bi-mensuel de la Société Linnéenne de Lyon* **6** (17): 132–133. DP: 10 November 1927 (issue cover).

Pic, M. 1927o: Descriptions de deux coléoptères d'Egypte. *Bulletin de la Société Royale Entomologique d'Egypte, Nouvelle Série* **11** (1): 1–2. DP: [31 December] 1927 (issue cover).

Pic, M. 1928a: Nouveaux Bruchidae. *Annals and Magazine of Natural History, Tenth Series* **1** (2): 297–299. DP: 1 February 1928 (Evenhuis 2003).

Pic, M. 1928b: Contribution à l'étude du genre Lema F. (Col.). *Bulletin de la Société Entomologique de France* **1928** (5): 95–96. DP: 16 April 1928 (*dates d'apparition* on p. 332).

Pic, M. 1928c: Notes et descriptions. *Mélanges Exotico-Entomologiques* **51**: 1–36. DP: 16 April 1928 (issue cover).

Pic, M. 1928d: Nouveaux coléoptères. *Revue Scientifique du Bourbonnais et du Centre de la France* **1928** (1): 15–21. DP: [30] April 1928 (issue cover).

Pic, M. 1928e: Six nouveaux coléoptères du Tonkin et mutation. *Bulletin Bi-mensuel de la Société Linnéenne de Lyon* 7 (10): 87–88. DP: 18 May 1928 (issue cover).

Pic, M. 1928f: Notes diverses, descriptions et diagnoses (Suite). *L'Échange, Revue Linnénne* **44** (433): 9–10. DP: 12 August 1928 (issue cover).

Pic, M. 1928g: Nouveautés diverses. *Mélanges Exotico-Entomologiques* **52**: 1–32. DP: 10 September 1928 (issue cover).

Pic, M. 1928h: Coléoptères asiatiques nouveaux. *Bulletin de la Société Entomologique de France* **1928** (16): 264. DP: 24 November 1928 (*dates d'apparition* on p. 332).

Pic, M. 1928i: Nouveaux coléoptères surtout Tonkinois. *Bulletin de la Société Zoologique de France* **53** (6): 377–379. DP: 10 December 1928 (issue cover).

Pic, M. 1929a: Nouveaux coléoptères Malachidae et Bruchidae. *Bulletin de la Société Entomologique de France* [**1928**] (20): 314–316. DP: 22 January 1929 (*dates d'apparition* on p. 332).

Pic, M. 1929b: Coléoptères exotiques en partie nouveaux (Suite). *L'Échange, Revue Linnéenne* **45** (435): 4. DP: 6 February 1929 (issue cover).

Pic, M. 1929c: Nouveaux coléoptères paléarctiques. *Časopis Československé Společnosti Entomologické* **25** [1928] (5/6): 118–120. DP: 25 February 1929 (issue cover).

Pic, M. 1929d: Lema femorata Guér. et formes voisines. *Entomologisches Nachrichtenblatt* (Troppau) **3** (1): 15. DP: [28] February 1929 (footer on p. 15).

Pic, M. 1929e: Coléoptères exotiques nouveaux ou peu connus. *Bulletin de la Société Zoologique de France* **54** (1): 43–46. DP: 28 February 1929 (issue cover).

Pic, M. 1929f: Diagnoses de coléoptères exotiques. *Bulletin Bi-mensuel de la Société Linnéenne de Lyon* **8** (5): 34–36. DP: 8 March 1929 (issue cover).

Pic, M. 1929g: Nouveautés diverses. *Mélanges Exotico-Entomologiques* **53**: 1–36. DP: 15 March 1929 (issue cover).

Pic, M. 1929h: Coléoptères phytophages nouveaux. *Bulletin de la Société Zoologique de France* **54** (2): 138–139. DP: 30 April 1929 (issue cover).

Pic, M. 1929i: Contribution a l'étude du genre Pseudometaxis Jac. (Col.). *Bulletin de la Société Zoologique de France* **54** (2): 140–141. DP: 30 April 1929 (issue cover).

Pic, M. 1929j: Coléoptères nouveaux de la Cyrénaique. *Bollettino della Società Entomologica Italiana* **61** (5/6): 94–96. DP: 3 June 1929 (issue cover and Poggi 2008).

Pic, M. 1929k: Phytophages exotiques nouveaux (Col.). *Bulletin de la Société Entomologique de France* **1929** (8): 147–148. DP: 5 June 1929 (*dates d'apparition* on p. 336).

Pic, M. 1929l: Notes diverses, nouveautés. *L'Échange, Revue Linnéenne* **45** (438): 13–14. DP: 4 November 1929 (issue cover).

Pic, M. 1930a: Nouveaux coléoptères d'Afrique et du Tonkin. *Bulletin de la Société Zoologique de France* **54** [1929] (6): 609–612. DP: 15 February 1930 (issue cover).

Pic, M. 1930b: Nouveaux coléoptères exotiques. *Bulletin Bi-mensuel de la Société Linnéenne de Lyon* **9** (6): 36–37. DP: 20 March 1930 (issue cover).

Pic, M. 1930c: Notes diverses, nouveautés. *L'Échange, Revue Linnénne* 46 (440): 5–7. DP: 3 May 1930 (issue cover).

Pic, M. 1930d: Nouveautés diverses. *Mélanges Exotico-Entomologiques* 56: 1–36. DP: 20 June 1930 (issue cover).

Pic, M. 1930e: Notes diverses, nouveautés. *L'Échange, Revue Linnénne* 46 (442): 13–14. DP: 12 November 1930 (issue cover).

Pic, M. 1930f: Coléoptères asiatiques nouveaux. *Acta Entomologica Musei Nationalis Pragae* 8: 58–59. DP: [31 December] 1930 (volume cover).

Pic, M. 1931a: Notes diverses, nouveautés. *L'Échange, Revue Linnéenne* 47 (443): 1–2. DP: 9 February 1931 (issue cover).

Pic, M. 1931b: Nouveautés diverses. *Mélanges Exotico-Entomologiques* 57: 1–36. DP: 25 May 1931 (issue cover).

Pic, M. 1931c: Notes diverses, nouveautés (Suite). *L'Échange, Revue Linnéenne* 47 (445): 9–10. DP: 8 August 1931 (issue cover).

Pic, M. 1931d: Nouveautés diverses. *Mélanges Exotico-Entomologiques* 58: 1–36. DP: 28 August 1931 (issue cover).

Pic, M. 1931e: Nouveaux coléoptères du globe. *Bulletin Bi-mensuel de la Société Linnéenne de Lyon* 10 (18): 138–139. DP: 22 November 1931 (issue cover).

Pic, M. 1932a: Nouveaux Bruchidae des Indes. *Annals and Magazine of Natural History, Tenth Series* 9 (51): 329–333. DP: 1 March 1932 (Evenhuis 2003).

Pic, M. 1932b: Coléoptères des Iles Mascareignes. Mission scientifique de P. Carié (1910–1913). Malacodermes, ptinides, anthicides et hylophilides. *Annales de la Société Entomologique de France* 101 (1): 40–54, pls. I–IV. DP: 31 March 1932 (*dates de publication* on p. 339).

Pic, M. 1932c: Nouveautés diverses. *Mélanges Exotico-Entomologiques* 59: 10–36. DP: 9 April 1932 (issue cover).

Pic, M. 1932d: Notes diverses, nouveautés (Suite.). *L'Échange, Revue Linnéenne* 48 (448): 21–23. DP: 18 May 1932 (issue cover).

Pic, M. 1932e: Notes diverses, nouveautés (Suite.). *L'Échange, Revue Linnéenne* 48 (449): 25–26. DP: 8 August 1932 (issue cover).

Pic, M. 1932f: Quatre nouveaux coléoptères de Chine et du Tonkin. *Bulletin de la Société Entomologique de France* 37 (13/14): 206–208. DP: 10 September 1932 (*dates d'apparition* on p. 320).

Pic, M. 1932g: Nouveautés diverses. *Mélanges Exotico-Entomologiques* 60: 1–36. DP: 14 September 1932 (issue cover).

Pic, M. 1933a: Nouveautés diverses. *Mélanges Exotico-Entomologiques* 61: 3–36. DP: 20 March 1933 (issue cover).

Pic, M. 1933b: Notes diverses, nouveautés (Suite). *L'Échange, Revue Linnéenne* 49 (453): 9–11. DP: 4 August 1933 (issue cover).

Pic, M. 1933c: Notes diverses, nouveautés (Suite). *L'Échange, Revue Linnéenne* 49 (454): 13–15. DP: 6 November 1933 (issue cover).

Pic, M. 1933d: Schwedisch-chinesische wissenschaftliche Expedition nach den nordwestlichen Provinzen Chinas, unter Leitung von Dr. Sven Hedin und Prof. Sü Ping-chang. Insekten gesammelt vom schwedischen Arzt der Expedition Dr. David Hummel 1927–1930. 16. Coleoptera. 2. Helmidae, Dermestidae, Anobiidae, Cleridae Malacodermata, Dascillidae, Heteromera (ex p.), Bruchidae, Cerambycidae, Phytophaga (ex p.). *Arkiv för Zoologi* 27A [1936] (2): 1–14. DP: 20 December 1933 (last page of the paper).

Pic, M. 1934a: Notes diverses, nouveautés (Suite). *L'Échange, Revue Linnéenne* 50 (455): 17–19. DP: 2 February 1934 (issue cover).

Pic, M. 1934b: Coléoptères exotiques en partie nouveaux. *L'Échange, Revue Linnéenne* 50 (455): 19–20. DP: 2 February 1934 (issue cover).

Pic, M. 1934c: Nouveautés diverses. *Mélanges Exotico-Entomologiques* 63: 1–36. DP: 4 April 1934 (issue cover).

Pic, M. 1934d: Notes diverses, nouveautés (Suite). *L'Échange, Revue Linnéenne* 50 (456): 21–22. DP: 8 May 1934 (issue cover).

Pic, M. 1934e: Coléoptères exotiques en partie nouveaux. *L'Échange, Revue Linnéenne* 50 (456): 22–23. DP: 8 May 1934 (issue cover).

Pic, M. 1934f: Notes variées sur les coléoptères paléarctiques. *Bulletin de la Société Entomologique de France* 39 (10): 156–158. DP: 24 July 1934 (*dates de parution* on p. 308).

Pic, M. 1934g: Nouveaux coléoptères de Chine. *Entomologisches Nachrichtenblatt* (Troppau) 8 (2): 84–87. DP: [31] July 1934 (footers on each page).

Pic, M. 1934h: Notes diverses, nouveautés (Suite). *L'Échange, Revue Linnéenne* 50 (457): 25–27. DP: 2 August 1934 (issue cover).

Pic, M. 1934i: Nouveautés diverses, notes. *Mélanges Exotico-Entomologiques* 64: 1–36. DP: 24 September 1934 (issue cover).

Pic, M. 1934j: Notes diverses, nouveautés (Suite). *L'Échange, Revue Linnéenne* 50 (458): 29–31. DP: 13 November 1934 (issue cover).

Pic, M. 1935a: Neue Bruchidae (Col.). *Entomologischer Anzeiger* 15 (9): 65–66. DP: 1 March 1935 (issue title page).

Pic, M. 1935b: Nouveautés diverses. *Mélanges Exotico-Entomologiques* 65: 1–36. DP: 8 April 1935 (issue cover).

Pic, M. 1935c: Coléoptères exotiques en partie nouveaux (Suite). *L'Echange, Revue Linnéenne* 51 (460): 8. DP: 7 May 1935 (issue cover).

Pic, M. 1935d: Notes diverses, nouveautés. *L'Échange, Revue Linnéenne* 51 (461): 11–12. DP: 2 August 1935 (issue cover).

Pic, M. 1935e: Nouveautés diverses. *Mélanges Exotico-Entomologiques* 66: 1–36. DP: 12 September 1935 (issue cover).

Pic, M. 1935f: Coléoptères exotiques en partie nouveaux (Suite). *L'Echange, Revue Linnéenne* 51 (462): 15–16. DP: 6 November 1935 (issue cover).

Pic, M. 1936a: Nouveautés diverses. *Mélanges Exotico-Entomologiques* 67: 1–36. DP: 25 February 1936 (issue cover).

Pic, M. 1936b: Nouveaux coléoptères exotiques. *Bulletin de la Société Zoologique de France* **61** (4): 298–300. DP: 5 July 1936 (issue cover).

Pic, M. 1936c: Notes diverses, nouveautés (Suite). *L'Échange, Revue Linnéenne* **52** (465): 25–27. DP: 4 August 1936 (issue cover).

Pic, M. 1936d: Nouveautés diverses. *Mélanges Exotico-Entomologiques* **68**: 10–36. DP: 28 August 1936 (issue cover).

Pic, M. 1936e: Descriptions de coléoptères. *Bulletin de la Société Entomologique de France* **41** (11): 213–216. DP: 23 September 1936 (*dates de parution* on p. 352).

Pic, M. 1936f: Notes diverses, nouveautés (Suite). *L'Échange, Revue Linnéenne* **52** (466): 29–30. DP: 4 November 1936 (issue cover).

Pic, M. 1937a: Nouveautés diverses. *Mélanges Exotico-Entomologiques* **69**: 1–36. DP: 18 June 1937 (issue cover).

Pic, M. 1937b: Coléoptères nouveaux de Chine. *Notes d'Entomologie Chinoise* **4** (7): 169–176. DP: 12 July 1937 (table of contents).

Pic, M. 1937c: Notes diverses, nouveautés (Suite.). *L'Échange, Revue Linnéenne* **53** (470): 13–15. DP: 4 November 1937 (issue cover).

Pic, M. 1938a: Notes diverses, nouveautés. *L'Échange, Revue Linnéenne* **54** (471): 1–3. DP: 4 February 1938 (issue cover).

Pic, M. 1938b: Coleoptera partim: Dermestidae – Phytophaga. Pp. 14–18. In: Sjöstedt, Y.: Insekten aus China im Naturhistorischen Reichsmuseum zu Stockholm. Heimgebracht von Direktor Kjell Kolthoff und anderen schwedischen Forschern und Reisenden. *Arkiv för Zoologi* **30A** (13): 1–19. DP: 19 April 1938 (last page of the paper).

Pic, M. 1938c: Nouveautés diverses, mutations. *Mélanges Exotico-Entomologiques* **70**: 1–36. DP: 10 June 1938 (issue cover).

Pic, M. 1938d: Descriptions de coléoptères et notes synonymiques. *Bulletin de la Société Entomologique de France* **43** (9/10): 121–124. DP: 13 July 1938 (*dates de parution* on p. 260).

Pic, M. 1938e: Trois Bruchus L. nouveaux des Indes (Coléoptères). *Annals and Magazine of Natural History, Eleventh Series* **2** (10): 399–400. DP: 1 October 1938 (Evenhuis 2003).

Pic, M. 1938f: Drei neue Coleopteren-Arten. *Arbeiten über Morphologische und Taxonomische Entomologie aus Berlin-Dahlem* **5** (4): 332–333. DP: 15 December 1938 (volume title page).

Pic, M. 1938g: Nouveaux coléoptères du Siam. *Bulletin de la Société Zoologique de France* **63** (6): 352–356. DP: 30 December 1938 (issue cover).

Pic, M. 1938h: Un nouveau coléoptère phytophage de Chine. *Entomologisches Nachrichtenblatt* (Troppau) **12** (3/4): 158. DP: [31] December 1938 (footers throughout the issue).

Pic, M. 1939a: Notes diverses, nouveautés (Suite.). *L'Échange, Revue Linnéenne* **55** (475): 17–19. DP: 2 February 1939 (issue cover).

Pic, M. 1939b: Nouveaux coléoptères de l'Indochine. *Bulletin de la Société Entomologique de France* **44** (1/2): 17–19. DP: 24 March 1939 (*dates de parution* on p. 240).

Pic, M. 1939c: Nouveautés et observations sur divers coléoptères. Pp. 342–344. In: Jordan, K. & Hering, E. M. (eds.): *Verhandlungen. VII. Internationaler Kongress für Entomologie. Band I.*

Weimar: Internationer Kongress für Entomologie, 617 pp., 29 pls. DP: [30] April 1939 (book title page).

Pic, M. 1939d: Mutations et nouveautés diverses. *Mélanges Exotico-Entomologiques* **71**: 1–36. DP: 24 May 1939 (issue cover).

Pic, M. 1939e: [new taxa]. In: Koch, C.: Die Käfer der libyschen Ausbeute des Herrn Georg Frey. *Mitteilungen der Münchner Entomologischen Gesellschaft* **29** (2/3): 216–293. DP: 1 July 1939 (volume pages).

Pic, M. 1939f: Notes diverses, nouveautés (Suite.). *L'Échange, Revue Linnéenne* **55** (477): 25–26. DP: 4 August 1939 (issue cover).

Pic, M. 1939g: Tenthredides (Hymén.) et coléoptères nouveaux. *L'Échange, Revue Linnéenne* **55** (478): 29–30. DP: 8 November 1939 (issue cover).

Pic, M. 1939h: Coléoptères nouveaux de l'Égypte et du Sinaï. *Bulletin de la Société Fouad 1er d'Entomologie* **23**: 143–149. DP: 31 December 1939 (verso of the volume title page).

Pic, M. 1940a: Diagnoses de coléoptères exotiques. *L'Échange, Revue Linnéenne* **56** (479): 2–4. DP: 2 February 1940 (issue cover).

Pic, M. 1940b: Diagnoses de coléoptères exotiques (Suite). *L'Échange, Revue Linnéenne* **56** (480): 6–8. DP: 6 May 1940 (footer on p. 5).

Pic, M. 1941a: Opuscula martialis II. *L'Échange, Revue Linnéenne* **Numéro Spécial 2**: 1–16. DP: 16 April 1941 (issue cover).

Pic, M. 1941b: Diagnoses de coléoptères exotiques (Suite). *L'Échange, Revue Linnéenne* **57** (484): 6–8. DP: 2 May 1941 (heading on p. 5).

Pic, M. 1941c: Opuscula martialis IV. *L'Échange, Revue Linnéenne* **Numéro Spécial 4**: 1–16. DP: 18 August 1941 (issue cover).

Pic, M. 1941d: Opuscula martialis V. *L'Échange, Revue Linnéenne* **Numéro Spécial 5**: 1–16. DP: 22 October 1941 (issue cover).

Pic, M. 1941e: Coléoptères du globe. *L'Échange, Revue Linnéenne* **57** (486): 13–16. DP: 6 November 1941 (heading on p. 13).

Pic, M. 1942a: Opuscula martialis VI. *L'Échange, Revue Linnéenne* **Numéro Spécial 6**: 1–16. DP: 20 February 1942 (issue cover).

Pic, M. 1942b: Coléoptères du globe (suite). *L'Échange, Revue Linnéenne* **58** (488): 5–8. DP: 8 May 1942 (heading on p. 5).

Pic, M. 1942c: Coléoptères nouveaux de l'Adrar des Iforas. *Bulletin de la Société Entomologique de France* **47** (5): 77–79. DP: [31] May 1942 (headings on p. 77 and 79).

Pic, M. 1942d: Opuscula martialis VII. *L'Échange, Revue Linnéenne* **Numéro Spécial 7**: 1–16. DP: 2 June 1942 (issue cover).

Pic, M. 1942e: Coléoptères du globe (suite). *L'Échange, Revue Linnéenne* **58** (489): 9–12. DP: 18 August 1942 (heading on p. 9).

Pic, M. 1943a: Coléoptères du globe (suite). *L'Échange, Revue Linnéenne* **59** (491): 1–4. DP: 4 February 1943 (heading on p. 1).

Pic, M. 1943b: Opuscula martialia IX. *L'Échange, Revue Linnéenne* **Numéro Spécial 9**: 1–16. DP: 22 February 1943 (issue cover).

Pic, M. 1943c: Opuscula martialia XI. *L'Échange, Revue Linnéenne* **Numéro Spécial 11**: 1–16. DP: 22 October 1943 (issue cover).

Pic, M. 1943d: Coléoptères du globe (suite). *L'Échange, Revue Linnéenne* **59** (493 [= 494]): 13–16. DP: 26 November 1943 (heading on p. 13).

Pic, M. 1944a: Coléoptères du globe (suite). *L'Échange, Revue Linnéenne* **60** (496): 5–8. DP: 4 May 1944 (heading on p. 5).

Pic, M. 1944b: Opuscula martialia XIII. *L'Échange, Revue Linnéenne* **Numéro Spécial 13**: 1–16. DP: 28 September 1944 (issue cover).

Pic, M. 1945a: Coléoptères du globe (suite). *L'Échange, Revue Linnéenne* **61** (499): 1–4. DP: 12 February 1945 (heading on p. 1).

Pic, M. 1945b: Coléoptères du globe (suite). *L'Échange, Revue Linnéenne* **61** (501): 10–12. DP: 18 August 1945 (heading on p. 9).

Pic, M. 1945c: Coléoptères du globe (suite). *L'Échange, Revue Linnéenne* **61** (502): 13–16. DP: 18 December 1945 (heading on p. 13).

Pic, M. 1946a: Coléoptères du globe (suite). *L'Échange, Revue Linnéenne* **62** (503): 1–4. DP: 5 February 1946 (issue cover).

Pic, M. 1946b: Coléoptères du globe (suite). *L'Échange, Revue Linnéenne* **62** (506): 13–16. DP: 16 November 1946 (heading on p. 13).

Pic, M. 1947a: Coléoptères du globe (suite). *L'Échange, Revue Linnéenne* **63** (508): 5–8. DP: [31] May 1947 (heading on p. 5).

Pic, M. 1947b: Diversités Entomologique. *Diversités Entomologiques* **2**: 1–16. DP: 10 September 1947 (issue cover).

Pic, M. 1947c: Coléoptères du globe (suite). *L'Échange, Revue Linnéenne* **63** (510): 13–16. DP: 28 November 1947 (heading on p. 13).

Pic, M. 1948: Coléoptères du globe (suite). *L'Échange, Revue Linnéenne* **64** (514): 13–16. DP: 29 November 1948 (heading on p. 13).

Pic, M. 1949a: Coléoptères du globe (suite). *L'Échange, Revue Linnéenne* **65** (515): 1–4. DP: 6 February 1949 (issue cover).

Pic, M. 1949b: Coléoptères du globe (suite). *L'Échange, Revue Linnéenne* **65** (517): 9–12. DP: 7 August 1949 (heading on p. 9).

Pic, M. 1949c: Nouveaux coléoptères exotiques et notes diverses. *Miscellanea Entomologica* **46** (6): 49–55. DP: [31] October 1949 (footer on the paper title page).

Pic, M. 1949d: Coléoptères du globe (suite). *L'Échange, Revue Linnéenne* **65** (518): 13–16. DP: 14 November 1949 (heading on p. 13).

Pic, M. 1950a: Nouveautés diverses, mutation. *Diversités Entomologiques* **6**: 15–16. DP: 25 January 1950 (issue cover).

Pic, M. 1950b: Coléoptères du globe (suite). *L'Échange, Revue Linnéenne* **66** (520): 5–8. DP: 12 May 1950 (heading on p. 5).

Pic, M. 1950c: Descriptions et notes variées. *Diversités Entomologiques* 7: 1–16. DP: 26 June 1950 (issue cover).

Pic, M. 1950d: Cinq nouveaux coléoptères du Maroc. *Comptes Rendus des Séances Mensuelles de la Société des Sciences Naturelles du Maroc* **1950** (6): 92–94. DP: 29 August 1950 (last page of the issue).

Pic, M. 1950e: Contribution a l'étude de l'Aïr (Mission L. Chopard et A. Villiers). Coléoptères Chrysomelidae. Descriptions d'espèces nouvelles. *Mémoires de l'Institut Français de l'Afrique Noire* 10: 203–207. DP: 12 December 1950 (last page of the volume).

Pic, M. 1950f: Contribution a l'étude de l'Aïr (Mission L. Chopard et A. Villiers). Coléoptères Bruchidae. *Mémoires de l'Institut Français de l'Afrique Noire* 10: 211. DP: 12 December 1950 (last page of the volume).

Pic, M. 1951a: Descriptions diverses et notes. *Diversités Entomologiques* 8: 7–16. DP: 1 February 1951 (issue cover).

Pic, M. 1951b: Notes et descriptions. *Diversités Entomologiques* 9: 4–16. DP: 14 April 1951 (issue cover).

Pic, M. 1951c: Coléoptères du globe (suite). *L'Échange, Revue Linnéenne* **67** (526): 13–16. DP: 6 November 1951 (heading on p. 13).

Pic, M. 1951d: Coléoptères rares ou nouveaux d'Afrique. *Bulletin de l'Institut Français d'Afrique Noire* **13** (4): 1099–1102. DP: [30] November 1951 (last page of the issue).

Pic, M. 1951e: Nouveaux bruchides (Coleoptera) de l'Afrique australe. *Annals of the Transvaal Museum* **21** [1948–1951] (4): 427–428. DP: 6 July 1951 (issue title page).

Pic, M. 1952: Coléoptères du globe (suite). *L'Échange, Revue Linnéenne* **68** (530): 13–16. DP: 8 November 1952 (heading on p. 13).

Pic, M. 1953a: Le variétisme de Cryptocephalus coryli L. *L'Échange, Revue Linnéenne* **69** (531): 1–2. DP: 12 February 1953 (issue cover).

Pic, M. 1953b: Coléoptères du globe (suite). *L'Échange, Revue Linnéenne* **69** (531): 2–4. DP: 12 February 1953 (issue cover).

Pic, M. 1953c: Quatre nouveaux coléoptères du Maroc. *Comptes Rendus des Séances Mensuelles de la Société des Sciences Naturelles du Maroc* **1953** (1): 18–20. DP: 27 April 1953 (date of publishing of the issue 2).

Pic, M. 1953d: Coléoptères du globe (suite). *L'Échange, Revue Linnéenne* **69** (532): 5–8. DP: 16 May 1953 (heading on p. 13).

Pic, M. 1953e: Coléoptères nouveaux de la région marocaine. *Comptes Rendus des Séances Mensuelles de la Société des Sciences Naturelles du Maroc* **1953** (4): 61–63. DP: 18 July 1953 (noticed on p. 66).

Pic, M. 1953f: Coléoptères du globe (suite). *L'Échange, Revue Linnéenne* **69** (533): 9–12. DP: 20 August 1953 (heading on p. 9).

Pic, M. 1953g: Coléoptères du globe (suite). *L'Échange, Revue Linnéenne* **69** (534): 13–16. DP: 10 November 1953 (heading on p. 13).

Pic, M. 1954a: Coléoptères du globe (suite). *L'Échange, Revue Linnéenne* **70** (535): 1–3. DP: 9 February 1954 (issue cover).

Pic, M. 1954b: Descriptions et notes. *Diversités Entomologiques* **13**: 4–12. DP: 10 March 1954 (issue cover).

Pic, M. 1954c: Coléoptères du globe (suite). *L'Échange, Revue Linnéenne* 70 (536): 5–8. DP: 6 May 1954 (heading on p. 5).

Pic, M. 1954d: Coléoptères nouveaux de Chine. *Bulletin de la Société Entomologique de Mulhouse* **1954**: 53–59. DP: [31] October 1954 (paper title page).

Pic, M. 1955a: Coléoptères du globe (suite). *L'Échange, Revue Linnéenne* **71** (539): 1–3. DP: 4 February 1955 (issue cover).

Pic, M. 1955b: Phytophages présumé nouveau du Maroc. *Comptes Rendus des Séances Mensuelles de la Société des Sciences Naturelle et Physiques du Maroc* 20 [1954] (8): 206. DP: [31] March 1955 (last page of the issue).

Pic, M. 1955c: Coléoptères nouveaux de Chine. *Bulletin de la Société Entomologique de Mulhouse* **1955**: 21–23. DP: [30] April 1955 (paper title page).

Pic, M. 1955d: Coléoptères du globe (suite). *L'Échange, Revue Linnéenne* **71** (540): 5–8. DP: 4 May 1955 (heading on p. 5).

Pic, M. 1955e: Descriptions diverses. *Diversités Entomologiques* **14**: 7–20. DP: 16 May 1955 (issue cover).

Pic, M. 1955f: Rectificatif. *Comptes Rendus des Séances de la Société des Sciences Naturelle et Physiques du Maroc* **21** (2): 41. DP: 21 April 1955 (noticed on p. 62).

Piolti, G. 1880: Descrizione di una nuova specie del genere Chrysomela Linn. *Atti della Reale Accademia delle Scienze di Torino* **15** [1879–1880]: 378–380. DP: 11 January 1880 (journal website).

Piton, L.-E. 1943: Aberrations nouvelles de Monolepta signata Ol. (coléoptères chrysomélides) du Yunnan. *Bulletin Mensuel de la Société Linnéenne de Lyon* **12** (9): 137–138. DP: [30] November 1943 (issue title page).

Poda von Neuhaus, N. 1761: *Insecta Musei Graecensis, que in ordines, genera et species juxta Systema Naturae Caroli Linnaei digessit.* Widmanstad: Graecii, 12 + 127 + 18 pp., 2 pls. DP: 3 September 1761 (Bousquet 2016).

Polentz, G. 1938: Beiträge zur schlesischen Käferfauna. *Mitteilungen aus der Entomologischen Gesellschaft zu Halle (Saale)* **16**: 48–60. DP: [31] August 1938 (table of contents).

Polentz, G. 1939: Beiträge zur schlesischen Käferfauna. *Zeitschrift für Entomologie* (Breslau) **18** (3): 4–11. DP: 31 July 1939 (issue title page).

Ponting, J. 2020: Replacement for the genus group name Anopsilus Jacobsen, 1917 (Coeleoptera: Chrysomelidae: Cryptocephalinae), junior homonym of Anopsilus Kirsch, 1870 (Coleoptera: Curculionidae). *Munis Entomology & Zoology* **15** (1): 298–300. DP: 1 January 2020 (headings throughtout the paper, see also the comments in the Dating of publications chapter).

Pontoppidan, E. L. 1763: *Den Danske atlas eller Konge-Riget Dannemark, med dets naturlige egenskaber, elementer, indbyggere, vaexter, dyr og andre affødninger, dets gamle tildragelser og naervaerende omstaendingheder i alle provintzer, staeder, kirker, slotte og herre-gaarde. Forestillet ved en udførlig lands-beskrivelse, saa og oplyst med dertil forfaerdigede land-kort over enhver provintz, samt ziret med staedernes prospecter, grund-ridser, og andre merkvaerdige kaaber-stykker. Efter Høy-Kongelig allernaadigst Befalning. Tomus 1.* Kiøbenhavn: A. H. Godiche, [6] + xl + [4] + 723 + [1] pp., 30 pls. DP: [31 December] 1763 (Bousquet 2016).

Ponza, L. 1805: Coleoptera salutiensia sive enumeratio methodica Coleopterorum, quae in agro salutiensi reperiuntur, locorum natalium indicatione, observationibus, novisque aliquot speciebus aucta. *Mémoires de l'Académie Impériale des Sciences, Littérature et Beaux-Arts de Turin, Sciences Physiques et Mathématiques* **14**: 29–94, 2 pls. DP: [31 December] 1805 (volume title page).

Porta, A. 1934: *Fauna Coleopterorum Italica. IV. – Heteromera – Phytophaga. Oedemeridae, Pythidae, Pyrochroidae, Hylophilidae, Anthicidae, Meloidae, Rhipiphoridae, Mordellidae, Melandryidae, Lagriidae, Alleculidae, Tenebrionidae, Cerambycidae, Chysomelidae, Lariidae.* Piacenza, 415 pp. DP: [31 December] 1934 (book title page).

Porta, A. 1949: *Fauna Coleopterorum Italica. Supplementum II.* San Remo: G. Gandolfi, 386 pp. DP: [31 December] 1949 (book title page).

Portevin, G. 1934: Histoire Naturelle des Coléoptères de France. Tome III. Polyphaga: Heteromera, Phytophaga. *Encyclopédie Entomologique, Série A – Travaux Généraux* 17: vi + 1–374, 15 pls. DP: [30] May 1934 (last page of the volume).

Power, J. A. 1863: Description of a new British Zeugophora. *Zoologist* **21**: 8735. DP: [31 December] 1863 (volume title page).

Prathapan, K. D. 2017: Identity of the citrus leaf mining flea beetle in northeast India and nomenclatural changes in Amphimela (Coleoptera: Chrysomelidae: Galerucinae: Alticini). *Florida Entomologist* 100 (2): 276–280. DP: 5 June 2017 (verso of the issue cover).

Prevett, P. F. 1965: The genus Caryedon in northern Nigeria, with descriptions of six new species (Col. Bruchidae). *Annales de la Société Entomologique de France, Nouvelle Série* 1 (3): 523–547. DP: 8 December 1965 (stamp on the Gallica copy).

Prevett, P. F. 1966a: A new genus and species of Pachymerinae (Coleoptera: Bruchidae) from South America. *Proceedings of the Royal Entomological Society of London, Series B Taxonomy* 35 (5/6): 81–83. DP: 8 July 1966 (issue cover).

Prevett, P. F. 1966b: che genus Pachymerus Thunberg (Coleoptera, Bruchidae). *Bulletin of Entomological Research* 57 [1966–1968] (1): 181–192. DP: 29 November 1966 (table of contents).

Raffray, A. 1873: [new taxa]. In: Fairmaire, L. & Raffray, A.: Coléoptères du nord de l'Afrique. *Revue et Magasin de Zoologie Pure et Appliquée, 3e Série* 1 (9): 331–385, pls. 15–16. DP: [31 December] 1873 (issue cover) [plate 16 published in issue 7, plate 15 in issue 8].

Rafinesque, C. S. 1815: *Analyse de la nature ou tableau de l'univers et des corps organisés.* Palermo: J. Barravecchia, 224 pp. DP: 15 July 1815 (Bousquet 2016).

Ragusa, E. 1871: Descrizione di due nuove specie di coleotteri trovate in Sicilia. *Bullettino della Società Entomologica Italiana* 2 [1870] (4): 315–316. DP: 28 February 1871 (Poggi 2008).

Ragusa, E. 1874: Di un nuovo Georyssus e Calomicrus trovati in Sicilia. *Bullettino della Società Entomologica Italiana* 5 [1873] (4): 233–234. DP: 20 February 1874 (Poggi 2008).

Ragusa, E. 1882: Coleotteri nuovi o poco conosciuti della Sicilia. *Il Naturalista Siciliano* 1 [1881–1882] (11): 248–251. DP: 1 August 1882 (issue cover).

Ragusa, E. 1887: Coleotteri nuovi o poco conosciuti della Sicilia. *Il Naturalista Siciliano* 6 [1886–1887] (11): 214–216. DP: 1 August 1887 (issue title page).

Ragusa, E. 1922: Coleotteri nuovi o poco conosciuti della Sicilia. *Bullettino della Società Entomologica Italiana* 53 [1921] (4): 121–130. DP: 15 February 1922 (Poggi 2008).

Ragusa, E. 1923: Coleotteri nuovi o poco conosciuti nella Sicilia. *Atti della Reale Accademia di Scienze, Lettere e Belle Arti di Palermo* 12 [1920–1922]: 1–18. DP: 14 February 1923 (*Bull. Soc. Ent. Fr.* 1923 (3): 44).

Randall, J. W. 1838: Description of new species of coleopterous insects inhabiting the state of Maine. *Boston Journal of Natural History* 2 [1838–1839] (1): 1–33. DP: [28] February 1838 (issue title page).

Rapilly, M. 1978: Contribution a la faune d'Iran (Col. Chrysomelidae Alticinae). *Nouvelle Revue d'Entomologie* 8 (3): 329–343. DP: 15 November 1978 (issue cover).

Rapilly, M. 1980: Contribution a la faune d'Iran. Coléoptères Chrysomelidae Cryptocephalinae. *Revue Française d'Entomologie, Nouvelle Série* 2 (2): 75–85. DP: 12 June 1980 (last page of the cover).

Rapilly, M. 1981: Une nouvelle espèce d'Hydrothassa pour la faune d'Iran (Coleoptera, Chrysomelidae, Chrysomelinae). *Revue Française d'Entomologie, Nouvelle Série* 3 (4): 143–145. DP: 10 December 1981 (last page of the cover).

Rapilly, M. 1982a: Description d'un nouveau Clytra d'Iran (Coleoptera, Chrysomelidae). *Revue Française d'Entomologie, Nouvelle Série* 4 (1): 18–20. DP: 31 March 1982 (last page of the cover).

Rapilly, M. 1982b: Description d'une nouvelle espèce de Pachybrachis de France (Col. Chrysomelidae). *L'Entomologiste* 38 (2): 53–56. DP: [30] April 1982 (issue cover).

Rapilly, M. 1984: Révision des Labidostomis d'Iran (Col. Chrysomelidae, Clytrinae). *Annales de la Société Entomologique de France, Nouvelle Série* 20 (2): 215–230. DP: 29 June 1984 (issue table of contents).

Ratzeburg, J. T. C. 1837: *Die Forst-Insecten oder Abbildung und Beschreibung der in den Wäldern Preussens und der Nachbarstaaten als schädlich oder nützlich bekannt gewordenen Insecten; in systematischer Folge und mit besonderer Rücksicht auf die Vertilgung der Schädlichen. Erster Theil. Die Käfer.* Berlin: Nicolai'schen Buchhandlung, x + [4] + 202 pp., 22 pls. DP: 3 November 1837 (Bousquet 2016).

Recalde Irurzun, J. I. & Ugarte San Vicente I. 2005: Los Cassidinae de Navarra y la comudidad autónoma Vasca: aproximación faunística y corológica (Coleoptera: Chrysomelidae). *Heteropterus Revista de Entomología* 5: 65–96. DP: 31 December 2005 (last page of the paper).

Redtenbacher, L. 1843: Bemerkungen über die in Syrien von Theodor Kotschy gesammelten Käfer. Coleopterorum Syriae genera et species novae. Pp. 971–990, 2 pls. In: Russegger, J.: *Reisen in Europa, Asien und Afrika mit besonderer Rücksicht auf die naturwissenschaftlichen Verhältnisse der betreffenden Länder, unternommen in den Jahren 1835 bis 1841. Erster Band. Reise in Griechenland, Unteregypten, im nördlichen Syrien und Kleinasien. Zweiter Theil.* Stuttgart: E. Schweitzerbart'sche Verlagshandlung, 471–1102 pp. DP (pp. 871–1102): 18 May 1843 (Bousquet 2016).

Redtenbacher, L. 1845: *Die Gattungen der deutschen Kaefer-Fauna nach der analytischen Methode berabeitet, nebst einem kurz gefassten Leitfaden, zum Studium dieses Zweiges der Entomologie*. Wien: Carl Ueberreuter, [12] + 177 + [1], 2 pls. DP: [30] September 1845 (Bousquet 2016).

Redtenbacher, L. 1848: *Fauna Austriaca. Die Käfer. Nach der analytischen Methode*. [1847–1849], Wien: Carl Gerold, xxvii + 883 pp., 2 pls. DP (pp. 481–640): 9 February 1848 (Bousquet 2016, 2017).

Redtenbacher, L. 1849: [new taxa]. In: Kollar, V. & Redtenbacher, L.: *Ueber den Character der Insecten-Fauna von Südpersien*. Wien: Hof- und Staatsdruckerei, 12 pp. DP: [31 December] 1849 (Bousquet 2016) [preprint of *Denkschriften der Kaiserlichen Akademie der Wissenschaften in Wien, Mathematisch-Naturwissenschaftliche Klasse* 1850, **1**: 42–53].

Redtenbacher, L. 1858: *Fauna Austriaca. Die Käfer. Nach der analytischen Methode bearbeitet. Zweite, gänzlich umgearbeitete, mit mehreren Hunderten von Arten und mit der Charakteristik sämmlicher europäischen Käfergattungen vermehrte Auflage*. [1856–1858], Wien: Carl Gerold's Sohn, cxxxvi + 1017 pp., 2 pls. DP: (pp. 769–1008): 5 May 1858, (pp. 1009–1017, 2 pls): 26 August 1858 (Bousquet 2016).

Redtenbacher, L. 1868: Coleopteren. In: *Reise der Oesterreichischen Fregatte Novara um die Erde in den Jahren 1857, 1858, 1859 unter dem Befehlen des Commodore B. von Wüllerstorf-Urbair. Zoologischer Teil. Zweiter Band. 1. Abtheilung A 2*. Wien: Karl Gerold's Sohn, iv + 249 pp., 5 pls. DP: 22 May 1868 (Bousquet 2016).

Redtenbacher, L. 1873: *Fauna Austriaca. Die Käfer, nach der analytischen Methode bearbeitet. Dritte, gänzlich umgearbeitete und bedeutend vermehrte Auflage. Zweiter Band*. [1873–1874], Wien: C. Gerold's Sohn, cliii + 571 + [1] + viii pp., 2 pls. DP (pp. ?225–?448): 15 August 1873 (Bousquet 2016).

Redtenbacher, L. 1874: *Fauna Austriaca. Die Käfer, nach der analytischen Methode bearbeitet. Dritte, gänzlich umgearbeitete und bedeutend vermehrte Auflage. Zweiter Band*. [1873–1874], Wien: C. Gerold's Sohn, cliii + 571 + [1] + viii pp., 2 pls. DP (pp. ?449–571, 2 pls): 1 January 1874 (Bousquet 2016).

Redtenbacher, W. [G.] 1842: *Dissertatio inauguralis entomologica sistens quaedam genera et species coleopterorum Archiducatus Austriae nondum descriptorum, quam consensu et auctoritate illustrissimi ac magnifici domini praesidis et directoris, perillustris ac spectabilis Domini Decani, nec non clarissimorum ac celeberrimorum D.D. Professorum pro doctoris medicinae laurea summisque in medicina honoribus et privilegiis rite et legitime consequendis in antiquissima ac celeberrima Universitate Vindobonensi publicae disquisitioni submittit Guilielmus Redtenbacher, Kirchdorfensis ex Austria superiori. Disputabitur publice in theses adnexas in Universitatis aedibus die mensis Januarii 1842*. Vindobona: C. Ueberreuter, 31 pp. DP: [31] January 1842 (Bousquet 1842).

Regalin, R. 1997: Nota sui crisomelidi di Creta (Grecia) e descrizione di Lachnaia (Lachnaia) zoiai n. sp. (Coleoptera Chrysomelidae). *Bollettino della Società Entomologica Italiana* **129** (1): 67–77. DP: 20 July 1997 (paper title page).

Regalin, R. & Medvedev, L. N. 2010: [new name]. In: New acts and comments. Chrysomelidae: Cryptocephalinae: Clytrini. Pp. 76–79. In: Löbl, I. & Smetana, A. (eds.): *Catalogue of*

Palaearctic Coleoptera. Volume 6. Chrysomeloidea. Stenstrup: Apollo Books, 924 pp. DP: 22 February 2010 (colophon).

Regalin, R. & Medvedev, L. N. 2011: [new name]. In: Errata for Volume 6. Pp. 35–61. In: Löbl, I. & Smetana, A. (eds.): *Catalogue of Palaearctic Coleoptera. Volume 7. Curculionoidea I*. Stenstrup: Apollo Books, 373 pp. DP: 1 February 2011 (colophon).

Reiche, L. J. 1850: Entomologie. Pp. 257–471. In: Ferret, P. V. A. & Galinier, J. G. (eds.): *Voyage en Abyssinie dans les provinces du Tigré, du Samen et de l'Ahmara. Tome Troisième.* [1847–1848], Paris: Paulin, 536 pp. DP: 19 December 1850 (Bousquet 2016).

Reiche, L. J. 1861a: Coléoptères de Sicile recueillis par M. E. Bellier de la Chavignerie et déterminés par M. L. Reiche, et description de dix espèce nouvelles. *Annales de la Société Entomologique de France, Troisième Série* 8 [1860] (4): 717–737. DP: 15 May 1861 (issue cover).

Reiche, L. 1861b: Sur quelques espèces de coléoptères du nord de l'Afrique. *Annales de la Société Entomologique de France, Quatrième Série* **1** (1): 87–92. DP: 14 August 1861 (Lefèvre 1885).

Reiche, L. J. 1862a: Espèces nouvelles de coléoptères découvertes en Corse par M. E. Bellier de la Chavignerie, en 1861. *Annales de la Société Entomologique de France, Quatrième Série* **2** (2): 293–300. DP: 8 October 1862 (Lefèvre 1885).

Reiche, L. J. 1862b: Espèces nouvelles de coléoptères appartenant à la faune circa-méditerranéenne (suite). *Annales de la Société Entomologique de France, Quatrième Série* **2** (3): 539–546. DP: 10 December 1862 (Lefèvre 1885).

Reiche, L. J. 1864: Espèces nouvelles de coléoptères d'Algérie. *Annales de la Société Entomologique de France, Quatrième Série* **4** (2): 233–246. DP: 12 October 1864 (Lefèvre 1885).

Reiche, L. J. & Saulcy, F. 1858a: Espèces nouvelles ou peu connues de coléoptères, requeillies par M. F. de Saulcy, membre de l'Institut, dans son voyage en Orient. *Annales de la Société Entomologique de France, Troisième Série* **5** [1857] (4): 649–694. DP: 24 February 1858 (Evenhuis 2022).

Reiche, L. & Saulcy, F. de 1858b: Espèces nouvelles ou peu connues de coléoptères, recueillies par M. F. de Saulcy. Membre de l'Institut, dans son voyage en Orient, et décrités par MM. L. Reiche et Félicien de Saulcy. *Annales de la Société Entomologique de France, Troisième Série* **6** (1): 5–60. DP: 12 May 1858 (Evenhuis 2022).

Reid, C. A. M. 2000: Spilopyrinae Chapuis: a new subfamily in the Chrysomelidae and its systematic placement (Coleoptera). *Invertebrate Taxonomy* **14** (6): 837–862. DP: [31 December] 2000 (paper title page).

Reid, C. A. M. & De Little, D. W. 2013: A new species of Paropsisterna Motschulsky, 1860, a significant pest of plantation eucalypts in Tasmania and Ireland (Coleoptera: Chrysomelidae: Chrysomelinae). *Zootaxa* **3681** (4): 395–404. DP: 24 June 2013 (footer on the paper title page).

Reid, C. A. M. 2014: Chrysomelinae Latreille, 1802. Pp. 243–251. In: Leschen, R. A. B. & Beutel, R. G. (eds.): *Handbook of Zoology. Coleoptera, beetles. Morphology and systematics. Volume 3.* Berlin/Boston: Walter de Gruyter, 675 pp. DP: [31 December] 2014 (colophon).

Reineck, G. 1908: Eine neue paläarktische Coptocephala-Art (Col.). *Deutsche Entomologische Zeitschrift* **1908** (3): 368. DP: 1 May 1908 (table of contents).

Reineck, G. 1910: Bemerkungen zu Chrysomela rufa Duft. und Phytodecta rufipes. *Zeitschrift für Wissenschaftliche Insektenbiologie* **6** (2): 65–66. DP: 15 February 1910 (issue cover).

Reineck, G. 1911a: Eine neue paläarktische Chrysomelide (Col.). *Deutsche Entomologische Zeitschrift* **1911** (3): 336. DP: 4 May 1911 (table of contents).

Reineck, G. 1911b: Phytodecta (Spartophila) variabilis Ol. (Col.). *Deutsche Entomologische Zeitschrift* **1911** (4): 479–480, pl. V. DP: 26 June 1911 (table of contents).

Reineck, G. 1915a: Beitrag zur Kenntnis von Chrysomela graminis L. und virgata Motsch. (Col.). *Stettiner Entomologische Zeitung* **75** [1914] (2): 240–242. DP: 15 January 1915 (noticed on p. 414).

Reineck, G. 1915b: Über die äthiopischen Vertreter der Gattungen Cryptocephalus und Melixanthus (Anteriscus) des Königl. Zoolog. Museums in Berlin und einiger anderen Museen und Sammlungen (Coleopt. Chrysomelidae). *Mitteilungen aus dem Zoologischen Museum in Berlin* **7** [1913–1915] (3): 391–469. DP: [31] March 1915 (issue cover).

Reineck, G. 1922a: Beitrag zur Kenntnis der paläarktischen Chrysomeliden. *Neue Beiträge zur Systematischen Insektenkunde* **2** [1920–1922] (8): 78–80. DP: 15 March 1922 (issue cover).

Reineck, G. 1922b: Beitrag zur Kenntnis asiatischer Chrysomeliden (Col.). *Deutsche Entomologische Zeitschrift* **1922** (4): 368–371. DP: 1 December 1922 (table of contents).

Reineck, G. 1923: Beitrag zur Kenntnis der asiatisch-malayischen Megalopodiden (Col. Chrysom.). *Deutsche Entomologische Zeitschrift* **1923** (6): 605–612. DP: 15 December 1923 (issue cover).

Reineck, G. 1937: Chrysomelidae. Pp. 183–187. In: Entomologische Sammelergebnisse der Deutschen Hindukusch-Expedition 1935 der Deutschen Forschungsgemeinschaft. II Teil. *Arbeiten über Morphologische und Taxonomische Entomologie aus Berlin-Dahlem* **4** (3): 177–191. DP: 13 September 1937 (volume title page).

Reitter, E. 1872: Neue Käferarten von Oran, gesammelt von Hans Leder. *Berliner Entomologische Zeitschrift* **16** (1): 167–176 & (2/4): 177–186. DP (pp. 167–176): [31] March 1872, (pp. 177–186): [30] November 1872 (table of contents).

Reitter, E. 1879: Coleopterologische Ergebnisse einer Reise nach Croatien und Slavonien. Unter Mitwirkung der Herren Dr. Eppelsheim und Dr. von Heyden. *Verhandlungen der Kaiserlich-Königlichen Zoologisch-Botanischen Gesellschaft in Wien* **29** (1): 35–56. DP: 1 October 1879 (*Ent. Nachricht.* 1879: 260).

Reitter, E. 1880a: [new taxa]. In: Leder, H.: Beitrag zur kaukasischen Käfer-Fauna. Unter Mitwirkung hervorragender Fachgenossen. *Verhandlungen der Kaiserlich-Königlichen Zoologisch-Botanischen Gesellschaft in Wien* **29** [1879] (2): 451–488. DP: 15 April 1880 (Evenhuis 2023).

Reitter, E. 1880b: Neun neue Clavicornier (Coleoptera). *Verhandlungen des Naturforschen Vereins in Brünn* **18** [1879]: 1–6. DP: [31 December] 1880 (volume title page; *Lotos* 1880, 30: XV).

Reitter, E. 1884a: [new taxon]. In: Brenske, E. & Reitter, E.: Neuer Beitrag zur Käferfauna Griechenlands. *Deutsche Entomologische Zeitschrift* **28** (1): 17–100, pls. I–II. DP: [31] January 1884 (table of contents).

Reitter, E. 1884b: Resultate einer coleopterologischen Sammel-Campagne während den Monaten Februar bis April 1883 auf den jonischen Inseln. *Deutsche Entomologische Zeitschrift* **28** (1): 101–122. DP: [31] January 1884 (table of contents).

Reitter, E. 1886: Neue Coleopteren aus Europa und den angrenzenden Ländern, mit Bemerkungen über bekannte Arten. Zweiter Theil. *Deutsche Entomologische Zeitschrift* 30 (1): 67–72. DP: [30] June 1886 (table of contents).

Reitter, E. 1889a: Neue Coleopteren aus Circassien, gesammelt von A. Starck. *Wiener Entomologische Zeitung* 8 (3): 97–104. DP: 31 March 1889 (footers throughout the issue).

Reitter, E. 1889b: Neue Coleopteren aus Europa, den angrenzenden Ländern und Sibirien, mit Bemerkungen über bekannte Arten. Sechster Theil. *Deutsche Entomologische Zeitschrift* 1889 (1): 17–44. DP: [31] May 1889 (table of contents).

Reitter, E. 1889c: Neue Coleopteren aus Europa, den angrenzenden Ländern und Sibirien, mit Bemerkungen über bekannte Arten. Siebenter Theil. *Deutsche Entomologische Zeitschrift* 1889 (2): 273–288. DP: [30] November 1889 (table of contents).

Reitter, E. 1889d: Neue Coleopteren aus Europa, den angrenzenden Ländern und Sibirien, mit Bemerkungen über bekannte Arten. Achter Theil. *Deutsche Entomologische Zeitschrift* 1889 (2): 369–376. DP: [30] November 1889 (table of contents).

Reitter, E. 1889e: Coleopterologische Ergebnisse der im Jahre 1886 und 1887 in Transcaspien von Dr. G. Radde, Dr. A. Walter und A. Konchin ausgeführten Expedition. *Verhandlungen des Naturforschenden Vereines in Brünn* 27 [1888]: 95–133. DP: [31 December] 1889 (volume title page).

Reitter, E. 1890a: Neue Coleopteren aus Europa, den angrenzenden Ländern und Sibirien, mit Bemerkungen über bekannte Arten. Neunter Theil. *Deutsche Entomologische Zeitschrift* 1890 (1): 145–164. DP: 28 May 1890 (*Bull. Soc. Ent. Fr.* 1890: c).

Reitter, E. 1890b: Neue Coleopteren aus Europa, den angrenzenden Ländern und Sibirien, mit Bemerkungen über bekannte Arten. Zehnter Theil. *Deutsche Entomologische Zeitschrift* 1890 (1): 165–176. DP: 28 May 1890 (*Bull. Soc. Ent. Fr.* 1890: c).

Reitter, E. 1890c: Beschreibungen neuer Coleopteren aus Europa, dem Kaukasus, Russisch-Armenien und Turkestan. *Wiener Entomologische Zeitung* 9 (7): 189–198. DP: 25 September 1890 (footers throughout the issue).

Reitter, E. 1890d: Neue Coleopteren aus Europa, den angrenzenden Ländern und Sibirien, mit Bemerkungen über bekannte Arten. Elfter Theil. *Deutsche Entomologische Zeitschrift* 1890 (2): 385–396. DP: [31] October 1890 (table of contents).

Reitter, E. 1890e: Coleopterologische Notizen. XXXIX. *Wiener Entomologische Zeitung* 9 (9): 264–267. DP: 30 November 1890 (footers throughout the issue).

Reitter, E. 1891: Neue Coleopteren aus Europa, den angrenzenden Ländern und Sibirien, mit Bemerkungen über bekannte Arten. Zwölfter Theil. *Deutsche Entomologische Zeitschrift* 1891 (1): 17–36. DP: [30] June 1891 (table of contents).

Reitter, E. 1892a: Siebenter Beitrag zur Coleopterenfauna des russischen Reiches. *Wiener Entomologische Zeitung* 11 (5): 151–154. DP: 31 May 1892 (footers throughout the issue).

Reitter, E. 1892b: Coleopterologische Notizen. XLVI. *Wiener Entomologische Zeitung* 11 (8): 239–240. DP: 31 October 1892 (footers throughout the issue).

Reitter, E. 1893: Ueber die Crioceris-Arten aus der Verwandtschaft der Crioceris merdigera L. mit rothem oder theilweise rothem Kopfe. *Wiener Entomologische Zeitung* **12** (9): 302. DP: 30 November 1892 (footers throughout the issue).

Reitter, E. 1894a: Coleopterologische Notizen. XLIX. *Wiener Entomologische Zeitung* **13** (1): 15–16. DP: 15 January 1894 (footers throughout the issue)..

Reitter, E. 1894b: Coleopterologische Notizen, LII. *Wiener Entomologische Zeitung* **13** (6): 191–192. DP: 25 July 1894 (footers throughout the issue).

Reitter, E. 1894c: Elfter Beitrag zur Coleopteren-Fauna des russischen Reiches. *Wiener Entomologische Zeitung* **13** (8): 238–248. DP: 31 October 1894 (footers throughout the issue).

Reitter, E. 1894d: Fünfter Beitrag zur Coleopteren-Fauna von Europa and den angrenzenden Ländern. *Wiener Entomologische Zeitung* **13** (10): 299–306. DP: 31 December 1894 (footers throughout the issue).

Reitter, E. 1895a: Beschreibung neuer oder wenig gekannter Coleopteren aus der Umgebung von Akbes in Syrien. *Wiener Entomologische Zeitung* **14** (3): 79–88. DP: 31 March 1895 (footers throughout the issue).

Reitter, E. 1895b: Zwölfter Beitrag zur Coleopteren-Fauna des russischen Reiches. *Wiener Entomologische Zeitung* **14** (5): 149–162. DP: 5 June 1895 (footers throughout the issue).

Reitter, E. 1896: Achter Beitrag zur Coleopteren-Fauna von Europa und den angrenzenden Ländern. *Wiener Entomologische Zeitung* **15** (9): 265–272. DP: 31 October 1896 (footers throughout the issue).

Reitter, E. 1897a: Coleopterologische Notizen. LXII. *Wiener Entomologische Zeitung* **16** (9): 241–242. DP: 30 November 1897 (footers throughout the issue).

Reitter, E. 1897b: Dreissig neue Coleopteren aus russisch Asien und der Mongolei. *Deutsche Entomologische Zeitschrift* **1897** (2): 209–228. DP: [31] December 1897 (table of contents).

Reitter, E. 1898a: Sechzehnter Beitrag zur Coleopteren-Fauna des russischen Reiches. *Wiener Entomologische Zeitung* **17** (1): 17–22. DP: 25 January 1898 (footers throughout the issue).

Reitter, E. 1898b: Siebzehnter Beitrag zur Coleopteren-Fauna des russischen Reiches. *Wiener Entomologische Zeitung* **17** (4/5): 109–127. DP: 31 May 1898 (footers throughout the issue).

Reitter, E. 1898c: Neue Coleopteren aus Europa und den angrenzenden Ländern. *Deutsche Entomologische Zeitschrift* **1898** (2): 337–360. DP: 28 December 1898 (*Bull. Soc. Ent. Fr.* 1898: 374).

Reitter, E. 1900a: Coleoptera, gesammelt im Jahre 1898 in Chin. Central-Asien von Dr. Holderer in Lahr. *Wiener Entomologische Zeitung* **19** (6/7): 153–166, pl. I. DP: 15 July 1900 (footers throughout the issue).

Reitter, E. 1900b: Beschreibung und Abbildung von neuen Coleopteren der palaearctischen Fauna. *Wiener Entomologische Zeitung* **19** (9): 225–232. DP: 5 November 1900 (footers throughout the issue).

Reitter, E. 1901a: Uebersicht der Arten der Coleopteren-Gattung Pachnephorus Redt. aus der palaearctischen Fauna. *Wiener Entomologische Zeitung* **20** (3): 53–54. DP: 10 March 1901 (footers throughout the issue).

Reitter, E. 1901b: Neue Uebersicht der Arten der Coleopteren-Gattung Ischyronota Weise. *Wiener Entomologische Zeitung* 20 (5): 103–108. DP: 29 June 1901 (footers throughout the issue).

Reitter, E. 1901c: Weitere Beiträge zur Coleopteren-Fauna des russischen Reiches. *Deutsche Entomologische Zeitschrift* **1901** (1): 65–84. DP: 10 July 1901 (*Bull. Soc. Ent. Fr.* 1901: 241).

Reitter, E. 1901d: Abbildungen und Beschreibungen neuer Coleopteren aus der palaearctischen Fauna. *Wiener Entomologische Zeitung* 20 (8/9): 157–164. DP: 25 November 1901 (footers throughout the issue).

Reitter, E. 1901e: Vierzehnter Beitrag zur Coleopteren-Fauna von Europa und den angrenzenden Ländern. *Wiener Entomologische Zeitung* 20 (10): 200–202. DP: 25 December 1901 (footers throughout the issue).

Reitter, E. 1902a: Eine Serie neuer Coleopteren aus dem russischen Reiche. *Deutsche Entomologische Zeitschrift* **1901** (2): 177–186. DP: 22 January 1902 (*Bull. Soc. Ent. Fr.* 1902: 30).

Reitter, E. 1902b: Neue Coleopteren der palaearctischen Fauna. *Wiener Entomologische Zeitung* **21** (6): 137–141. DP: 15 July 1902 (footers throughout the issue).

Reitter, E. 1902c: Coleopterologische Studien. Uebersicht der Arten der Coleopteren-Gattung Malacosoma Chevrl., aus der palaearctischen Fauna. *Wiener Entomologische Zeitung* **21** (8/9): 217. DP: 15 October 1902 (footers throughout the issue).

Reitter, E. 1903: Uebersicht der mir bekannten Coleopteren-Arten der Gattung Galeruca Geoffr. *Wiener Entomologische Zeitung* **22** (4/5): 133–139. DP: 30 May 1903 (footers throughout the issue).

Reitter, E. 1904: Über vier Coleopteren aus der palaearktischen Fauna. *Wiener Entomologische Zeitung* **23** (5): 81–82. DP: 31 May 1904 (footers throughout the issue).

Reitter, E. 1905a: Acht neue Coleopteren aus der palaearktischen Fauna. *Wiener Entomologische Zeitung* **24** (3/4): 90–94. DP: 10 April 1905 (footers throughout the issue).

Reitter, E. 1905b: Sechzehn neue Coleopteren aus der palaearktischen Fauna. *Wiener Entomologische Zeitung* **24** (7/8): 241–251. DP: 15 August 1905 (footers throughout the issue).

Reitter, E. 1906a: Neue Coleopteren aus der palaearktischen Fauna. *Wiener Entomologische Zeitung* **25** (1): 31–37. DP: 20 January 1906 (footers throughout the issue).

Reitter, E. 1906b: Einige neue Coleopteren. *Deutsche Entomologische Zeitschrift* **1906** (2): 449–451. DP: 10 September 1906 (table of contents).

Reitter, E. 1908: Zwei neue Coleopteren aus Adana in Kleinasien. *Wiener Entomologische Zeitung* **27** (4/5): 133–136. DP: 15 April 1908 (footers throughout the issue).

Reitter, E. 1909a: Einige neue Coleopteren aus der paläarktischen Fauna. *Wiener Entomologische Zeitung* **28** (2): 53–58. DP: 1 March 1909 (footers throughout the issue).

Reitter, E. 1909b: Neun neue Coleopterenarten und -Varietäten aus der paläarktischen Fauna. *Wiener Entomologische Zeitung* **28** (4): 99–103. DP: 15 May 1909 (footers throughout the issue).

Reitter, E. 1909c: Due nuovi coleotteri italiani. *Rivista Coleotterologica Italiana* 7 (5): 87–89. DP: 15 May 1909 (issue cover).

Reitter, E. 1910: Due nuovi coleotteri di Sardegna. *Rivista Coleotterologica Italiana* 8 (5/6): 115–116. DP: [30] June 1910 (issue cover).

Reitter, E. 1911: Vier neue Koleopteren-Arten aus der palaearctischen Fauna. *Wiener Entomologische Zeitung* 30 (8): 234–236. DP: 1 November 1911 (footers throughout the issue).

Reitter, E. 1912: Übersicht der mir bekannten Malegia-Arten aus dem russischen Reiche (Col. Chrysomelidae.). *Wiener Entomologische Zeitung* 31 (2): 91–92. DP: 15 March 1912 (footers throughout the issue).

Reitter, E. 1913a: Coleopterologische Novitäten der palaearktischen Fauna. *Coleopterologische Rundschau* 2 (8/9): 121–125. DP: 1 August 1913 (*Ent. Litteraturbl.* 1913, 13 (10): 208).

Reitter, E. 1913b: Neue paläarktische Coleopteren. *Entomologische Blätter* 9 (11/12): 289–292. DP: 10 November 1913 (issue title page).

Reitter, E. 1913c: *Fauna Germanica. Die Käfer des Deutschen Reiches. IV. Band.* [1912], Stuttgart: K. G. Lutz, 236 pp., pl. 129–152. DP: [31] December 1913 (*Natur. Novit.* 1913, 35 (24): 612, see also *Ent. Bl.* 1914, 10: 127).

Reitter, E. 1914a: Über die mir bekannten Varietäten von Phytodecta (Spartophila) pallida Lin. *Wiener Entomologische Zeitung* 33 (1/2): 46. DP: 1 January 1914 (footers throughout the issue).

Reitter, E. 1914b: [new taxon]. In: Heikertinger, F.: Klarheit über die europäischen Crepidodera-Arten mit dunklem Thorax. *Deutsche Entomologische Zeitschrift* 1914 (2): 103–112. DP: 1 April 1914 (table of contents).

Reitter, E. 1916: Spartophila fornicata Brügg. ab. nov. Sequensi. *Wiener Entomologische Zeitung* 35 (8/10): 234. DP: 28 December 1916 (footers throughout the issue).

Reitter, E. 1917: Über Chrysochloa viridis Dfschm. und einige damit verwandte Formen. *Coleopterologische Rundschau* 6 (1/2): 6–7. DP: 18 April 1917 (issue cover).

Reitter, E. 1918: Einige neue Coleopteren aus Albanien. *Entomologische Blätter* 14 (1/3): 42–45. DP: 15 March 1918 (issue cover).

Reitter, E. 1920: Bestimmungs-Tabelle der europäischen Donaciini, mit Berücksichtigung der Arten aus der paläarktischen Region. *Wiener Entomologische Zeitung* 38 (1/3): 21–43. DP: 10 November 1920 (footers throughout the issue).

Reitter, E. 1926: [new taxa]. In: Spaeth, F. & Reitter, E.: *Bestimmungs-Tabellen der europäischen Coleopteren. 95 Heft. Cassidinae der palaearktischen Region.* Troppau: E. Reitter, 68 pp. DP: 9 June 1926 (stamp on L. Sekerka's copy).

Ren, G.-D. 2019: *Fauna of the beetles from Ningxia, China.* Beijing: Publishing House of Electronic Industry, 664 pp. DP: [31] August 2019 (colophon).

Reuter, O. M. 1875: En ny Haemonia-art. *Notiser ur Sällskapets pro Fauna et Flora Fennica Förhandlingar, Ny Serie* 14: 326–327. [31 December] 1875 (volume title page).

Rey, C. 1883a: Révision des genres Disopus, Pachybrachys et Stylosomus. *Revue d'Entomologie* 2 (10): 257–285. DP: [31] October 1883 (footer on p. 285).

Rey, C. 1883b: Révision des genres Disopus, Pachybrachys et Stylosomus (Suite). *Revue d'Entomologie* 2 (11): 289–306. DP: [30] November 1883 (footer on p. 289).

Rey, C. 1883c: Révision des genres Disopus, Pachybrachys et Stylosomus (Suite et fin). *Revue d'Entomologie* 2 (12): 313–326. DP: [31] December 1883 (footer on p. 313).

Rey, C. 1885: Supplément a la révision des Stylosomus. *Revue d'Entomologie* 4 (9): 274–275. DP: [30] September 1883 (footer on p. 269).

Rey, C. 1893a: Remarques et passant. Famille des bruchides. *L'Échange, Revue Linnéenne* **9** (97): 3. DP: [31] January 1893 (issue title page).

Rey, C. 1893b: Remarques et passant. Famille des bruchides. *L'Échange, Revue Linnéenne* **9** (99): 25–26. DP: [31] March 1893 (issue title page).

Rey, C. 1893c: Remarques et passant. Famille des bruchides. *L'Échange, Revue Linnéenne* **9** (100): 37. DP: [30] April 1893 (issue title page).

Rheinheimer, J. & Hassler, M. 2018: *Die Blattkäfer Baden-Würtembergs*. Karlsruhe: Kleinsteuber Books, 928 pp. DP: [31 December] 2018 (colophon).

Ricci, M. S. & Zampetti, M. F. 2006: Contributo alla conoscenza dei bruchidi dell'Europa (Coleoptera Bruchidae). *Entomologica* (Bari) **39** [2005]: 121–167. DP: [31] October 2006 (last page of the volume).

Ricci, M. S. & Zampetti, M. F. 2007: Contributo alla conoscenza dei bruchidi del Nord Africa e delle Isole Canarie (Coleoptera Bruchidae). *Bollettino della Società Entomologica Italiana* **139** (1): 43–57. DP: 30 April 2007 (issue cover).

Richter, C. F. W. 1820: *Supplementa faunae insectorum Europae. Fasciculus 1*. Vratislaviae: R. F. Schoene, [3] + 12 pp., 12 pls. DP: [31 December] 1820 (Bousquet 2016).

Riley, C. V. 1871: *Third annual report on the noxious, beneficial and other insects, of the state of Missouri*. Jefferson City: Horace Wilcox, 175 pp. DP: [31 December] 1871 (book title page).

Riley, C. V. 1875: *Seventh annual report on the noxious, beneficial and other insects of the state of Missouri*. Jefferson City: Regan & Carter, 196 + iii pp. DP: [31 December] 1875 (book title page).

Riley, E. G., Clark, S. M. & Seeno, T. N. 2003: *Catalogue of the leaf beetles of America north of Mexico*. Special publication No. 1. Sacramento: The Coleopterists Society, 290 pp. DP: 31 July 2003 (colophon).

Ritsema, C. 1876: Bijdrage tot de kennis der insecten-fauna van het noordelijkste gedeelte van Sumatra. *Tijdschrift voor Entomologie* **19** (1): 43–50. DP: [31] January 1876 (Barendrecht 1957).

Rizvi, S. G. & Kamaluddin, S. 2011: Oides neobengalensis: a new species of the genus Oides Weber (Coleoptera: Chrysomelidae: Galerucinae) from Pakistan. *Pakistan Journal of Entomology Karachi* **26** (1): 57–60. DP: [30] June 2011 (issue cover).

Rodríguez-Mirón, G. M., Zaragoza-Caballero, S. & Morrone, J. J. 2021: Phylogenetic analysis of the family Megalopodidae (Coleoptera: Chrysomeloidea): better taxon-sampling facilitates detection of new relationships and new taxa. *Cladistics* **37** (6): 677–716. DP: [31] December 2021 (issue cover).

Roelofs, W. 1879: Diagnoses de nouvelles espèces de curculionides, brenthides, anthribides et bruchides du Japon. *Comptes-Rendus des Séances de la Société Entomologique de Belgique* **22**: liii–lv. DP: 14 May 1879 (*Bull. Soc. Ent. Fr.* 1879: 23).

Roemer, J. J. 1789: *Genera insectorum Linnaei et Fabricii iconibus illustrata*. Vitoduri Helvetorum [= Winterthur]: Henric. Steiner et Socios, viii + 86 + [3 (Index tabularum)] pp. + 37 pls. DP: [31 December] 1789 (Bousquet 2016).

Rogers, W. F. 1856: Synopsis of species of Chrysomela and allied genera inhabiting the United States. *Proceedings of the Academy of Natural Sciences of Philadelphia* **8** (1): 29–39. DP: 26 March 1856 (Nolan 1913).

Romantsov, P. V. 2003: Opisanie novogo vida zhukov-listoedov roda Thelyterotarsus Wse. (Coleoptera, Chrysomelidae) iz Kazakhstana. (A description of a new species of the chrysomelid genus Thelyterotarsus Wse. (Coleoptera, Chrysomelidae) from Kazakhstan). *Entomologicheskoe Obozrenie* 82 (3): 698–700. SFP: 22 August 2003, DP: 26 November 2003 (registered at ZIN library).

Romantsov, P. V. 2004: Opisanie novogo vida zhukov-listoedov roda Luperus Geoffr. (Coleoptera, Chrysomelidae) iz vostochnogo Kazakhstana i zametki o blizkikh k nemu vidakh. (A description of a new species of the leaf-beetle genus Luperus Geoffr. (Coleoptera, Chrysomelidae) from Eastern Kazakhstan and notes on some related species). *Entomologicheskoe Obozrenie* 83 (3): 622–624. SFP: 26 August 2004, DP: 5 November 2004 (registered at ZIN library).

Romantsov, P. V. 2005: Novyy vid zhukov-listoedov roda Chrysolina Motsch. (Coleoptera, Chrysomelidae) iz severnogo Tyan'-Shanya. (A new species of the leaf-beetle genus Chrysolina Motsch. (Coleoptera, Chrysomelidae) from northern Tien Shan). *Entomologicheskoe Obozrenie* 84 (4): 881–883. SFP: 30 November 2005, DP: [31 December] 2005 (issue cover).

Romantsov, P. V. 2008: Obzor dzhungaro-tyan'shanskikh vidov zhukov-listoedov podroda Taeniosticha Motsch. roda Chrysolina Motsch. (Coleoptera, Chrysomelidae) s opisaniem dvukh novykh vidov. (Review of the Dzhungaro-Tien-Shanian species of the subgenus Taeniosticha Motsch., genus Chrysolina Motsch. (Coleoptera, Chrysomelidae), with description of two new species). *Entomologicheskoe Obozrenie* 87 (4): 842–853. SFP: 21 November 2008, DP: 31 December 2008 (based on invoice deposited in Russian Entomological Society, A. G. Moseyko 2022, pers. comm.).

Romantsov, P. V. 2011: A new species of the genus Otiothraea Warchałowski, 1990 (Coleoptera: Chrysomelidae: Cryptocephalinae: Clytrini) from Morocco. *Caucasian Entomological Bulletin* 7 (2): 145–146, pl. 7. DP: 9 December 2011 (journal website).

Romantsov, P. V. 2012: To the knowledge of Palaearctic Cryptocephalinae (Coleoptera: Chrysomelidae) with description of two new species and with some taxonomical remarks. *Caucasian Entomological Bulletin* 8 (1): 63–69. DP: 5 July 2012 (journal website).

Romantsov, P. V. 2017: Novye dannye o zhukakh-listoedakh podsemeystva Cryptocephalinae (Coleoptera: Chrysomelidae) fauny Sredney Azii i Afghanistana. (New data on the leaf beetles of the subfamily Cryptocephalinae (Coleoptera: Chrysomelidae) from Middle Asia and Afghanistan. *Entomologicheskoe Obozrenie* 96 (2): 350–355, 4 pls. DP: 31 July 2017 (colophon).

Romantsov, P. V. 2019: A new Galerucinae species from Iran and new records for Iranian leaf beetle fauna (Coleoptera, Chrysomelidae). *Zoosystematica Rossica* 28 (1): 19–23 DP: 18 April 2019 (paper title page).

Romantsov, P. V. 2021: K poznaniyu fauny zhukov-listoedov (Coleoptera, Chrysomelidae) vostochnoy Sibiri i yuga Dalnego Vostoka Rossii. (To the knowledge of the fauna of leaf-beetles (Coleoptera, Chrysomelidae) of Siberia and the Russian Far East. *Entomologicheskoe Obozrenie* 100 (1): 153–180. DP: 26 March 2021 (based on invoice deposited in Russian Entomological Society, A. G. Moseyko 2022, pers. comm.).

Romantsov, P. V. & Bukejs, A. 2009: New leaf-beetle species of Acolastus Gerstaecker, 1855 (Coleoptera: Chrysomelidae: Cryptocephalinae) from Iran and notes on close species. *Acta Biologica Universitatis Daugavpiliensis* 9 (1): 89–96. DP: 1 July 2009 (verso of the issue cover).

Romantsov, P. V. & Moseyko, A. G. 2019: New and little known species of Eumolpinae (Coleoptera: Chrysomelidae) from Northern Vietnam. *Zootaxa* **4609** (2): 321–334. DP: 23 May 2019 (footer of paper title page).

Romantsov, P. V. & Moseyko, A. G. 2020: A new subgenus and species of Eumolpinae (Coleoptera: Chrysomelidae) from Azerbaijan, with faunistic notes on adjacent countries. *Zootaxa* **4853** (1): 139–145. DP: 21 September 2020 (footer on the paper title page).

Romantsov, P. V. & Moseyko, A. G. 2022: Obzor skrytoglavov podroda Burlinius Lopatin, 1965 iz evropeyskoy chasti Rossii i Zakavkazya s tselikom ili chastichno zheltymi nadkrylyami (Coleoptera, Chrysomelidae: Cryptocephalinae). (A review of cryptocephaline of the subgenus Burlinius Lopatin, 1965 with entirely or partly yellow elytra from European Russia and Transcaucasia (Coleoptera, Chrysomelidae: Cryptocephalinae)). *Entomologicheskoe Obozrenie* 101 (3): 597–626. DP: 26 September 2022 (based on invoice deposited in Russian Entomological Society, A. G. Moseyko 2022, pers. comm.).

Romantsov, P. V. & Schöller, M. 2023: Contribution to the knowledge of the subgenus Homalopus Chevrolat, 1837 of Cryptocephalus Geoffroy, 1762 from China (Coleoptera, Chrysomelidae, Cryptocephalinae). *Linzer Biologische Beiträge* 55 (2): 705–711. DP: 29 December 2023 (paper title page).

Ronchetti, V. 1922: Al Monte Cevedale (m 3774) per la parete Sud-Ovest. *Rivista Mensile del Club Alpino Italiano* **41** (3/4): 86–89. DP: [30] April 1922 (issue cover).

Rondani, C. 1877: Vesparia parasita non vel minus cognita. *Bullettino della Società Entomologica Italiana* 9 (2): 166–213, pls. I–IV. DP: 22 August 1877 (*Bull. Soc. Ent. Fr.* 1877 (16): 183).

Rosenhauer, W. G. 1847: *Beiträge zur Insekten-Fauna Europas. Erstes Bändchen; enthält die Beschreibung von sechzig neuen Käfern aus Bayern, Tyrol, Ungarn etc., sowie die Käfer Tyrols nach dem Ergebnisse von vier Reisen.* Erlangen: Theodor Blaesing, x + 159 + [1] pp., 1 pl. DP: [31 December] 1847 (Bousquet 2016).

Rosenhauer, W. G. 1856: *Die Thiere Andalusiens nach dem Resultate einer Reise zusammengestellt. nebst den Beschreibungen von 249 neuen oder bis jetzt noch unbeschriebenen Gattungen und Arten.* Erlangen: Theodor Blaesing, viii + 429 pp., 3 pls. DP: [30] November 1856 (Bousquet 2016).

Rossi, G. de 1882: Die Käfer der Umgegend von Neviges. *Verhandlungen des Naturhistorischen Vereins der Preussischen Rheinlande, Westfalens* 39 (2): 196–215. DP: [31 December] 1982 (issue cover).

Rossi, P. 1790: *Fauna Etrusca sistens insecta quae in provinciis Florentina et Pisana praesertim collegit. Tomus primus.* Liburni: Thomae Masi & Sociorum, xxii + 272 pp., 10 pls. DP: 3 October 1790 (Bousquet 2017).

Rossi, P. 1792: *Mantissa insectorum exhibens species nuper in Etruria collectas a Petro Rossio adiectis Faunae Etruscae illustrationibus, ac emendationibus.* Pisis: Polloni, 148 pp. DP: [31 December] 1792 (Bousquet 2016).

Rossi, P. 1794: *Mantissa insectorum exhibens species nuper in Etruria collectas a Petro Rossio adiectis Faunae Etruscae illustrationibus, ac emendationibus. Tom. II.* Pisis: Prosperi, 154 pp., 8 pls. DP: [31 December] 1794 (Bousquet 2016).

Roth von Schreckenstein, F. 1801: *Verzeichniss der Kaefer, welche um den Ursprung der Donau und des Nekars, dann um den untern Theil des Bodensees vorkommen.* Tübingen: J. G. Cotta, 67 pp. DP: [31 December] 1801 (Bousquet 2016).

Rottenberg, A. L. A. M. von 1872a: Beiträge zur Coleopteren-Fauna von Sicilien (Drittes Stück). *Berliner Entomologische Zeitschrift* 15 [1871] (2/3): 225–247. DP: [29] February 1872 (table of contents).

Rottenberg, A. L. A. M. von 1872b: Cryptocephalus princeps n. sp. *Berliner Entomologische Zeitschrift* 15 [1871] (2/3): 248. DP: [29] February 1872 (table of contents).

Roubal, J. 1910: Beschreibungen dreier neuer Käferformen. *Verhandlungen der Kaiserlich-Königlichen Zoologisch-Botanischen Gesellschaft in Wien* 59 [1909] (10): (373)–(374). DP: 20 January 1910 (volume title page).

Roubal, J. 1911: Koleopterologické výsledky mé cesty na Kavkaz v červenci r. 1910. – Quid novi de Coleopterorum Caucasi ad occidentem vergentis fauna in meo itinere Julio mense anni 1910 suscepto cognoverim. *Časopis České Společnosti Entomologické* 8 (3): 101–104. DP: 1 November 1911 (*Bibl. Zool.* 1912, 22: 127 [cited another paper from the same issue]).

Roubal, J. 1912a: Nuova specie di Cryptocephalus della Sicilia. *Rivista Coleotterologica Italiana* 10 (1): 1–2. DP: [31] January 1912 (issue title page).

Roubal, J. 1912b: Koleopterologické výsledky mé cesty na Kavkaz v červenci r. 1910. Quid novi de Coleopterorum Caucasi ad occidentem vergentis fauna in meo itinere Julio mense anni 1910 suscepto cognoverim. *Časopis České Společnosti Entomologické* 9 (1): 25–28. DP: [31] December 1912 (*Bibl. Zool.* 1913, 23: 479).

Roubal, J. 1913a: Zwei neue paläarktische Coleopteren. *Entomologische Mitteilungen* 2 (1): 21–22. DP: 1 January 1913 (issue cover).

Roubal, J. 1913b: Pět nových brouků z Čech. Quinque Coleoptera nova de Bohemia. *Časopis České Společnosti Entomologické* 10 (3): 109–112. DP: [31 December] 1913 (issue cover).

Roubal, J. 1914: Verschiedene coleopterologische Notizen. *Coleopterologische Rundschau* 3 (5): 90–92. DP: 8 May 1914 (issue cover).

Roubal, J. 1915: Vier Coleopteren-Neuheiten. *Entomologische Blätter* 11 (7/9): 169–171. DP: 25 July 1915 (issue title page).

Roubal, J. 1916a: Trois formes nouvelles des coléoptères de la Bohême. *Societas Entomologica* 31 (3): 14. DP: 3 March 1916 (issue title page).

Roubal, J. 1916b: Nová odrůda a nový druh rodu Pachybrachys Redtenbacher. (Coleopterorum generis Pachybrachys Redtenbacher aberratio ac species nova.). *Časopis Moravského Zemského Musea* 15 (2): 239–241. DP: [31 December] 1916 (issue cover).

Roubal, J. 1917: Beschreibung drei neuer Chrysomela-Formen. *Entomologische Rundschau* 34 (1): 3. DP: 19 January 1917 (issue title page).

Roubal, J. 1922: Sechs Coleopteren-Neuheiten nebst Bemerkungen. *Entomologische Blätter* 18 (3): 115–116. DP: 30 September 1922 (issue cover).

Roubal, J. 1923: Novinky zvířeny broučí Malých Tater. [News of the beetle fauna of the Low Tatras]. *Časopis Československé Společnosti Entomologické* 19 (3/6): 64–66. DP: 1 March 1923 (issue cover).

Roubal, J. 1926: Sieben neue Coleopteren aus der paläarktischen Region. *Coleopterologisches Centralblatt* 1 [1926–1927] (3/4): 244–249. DP: 25 September 1926 (table of contents).

Roubal, J. 1927a: Eine coleopterologische Reise nach Mittel-Dalmatien im Juli 1924. *Entomologischer Anzeiger* 7 (6): 74–75. DP: 20 March 1927 (issue title page).

Roubal, J. 1927b: Sex nova Coleoptera palaearctica. *Boletín de la Real Sociedad Española de Historia Natural* 27 (3): 134–136 DP: 25 March 1927 (*Advertencia* at the end of volume).

Roubal, J. 1929a: Coleoptera nova asiatica. *Bollettino della Società Entomologica Italiana* 61 (5/6): 96–98. DP: 3 June 1929 (issue cover and Poggi 2008).

Roubal, J. 1929b: Popis šesti nových coleopter. Beschreibung von sechs neuen Coleopteren. *Časopis Československé Společnosti Entomologické* 26 (3/4): 56–57. DP: 9 November 1929 (issue cover).

Roubal, J. 1931a: Aphthona Kuntzei sp. n. *Coleopterologisches Centralblatt* 5 (2/5): 116–117. DP: 30 April 1931 (issue title page).

Roubal, J. 1931b: Fragmente zur Koleopterenfaunistik des balkanischen Festlandes. *Entomologischer Anzeiger* 11 (23): 453–454. DP: [31 December] 1931 (volume title page) [the issue cover was not examined].

Roubal, J. 1931c: Einige neue palaearktische Koleopteren. *Entomologische Blätter* 27 (4): 183–185. DP: 31 December 1931 (issue cover).

Roubal, J. 1932: Fragmente zur Koleopterenfaunistik des balkanischen Festlandes. *Entomologischer Anzeiger* 12 (7): 129–130. DP: [31 December] 1932 (volume title page) [the issue cover was not examined].

Roubal, J. 1934: Popisy čtyř nových coleopter z Československé republiky. Descriptiones quattuor Coleopterorum novorum e Republica ČS. *Časopis Československé Společnosti Entomologické* 31 (1): 36–38. DP: 15 March 1934 (issue cover).

Roubal, J. 1936a: Chrysomela Tesaři sp. n. *Entomologisches Nachrichtenblatt* (Troppau) 10 (2): 68–69. DP: [31] July 1936 (footer on the paper title page).

Roubal, J. 1936b: Přírodovědecký výzkum státní reservace „Pieniny" na hranicích Polsko-Českoskovenských. Skarby przyrody w parku narodowym Pienińskim. L'exploration zoologique dans le parc national limitrophe „Pieniny" sur la rive du Dunajec entre la Pologne et la Tchécoslovaquie. *Acta Entomologica Musei Nationalis Pragae* 14: 180–199. DP: 15 October 1936 (volume title page).

Roubal, J. 1939: Orestia alpina Germ. ssp. mařani nov. *Časopis České Společnosti Entomologické* 36 (1/2): 34–35. DP: 15 March 1939 (issue cover).

Roubal, J. 1940a: Quinque Coleoptera palaearctica nova descripta. *Folia Zoologica et Hydrobiologica* 10 (1): 281–284. DP: 30 March 1940 (reprint cover).

Roubal, J. 1940b: [new name]. In: Heikertinger, F. & Csiki, E: Chrysomelidae: Halticinae. In: Schenkling, S. (ed.): *Coleopterorum Catalogus. Pars 169*. s-Gravenhage: W. Junk, 337–635 pp. DP: 15 August 1940 (book cover).

Roubal, J. 1940c: O dvou kritických, v Tatrách společně žijících chrysochloách, Ch. bifrons obenbergeri Marchand a viridis Duft. De duabus, simillimis, coniuncte viventibus chrysochlois. *Časopis České Společnosti Entomologické* 37 (3/4): 91–93 DP: 1 December 1940 (issue cover).

Roubal, J. 1941: *Katalog Coleopter (brouků) Slovenska a Východních Karpat. III. (Katalog der Coleopteren der Slowakei und der Ost-Karpathen. III.).* [1937–1941], Praha: Orbis, 363 pp. DP: [31 December] 1941 (book cover).

Roubal, J. 1942a: Přehled známých forem Cryptocephalus schaefferi Schrank. Conspectus formarum Cryptocephali schaefferi Schrank. *Časopis České Společnosti Entomologické* 39 (1): 33. DP: 8 June 1942 (issue cover).

Roubal, J. 1942b: Variabilita mandelinky Chrysomela analis L. Variabilitas Chrysomelae analis L. *Časopis České Společnosti Entomologické* 39 (1): 63. DP: 8 June 1942 (issue cover).

Roubal, J. 1943: Popisy nových coleopter z Čech a Slovenska. Col. nova de Bohemia et Slovakia. *Časopis České Společnosti Entomologické* 40 (1): 57–59. DP: 15 April 1943 (issue title page).

Roubal, J. 1946: Několik nových coleopter z mé sbírky. Quelques coléopteres nouveaux de ma collection. *Časopis Československé Společnosti Entomologické* 43 (1/4): 41–44. DP: [31 December] 1946 (issue cover).

Roubal, J. 1948a: Systematická, zoogeografická, bibliografická studie o červených druzích rodu Coptocephala Dejean (Chrysomelidae). (De chrysomelidarum genere Coptocephala Dejean adnotationes systematicae, zoogeographicae, necnon bibliographicae). *Entomologické Listy* 11 (2): 33–34. DP: 31 August 1948 (issue cover).

Roubal, J. 1948b: Popisy několika nových broučích forem z Československa (II). Coleopterorum formae novae e Čechoslovenia descriptae (II). *Časopis Československé Společnosti Entomologické* 45 (3/4): 123–125. DP: 1 October 1948 (issue cover).

Roubal, J. 1949: Popisy nových forem coleopter z ČSR. Coleopterorum formae novae de RČS descriptae. *Časopis Československé Společnosti Entomologické* 46 (1/2): 44–46. DP: 1 February 1949 (issue cover).

Ruan, Y.-Y. & Konstantinov, A. S. 2017: [new taxa]. In: Ruan, Y.-Y., Konstantinov, A. S., Prathapan, K. D. & Yang, X.-K.: New contributions to the knowledge of Chinese flea beetle fauna (I): Gansuapteris new genus and Primulavorus new genus (Coleoptera: Chrysomelidae: Galerucinae). *Zootaxa* 4282 (1): 111–122. DP: 23 June 2017 (footer on the paper title page).

Ruan, Y.-Y. & Konstantinov, A. S. 2023: [new taxa]. In: Ruan, Y.-Y., Konstantinov, A. S., Damaška, A. F., Zheng, L.-H., Chen, J. & Meng, Z.-Y.: Description of three new species of Benedictus (Coleoptera, Chrysomelidae, Galerucinae, Alticini) from China, with comments on their biology and modified ethanol traps for collecting flea beetles. *ZooKeys* 1177: 147–165. DP: 30 August 2023 (Paper title page).

Ruan, Y.-Y., Konstantinov, A. S. & Damaška, A. F. 2020: The biology and immature stages of the moss-eating flea beetle Cangshanaltica fuanensis sp. nov. (Coleoptera, Chrysomelidae, Galerucinae, Alticini), with description of a fan-driven high-power Berlese funnel. *Insects* 11 (9) 571: 1–27. DP: 26 August 2020 (paper title page).

Ruan, Y.-Y., Konstantinov, A. S. & Damaška, A. F. 2022: Description of the moss-eating flea beetle Cangshanaltica fuanensis sp. nov. (Coleoptera, Chrysomelidae, Galerucinae, Alticini) from

China. *Journal of Insect Biodiversity* **34** (2): 28–32. DP: 4 August 2022 (footer on the paper title page).

Ruan, Y.-Y., Konstantinov, A. S. & Prathapan, K. D. 2019: [new taxon]. In: Ruan, Y.-Y., Konstantinov, A. S., Prathapan, K. D., Zhang, M.-N. & Yang, X.-K.: A review of the genus Lankaphthona Medvedev, 2001, with comments on the modified phallobase and the unique abdominal appendage of L. binotata (Baly) (Coleoptera, Chrysomelidae, Galerucinae, Alticini). *ZooKeys* **857**: 29–58 DP: 25 June 2019 (paper title page).

Ruan, Y.-Y., Konstantinov, A. S., Prathapan, K. D., Ge, S.-Q. & Yang, X.-K. 2015: Penghou, a new genus of flea beetles from China (Coleoptera: Chrysomelidae: Galerucinae: Alticini). *Zootaxa* **3973** (2): 300–308. DP: 17 June 2015 (footer on the paper title page).

Ruan, Y.-Y., Konstantinov, A. S. & Yang, X.-K. 2014a: [new taxa]. In: Ruan, Y.-Y., Konstantinov, A. S., Ge S.-Q. & Yang, X.-K.: Revision of the Chaetocnema picipes species-group (Coleoptera, Chrysomelidae, Galerucinae, Alticini) in China, with descriptions of three new species. *ZooKeys* **387**: 11–32. DP: 10 March 2014 (paper title page).

Ruan, Y.-Y., Konstantinov, A. S. & Yang, X.-K. 2014b: [new taxa]. In: Ruan, Y.-Y., Konstantinov, A. S., Ge, S.-Q. & Yang, X.-K.: Revision of Chaetocnema semicoerulea species-group (Coleoptera, Chrysomelidae, Galerucinae, Alticini) in China, with descriptions of three new species. *ZooKeys* **463**: 57–74. DP: 12 December 2014 (paper title page).

Ruan, Y.-Y., Konstantinov, A. S. & Yang, X.-K. 2019: [new taxa]. In: Ruan, Y.-Y., Yang, X.-K., Konstantinov, A. S., Prathapan, K. D. & Zhang, M.-N.: Revision of the Oriental Chaetocnema species (Coleoptera, Chrysomelidae, Galerucinae, Alticini). *Zootaxa* **4699** (1): 1–206. DP: 15 November 2019 (footer on the paper title page).

Ruan, Y.-Y., Yang, X.-K., Konstantinov, A. S., Prathapan, K. D. & Zhang, M.-N. 2019: Revision of the Oriental Chaetocnema species (Coleoptera, Chrysomelidae, Galerucinae, Alticini). *Zootaxa* **4699** (1): 1–206. DP: 15 November 2019 (footer on the paper title page).

Rudd, G. T. 1846: Description of a new species of Haltica. *Zoologist* **4**: 1517–1518. DP: [31 December] 1846 (volium title page).

Ruffo, S. 1946: Studi sui crisomelidi II. Nota su alcune specie italiane dei generi Chrysomela L. e Chrysochloa Hope. *Bollettino dell'Instituto di Entomologia della Università di Bologna* **15** [1943–1946]: 171–183. DP: 15 September 1946 (reprint cover).

Ruffo, S. 1964: Contributi alla conoscenza della distribuzione dei coleotteri crisomelidi nella regione Appeninica. 1. Orsodacnini, Donaciini, Criocerini. *Memorie del Museo Civico di Storia Naturale di Verona* **12**: 41–96. DP: 31 December 1964 (reprint cover).

Ruffo, S. 1965: Resultati delle missioni entomologiche dell'Istituto di Entomologia dell'Università di Bologna nel Nord-Africa compiute dai Professori G. Fiori ed E. Mellini. XIV. Un nuovo Chloropterus (Coleoptera Chrysomelidae-Eumolpini) della Libia. *Bolletino dell'Instituto di Entomologia della Università degli studi di Bologna* **27** [1964]: 245–248. DP: 2 April 1965 (reprint cover).

Rüschkamp, F. 1935: Zur Rheinischen Käferfauna XV. *Entomologische Blätter* **31** (2): 61–69. DP: 30 April 1935 (issue cover).

Rybakow, G. 1884a: Neue Käfer-Art aus Turkestan. *Horae Societatis Entomologicae Rossicae* 18 [1883–1884] (1/2): 135–136. DP: 13 May 1884 (Kerzhner 1984, Standfuss & Kerzhner 2004).

Rybakow, G. 1884b: Neue Cassida-Art von Ost-Sibirien. *Horae Societatis Entomologicae Rossicae* 18 [1883–1884] (1/2): 136. DP: 13 May 1884 (Kerzhner 1984, Standfuss & Kerzhner 2004).

Rybakow, G. 1889: Insecta in itinere Cl. N. Przewalski in Asia centrali novissime lecta. XIV. Chrysomelidae et Coccinellidae. *Horae Societatis Entomologicae Rossicae* 23 (1/2): 286–290. DP: 18 February 1889 (Kerzhner 1984, Standfuss & Kerzhner 2004).

Rye, E. C. 1868: Description of a new species of Thyamis. *Entomologist's Monthly Magazine* 5 [1868–1869] (54): 133–134. DP: 2 November 1868 (*Trans. Ent. Soc. London* (*Proc.*) 1868: xxxvii).

Rye, E. C. 1872a: A list of the species of Coleoptera recorded as new to Britain in the Entomologist's Annuals, 1855–72; with the more important corrections in nomenclature, &c. *Entomologist's Annual* 1872: 125–201. DP: 1 February 1872 (*Pet. Nouv. Ent.* 1872, 4 (45): 181).

Rye, E. C. 1872b: Additions to the list of British Coleoptera, &c., including description of a new species of Thyamis. *Entomologist's Monthly Magazine* 9 (103): 156–158. DP: 15 December 1872 (*Pet. Nouv. Ent.* 1872, 4 (66): 265).

Sahlberg, C. R. 1823a: *Periculum entomographicum, species insectorum nondum descriptas proponens. Publicae censurae submittit pro laurea Johannes Fridericus Homén, stipend. publicus satacundensis. In auditorio philos. die XIV Junii 1823. h. p. m. s. P. IV.* Aboae: Typis Frenckelliorum, pp. 49–64. DP: 14 June 1823 (Bousquet 2016).

Sahlberg, C. R. 1823b: *Periculum entomographicum, species insectorum nondum descriptas proponens. Publicae censurae submittit pro laurea Carolus Tengström, Ostrobottn. In auditorio philos. die XVI Junii 1823. h. a. m. c. P. V.* Aboae: Typis Frenckelliorum, pp. 65–82, 4 pls. DP: 16 June 1823 (Bousquet 2016).

Sahlberg, C. R. 1823c: *Periculi entomographici, species insectorum nondum descriptas proposituri, fasciculus.* Aboae: Typis Frenckelliorum, 82 pp., 4 pls. DP: [31 December] 1823 (Bousquet 2016).

Sahlberg, C. R. 1829: Periculi entomographici, species insectorum nondum descriptas proposituri, fasciculus. *Entomologisches Archiv* [Thon] 2 (1): 12–31. DP: 10 May 1829 (*Allg. Verz. Bücher, Ostermesse* 1829: 174 and Evenhuis 2014).

Sahlberg, C. R. 1838: *Dissertatio entomologica insecta Fennica enumerans. Cujus particulam decimam quartam partis secundae, cons. ampl. facult. philos. ad Imper. Univers. Alexandr. in Fennia, publico offert examini Carolus Reginaldus Sahlberg, respondente Zacharia Topelio, stip. Bilmark. Borea-Ostrobottniensi. In auditorio philos. die 16 Maji 1838. H.p.m.s.* Aboae: J. C. Frenckel, pp. 209–224. DP: 16 May 1838 (Bousquet 2016).

Sahlberg, C. R. 1839: *Dissertatio entomologica insecta Fennica enumerans. Cujus particulam decimam septimam partis secundae, cons. ampl. facult. philos. ad Imper. Univers. Alexandr. in Fennia, publicae submittit censurae Carolus Reginaldus Sahlberg, respondente Joachimo Wilhelmo Pipping, Boreali. In auditorio philos. die 10 Aprilis 1839. H.a.m.s.* Aboae: J. C. Frenckel, pp. 257–273. DP: 10 April 1839 (Bousquet 2016).

Sahlberg, J. R. 1870a: Finska arter af coleopter-slägtet Haemonia Latreille. *Öfversigt af Finska Vetenskaps-Societetens Förhandlingar* 12 [1869–1870]: 64–69. DP: 27 October 1870 (*Monatsber. Königl. Preuss. Akad. Wiss. Berlin* 1870: 801).

Sahlberg, J. R. 1870b: Anteckningar till Lapplands coleopter-fauna. *Notiser ur Sällskapets pro Fauna et Flora Fennica Förhandlingar* 11 [1871]: 385–440. DP (preprint): [31 December] 1870 (preprint title page).

Sahlberg, J. R. 1885a: Bidrag till Tschuktsch-halföns insektfauna. Coleoptera och Hemiptera, insamlade under Vega-expeditionen vid halföns norra och östra kust, 1878–1879. Stockholm: F. & G. Beijers, pp. 1–42. DP: 1 May 1885 (Bousquet 2016) [preprint of paper published in Nordeskiöld, N. A. E. von. 1887: *Vega-Expeditionen Vetenskapliga Jakttagelser bearbetade af deltagare i resan och andra forskare. Fjerte Bandet.* Stockholm: Beijers, 582 pp.].

Sahlberg, J. R. 1885b: Coleoptera och Hemiptera, insamlade af Vega-expeditionens medlemmar å Berings sunds amerikanska kust uti omgifningarna af Port Clarence, vid Grantley Harbour och sjön Iman-Ruk den 23–26 juli 1879. Stockholm: F. & G. Beijers, pp. 43–57. DP: 1 May 1885 (Bousquet 2016) [preprint of paper published in Nordeskiöld, N. A. E. von. 1887: *Vega-Expeditionen Vetenskapliga Jakttagelser bearbetade af deltagare i resan och andra forskare. Fjerte Bandet.* Stockholm: Beijers, 582 pp.].

Sahlberg, J. R. 1893: Anmärkningsvärda skalbaggar och getingar. *Meddelanden af Societas pro Fauna et Flora Fennica* 19 [1892–1893]: 19–20, 167. DP: [31 December] 1893 (volume title page).

Sahlberg, J. R. 1903: Coleoptera levantina mensibus februario et martio 1896 in Palaestina et Aegypto inferiore collecta. *Öfversigt af Finska Vetenskaps-Societetens Förhandlingar* 45 [1902–1903] (18): 1–36. DP: [31 December] 1903 (volume title page).

Sahlberg, J. R. 1913a: Coleoptera mediterranea et rosso-asiatica nova et minus cognita maxima ex parte itineribus annis 1895–1896, 1898–1899 et 1903–1904 collecta. *Öfversigt af Finska Vetenskaps-Societetens Förhandlingar* 55 A [1912–1913] (8): 1–88. DP: [31 December] 1913 (volume title page).

Sahlberg, J. R. 1913b: Coleoptera balcanica quae mensibus Octobri et Decembri 1903 atque Martis et Aprili 1906 in peninsula balcanica collegerunt John Sahlberg et Unio Saalas. *Öfversigt af Finska Vetenskaps-Societetens Förhandlingar* 55 A [1912–1913] (15): 1–108. DP: [31 December] 1913 (volume title page).

Sahlberg, J. R. 1913c: Coleoptera mediterranea orientalia, quae in Aegypto, Palaestina, Syria, Caramania atque in Anatolia occidentali anno 1904 collegerunt John Sahlberg et Unio Saalas. *Öfversigt af Finska Vetenskaps-Societetens Förhandlingar* 55 A [1912–1913] (19): 1–281. DP: [31 December] 1913 (volume title page).

Sahlberg, J. R. 1921: Coleoptera Fennica nova II. *Notulae Entomologicae* 1 (2): 33–37. DP: 15 June 1921 (table of contents).

Sahlberg, R. F. 1834: Novae coleopterorum fennicorum species. *Bulletin de la Société Impériale des Naturalistes de Moscou* 7: 266–280. DP: [30] September 1834 (Bousquet & Davies 2021).

Sainte-Claire, Deville J. 1911: *Catalogue critique des coléoptères de la Corse. Revue d'Entomologie* 28 [1909–1910] (9/10): 353–400. DP: 21 June 1911 (registered at ZIN library, 1911 also written on the issue cover).

Salvi, D., D'Alessandro, P. & Biondi, M. 2019: Host plant associations in Western Palaearctic Longitarsus flea beetles (Chrysomelidae, Galerucinae, Alticini): a preliminary phylogenetic assessment. *ZooKeys* 856: 101–114. DP: 17 June 2019 (paper title page).

Samin, N., Ghahari, H. & Warchałowski, A. 2019: New records of Chrysomelinae and Cryptocephalinae (Coleoptera: Chrysomelidae) from Iran. *Entomological News* **128** (3): 257–266. DP: 30 April 2019 (paper title page).

Samoderzhenkov, E. V. 1988: Zhuki-listoedy triby Galerucini (Chrysomelidae, Galerucinae) fauny V'etnama. [Leaf-beetles of the tribe Galerucini (Chrysomelidae, Galerucinae) of Vietnam fauna]. Pp. 70–95. In: Medvedev, L. N. & Striganova, B. R. (eds.): *Fauna i ekologiya nasekomykh V'etnama.* [*Fauna and ecology of insects of Vietnam*]. Moskva: Nauka, 197 + [3] pp. SFP: 20 April 1988, DP: 26 September 1988 (registered at ZIN library).

Samouelle, G. 1819: *The entomologist's useful compendium; or an introduction to the knowledge of British insects, comprising the best means of obtaining and preserving them, and a description of the apparatus generally used; together with the genera of Linné, and the modern method of arranging the classes Crustacea, Myriapoda, spiders, mites and insects, from their affinities and structure, according to the views of Dr. Leach. Also an explanation of the terms used in entomology; a calendar of the times of appearance and usual situations of near 3,000 species of British insects; with instructions for collecting and fitting up objects for the microscope.* London: Thomas Boys, 496 pp., 12 pls. DP: [31] May 1819 (Bousquet 2016).

Samuelson, G. A. 1984: Plant associated Alticinae from the Bismarck range, Papua New Guinea (Coleoptera: Chrysomelidae). *Esakia* **21**: 31–47. DP: 15 February 1984 (colophon).

Sassi, D. 1993: Contribution to the knowledge of Palearctic Cassidinae. The species near Cassida pusilla Waltl (Coleoptera Chrysomelidae). *Natura Bresciana, Annali del Museo Civico di Scienze Naturali di Brescia* **28** [1992]: 331–336. DP: [31 December] 1993 (paper title page).

Sassi, D. 1995: Cryptocephalus etruscus Suffrian dell'Italia peninsulare, specie vicina a Cryptocephalus ilicis Olivier (Coleoptera Chrysomelidae Cryptocephalinae). *Giornale Italiano di Entomologia* 7: 187–193. DP: 30 April 1995 (paper title page).

Sassi, D. 1997: Cryptocephalus (Homalopus) loebli n. sp. from Turkey (Coleoptera Chrysomelidae Cryptocephalinae). *Revue Suisse de Zoologie* **104** (1): 57–60. DP: 21 March 1997 (stamp on the BHL copy).

Sassi, D. 2001a: Nuove specie del genere Cryptocephalus vicine a Cryptocephalus marginellus (Coleoptera Chrysomelidae). *Memorie della Società Entomologica Italiana* 80: 107–138. DP: 30 November 2001 (paper title page).

Sassi, D. 2001b: Cryptocephalus convergens, nuova specie dell'Europa sud occidentale (Coleoptera Chrysomelidae). *Atti della Società Italiana di Scienze Naturali e del Museo Civile di Storia Naturale in Milano* **142** (1): 135–146 DP: [30] November 2001 (paper title page).

Sassi, D. 2011: A new species of the Cryptocephalus marginellus complex from Italian Western Alps (Coleoptera: Chrysomelidae: Cryptocephalinae). *Genus* **22** (1): 123–132. DP: 30 April 2011 (paper title page).

Sassi, D. & Borowiec, L. 2006: Cassida inopinata, a new species from Italy and Balkan Region (Coleoptera: Chrysomelidae: Cassidinae). *Genus* 17 (4): 545–560. DP: 28 December 2006 (paper title page).

Sassi, D. & Kismali, S. 2000: The Cryptocephalinae of Turkey, with informations on their distribution and ecology (Coleoptera Chrysomelidae). *Memorie della Società Entomologica Italiana* 78 [1999] (1): 71–129. DP: 30 April 2000 (paper title page).

Sassi, D. & Regalin, R. 1998: Cryptocephalus (Burlinius) biondii n. sp. dell'arcipelago sardo-corso e dell'isola di Capraia (Toscana) (Coleoptera Chrysomelidae). *Bollettino della Società Entomologica Italiana* 130 (2): 135–146. DP: 31 August 1998 (paper title page).

Sassi, D. & Schöller, M. 2003: Two new species of Pachybrachis Chevrolat, 1837 closely related to P. limbatus (Ménétriés, 1836) (Insecta, Coleoptera, Chrysomelidae, Cryptocephalinae). *Senckenbergiana Biologica* 82 (1/2): 135–156. DP: 5 September 2003 (paper title page).

Savitskiy, M. Y. 2000: Novye i maloizvestnye vidy zhukov-zernovok roda Spermophagus (Coleoptera: Bruchidae). (New and little-known species of seed-beetles from the genus Spermophagus (Coleoptera: Bruchidae)). *Zoologicheskiy Zhurnal* 79 (5): 556–563. SFP: 18 April 2000, DP: [31] May 2000 (issue cover).

Say, T. 1824a: Descriptions of coleopterous insects collected in the late Expedition to the Rocky Mountains, performed by order of Mr. Calhoun, Secretary of War, under the command of Major Long. *Journal of the Academy of Natural Sciences of Philadelphia* 3 [1823–1824] (13): 403–416 & (14): 417–462. DP: (pp. 403–416): [31] May 1824, (pp. 417–462): [30] June 1824 (Nolan 1913).

Say, T. 1824b: Appendix. Part I. – Natural History. §1. Zoology. Pp 253–378. In: Keating, W.H. (ed.): *Narrative of an expedition to the source of St. Peter's River, Lake Winnipeek, Lake of the Woods, &c &c performed in the year 1823, by order of the Hon. J. C. Calhoun, Secretary of war, under the command of Stephen H. Long, Major U.S.T.E.* Philadelphia: H.C. Carey & I. Lea, 459 pp. DP: [31 December] 1824 (Bousquet 2016).

Say, T. 1826: Descriptions of new species of coleopterous insects inhabiting the United States. *Journal of the Academy of Natural Sciences of Philadelphia* 5 [1825–1827] (8): 293–304. DP: [31] December 1826 (Nolan 1913).

Say, T. 1831: *Descriptions of new species of curculionites of North America, with observations on some of the species already known.* [1831–1832], New Harmony: School Press, 30 pp. DP (pp. 1–8): [31] July 1831 (Bousquet 2016).

Schaefer, L. 1958: La faune entomologique de St.-Guilhem-le-Désert (Hérault) (2me note). *Bulletin Mensuelle de la Société Linnéenne de Lyon* 27 (3): 68–74. DP: [31] March 1958 (issue title page).

Schaeffer, C. 1907: New Bruchidae with notes on known species and list of species known to occur at Brownsville, Texas, and in the Huachuca Mountains, Arizona. *Science Bulletin, The Museum of the Brooklyn Institute of Arts and Sciences* 1 (10): 291–306. DP: 21 February 1907 (footer on the paper title page).

Schaeffer, C. 1920a: Synonymical and other notes on some species of the family Chrysomelidae and descriptions of new species. *Journal of the New York Entomological Society* 27 [1919] (4): 307–340. DP: 17 April 1920 (mailing date on p. 349).

Schaeffer, C. 1920b: Change of names in Coleoptera. *Bulletin of the Brooklyn Entomological Society, New Series* 15 (4): 117. DP: 22 October 1920 (issue cover).

Schaller, J. G. 1783: Neue Insekten. *Abhandlungen der Hallischen Naturforschenden Gesellschaft* 1: 217–332, pl. 1. DP: 10 May 1783 (*Allg. Verz. Bücher, Ostermesse* 1783: 498 and Evenhuis 2014).

Schatzmayr, A. 1923: Risultati scientifici della spedizione Ravasini-Lona in Albania. IV. Nuovi coleotteri. *Bollettino della Società Entomologica Italiana* 55 (1): 7–8. DP: 22 January 1923 (Poggi 2008).

Schatzmayr, A. 1927: Appunti coleotterologici. *Bollettino della Società Adriatica di Scienze Naturali in Trieste* 29 (2): 149–154. DP: [31 December] 1927 (reprint cover).

Schatzmayr, A. 1928: Coleopterologische Notizen. *Coleopterologisches Centralblatt* 3 [1928–1929] (1/2): 45–48. DP: 20 August 1928 (table of contents).

Schatzmayr, A. 1942a: Appunti coleotterologici VI. *Natura, Rivista di Scienze Naturali* (Milano) **32** [1941] (4): 161–165. DP: 2 January 1942 (last page of reprint).

Schatzmayr, A. 1942b: Coleoptera. Pp. 59–119. In: Castellani, O., Illuminati, G., Ghidini, G., Giordani Soika, A., Guareschi, C., Menozzi, C., Romei, L., Salfi, M., Schatzmayr, A. & Topi, N.: *Prima esplorazione entomologica del Parco Nazionale del Circeo*. Salerno: Istituto Nazionale di Biologia del Consiglio Nazionale delle Richerche, 201 pp. DP: [31 December] 1942 (book title page).

Schatzmayr, A. 1943: Appunti coleotterologici VIII. *Natura, Rivista di Scienze Naturali* (Milano) **34** (1): 25–29. DP: [31] March 1943 (issue cover).

Schaufuss, L. W. 1862a: Diagnoses de coléoptères nouveaux. *Annales de la Société Entomologique de France, Quatrième Série* 2 (2): 309–312. DP: 8 October 1862 (Lefèvre 1885).

Schaufuss, L. W. 1862b: [new taxon]. *Sitzungs-Berichte der Naturwissenschaftlichen Gesellschaft Isis zu Dresden* [**1861**] (10/12): 91–92. DP: 8 April 1862 (*Mitth. Geogr. Ges.* 1862, 6 : 87).

Schaufuss, L. W. 1863: Coléoptères nouveaux appartenant à la faune d'Espagne. *Revue et Magasin de Zoologie Pure et Appliquée, 2e Série* 15 (3): 120–121. DP: [30] April 1863 (Dickinson et al. 2011).

Schaufuss, L. W. 1871a: Ueber die Gattung Pseudocolaspis Casteln. *Nunquam Otiosus* 1 [1870–1871]: 197–200. DP: 31 August 1871 (table of contents).

Schaufuss, L. W. 1871b: Ueber die Gattung Pseudocolaspis Casteln. (Fortsetzung.). *Nunquam Otiosus* 1 [1870–1871]: 201–204. DP: 20 October 1871 (table of contents).

Schaufuss, L. W. 1877: Ueber Argopus F. v. W. *Nunquam Otiosus* 2 [1872–1877]: 423–424. DP: 8 January 1877 (table of contents).

Schenkling, S. 1917: *Erklärung der wissenschaftlichen Käfernamen: aus Reitter's Fauna Germanica*. Stuttgart: K. G. Lutz, 80 pp. DP: [31 December] 1917 (book title page).

Scherer, G. 1969: Die Alticinae des indischen Subkontinentes (Coleoptera-Chrysomelidae). *Pacific Insects Monograph* 22: 1–251. DP: 30 July 1969 (paper title page).

Scherer, G. 1971: Das Genus Livolia und seine umstrittene Stellung im System. Eine taxonomische-zoogeographische-evolutionistische Studie (Coleoptera – Chrysomelidae – Alticinae). *Entomologische Arbeiten aus dem Museum G. Frey* 22: 1–37. DP: 1 July 1971 (table of contents).

Scherer, G. 1974: Clavicornaltica a new genus from Ceylon (Coleoptera-Chrysomelidae-Alticinae). *Revue Suisse de Zoologie* 81 (1): 57–68. DP: [30] April 1974 (paper title page).

Scherer, G. 1978: Podontia affinis (Gröndal) a polytypic species (Coleoptera, Chrysomelidae, Alticinae). *Spixiana* 1 [1977–1978] (3): 275–280. DP: 15 December 1978 (paper title page).

Scherer, G. 1979a: Ergebnisse der Bhutan-Expedition 1972 des Naturhistorischen Museums in Basel. Coleoptera: Fam. Chrysomelidae, Subfam. Alticinae (1. Teil). *Entomologica Basiliensia* 4: 127–139. DP: 10 October 1979 (table of contents).

Scherer, G. 1979b: Zwei neue Zipangia-Arten aus Nordindien. Coleoptera – Chrysomelidae – Alticinae. *Entomologica Basiliensia* 4: 479–482. DP: 10 October 1979 (table of contents).

Scherer, G. 1983: Beitrag zur Gattung Xuthea Baly (Coleoptera – Chrysomelidae – Alticinae). *Entomologische Arbeiten aus dem Museum G. Frey* **31/32**: 157–159. DP: 1 November 1983 (table of contents).

Scherer, G. 1985: Yemenaltica, eine neue Gattung aus dem Yemen (Coleoptera Chrysomelidae Alticinae). *Atti della Società Italiana di Scienze Naturali e del Museo Civile di Storia Naturale di Milano* **126** (1/2): 85–88. DP: 15 June 1985 (paper title page).

Scherer, G. 1987: The Genus Torodera Weise (Coleoptera – Chrysomelidae – Alticinae). *Entomologische Arbeiten aus dem Museum G. Frey* **35/36**: 67–71. DP: 15 December 1987 (table of contents).

Scherer, G. 1989: Ground living flea beetles from the Himalayas (Coleoptera, Chrysomelidae, Alticinae). *Spixiana* **12** (1): 31–55. DP: 30 July 1989 (paper title page).

Schilling, P. S. 1838: [new taxon]. In: Schummel, T. E.: Bericht über die Arbeiten der entomologischen Section im Jahre 1837. *Uebersicht der Arbeiten und Veränderungen der Schlesischen Gesellschaft für Vaterländische Kultur* (Breslau) [**1837**]: 99–112. [31 December] 1838 (volume title page).

Schilsky, J. 1888: Berichtigungen und Ergänzungen zu Schilsky's Verzeichniss der Deutschen Käfer. *Deutsche Entomologische Zeitschrift* **1888** (2): 321–323. DP: 28 November 1888 (*Bull. Soc. Ent. Fr.* 1888: clxxxv).

Schilsky, J. 1905: *Die Käfer Europa's. Nach der Natur beschrieben von Dr. H. C. Küster & Dr. G. Kraatz. Einundvierzigstes Heft*. Nürnberg: Bauer & Raspe, iv + 41a–41MM + 100 nos. DP: [31] May 1905 (Bousquet 2016).

Schilsky, J. 1906: *Die Käfer Europa's. Nach der Natur beschrieben von Dr. H. C. Küster & Dr. G. Kraatz. Zweiundvierzigstes Heft*. Nürnberg: Bauer & Raspe, vi + 100 nos. DP: 14 March 1906 (stamp on the BHL copy).

Schilsky, J. 1908: Neue märkische Käfer und Varietäten aus der Gegend von Luckenwalde. *Deutsche Entomologische Zeitschrift* **1908** (5): 599–604. DP: 1 September 1908 (table of contents).

Schleicher, H. 1924: Einige neue Aberrationen meiner Sammlung. *Entomologische Blätter* **20** (1): 57–59. DP: 31 March 1924 (issue cover).

Schlich, C. & Wagner, T. 2010: Species of Monolepta Chevrolat, 1836 from the Arabian Peninsula (Coleoptera, Chrysomelidae, Galerucinae). *Entomologica Basiliensia et Collectionis Frey* **32**: 307–322. DP: 20 December 2010 (table of contents on the cover).

Schmidt, F. J. 1841: Ueber Clythra quadripunctata Lin. und ihre nächsten Verwandten. *Entomologische Zeitung* (Stettin) **2** (10): 146–155. DP: [31] October 1841 (issue title page).

Schmidt, F. J. 1842: Lema suffrianii. Eine neue deutsche Art. *Entomologische Zeitung* (Stettin) **3** (2): 27–29. DP: [28] February 1842 (issue title page).

Schmidt, F. J. 1850: [new taxon]. *Berichte über die Mittheilungen von Freunden der Naturwissenschaften in Wien* 6 (9): 182–184. DP: 22 March 1850 (volume cover).

Schmitt, M., Hübner, H. & Gaedike, R. 1998: Nomina Auctorum – Auflösung von Abkürzungen taxonomischer Autoren-Namen. *Nova Supplementa Entomologica* **11**: 3–189. DP: 18 June 1998 (paper title page).

Schmitt, M. & Rönn, T. 2011: Types of geographical distribution of leaf beetles (Chrysomelidae) in Central Europe. *ZooKeys* **157**: 131–158. DP: 21 December 2011 (paper title page).

Schneider, D. H. 1791: Rezensionen. *Neuestes Magazin für die Liebhaber der Entomologie* **1** (1): 90–109. DP: 2 October 1791 (*Allg. Verz. Bücher, Michaelismesse* 1791: 246 and Evenhuis 2014).

Schneider, D. H. 1792: Verzeichniss und Beschreibung der in der Sammlung des Herausgebers befindlichen zur Gattung Cryptocephalus Fabr. gehörigen europäischen Käfer. Sägekäfer und Fallkäfer. *Neuestes Magazin für die Liebhaber der Entomologie* **1** (2): 186–220. DP: 29 April 1792 (*Allg. Verz. Bücher, Ostermesse* 1792: 81 and Evenhuis 2014).

Schneider, D. H. 1794: [new taxon]. In: Illiger, J. C. W.: Beschreibung einiger neuen Käferarten aus der Sammlung des Herrn Professors Hellwig in Braunschweig. *Neuestes Magazin für die Liebhaber der Entomologie* **1** (5): 593–620. DP: [31 December] 1794 (see the comments in Dating of publications).

Schneider, J. S. 1886: Nogle bemærkninger om en ny art af slægten Phyllodecta fra det arktiske Norge. *Tromsø Museums Aarshefter* 9: 66. DP: [31] December 1886 (*Natur. Novit.* 1886, 8 (25): 280).

Schneider, O. 1893: San Remo und seine Thierwelt im Winter. *Sitzungsberichte und Abhandlungen der Naturwissenschaftlichen Gesselschaft Isis in Dresden* **1893**: 3–62. DP: [30] September 1893 (*Natur. Novit.* 1893, 15 (17): 353).

Schoenherr [= Schönherr], C. J. 1808: *Synonymia insectorum, oder: Versuch einer Synonymie aller bisher bekannten Insecten; nach Fabricii Systema Eleutheratorum &c. geordnet. Erster Band. Eleutherata oder Käfer. Zweiter Theil. Spercheus---Cryptocephalus*. Stockholm: C. F. Marquard, ix + 424 pp., 1 pl. DP: [31 December] 1808 (Bousquet 2016).

Schoenherr, C. J. 1823: Curculionides. *Isis von Oken* **1823** (10): 1132–1146. DP: [31] October 1823 (published monthly).

Schoenherr, C. J. 1832: [new taxon]. In: Ménétriés, E.: *Catalogue raisonné des objets de zoologie recueillis dans un voyage au Caucase et jusqu'aux frontières actuelles de la Perse entrepris par l'ordre de S. M. l'Empereur*. St.-Pétersbourg: Académie Impérial des Sciences, xxxiii + 272 + iv + [1] pp., 5 pls. DP: [30] September 1832 (Bousquet 2016).

Schoenherr, C. J. 1833: *Genera et species curculionidum, cum synonymia hujus familiae. Tomus primus. Pars prima*. Paris: A. Roret, xv + 381 pp. DP: [30] June 1833 (Bousquet 2016).

Schöller, M. 1999: Nomenklatorische Änderungen in der Gattung Coenobius Suffrian (Coleoptera: Chrysomelidae: Cryptocephalinae). *Mitteilungen des Internationalen Entomologischen Vereins e. V.* (Frankfurt am Main) **24** (1/2): 73–75. DP: 31 July 1999 (paper title page).

Schöller, M. 2000: A new species of Acolastus Gerstäcker from Tibet (Coleoptera: Chrysomelidae: Cryptocephalinae). *Mitteilungen des Internationalen Entomologischen Vereins e. V.* (Frankfurt am Main) **25** (1/2): 25–30. DP: 31 July 2000 (paper title page).

Schöller, M. 2005: New records and a new species of the genus Cryptocephalus Geoffroy 1762 from Yemen (Coleoptera, Chrysomelidae). *Linzer Biologische Beiträge* **37** (2): 1555–1558. DP: 16 December 2005 (paper title page).

Schöller, M. 2006: Two new species of Cryptocephalus Geoffroy, 1762 from Yemen and re-examination of C. oblitus Suffrian and C. saudiensis Lopatin (Coleoptera: Chrysomelidae: Cryptocephalinae). *Mitteilungen des Internationalen Entomologischen Vereins e. V.* (Frankfurt am Main) **31** (3/4): 145–155. DP: 24 November 2006 (paper title page).

Schöller, M. 2007: A new species of Stylosomus Suffrian from Afghanistan (Coleoptera: Chrysomelidae: Cryptocephalinae). *Genus* **18** (4): 567–570. DP: 28 December 2007 (paper title page).

Schöller, M. 2008: Comparative morphology of sclerites used by Camptosomatan leaf beetles for formation of the extrachorion (Chrysomelidae: Cryptocephalinae, Lamprosomatinae). Pp. 87–120. In: Jolivet, P., Santiago-Blay, J. & Schmitt, M. (eds.): *Research on Chrysomelidae. Volume 1.* Leiden: Brill, 430 pp. DP: 10 July 2008 (publisher website).

Schöller, M. 2009a: A new species of the genus Pachybrachis Chevrolat, 1832 from Pakistan and notes on similar Asian species (Coleoptera: Chrysomelidae: Cryptocephalinae). *Mitteilungen des Internationalen Entomologischen Vereins e. V.* (Frankfurt am Main) **33** (3/4): 91–104. DP: 2 March 2009 (paper title page).

Schöller, M. 2009b: Nomenklatorische Änderungen in der Gattung Cryptocephalus Geoffroy, 1762 (Coleoptera: Chrysomelidae: Cryptocephalinae). *Mitteilungen des Internationalen Entomologischen Vereins e. V.* (Frankfurt am Main) **33** (3/4): 105–108. DP: 2 March 2009 (paper title page).

Schöller, M. 2010a: New acts and comments. Chrysomelidae: Cryptocephalini. P. 80. In: Löbl, I. & Smetana, A. (eds.): *Catalogue of Palaearctic Coleoptera. Volume 6. Chrysomeloidea.* Stenstrup: Apollo Books, 924 pp. DP: 22 February 2010 (colophon).

Schöller, M. 2010b: Pachybrachis salsuginosus n. sp. from Iran (Coleoptera: Chrysomelidae: Cryptocephalinae). *Mitteilungen des Internationalen Entomologischen Vereins e. V.* (Frankfurt am Main) **35** (3/4): 149–154. DP: 30 December 2010 (paper title page).

Schöller, M. 2011: Cryptocephalus (Burlinius) pseudopopuli n. sp. from South Korea (Coleoptera: Chrysomelidae: Cryptocephalinae). *Mitteilungen des Internationalen Entomologischen Vereins e. V.* (Frankfurt am Main) **36** (1/2): 25–32. DP: 30 June 2011 (paper title page).

Schöller, M. 2014a: Revision of the Cryptocephalus vitellinus species group (Coleoptera: Chrysomelidae: Cryptocephalinae). *Acta Entomologica Musei Nationalis Pragae* **54** (2): 645–655. DP: 15 December 2014 (paper title page).

Schöller, M. 2014b: Cryptocephalus socotrensis sp. nov., the first representative of the genus from Socotra Island (Coleoptera: Chrysomelidae: Cryptocephalinae). Pp. 269–275. In: Hájek, J. & Bezděk, J. (eds.): Insect biodiversity of the Socotra Archipelago II. *Acta Entomologica Musei*

Nationalis Pragae **54** (supplementum): i–vi + 1–440. DP: 30 December 2014 (colophon and paper title page).

Schöller, M. 2018: Two new species and new records of Acolastus Gerstaecker from the Arabian Peninsula (Coleoptera, Chrysomelidae, Cryptocephalinae). *Entomologische Blätter und Coleoptera* **114**: 407–413. DP: [30] November 2018 (volume cover).

Schöller, M. 2020: New country records, taxonomical changes and a new species in Palaearctic Cryptocephalinae (Coleoptera: Chrysomelidae). *Mitteilungen des Internationalen Entomologischen Vereins e. V.* (Frankfurt am Main) **43** (1/2): 75–90. DP: 30 January 2020 (paper title page).

Schöller, M. 2021a: Cryptocephalus from Oman and Yemen, with description of a new species and notes on subgenera (Coleoptera: Chrysomelidae). *Mitteilungen des Internationalen Entomologischen Vereins e. V.* (Frankfurt am Main) **44** (1/2): 13–32. DP: 1 June 2021 (paper title page).

Schöller, M. 2021b: Three new species of Cryptocephalini from China (Coleoptera: Chrysomelidae: Cryptocephalinae). *Mitteilungen des Internationalen Entomologischen Vereins e. V.* (Frankfurt am Main) **44** (3/4): 135–145. DP: 31 December 2021 (paper title page).

Schöller, M., Smetana, A. & Lopatin, I. K. 2010: [new name]. In: Schöller, M. & Smetana, A.: New acts and comments. Chrysomelidae: Cryptocephalini. P. 80. In: Löbl, I. & Smetana, A. (eds.): *Catalogue of Palaearctic Coleoptera. Volume 6. Chrysomeloidea*. Stenstrup: Apollo Books, 924 pp. DP: 22 February 2010 (colophon).

Scholz, R. 1926: Eine übersehene Chrysomelidae (Col.). Lamprosoma Kolbei n. sp. *Entomologische Blätter* **22** (3): 104–109. DP: 30 September 1926 (issue cover).

Schönfeldt, H. von 1887: Catalog der Coleopteren von Japan mit Angabe der bezüglichen Beschreibungen und der sicher bekannten Fundorte. *Jahrbücher des Nassauischen Vereins für Naturkunde* 40: 29–204. DP: [30] November 1887 (*Ent. Nachricht.* 1887, 13 (22): 350).

Schönfeldt, H. von 1888: Eine neue Donacia aus Japan. *Entomologische Nachrichten* **14** (3): 33–34. DP: [29] February 1888 (issue title page).

Schönfeldt, H. von 1890: Ein Beitrag zur Coleopterenfauna der Liu-Kiu-Inseln. *Entomologische Nachrichten* **16** (11): 168–175. DP: [30] June 1890 (issue title page).

Schrank von Paula, F. 1776: *Beyträge zur Naturgeschichte. Mit sieben von Verfasser selbst gezeichneten, und in Kupfer gestochenen Tabellen.* Augsburg: Gebrüdern Veith, [6] + 137 + [3] pp., 7 pls. DP: 21 April 1776 (Bousquet 2016).

Schrank von Paula, F. 1781: *Enumeratio insectorum Austriae indigenorum.* Augustae Vindelicorum [= Augsburg]: E. Klett et Franck, [22] + 548 + [2 (Errata)], 4 pls. DP: [30] June 1781 (Bousquet 2016).

Schrank von Paula, F. 1782: Kritische Revision des österreichischen Insectenverzeichnisses. *Neues Magazin für die Liebhaber der Entomologie* (Fuessly) **1** (2): 135–168. DP: 21 April 1782 (*Allg. Verz. Bücher, Ostermesse* 1782: 311 and Evenhuis 2014).

Schrank von Paula F. 1785: Verzeichniss beobachteter Insecten im Fürstenthume Berchtesgaden. *Neues Magazin für die Liebhaber der Entomologie* (Fuessly) **2** (4): 313–345. DP: 17 April 1785 (*Allg. Verz. Bücher, Ostermesse* 1785: 62 and Evenhuis 2014).

Schrank von Paula, F. 1789: Entomologische Beobachtungen. *Der Naturforscher* **24**: 60–90. DP: 3 May 1789 (*Allg. Verz. Bücher, Ostermesse* 1789: 90 and Evenhuis 2014).

Schrank von Paula, F. 1798: *Fauna Boica. Durchgedachte Geschichte der in Baiern einheimischen und zahmen Thiere. Erster Band. Zweyte Abtheilung.* Nürnberg: Stein'schen Buchhadlung, 295–720 pp. DP: 29 April 1798 (Bousquet 2017).

Schrank von Paula, F. 1837: [new taxon]. Gistel, J. N. F. X.: Kritische Revisionen und Ergänzungen zu Schrank's "Enumeratio Insectorum Austriae, Fauna boica u.s.w.". *Faunus, Neue Folge* **1837** (1): 5–19. DP: [31] January 1837 (dated review in *Isis von Oken* 1837 (4)).

Schulz, H. 1909: Die Varietäten von Macrolenes bimaculata. *Internationale Entomologische Zeitschrift* (Guben) **2** [1908/1909] (52): 323. DP: 27 March 1909 (issue title page).

Schummel, T. E. 1844: [new taxa]. *Uebersicht der Arbeiten und Veränderungen der Schlesischen Gesellschaft für Vaterländische Kultur* (Breslau) [**1843**]: 184–199. DP: 24 October 1844 (*Ber. Bekanntmach. Verh. Akad. Wiss. Berlin* 1844: 347).

Schuster, W. 1905: Die typischen stehenden Formen von Crioceris asparagi L. (Spargelhähnchen) im Mainzer Becken. *Der Zoologische Garten* **46** (7): 211–214. DP: [31] July 1905 (issue title page).

Schuster, W. 1906: Crioceris asparagi var. Linnei, anticeconjuncta, schusteri, impupillata, apiceconjuncta, quadripunctata, cruciata, (incrucifer, pupillata) und muguntiaca im Mainzer Becken. Biologisches. *Jahrbücher des Nassauischen Vereins für Naturkunde* (Wiesbaden) **59**: 145–152. DP: 10 June 1906 (*Bibl. Zool.* 1906, 13: 294).

Scopoli, J. A. 1763: *Entomologia Carniolica exhibens insecta Carnioliae indigena et distributa in ordines, genera, species, varietates. Methodo linnaeana.* Vindobonae: Ioannis Thomae Trattner, xxxii + 420 + [4] pp., 3 pls. DP: 23 June 1763 (Bousquet 2016).

Scopoli, J. A. 1772: *Annus V. Historico-Naturalis.* Lipsiae: Christ. Gottlob Hilscheri, 128 pp. DP: 10 May 1772 (Bousquet 2017).

Scriba, L. G. 1791: Dritte Fortsetzung des Verzeichniss der Insekten in der Darmstädter Gegend. *Journal für die Liebhaber der Entomologie* **1** (3): 275–296. DP: [31 December] 1791 (issue cover, see also Ziani et al. 2021).

Seeno, T. N. & Wilcox, J. A. 1982: Leaf beetle genera (Coleoptera: Chrysomelidae). *Entomography* **1**: 1–221. DP: [31 December] 1982 (paper title page).

Seidlitz, G. C. M. von 1867: Einige entomologische Excursionen in den Castilischen Gebirgen im Sommer 1865. *Berliner Entomologische Zeitschrift* **11** (1/2): 167–193. DP: [31] May 1867 (table of contents).

Seidlitz, G. C. M. von 1872: Geänderte Namen. *Coleopterologische Hefte* **9/10**: 254. DP: [31 December] 1872 (volume title page).

Seidlitz, G. C. M. von 1875: *Fauna Baltica. Die Käfer (Coleoptera) der Ostseeprovinzen Russlands.* [1872–1875], Dorpat: H. Laakmann, xlii + 142 + 560 pp. DP: (pp. xxi–xlii + Gattungen 81–142 + Arten 341–560): [31 December] 1875 (Bousquet 2016).

Seidlitz, G. C. M. von 1891a: *Fauna Baltica. Die Kaefer (Coleoptera) der deutschen Ostseeprovinzen Russlands. Zweite neu bearbeitete Auflage.* [1887–1891], Königsberg: Hartungsche Verlagsdruckerei, [10] + lvi + 192 + 818 pp., 1 pl. DP: (pp. xlix–lvi, Gattungen 161–192, Arten 609–818): [28] February 1891 (Bousquet 2016).

Seidlitz, G. C. M. von 1891b: *Fauna Transsylvanica. Die Kaefer (Coleoptera) Siebenbürgens.* [1888–1891], Königsberg: Hartungsche Verlagsdruckerei, [12] + lvi + 192 + 914 pp., 1 pl. DP: (pp. xlix–lvi, Gattungen 129–192, Arten 545–914): [28] February 1891 (Bousquet 2016).

Sekerka, L. 2004: Species of Cassidinae and Hispinae contained in the Moravian Museum collection in Brno (Coleoptera, Chrysomelidae). *Acta Musei Moraviae, Scientiae Biologicae* 89 (1/2): 117–165. DP: [31 December] 2004 (issue title page).

Sekerka, L. 2005: Cassida olympica, a new species from Greece (Coleoptera: Chrysomelidae: Cassidinae). *Genus* 16 (2): 285–289. DP: 30 June 2005 (paper title page).

Sekerka, L. 2006a: A new species of Cassida Linné, 1758 from Cyprus (Coleoptera: Chrysomelidae: Cassidinae). *Genus* 17 (2): 253–262. DP: 30 June 2006 (paper title page).

Sekerka, L. 2006b: Cassidinae and Hispinae preserved in the East Bohemian Museum in Hradec Králové (Coleoptera: Chrysomelidae). *Acta Musei Reginaehradensis, Series A – Scientiae Naturales* 31: 89–102. DP: 30 August 2006 (colophon).

Sekerka, L. 2006c: A new species of Cassida undecimnotata group from Turkey (Coleoptera: Chrysomelidae: Cassidinae). *Genus* 17 (4): 561–566. DP: 28 December 2006 (paper title page).

Sekerka, L. 2008: Review of the genus Macromonycha (Coleoptera: Chrysomelidae: Cassidinae). *Acta Entomologica Musei Nationalis Pragae* 48 (1): 95–102. DP: 15 August 2008 (paper title page).

Sekerka, L. 2010: Icones Insectorum Europae Centralis. Coleoptera: Chrysomelidae: Cassidinae. *Folia Heyrovskyana, Series B* 13: 1–24. DP: 30 April 2010 (paper title page).

Sekerka, L. 2011: Cassida stevensi, a new species from India (Coleoptera: Chrysomelidae: Cassidinae: Cassidini). *Genus* 22 (3): 499–504. DP: 30 November 2011 (paper title page).

Sekerka, L. 2015: Wallacea, Pistosia and Neodownesia: three distinct genera and their tribal placement (Coleoptera: Chrysomelidae: Cassidinae). *Acta Entomologica Musei Nationalis Pragae* 55 (2): 713–743. DP: 31 December 2015 (paper title page).

Sekerka, L. in prep.: Contribution to the knowledge of Palaearctic Cassidinae (Coleoptera: Chrysomelidae).

Sekerka, L. & Barclay, M. V. L. 2014: Fabrician types of Cassidinae (Coleoptera: Chrysomelidae) deposited in the Natural History Museum, London. *Acta Entomologica Musei Nationalis Pragae* 54 (2): 657–684. DP: 15 December 2014 (paper title page).

Sekerka, L. & Borowiec, L. 2008: Three new species of Cassida Linné, 1758 from India and note on Thlaspida obenbergeri Spaeth, 1928 (Coleoptera: Chrysomelidae: Cassidinae: Cassidini). *Annales Zoologici* (Warszawa) 58 (3): 611–620. DP: 30 September 2008 (verso of the issue cover).

Selman, B. J. 1969: Atomyria Jacobson (Coleoptera, Eumolpinae), a genus new to Africa. *Bulletin du Muséum National d'Histoire Naturelle, 2e Série* (Paris) 41 (1): 200–202. DP: 28 November 1969 (issue cover).

Selman, B. J. & Vogt, G. B. 1971: Lectotype designations in the South American genus Agasicles (Coleoptera: Chrysomelidae), with description of a new species important as a suppressant of alligatorweed. *Annals of the Entomological Society of America* 64 (5): 1016–1020. DP: 15 September 1971 (table of contents).

Semenov [= Semenow], A. P. 1891: Diagnoses coleopterorum novorum ex Asia centrali et orientali. III. *Horae Societatis Entomologicae Rossicae* **25** [1890–1891] (1/2): 262–382. DP: 1 April 1891 (Kerzhner 1984, Standfuss & Kerzhner 2004).

Semenov, A. P. 1894: Coleoptera asiatica nova. II. *Horae Societatis Entomologicae Rossicae* **28** [1893–1894] (3/4): 526–547. DP (reprint): 12 November 1894 (Kerzhner 1984, Standfuss & Kerzhner 2004).

Semenov, A. P. 1895a: Coleoptera asiatica nova. III. *Horae Societatis Entomologicae Rossicae* **29** [1894–1895] (1/2): 125–144. DP (reprint): 12 March 1895 (Kerzhner 1984, Standfuss & Kerzhner 2004).

Semenov [= Semenow], A. P. 1895b: Coleoptera asiatica nova. V. *Horae Societatis Entomologicae Rossicae* **29** [1894–1895] (1/2): 251–270. DP (reprint): 12 April 1895 (Kerzhner 1984, Standfuss & Kerzhner 2004).

Semenov, A. P. 1896a: [new taxon]. In: Jacobson, G. G.: Chrysomelidae palaearcticae novae vel parum cognitae. II. *Horae Societatis Entomologicae Rossicae* **29** [1894–1895] (3/4): 529–558. DP: 1 January 1896 (Kerzhner 1984, Standfuss & Kerzhner 2004).

Semenov, A. P. 1896b: Insectorum quorundam novorum faunae transcaspicae diagnoses. *Annuaire du Musée Zoologique de l'Académie Impériale des Sciences de St.-Pétersbourg* **1** (4): 378–386. DP: [31] December 1896 (Asanovich et al. 2007).

Semenov, A. P. 1902: Generica quaedam nomina mutanda vel emendanda. *Russkoe Entomologicheskoe Obozrenie* **2** (6): 353. DP: [31] December 1902 (footer on the paper title page).

Semenov, A. P. 1903a: Rod Nyctiphantus Sem. (Coleoptera, Chrysomelidae) i ego vidy. *Horae Societatis Entomologicae Rossicae* **36** (1/2): 176–182. DP (reprint): 13 January 1903 (Kerzhner 1984, Standfuss & Kerzhner 2004).

Semenov, A. P. 1903b: Analecta coleopterologica. III. *Russkoe Entomologicheskoe Obozrenie* **3** (3/4): 169–173. DP: [31] August 1903 (footers throughout the paper).

Semenov, A. P. 1904: Dopolnitelnaya zametka o vidakh roda Nyctiphantus Sem. (Coleoptera, Chrysomelidae). *Horae Societatis Entomologicae Rossicae* **37** [1904–1905] (1/2): 194–196. DP (reprint): 13 September 1904 (Kerzhner 1984, Standfuss & Kerzhner 2004).

Semenov, A. P. 1909: Coleoptera nova faunae turanicae. III. *Russkoe Entomologicheskoe Obozrenie* **8** [1908] (3/4): 295–304. DP: [31] January 1909 (footers throughout the paper).

Semenov [= Semenov-Tian-Shanskyi], A. P. & Reichardt, A. 1927: De nova Donaciae specie e Transcaucasia (Coleoptera, Chrysomelidae). *Russkoe Entomologicheskoe Obozrenie* **21** (3/4): 218–219. DP: [31 December] 1927 (issue cover).

Şen, I. 2015: A new species of the genus Chrysolina Motschulsky, 1860 from Turkey (Coleoptera: Chrysomelidae: Chrysomelinae). *Zoology in the Middle East* **61** (3): 252–255. DP: 24 July 2015 (paper title page).

Sengupta, T. & Basu, C. 1977: A revision of Xuthea (Coleoptera: Chrysomelidae: Alticinae) from India with descriptions of two new species. *Oriental Insects* **11** (2): 195–204. DP: [31] August 1977 (footer on the paper title page).

Serafim, R. & Maican, S. 2011: Catalogue of Cerambycidae, Megalopodidae and Chrysomelidae, Chrysomelidae (Coleoptera: Chrysomeloidea) recently entered in the patrimony of "Grigore Antipa" National Museum of Natural History (Bucharest). "Igor Ceianu" collection. *Travaux du Muséum National d'Histoire Naturelle « Grigore Antipa »* 54 (2): 425–460. DP: 30 December 2011 (paper title page).

Sergeeva, E. V. & Dedyukhin, S. V. 2021: Novye dannye po faune zhukov-listoedov (Coleoptera, Chrysomelidae) Tyumenskoy oblasti. (New records of leaf-beetles (Coleoptera, Chrysomelidae) from Tyumenskaya Oblast, Russia). *Euroasian Entomological Journal* 20 (1): 9–14. DP: 3 March 2021 (journal website).

Shapiro, D. S. 1956: Opisanie novogo vida zemlyanoy bloshki (Coleoptera, Chrysomelidae) iz Ukrainskoy SSR. (New halticin beetle (Coleoptera, Chrysomelidae) from Ukraine). *Entomologicheskoe Obozrenie* 35 (1): 152–153. SFP: 17 May 1956, DP: [31 December] 1956 (issue cover).

Shapiro, D. S. 1958: Novyy vid zemlyanoy bloshki (Coleoptera, Chrysomelidae) iz Kazakhstana. (A new flea-beetle (Coleoptera, Chrysomelidae) from Kazakhstan). *Entomologicheskoe Obozrenie* 37 (2): 432–433. SFP: 9 June 1958, DP: 9 July 1958 (registered at ZIN library).

Shapiro, D. S. 1960: Novye i maloizvestnye palearkticheskie vidy listoedov iz rodov Haltica Müll. i Phyllotreta Foudras (Coleoptera, Chrysomelidae). (New and little known Palearctic species of the genera Haltica Müll. and Phyllotreta Foudras (Coleoptera, Chrysomelidae)). *Entomologicheskoe Obozrenie* 39 (1): 187–188. SFP: 22 February 1960, DP: 19 March 1960 (registered at ZIN library).

Shapiro, D. S. 1970: Novyy vid listoeda roda Haltica (Coleoptera, Chrysomelidae) iz Dagestana. (A new species of leaf beetles of the genus Haltica (Coleoptera, Chrysomelidae) from Daghestan). *Entomologicheskoe Obozrenie* 49 (3): 656. SFP: 10 September 1970, DP: 7 October 1970 (registered at ZIN library).

Sharp, D. 1885: Tribe Bruchides. Fam. Bruchidae. Pp. 437–504, pl. 26. In: Godman, F. D. & Salvin, O. (eds.): *Biologia Centrali-Americana. Insecta. Coleoptera. Vol. V.* [1879–1886]. London: Taylor and Francis, xii + 525 + [1] pp. DP (pp. 437–440): [31] October 1885, (pp. 441–464, pl. 26): [30] November 1885, (pp. 465–504): [31] December 1885 (Bousquet 2016).

Sharp, D. 1886a: On the Bruchidae of Japan. *Annals and Magazine of Natural History, Fifth Series* 17 (97): 34–38. DP: 1 January 1886 (Evenhuis 2003).

Sharp, D. 1886b: On New Zealand Coleoptera, with descriptions of new genera and species. *Scientific Transactions of the Royal Dublin Society (Series II)* 3 [1883–1887]: 351–456, pls. XII–XIII. DP: [30] November 1886 (issue cover).

Sharp, D. 1904: Description of a new genus and species of Coleoptera (Fam. Hispidae) from New Britain. *Proceedings of the Linnean Society of New South Wales* 28 [1903–1904] (4): 924–925. DP: 28 April 1904 (contents p. ix).

Sharp, D. 1910: On the British species of Phaedon. *Entomologist's Monthly Magazine, Second Series* 46 (548): 4–6. DP: 4 January 1910 (registered at NHMUK library).

Sharp, D. 1914: Notes on some British species of Haltica. *Entomologist's Monthly Magazine, Second Series* 50 (606): 259–264. DP: 4 November 1914 (registered at NHMUK library).

Sharp, D. 1917: Cryptocephalus bipunctatus L., and C. biguttatus Scop. (= bipustulatus F.). *Entomologist's Monthly Magazine, Third Series* **53** (28): 76–79. DP: 5 April 1917 (registered at NHMUK library).

Shavrov, V. B. 1946: Novye formy i mestonakhozhdeniya donaciy (Coleoptera, Chrysomelidae, subfam. Donaciinae) fauny SSSR. I. *Byulleten' Moskovskogo Obshchestva Ispytateley Prirody, Otdel Biologicheskiy, Novaya Seriya* **51** (6): 36–39. SFP: 28 December 1946, DP: [31 December] 1946 (issue title page).

Shavrov, V. B. 1948: Novye formy i mestonakhozhdeniya donaciy (Coleoptera, Chrysomelidae, Subfam. Donaciinae) fauny SSSR. II. *Byulleten' Moskovskogo Obshchestva Ispytateley Prirody, Otdel Biologicheskiy, Novaya Seriya* **53** (1): 49–52. SFP: 2 February 1948, DP: 23 March 1948 (registered at ZIN library).

Shi, J. 2008: [*Common forest insects and spiders in the middle and lower reaches of the Yangtze River*]. Harbin: Northeast Forestry University, 199 pp. DP: [30] November 2008 (colophon).

Shigetoh, H. & Suenaga, H. 2022: Description of a new species of the genus Argopistes Motschulsky (Coleoptera, Chrysomelidae, Galerucinae, Alticini) from the Ryukyu Islands and Daitô Islands, southwestern Japan, and redescription of A. coccinelliformis Csiki. *Elytra, New Series* **12** (1): 1–13. DP: 15 September 2022 (paper title page).

Shin, C., Jin, H. & Chaboo, C. S. 2012: Biology and morphology of Neochlamisus gibbosus (Fabricius, 1777) (Coleoptera: Chrysomelidae: Cryptocephalinae: Fulcidacini). *Journal of the Kansas Entomological Society* **85** (2): 116–134. DP: [31 December] 2012 (paper title page).

Shirôzu, T. 1957: Three new species of the angulosa-group of the genus Dactylispa Weise from Japan, Manchuria and Formosa (Coleoptera: Chrysomelidae). *Sieboldia* **2** (1): 53–56, pl. 6. DP: [30] June 1957 (issue title page).

Shukla, S. P. 1960: Entomological survey of Himalaya. Part XXX – On some Chrysomelidae (Coleoptera) from the North-West (Punjab) Himalaya. *Agra University Journal of Research (Science)* **9** (1): 65–88. DP: [31] January 1960 (issue cover).

Siede, D. 1998: Bestimmungshilfe für die mitteleuropäischen Altica-Arten (Coleoptera, Chrysomelidae). *Entomologische Blätter* **94** (1/2): 77–90. DP: [31] July 1998 (issue cover).

Silfverberg, H. 1977: Nomenclatoric notes on Coleoptera Polyphaga. *Notulae Entomologicae* **57** (3): 91–94. DP: 30 September 1977 (volume contents).

Silfverberg, H. 2004: Enumeratio nova Coleopterorum Fennoscandiae, Daniae et Baltiae. *Sahlbergia* **9**: 1–111. DP: [31 December] 2004 (paper title page).

Silfverberg, H. 2010: Enumeratio renovata Coleopterorum Fennoscandiae, Daniae et Baltiae. *Sahlbergia* **16** (2): 1–144. DP: [31 December] 2010 (paper title page).

Singal, S. K. & Pajni, H. R. 1986: Conicobruchus of India (Bruchidae: Coleoptera). *Geobios New Reports* **5** (2): 97–100. DP: [31 December] 1986 (paper title page).

Singal, S. K. & Pajni, H. R. 1990: Six new species of Callosobruchus Pic from India (Coleoptera, Bruchidae). *Polskie Pismo Entomologiczne* **59** [1989] (4): 761–782. DP: 30 March 1990 (paper title page).

Singh, T. & Saini, M. S. 1978: Three new species of Bruchidae (Coleoptera) from Patiala (Panjab – India). *Journal of Entomological Research* **2** (2): 129–135. DP: [31] December 1978 (issue cover).

Sirgue, P. 1929: Deuxième note sur les chrysomèles. *Miscellanea Entomologica* **32** (4): 35. DP: 20 October 1929 (footer on p. 33).

Solier, A. J. J. 1837: Description d'une nouvelle espèce du genre Cryptocephalus. *Annales de la Société Entomologique de France* **5** [1836] (4): 687–688. DP: 6 March 1837 (Evenhuis 2022).

Solsky [= Solskij], S. M. 1866: Matériaux pour servir a l'étude des insectes de la Russie. I. Notes sur quelques coléoptères nouveaux ou peu connus. *Horae Societatis Entomologicae Rossicae* **4** [1866–1867] (2): 79–96. DP: [31 December] 1866 (Kerzhner 1984, Standfuss & Kerzhner 2004).

Solsky [= Solskij], S. M. 1867: Matériaux pour servir a l'étude des insectes de la Russie. II. Insectes nouveaux et remarques sur des espèces connues. *Horae Societatis Entomologicae Rossicae* **4** [1866–1867] (3/4): 179–185. DP: 3 October 1867 (Kerzhner 1984, Standfuss & Kerzhner 2004).

Solsky, S. M. 1870: Deux coléopteres nouveaux de la Sibérie orientale. *Horae Societatis Entomologicae Rossicae* **6** [1869] (4): 310–314. DP: 6 August 1870 (Kerzhner 1984, Standfuss & Kerzhner 2004).

Solsky, S. M. 1871: Coléoptères de la Sibérie orientale. *Horae Societatis Entomologicae Rossicae* **8** (3): 232–272 & (4): 273–277. DP (pp. 232–272): 22 December 1871, (pp. 273–277): 13 November 1872 (Kerzhner 1984, Standfuss & Kerzhner 2004).

Solsky, S. M. 1876: Matériaux pour l'entomographie des provinces asiatiques de la Russie. *Horae Societatis Entomologicae Rossicae* **11** [1875–1876] (3): 253–272 & (4): 273–299. DP (pp. 253–272): 7 October 1875, (pp. 273–299): 13 August 1876 (Kerzhner 1984, Standfuss & Kerzhner 2004).

Solsky, S. M. 1881: Novyya ili maloizvestnyya zhestkokrylya okrain Rossiyskoy imperii i prilegayushchikh k ney stran. [New and little known beetles from Russian imperium and neighboring countries]. *Trudy Russkago Entomologicheskago Obshchestva* **13** [1881–1882] (1/2): 31–84. DP: 13 October 1881 (Kerzhner 1984, Standfuss & Kerzhner 2004).

Sota, T., Bocák, L. & Hayashi, M. 2007: Molecular phylogeny and historical biogeography of the Holarctic wetland leaf beetle of the genus Plateumaris. *Molecular Phylogenetics and Evolution* **46** [2008]: 183–192. DP: 25 October 2007 (paper title page).

Southgate, B. J. 1971: On the identity of Caryedon pallidus (Olivier) (Col., Bruchidae), and the description of two new Caryedon spp. *Bulletin of Entomological Research* **60** [1970–1971] (3): 409–414. DP: 19 February 1971 (table of contents).

Southgate, B. J. 1977: A new subspecies of Caryedon (Coleoptera: Bruchidae) from the Middle East. *Israel Journal of Zoology* **25** [1976] (4): 194–198. DP: [30] April 1977 (paper title page).

Southgate, B. J., Howe, R. W. & Brett, G. A. 1957: The specific status of Callosobruchus maculatus (F.) and Callosobruchus analis (F.). *Bulletin of Entomological Research* **48** (1): 79–89, pls. III–IV. DP: 19 March 1957 (table of contents).

Spaeth, F. 1901a: Neue Cassiden aus Sumatra gesammelt von Dr. H. Dohrn. *Entomologische Zeitung* (Stettin) **62** (1/2): 3–13. DP: [31] May 1901 (last page of the issue).

Spaeth, F. 1901b: Ueber Chelysida und Oxylepus als zwei verschiedene Cassiden-Gattungen. *Verhandlungen der Kaiserlich-Königlichen Zoologisch-Botanischen Gesellschaft in Wien* **51** (10): 750–756. DP: 28 December 1901 (issue cover).

Spaeth, F. 1902: Beschreibung neuen Centralafrikanischer Cassiden aus dem Museum zu Brüssel. *Annales de la Société Entomologique de Belgique* **46** (11): 446–461. DP: 4 December 1902 (footer on p. 445).

Spaeth, F. 1903a: Eine neue Casside aus Birma. *Entomologisk Tidskrift* **24** (2/3): 111–112. DP: 4 August 1903 (verso of the volume title page).

Spaeth, F. 1903b: Zusammenstellung der bisher von Neu-Guinea bekannt gewordenen Cassiden mit besonderer Berücksichtigung der Sammlungen des Ungarischen National-Museums und des Museo Civico von Genua. *Annales Historico-Naturales Musei Nationalis Hungarici* **1** (1): 109–160. DP: 31 August 1903 (table of contents, Merkl et al. 2008).

Spaeth, F. 1904: Zur Kenntnis der Cassiden des Ostindischen Archipels. *Annali del Museo Civico di Storia Naturale di Genova* **41**: 69–79. DP: 24 May 1904 (Poggi 2010).

Spaeth, F. 1905: Beschreibung neuer Cassididen nebst synonymischen Bemerkungen. V. *Verhandlungen der Kaiserlich-Königlichen Zoologisch-Botanischen Gesellschaft in Wien* **55** (1): 79–118. DP: 28 January 1905 (volume title page).

Spaeth, F. 1911: Beschreibung neuer Cassididen nebst synonymischen Bemerkungen. VIII. *Verhandlungen der Kaiserlich-Königlichen Zoologisch-Botanischen Gesellschaft in Wien* **61** (5/6): 239–240 & (7/8): 241–277. DP (pp. 239–240): 28 August 1911, (pp. 241–277): 28 November 1911 (volume title page).

Spaeth, F. 1912a: One new genus and some new species of Cassidae from Borneo, with a list of all the species at present known from that Island. *Sarawak Museum Journal* **1** [1911–1913] (2): 113–122. DP: [29] February 1912 (table of contents).

Spaeth, F. 1912b: Ostafrikanische Cassiden. *Annales Historico-Naturales Musei Nationalis Hungarici* **10** (2): 496–508. DP: 29 December 1912 (table of contents, Merkl et al. 2008).

Spaeth, F. 1913a: H. Sauter's Formosa-Ausbeute: Cassidinae. *Annales Historico-Naturales Musei Nationalis Hungarici* **11** (1): 46–48. DP: 30 July 1913 (table of contents, Merkl et al. 2008).

Spaeth, F. 1913b: Studien über die Gattung Hoplionota Hope und Beschreibung einer verwandten neuen Gattung. *Verhandlungen der Kaiserlich-Königlichen Zoologisch-Botanischen Gesellschaft in Wien* **63** (9/10): 381–534. DP: 10 December 1913 (volume title page).

Spaeth, F. 1914a: H. Sauter's Formosa-Ausbeute. Cassidinae (Col.) II. *Supplementa Entomologica* **3**: 14–19. DP: 27 January 1914 (reprint cover).

Spaeth, F. 1914b: Neue Cassidinen aus Yünnan (Col.). *Entomologische Mitteilungen* **3** (7/8): 226–230. DP: 1 July 1914 (issue cover).

Spaeth, F. 1914c: Über die paläarktischen Cassiden mit besonderer Berücksichtigung jener von Asien. *Verhandlungen der Kaiserlich-Königlichen Zoologisch-Botanischen Gesellschaft in Wien* **64** (5/6): (128)–(147). DP: 5 September 1914 (volume title page).

Spaeth, F. 1914d: Zur Kenntnis der indischen Cassidinen. *Deutsche Entomologische Zeitschrift* **1914** (5): 542–568. DP: 30 September 1914 (table of contents).

Spaeth, F. 1914e: Chrysomelidae: 16. Cassidinae. In: Schenkling, S. (ed.): *Coleopterorum Catalogus, Pars 62*. Berlin: W. Junk, 182 pp. DP: 10 November 1914 (book cover).

Spaeth, F. 1914f: Die Hoplionota-Arten des königl. Museums in Berlin. Ein Nachtrag zu den Studien über die Gattung Hoplionota. *Verhandlungen der Kaiserlich-Königlichen Zoologisch-Botanischen Gesellschaft in Wien* 64 (7/8): 290–298. DP: 27 November 1914 (volume title page).

Spaeth, F. 1915: Hebdomecosta, eine neue Cassidinen-Gattung aus China, und Mitteilungen über Metriona sigillata Gorh. aus Japan. *Wiener Entomologische Zeitung* 34 (8/10): 361–364. DP: 22 October 1915 (footers throughout the paper).

Spaeth, F. 1917: Neuer Beitrag zur Kenntnis der Ost- und Zentralafrikanischen Cassidinen. *Annales Historico-Naturales Musei Nationalis Hungarici* 15 (2): 422–444. DP: 31 December 1917 (table of contents, Merkl et al. 2008).

Spaeth, F. 1919: Neue Cassidinen aus der Sammlung von Dr. K. Brancsik, dem Ungarischen National-Museum und meiner Sammlung. *Annales Historico-Naturales Musei Nationalis Hungarici* 17: 184–204. DP: 29 December 1919 (table of contents, Merkl et al. 2008).

Spaeth, F. 1921: Zwei neue Cassiden aus dem Ussurigebiet. *Koleopterologische Rundschau* 9 (4/9): 84–86. DP: 30 September 1921 (issue cover).

Spaeth, F. 1922: Chrysomélides. Cassidines. Pp. 997–1004. In: Rothschild, M. de (ed.): *Voyage de M. le Baron Maurice de Rothschild en Éthiopie et en Afrique Orientale Anglaise (1904–1905). Résultats scientifiques. Animaux articulés. Deuxième partie.* Paris: Imprimerie Nationale, 483–1041 pp. DP: 1 December 1922 (last page of the book).

Spaeth, F. 1925: Die Gattung Prioptera Hope (Coleoptera, Cassididae). *Philippine Journal of Science* 28 (3): 379–437, 2 pls. DP: 6 November 1925 (volume table of contents).

Spaeth, F. 1926a: Neue Cassidinen des Rijksmuseums in Leiden, des British Museums und meiner Sammlung. *Zoologische Mededeelingen* 9 (1): 1–15. DP: 30 January 1926 (volume table of contents).

Spaeth, F. 1926b: [new taxa]. In: Spaeth, F. & Reitter, E.: *Bestimmungs-Tabellen der europäischen Coleopteren. 95 Heft. Cassidinae der palaearktischen Region.* Troppau: E. Reitter, 68 pp. DP: 9 June 1926 (stamp on L. Sekerka's copy).

Spaeth, F. 1926c: Beschreibung neuer Cassidinen. *Bulletin Mensuel de la Société des Naturalistes Luxembourgeois, Nouvelle Série* 20 (6/10): 47–60. DP: 29 September 1926 (issue title page).

Spaeth, F. 1926d: Fauna sumatrensis (Beitrag Nr. 27). Cassidinae (Col.). *Supplementa Entomologica* 14: 116–119. DP: 20 December 1926 (volume cover).

Spaeth, F. 1926e: Mitteilungen ueber die Cassidinen des Prager Nationalmuseums (Col. Chrysomelidae). *Acta Entomologica Musei Nationalis Pragae* 4: 81–96. DP: [31 December] 1926 (issue cover).

Spaeth, F. 1928: Mitteilungen ueber die Cassidinen des Nationalmuseum in Prag. II. (Col., Chrysomelidae). *Acta Entomologica Musei Nationalis Pragae* 6: 29–47. DP: [31 December] 1928 (issue cover).

Spaeth, F. 1932a: Neue Beiträge zur Kenntnis der afrikanischen Cassidinen (Col. Chrys.). *Revue de Zoologie et de Botanique Africaines* 22 (1): 2–22. DP: 1 August 1932 (issue cover).

Spaeth, F. 1932b: Neue Cassidinen (Col. Chrysom.). *Stettiner Entomologische Zeitung* 93 (2): 182–204. DP: [30] November 1932 (table of contents).

Spaeth, F. 1933a: Neue Cassidinen von den Philippinen-Inseln (Coleoptera; Chrysomelidae). *Philippine Journal of Science* **51** (4): 495–506. DP: 25 October 1933 (volume table of contents).

Spaeth, F. 1933b: Dactylispa reitteri nov. spec. *Entomologisches Nachrichtenblatt* (Troppau) 7 (4): 136–137. DP: [31] December 1933 (issue cover, footers throughout the issue).

Spaeth, F. 1938: Über die von Herrn R. Malaise 1934 in Burma gesammelten Cassidinen. *Entomologisk Tidskrift* **59** (3/4): 228–236. DP: 5 December 1938 (verso of the volume title page).

Spaeth, F. 1940: Eine neue Cassida aus China. *Koleopterologische Rundschau* **26** (1/2): 37–38. DP: 1 March 1940 (issue cover).

Spaeth, F. 1941: Neue Cassidinen aus Italienish-Ost-Afrika des Museo Civico di Storia Naturale-Trieste. *Atti del Museo Civico di Storia Naturale di Trieste* **14** [1937–1941] (22): 315–318. DP: 26 April 1941 (footer on the paper title page).

Spaeth, F. 1942: Zur Kenntnis der Insekten von Mandschukuo. 12. Beitrag. Über einige Cassidinen aus Mandschukuo (Coleoptera: Chrysomelidae). *Arbeiten über Morphologische und Taxonomische Entomologie aus Berlin-Dahlem* 9 (1): 12–13. DP: 25 March 1942 (volume title page).

Spaeth, F. 1952: [new taxa]. In: Hincks, W. D.: The genera of the Cassidinae (Coleoptera: Chrysomelidae). *Transactions of the Royal Entomological Society of London* 103 (10): 327–358. DP: 31 December 1952 (issue cover).

Spaeth, F. & Reitter, E. 1926: *Bestimmungs-Tabellen der europäischen Coleopteren. 95 Heft. Cassidinae der palaearktischen Region.* Troppau: E. Reitter, 68 pp. DP: 9 June 1926 (stamp on L. Sekerka's copy).

Sprecher-Uebersax, E. 2000: Trachyaphthona hiunchulii, eine neue Alticinen-Art in Nepal (Col. Chrysomelidae, Alticinae). *Entomologica Basiliensia* **22**: 203–207. DP: 1 September 2000 (table of contents).

Sprecher-Uebersax, E. & Daccordi, M. 2012: New species of Phaedon Latreille, 1829 (Insecta: Coleoptera: Chrysomelidae) from the Himalaya. Pp. 415–421. In: Hartmann, M. & Weipert, J. (eds.): *Biodiversität und Naturausstattung im Himalaya IV.* Erfurt: Verein der Freunde & Förderer des Naturkundemuseums Erfurt e. V., 492 pp. DP: 15 August 2012 (editor of the book M. Hartmann 2019, pers. comm.).

Sprecher-Uebersax, E. & Daccordi, M. 2016: Leaf-beetles of the subfamily Chrysomelinae of Laos (Coleoptera: Chrysomelidae). *Entomologica Basiliensia et Collectionis Frey* **35**: 455–485. DP: 10 May 2016 (colophon).

Sprecher-Uebersax, E., Konstantinov, A. S., Prathapan, K. D. & Döberl, M. 2009: Revision of the genus Benedictus Scherer (Insecta: Coleoptera: Chrysomelidae: Galerucinae). Pp. 475–514. In: Hartmann, M. & Weipert, J. (eds.): *Biodiversität und Naturausstattung im Himalaya III.* Erfurt: Verein der Freunde & Förderer des Naturkundemuseums Erfurt e. V., 477 pp. DP: 2 July 2009 (editor of the book M. Hartmann 2019, pers. comm.).

Sprecher-Uebersax, E. & Zoia, S. 2002: Pyrrhalta medvedevi sp. nov., a new species from the Nepal Himalayas (Coleoptera, Chrysomelidae, Galerucinae). *Mitteilungen der Schweizerischen*

Entomologischen Gesellschaft 75 (3/4): 161–167. DP: [31] December 2002 (last page of the cover).

Springer, G. 1953: [new taxon]. In: Müller, G.: *I coleotteri della Venezia Giulia. Catalogo ragionato con tabelle dicotomiche per la classificazione delle specie della regione Adriatica orientale, del Veneto e della Pianura Padana. Volume II: Coleoptera Phytophaga (Cerambycidae, Chrysomelidae, Bruchidae)*. [1949–1953], Trieste: Centro sperimentale agrario e forestale, 685 + [1] pp. DP (pp. 481–592): [30] June 1953 (last page of the book).

Špringlová, B. 1945: Přehled forem forem druhu Chrysolina geminata Payk. Formarum Chrysolinae geminatae Payk. conspectus (Col. Phytoph. Chrysomelidae). *Časopis Československé Společnosti Entomologické* 42 (1/4): 142–144. DP: 1 September 1945 (issue cover).

Staines, C. L. 2007: A review of the genus Chaeridiona Baly, 1869 (Coleoptera: Chrysomelidae: Cassidinae). *Zootaxa* 1521: 19–29. DP: 15 July 2007 (footer on the paper title page).

Staines, C. L. 2010: Some nomenclatural notes on Hispines (Coleoptera: Chrysomelidae: Hispinae). *Proceedings of the Entomological Society of Washington* 112 (1): 172–173. DP: 19 February 2010 (verso of the issue cover).

Stål, C. 1857: Till kännedomen af Phytophaga. *Öfversigt af Kongliga Vetenskaps-Akademiens Förhandlingar* 14 (2): 53–63. DP: [31 December] 1857 (last page of the issue, see also the comments in Dickinson et al. 2011: 227).

Stål, C. 1858: Nya genera och arter bland Phytophaga. *Öfversigt af Kongliga Vetenskaps-Akademiens Förhandlingar* 15 (4/5): 250–258. DP: [31 December] 1858 (footer on last page of the issue, see also the comments in Dickinson et al. 2011: 227).

Stål, C. 1859: Bidrag till kännedomen om Amerikas Chrysomeliner. *Öfversigt af Kongliga Vetenskaps-Akademiens Förhandlingar* 16 (7): 305–326. DP: [31 December] 1859 (last page of the issue, see also the comments in Dickinson et al. 2011: 227).

Stål, C. 1861: Till kännedomen om Chrysomelidae. *Öfversigt af Kongliga Vetenskaps-Akademiens Förhandlingar* 17 [1860] (8): 455–470. DP: [31 December] 1861 (last page of the issue, see also the comments in Dickinson et al. 2011: 227).

Stål, C. 1862: Monographie des chrysomélides de l'Amérique. I. *Nova Acta Societatis Scientiarum Upsaliensis, Seriei Tertiae* 4 [1862–1863] (1): 1–86. DP: 10 December 1862 (*Öfv. Kongl. Vetens.-Akad. Forhand. 1862, 19: 598*).

Steinhausen, W. R. 1993: Phratora polaris leederi, eine boreoalpine Rasse in Mitteleuropa (Coleoptera Chrysomelidae). *Acta Coleopterologica* 9 (1): 27–29. DP: 1 March 1993 (paper title page).

Steinhausen, W. R. 1994: 116. Familie: Chrysomelidae. Pp. 231–314. In: Klausnitzer, B. (ed.): *Die Larven der Käfer Mitteleuropas. 2. Band. Myxophaga, Polyphaga Teil 1.* Krefeld: Goecke & Evers, 325 pp. DP: [31 December] 1994 (colophon).

Steinhausen, W. R. 1996: Versuch einer Ubereinstimmung imaginaler und larvaler Systeme (Coleoptera, Chrysomelidae, Chrysomelinae). Pp. 261–263. In: Gerstmeier, R. & Scherer, G. (eds.): *Verhandlungen des 14. Internationalen Symposiums über Entomofaunistik in*

Mitteleuropa (SIEEC) 4–9. September 1994 München. München, 449 pp. DP: [31 December] 1996 (book cover).

Steinhausen, W. R. 1998: Comparative morphology of pupae of the leaf beetle subfamilies Chrysomelinae and Galerucinae. Pp. 311–318. In: Biondi, M., Daccordi, M. & Furth, D. G. (eds.): *Proceedings of XX International Congress of Entomology, Firenze, 1996. Volume 3: Proceedings of the Fourth International Symposium on Chrysomelidae.* Torino: Museo Regionale di Scienze Naturali, 327 pp. DP: [31 December] 1998 (book cover).

Steinhausen, W. R. 2001: Die Puppen mitteleuropäischer Blattkäfer – Eine vorläufige Bestimmungstabelle. 1. Teil (Coleoptera: Chrysomelidae). *Mitteilungen der Münchner Entomologischen Gesellschaft* 91: 35–63. DP: 1 October 2001 (paper title page).

Steinhausen, W. R. 2002: Die Puppen mitteleuropäischer Blattkäfer – Eine vorläufige Bestimmungstabelle. 2. Teil (Coleoptera: Chrysomelidae). *Mitteilungen der Münchner Entomologischen Gesellschaft* 92: 5–36. DP: 1 October 2002 (paper title page).

Steinhausen, W. R. 2007: Die Blattkäfergattung Cryptocephalus Geoffroy (1768) und ihre Untergattungen in Mitteleuropa nach larvaler Morphologie mit einer Revision der Larven-Bestimmungstabelle (Steinhausen, 1994) (Coleoptera, Chrysomelidae, Cryptocephalinae). *Mitteilungen der Münchner Entomologischen Gesellschaft* 97: 23–32. DP: 31 October 2007 (paper title page).

Stephens, J. F. 1829a: *Illustrations of British entomology; or, a synopsis of indigenous insects: containing their generic and specific distinctions; with an account of their metamorphoses, times of appearance, localities, food, and economy, as far as practicable. Mandibulata. Volume II.* London: Baldwin & Cradock, 200 + [1] pp., pls. X–XV. DP (pp. 129–160): 1 February 1829 (Bousquet 2016).

Stephens, J. F. 1829b: *A systematic catalogue of British insects: being an attempt to arrange all the hitherto discovered indigenous insects in accordance with their natural affinities. Containing also the references to every English writer on entomology, and to the principal foreign authors. With all the published British genera to the present time.* London: Baldwin & Cradock, xxxiv + [1] + 416 pp. [Part I. Insecta Mandibulata] + 388 pp. [Part II. Insecta Haustellata]. DP: 15 July 1829 (Bousquet 2016).

Stephens, J. F. 1831: *Illustrations of British entomology; or, a synopsis of indigenous insects: containing their generic and specific distinctions; with an account of their metamorphoses, times of appearance, localities, food, and economy, as far as practicable. Mandibulata. Volume IV.* London: Baldwin & Cradock, 413 + [1] pp., pls. XX–XXIII. DP (pp. 143–222): 31 October 1831, (pp. 223–270): 30 November 1831, (pp. 271–286): 1 December 1831, (pp. 287–366): 31 December 1831 (Bousquet 2016).

Stephens, J. F. 1832: *Illustrations of British entomology; or, a synopsis of indigenous insects: containing their generic and specific distinctions; with an account of their metamorphoses, times of appearance, localities, food, and economy, as far as practicable. Mandibulata. Volume IV.* London: Baldwin & Cradock, 413 + [1] pp., pls. XX–XXIII. DP (pp. 367–423): 31 January 1832 (Bousquet 2016).

Stephens, J. F. 1835: *Illustrations of British entomology; or, a synopsis of indigenous insects: containing their generic and specific distinctions; with an account of their metamorphoses, times of appearance, localities, food, and economy, as far as practicable. Mandibulata. Volume V.* [1832–1835]. London: Baldwin and Cradock, 447 pp. + list of illustrations, pls. XXIV–XXVII. DP (pp. 385–432): 31 March 1835 (Bousquet 2016).

Stephens, J. F. 1839: *A manual of British Coleoptera, or beetles; containing a brief description of all the species of beetles hitherto ascertained to inhabit Great Britain and Ireland; together with a notice of their chief localities, times and places of appearances, etc.* London: Longman, Orme, Brown, Green, and Longmans, xii + 443 pp., 12 pls. DP: 15 October 1839 (Bousquet 2016).

Steven, C. 1806: Déscription de dix insectes de la Russie méridionale. *Mémoires de la Société des Naturalistes de l'Université Impériale de Moscou* 1: 155–167, pl. IX. DP: [31] December 1806 (Bousquet & Davies 2021).

Stierlin, G. 1863a: Ueber einige neue oder wenig gekannte Insekten der Gegend von Sarepta. *Bulletin de la Société Impériale des Naturalistes de Moscou* **36** (4): 489–502. DP: 4 March 1863 (Bousquet & Davies 2021).

Stierlin, W. G. 1863b: Verzeichniss der Während einer entomologischen Excursion nach dem Engadin im Juni 1862 gesammelten Käfer. *Mittheilungen der Schweizerischen Entomologischen Gesellschaft* 1 [1862–1865] (3): 57–66. DP: [31] March 1863 (issue title page).

Stierlin, W. G. 1866: Coleopterologisches. *Mittheilungen der Schweizerischen Entomologischen Gesellschaft* 2 [1866–1868] (1): 30–32. DP: [31] January 1866 (issue title page).

Stierlin, W. G. 1867: Beschreibung einiger neuen Käfer-Arten. *Mittheilungen der Schweizerischen Entomologischen Gesellschaft* 2 [1866–1868] (5): 218–228. DP: [28] February 1867 (issue title page).

Stierlin, W. G. 1887: Ueber eine neue Varietät der Oreina luctuosa Ol. (tristis F.). *Societas Entomologica* 1 [1886–1887] (18): 137–138. DP: 1 January 1887 (issue title page).

Stierlin, W. G. 1890: Sammelnotiz. *Mittheilungen der Schweizerischen Entomologischen Gesellschaft* 8 [1888–1893] (6): 250–252. DP: [31] December 1890 (issue title page).

Stierlin, W. G. 1896a: Coleoptera Helvetiae. *Mittheilungen der Schweizerischen Entomologischen Gesellschaft* 9 (7): 513–544. DP: [30] April 1896 (issue cover).

Stierlin, W. G. 1896b: Coleoptera Helvetiae. *Mittheilungen der Schweizerischen Entomologischen Gesellschaft* 9 (8): 545–576. DP: [31] July 1896 (issue cover).

Stierlin, W. G. 1897: Coleoptera Helvetiae. *Mittheilungen der Schweizerischen Entomologischen Gesellschaft* 10 (1): 577–624. DP: [30] November 1897 (issue cover).

Stiller, V. 1923: A Fiumei tengermellék és a mögöttes Karszt-hegységet összekötő táj bogárfaunája. (Die Käferfauna der Kontaktzone zwischen dem Litorale von Fiume und dem dahinterliegenden Karstzgebiete). *Rovartani Lapok* **26** (7/12): 131–140. DP: 1 December 1923 (issue title page).

Stojanova, A. & György, Z. 2011: Checklist of the Bulgarian Bruchinae (Coleoptera: Chrysomelidae). *ZooNotes* **25**: 1–7. DP: 30 August 2011 (footer on the paper title page).

Strand, A. 1936: Phytodecta norvegicus n. sp. (Col. Chrysomelidae). *Norsk Entomologisk Tidsskrift* 4 (3): 104–105. DP: [30] June 1936 (verso of the volume title page).

Strand, E. 1917: Übersicht der in Gistel's "Achthundert und zwanzig neue oder unbeschriebene wirbellose Thiere" (1857) behandelten Insekten. *Archiv für Naturgeschichte, Abteilung A* 82 [1916] (5): 75–101. DP: [31] August 1917 (issue cover).

Strand, E. 1928: Nomenklatorische Bemerkungen über einige Coleopteren-Gattungen. *Entomologisches Nachrichtsblatt* (Troppau) 2 (1): 2–3. DP: [31] March 1928 (footer on the paper title page).

Strand, E. 1935: Revision von Gattungsnamen palaearktischer Coleoptera. *Folia Zoologica et Hydrobiologia* 7 [1934–1935] (2): 282–299. DP: 20 April 1935 (issue cover).

Strand, E. 1942: Miscellanea nomenclatorica zoologica et palaeontologica X. *Folia Zoologica et Hydrobiologica* 11 (2): 386–402. DP: 15 September 1942 (issue cover).

Strejček, J. 1990: *Brouci čeledí Bruchidae, Urodonidae a Anthribidae. [Beetles of the families Bruchidae, Urodonidae and Anthribidae]*. Praha: Academia, 88 pp., 14 pls. DP: [31 December] 1990 (book title page).

Strejček, J. 1993: Chrysomelidae. Pp. 123–132. In: Jelínek, J. (ed.): Check-list of Czechoslovak Insects IV (Coleoptera). Seznam československých brouků. *Folia Heyrovskyana* **Supplementum** 1: 3–172. DP: [31 December] 1990 (supplement cover).

Strejček, J. 2012: Icones Insectorum Europae Centralis. Coleoptera: Bruchidae, Urodontidae. *Folia Heyrovskyana, Series B* 15: 1–24. DP: 31 October 2012 (paper title page).

Strouhal, H. 1955: Franz Heikertinger †. *Annalen des Naturhistorischen Museums in Wien* 60 [1954–1955]: 20–35. DP: [31] December 1955 (volume title page).

Sturm, J. 1807: *Deutschlands Fauna in Abbildungen nach der Natur mit Beschreibungen. V. Abtheilung. Die Insecten. Zweites Bändchen. Käfer.* Nürnberg: J. Sturm, iv + 279 pp. + pls. 21–52. DP: 19 April 1807 (Bousquet 2017).

Sturm, J. 1826: *Catalog meiner Insecten-Sammlung. Erster Theil. Käfer.* Nürnberg, viii + 208 + 16 pp., 4 pls. DP: 15 April 1826 (Bousquet 2017).

Sturm, J. 1843: *Catalog der Kaefer-Sammlung von Jacob Sturm.* Nürnberg: J. Sturm: xii + 386 pp. + 6 pls. DP: 27 July 1843 (Bousquet 2016).

Su, L. & Zhou, H.-Z. 2017: Taxonomy of the genus Chlamisus Rafinesque (Coleoptera: Chrysomelidae) from China with description of three new species. *Zootaxa* 4233 (1): 1–138. DP: 23 February 2017 (footer on the paper title page).

Suenaga, H. 2012: A new species of the genus Gonioctena (Coleoptera: Chrysomelidae: Chrysomelinae) from Shikoku, Japan, with description of its immature stages. *Zootaxa* 3268: 29–39. DP: 13 April 2012 (footer on the paper title page).

Suenaga, H. 2013: Notes on Cassida ferruginea and Cassida mongolica in Japan, with descriptions of their reproductive systems (Coleoptera: Chrysomelidae: Cassidinae). *Genus* 24 (3/4): 325–333. DP: 28 December 2013 (paper title page).

Suenaga, H. 2020: A revision of the genus Altica (Coleoptera: Chrysomelidae: Galerucinae) of Japan. *Japanese Journal of Systematic Entomology, Supplementary Series* **2**: 163–258. DP: 30 April 2020 (paper title page).

Suenaga, H. 2023: Two new species of Clavicornaltica Scherer (Coleoptera, Chrysomelidae, Galerucinae) from the Ryukyu Islands, Southwestern Japan, with a redescription of C. sakishimana Suenaga and Yoshida. *Zootaxa* **5301** (4): 447–460. DP: 13 June 2023 (footer on the paper title page).

Suenaga, H. & Lee, C.-F. 2023: Taxonomic studies on the genus Altica Geoffroy (Coleoptera, Chrysomelidae, Galerucinae, Alticini) in Taiwan: Re-evaluation of A. himalayensis (Chen), A. cyanea (Weber), and A. caerulea (Olivier). *Formosan Entomologist* **43**: 49–65. DP: 17 November 2023 (paper title page)

Suenaga, H., Lee, C.-L. & Bezděk, J. 2017: Two new species of the genus Menippus Clark (Coleoptera: Chrysomelidae: Galerucinae) from islands of Ryukyu-rettô, Japan, and redescription of M. issikii (Chûjô). *Entomologische Blätter und Coleoptera* **113** (1): 231–240. DP: [30] November 2017 (issue cover).

Suenaga, H. & Yoshida, T. 2016: Two new species of the genus Clavicornaltica (Coleoptera, Chrysomelidae, Galerucinae) from Taiwan and Ishigaki-jima Island, Japan. *Elytra, New Series* **6** (1): 1–9. DP: 25 May 2016 (paper title page).

Suffrian, E. 1840: Fragmente zur genauern Kenntniss deutscher Käfer. *Entomologische Zeitung* (Stettin) **1** (7): 98–104. DP: [31] July 1840 (issue title page).

Suffrian, E. 1841a: Fragmente zur genauern Kenntniss deutscher Käfer. *Entomologische Zeitung* (Stettin) **2** (3): 38–47. DP: [31] March 1841 (issue title page).

Suffrian, E. 1841b: Fragmente zur genauern Kenntniss deutscher Käfer. *Entomologische Zeitung* (Stettin) **2** (7): 97–106. DP: [31] July 1841 (issue title page).

Suffrian, E. 1844a: Fragmente zur genaueren Kenntniss deutscher Käfer. 4. Cassida L. *Entomologische Zeitung* (Stettin) **5** (3): 89–107. DP: [31] March 1844 (issue title page).

Suffrian, E. 1844b: Fragmente zur genaueren Kenntniss deutscher Käfer. 4. Cassida L. *Entomologische Zeitung* (Stettin) **5** (4): 135–148. DP: [30] April 1844 (issue title page).

Suffrian, E. 1844c: Fragmente zur genaueren Kenntniss deutscher Käfer. 4. Cassida L. *Entomologische Zeitung* (Stettin) **5** (5): 186–192. DP: [31] May 1844 (issue title page).

Suffrian, E. 1844d: Fragmente zur genaueren Kenntniss deutscher Käfer. 4. Cassida L. *Entomologische Zeitung* (Stettin) **5** (6): 206–224. DP: [30] June 1844 (issue title page).

Suffrian, E. 1844e: Fragmente zur genaueren Kenntniss deutscher Käfer. 4. Cassida L. *Entomologische Zeitung* (Stettin) **5** (7): 241–257. DP: [31] July 1844 (issue title page).

Suffrian, E. 1844f: Fragmente zur genaueren Kenntniss deutscher Käfer. 4. Cassida L. *Entomologische Zeitung* (Stettin) **5** (8): 270–287. DP: [31] August 1844 (issue title page).

Suffrian, E. 1845: Bemerkungen zu den in “Lacordaire’s Monographie des coléoptères subpentamères de la famille des Phytophages” vorkommenden deutschen Arten. *Entomologische Zeitung* (Stettin) **6** (11): 359–367. DP: [30] November 1845 (issue title page).

Suffrian, E. 1846: Bemerkungen zu den in "Lacordaire's Monographie des coléoptères subpentamères de la famille des Phytophages" vorkommenden deutschen Arten. *Entomologische Zeitung* (Stettin) 7 (3): 80–92. DP: [31] March 1846 (issue title page).

Suffrian, E. 1847a: Zur Kritik einiger Käferarten, nach Vergleich der Typen aus der Fabricius'schen Sammlung. *Entomologische Zeitung* (Stettin) 8 (4): 98–102. DP: [30] April 1847 (issue title page).

Suffrian, E. 1847b: Revision der europäischen Arten der Gattung Cryptocephalus. *Linnaea Entomologica* 2: 1–194. DP: [31] July 1847 (*Ent. Zeit.* 1847, 8 (7): 224).

Suffrian, E. 1848: Revision der europäischen Arten der Gattung Cryptocephalus. (Fortsetzung und Schluss). *Linnaea Entomologica* 3: 1–152. DP: [30] April 1848 (*Ent. Zeit.* 1848, 9 (4): 128).

Suffrian, E. 1849a: Drei neue europäische Cryptocephali. *Entomologische Zeitung* (Stettin) 10 (1): 17–21. DP: [31] January 1849 (issue title page).

Suffrian, E. 1849b: Abermals zwei neue europäische Cryptocephali. *Entomologische Zeitung* (Stettin) 10 (10): 290–293. DP: [31] October 1849 (issue title page).

Suffrian, E. 1851a: Zur Kenntniss der europäischen Chrysomelen. *Linnaea Entomologica* 5: 1–280. DP: 6 February 1851 (*Allg. Bibl. Deutsch.* 1851 (5/6): 36).

Suffrian, E. 1851b: Bemerkungen zu den in Lacordaire's "Monographie des Coleoptères subpentamères de la famille des Phytophages" vorkommenden europäischen Arten. *Entomologische Zeitung* (Stettin) 12 (7): 194–220. DP: [31] July 1851 (issue title page).

Suffrian, E. 1852a: [new taxa]. In: Kiesenwetter, E. A. H. von.: Énumération des coléoptères trouvés dans le midi de la France et en Catalogne (IIe Partie). *Annales de la Société Entomologique de France, Deuxième Série* 9 [1851] (4): 577–656. DP: 28 January 1852 (Evenhuis 2022).

Suffrian, E. 1852b: Zur Kenntniss der nordamerikanischen Cryptocephalen. *Linnaea Entomologica* 7: 1–238. DP: 25 November 1852 (*Allg. Bibl. Deutsch.* 1852 (47): 441).

Suffrian, E. 1853a: Monographie des chrysomèles d'Europe. Traduction de M. L. Fairmaire. *Annales de la Société Entomologique de France, Troisième Série* 1 (1): 91–126. DP: 8 June 1853 (Evenhuis 2022).

Suffrian, E. 1853b: Berichtigtes Verzeichniss der europäischen Cryptocephalen. *Linnaea Entomologica* 8: 88–153. DP: 20 October 1853 (*Allg. Bibl. Deutsch.* 1853 (42): 373).

Suffrian, E. 1854a: Monographie des chrysomèles d'Europe (Suite). Traduction de M. L. Fairmaire. *Annales de la Société Entomologique de France, Troisième Série* 2 (2): 313–328. DP: 13 September 1854 (Evenhuis 2022).

Suffrian, E. 1854b: Verzeichniss der bis jetzt bekannt gewordenen asiatischen Cryptocephalen. *Linnaea Entomologica* 9: 1–169. DP: 14 December 1854 (*Allg. Bibl. Deutsch.* 1854 (50): 480).

Suffrian, E. 1855: Synonymische Miscellaneen. *Entomologische Zeitung* (Stettin) 16 (5/6): 142–150. DP: [30] June 1855 (issue title page).

Suffrian, E. 1857: Zur Kenntniss der afrikanischen Cryptocephalen. *Linnaea Entomologica* 11: 57–260. DP: 5 February 1857 (*Allg. Bibl. Deutsch.* 1857 (5/6): 33).

Suffrian, E. 1858a: Monographie des chrysomèles d'Europe (suite). Traduction de M. L. Fairmaire. *Annales de la Société Entomologique de France, Troisième Série* **6** (3): 531–567. DP: 24 November 1858 (Evenhuis 2022).

Suffrian, E. 1858b: [new taxa]. In: Rogers, F. W.: Uebersicht der in den Verein. Staaten von Nord-Amerika einheimischen Chrysomelen. *Entomologische Zeitung* (Stettin) **19** (10/12): 381–400. DP: [31] December 1858 (issue title page).

Suffrian, E. 1860: Berichtigtes Verzeichniss der bis jetzt bekannt gewordenen asiatischen Cryptocephalen. *Linnaea Entomologica* **14**: 1–72. DP: [30] April 1860 (Evenhuis 2022).

Suffrian, E. 1861a: Synonymische Miscellaneen. *Entomologische Zeitung* (Stettin) **22** (1/3): 86–92. DP: [31] March 1861 (issue title page).

Suffrian, E. 1861b: Synonymische Miscellaneen. *Entomologische Zeitung* (Stettin) **22** (10/12): 429–437. DP: [31] December 1861 (issue title page).

Suffrian, E. 1864: Synonymische Miscellaneen. *Entomologische Zeitung* (Stettin) **25** (7/9): 263–265. DP: [30] September 1864 (issue title page).

Suffrian, E. 1866: Synonymiche [sic] Miscellaneen. *Entomologische Zeitung* (Stettin) **27** (4/6): 205–210. DP: [30] June 1866 (issue title page).

Suffrian, E. 1867a: Rhaebus beckeri m., ein neuer europäischer Käfer. *Entomologische Zeitung* (Stettin) **28** (4/6): 141–144. DP: [31] March 1867 (last page of the issue).

Suffrian, E. 1867b: Cryptocephalus astracanicus n. sp. *Entomologische Zeitung* (Stettin) **28** (7/9): 309–311. DP: [30] June 1867 (last page of the issue).

Suffrian, E. 1868: Synonymische Miscellaneen. *Entomologische Zeitung* (Stettin) **29** (4/6): 170–176. DP: [30] April 1868 (last page of the issue).

Suffrian, E. 1870a: Verzeichniss der von Dr. Gundlach auf der Insel Cuba gesammelten Rüsselkäfer. *Archiv für Naturgeschichte* **36** [1] (2): 150–234. DP: 12 November 1870 (*Lit. Centralbl. Deutschl.* 1870 (47): 1257).

Suffrian, E. 1870b: Synonymische Miscellaneen. *Entomologische Zeitung* (Stettin) **32** [1871] (1/3): 21–24. DP: [30] November 1870 (last page of the issue).

Sulzer, J. H. 1776: *Abgekürzte Geschichte der Insecten. Nach dem Linaeischen System. Erster Theil.* Winterthur: Steiner u. Co., xxviii + 274 + 72 pp. + 32 pl. DP: 21 April 1776 (Bousquet 2016).

Sumakov [= Sumakow], G. G. 1901a: Novyy vid iz roda Malegia Lef. (Coleoptera, Chrysomelidae). *Russkoe Entomologicheskoe Obozrenie* **1** (4/5): 179–180. DP: [31] October 1901 (footer on the paper title page).

Sumakov, G. G. 1901b: Species nova generis Donacia Fab. *Sitzungsberichte der Naturforscher-Gesellschaft bei der Universität Jurjew (Dorpat)* **12** [1898–1901] (3): 454–459. DP: 19 November 1901 (*Sitzungsber. Ges. Naturforsch. Freunde Berlin* 1901: 256).

Sun, C.-H. 1985: A new species of the genus Javeta from China (Coleoptera: Hispidae). *Entomotaxonomia* **7** (1): 7–8. DP: [31] March 1985 (paper title page).

Suzuki, K. 2005: Description of a new species of the genus Lema (Coleoptera, Chrysomeliae, Criocerinae) from Honshu, Japan. *Elytra* **33** (1): 86–94. DP: 30 May 2005 (paper title page).

Suzuki, K. 2006: Description of a new species of the genus Crepidodera (Coleoptera, Chrysomelidae, Alticinae) from Honshu, Japan. *Elytra* **34** (1): 189–198. DP: 20 May 2006 (paper title page).

Suzuki, K. & Saitoh, S. 2011: Description of a new species of the genus Chrysolina (Coleoptera, Chrysomelidae) from central Honshu, Japan. *Elytra, New Series* **1** (1): 131–140. DP: 31 July 2011 (paper title page).

Swartz, O. 1808: [new taxa]. In: Schoenherr, C. J.: *Synonymia insectorum, oder: Versuch einer Synonymie aller bisher bekannten Insecten; nach Fabricii Systema Eleutheratorum &c. geordnet. Erster Band. Eleutherata oder Käfer. Zweiter Theil. Spercheus---Cryptocephalus*. Stockholm: C. F. Marquard, ix + 424 pp., 1 pl. DP: [31 December] 1808 (Bousquet 2016).

Świętojańska, J. 2001a: A revision of the tribe Aspidimorphini of the Oriental region (Coleoptera: Chrysomelidae: Cassidinae). *Genus* **Supplement** [11]: 1–318, 18 pl. DP: 28 December 2001 (colophon).

Świętojańska, J. 2001b: Hispa tarsata, a new species from Iran (Coleoptera: Chrysomelidae: Hispinae). *Genus* **12** (4): 479–482. DP: 28 December 2001 (paper title page).

Świętojańska, J. 2005: Description of first instar larvae of Cassida denticollis Suffrian, 1844, C. prasina Illiger, 1798, C. sanguinolenta Müller, 1776, and C. sanguinosa Suffrian, 1844 (Coleoptera: Chrysomelidae: Cassidinae). *Genus* **16** (4): 583–610. DP: 28 December 2005 (paper title page).

Świętojańska, J. & Borowiec, L. 2006: Aspidimorpha (s. str.) tibetana, a new species from China (Coleoptera: Chrysomelidae: Cassidinae). *Genus* **17** (4): 541–544. DP: 28 December 2006 (paper title page).

Świętojańska, J. & Sekerka, L. 2007: Notosacantha warchalowskii, a new species from India (Coleoptera: Chrysomelidae: Cassidinae: Notosacanthini). *Genus* **18** (4): 681–685. DP: 28 December 2007 (paper title page).

Székessy, V. 1941a: Die zur Gattung Haemonia Latr. gestellten Arten aus Amerika (Coleopt., Chrysomelidae). *Mitteilungen der Münchner Entomologischen Gesellschaft* **31** (1): 148–154. DP: 15 January 1941 (volume table of contents).

Székessy, V. 1941b: Haemonia mutica balatonica nov. subsp. (Coleopt., Chrysom.). *Fragmenta Faunistica Hungarica* **4** (2): 21–22. DP: 1 April 1941 (issue cover).

Székessy, V. 1943: Die Koleopteren-Fauna der Halbinsel Tihany. *A Magyar Biologiai Kutatóintézet Munkái* **15**: 358–398. DP: [31 December] 1943 (volume cover).

Takei, T. 1916: [New insects collected by myself]. *Insect World* (Gifu) **20** (6): 254–256 (in Japanese). DP: 15 June 1916 (issue title page).

Takemoto, T. 2019: Revision of the genus Zeugophora (Coleoptera, Megalopodidae, Zeugophorinae) in Japan. *Zootaxa* **4644** (1): 1–62. DP: 22 July 2019 (paper title page).

Takemoto, T. 2022: Description of a new species of the genus Chrysolina (Coleoptera: Chrysomelidae) from Honshu, and notes on records of Chrysolina aeruginosa in Japan. *Acta Entomologica Musei Nationalis Pragae* **62** (2): 457–474. DP: 31 December 2022 (paper title page).

Takemoto, T. & Suenaga, H. 2021: Four new species of the genus Ivalia Jacoby and a new species of Cangshanaltica Konstantinov, Chamorro, Prathapan, Ge et Yang (Coleoptera, Chrysomelidae, Galerucinae) from the Ryukyus, southwestern Japan. *Elytra, New Series* **11** (1): 129–153. DP: 16 August 2021 (paper title page).

Takizawa, H. 1970: Descriptions of five new species of the genus Chrysolina Motschulsky in Japan (Coleoptera: Chrysomelidae). *Kontyû* **38** (2): 117–125. DP: 30 June 1970 (heading on p. 93).

Takizawa, H. 1971: A new species of Cryptocephalus from Sakhalin (Coleoptera: Chrysomelidae). *Kontyû* **39** (1): 137–138. DP: 30 July 1971 (heading on p. 77).

Takizawa, H. 1975a: Descriptions of new chrysomelid beetles from Ryukyu Archipelago (Coleoptera, Chrysomelidae). *Entomological Review of Japan* 28 (1/2): 56–62. DP: 30 August 1975 (General index 1974–1977).

Takizawa, H. 1975b: A review of the approximatus-group of Cryptocephalus (Coleoptera, Chrysomelidae) in Japan, with description of new species. *Kontyû* **43** (4): 422–436. DP: 25 December 1975 (paper title page).

Takizawa, H. 1978a: Notes on Taiwanese Chrysomelidae, I. *Kontyû* **46** (1): 123–134. DP: 25 March 1978 (paper title page).

Takizawa, H. 1978b: Notes on Taiwanese chrysomelid-beetles, 2. *Kontyû* **46** (4): 596–602. DP: 25 December 1978 (paper title page).

Takizawa, H. 1979: Notes on Taiwanese Chrysomelidae, III. *Kontyû* **47** (3): 340–351. DP: 25 September 1979 (paper title page).

Takizawa, H. 1980a: Notes on Korean Chrysomelidae. *Nature and Life* (Taegu) 10 (2): 1–13. DP: [31] December 1980 (issue cover).

Takizawa, H. 1980b: Notes on Indian Chrysomelinae based on the collection of the Forest Research Institute, Dehra Dun (Coleoptera: Chrysomelidae). *Insecta Matsumurana, New Series* **21**: 49–55. DP: [31] July 1980 (paper title page).

Takizawa, H. 1982: Notes on Japanese Chrysomelidae (Coleoptera). *Entomological Review of Japan* **37** (1): 55–60. DP: 30 June 1982 (General index 1981–1985).

Takizawa, H. 1985a: Notes on Korean Chrysomelidae, part 2. *Nature and Life* (Taegu) **15** (1): 1–18. DP: [31] August 1985 (issue cover).

Takizawa, H. 1985b: Notes on chrysomelid-beetles of India and its neighboring areas, part II (Coleoptera, Chrysomelidae). *Entomological Review of Japan* 40 (2): 95–114, pls. 4–5. DP: 20 December 1985 (General index 1981–1985).

Takizawa, H. 1986: Chrysomelidae collected by the Japan-India Cooperative Survey in India, 1978. Part III. *Entomological Review of Japan* **41** (2): 93–105, pls. 5–6. DP: 26 December 1986 (verso of the issue cover).

Takizawa, H. 1987: Notes on chrysomelid beetles (Coleoptera, Chrysomelidae) of India and its neigbouring areas. Part 5. *Proceedings of the Japanese Society of Systematic Zoology* **35**: 40–58. DP: 25 June 1987 (paper title page).

Takizawa, H. 1988a: Chrysomelid beetles of Nepal, collected by the Hokkaido University Scientific Expeditions to Nepal Himalaya Part IV (Coleoptera: Chrysomelidae). *Entomological Review of Japan* **43** (1): 1–16, pls. 1–2. DP: 30 June 1988 (verso of the issue cover).

Takizawa, H. 1988b: Notes on chrysomelid beetles (Coleoptera, Chrysomelidae) of India and its neighboring areas. Part 7. *Kontyû* 56 (3): 534–552. DP: 25 September 1988 (paper title page).

Takizawa, H. 1989a: Chrysomelid beetles of Nepal, northeastern India and western Sikkim collected by the Hamalaya expeditions of the National Science Museum, Tokyo (Part 1). *Japanese Journal of Entomology* 57 (2): 319–332. DP: 25 June 1989 (paper title page).

Takizawa, H. 1989b: A new species of the genus Gonioctena (Coleoptera: Chrysomelidae) from Japan. *Akitu, New Series* 109: 1–5. DP: 1 October 1989 (each even page of the paper).

Takizawa, H. 1990a: Chrysomelid beetles of Nepal, northeastern India and western Sikkim collected by the Himalaya expeditions of the National Science Museum, Tokyo (Part 2). *Japanese Journal of Entomology* 58 (2): 275–291. DP: 25 June 1990 (paper title page).

Takizawa, H. 1990b: Notes on Japanese Chrysomelidae (Coleoptera), with descriptions of two new species. Part 2. *Akitu, New Series* 114: 1–7. DP: 10 September 1990 (each even page of the paper).

Takizawa, H. 2002: A revision of the genus Neocrepidodera Heikertinger in Japan (Chrysomelidae: Alticinae). *Insecta Matsumurana, New Series* 59: 39–53. DP: [31] December 2002 (paper title page).

Takizawa, H. 2007: A revision of the genus Gonioctena Chevrolat in Japan (Coleoptera: Chrysomelinae). *Insecta Matsumurana, New Series* 63: 35–50. DP: [28] February 2007 (paper title page).

Takizawa, H. 2009: Description of a new alticine genus (Coleoptera, Chrysomelidae) from Japan. *Elytra* 37 (1): 149–154. DP: 29 May 2009 (paper title page).

Takizawa, H. 2012: Description of a new Aphthona species from Japan (Coleoptera, Chrysomelidae, Alticinae). *Elytra, New Series* 1 [2011] (2): 321–326. DP: 19 January 2012 (editorial note at the journal website).

Takizawa, H. 2015: Notes on Japanese Chrysomelidae (Coleoptera), III. *Elytra, New Series* 5 (1): 233–250. DP: 25 May 2015 (paper title page).

Takizawa, H. 2017: Descriptions of two new Asian species of the genus Aphthona Chevrolat (Chrysomelidae, Alticinae), feeding on Euphorbiaceae. *Special Bulletin of the Coleopterological Society of Japan* 1: 287–290. DP: 22 October 2017 (paper title page).

Takizawa, H. 2021a: Descriptions of four new alticine species from Japan (Coleoptera: Chrysomelidae). *Elytra, New Series* 11 (1): 155–165. DP: 16 August 2021 (paper title page).

Takizawa, H. 2021b: A taxonomic note on Sphaeroderma species in Japan (Coleoptera, Chrysomelidae, Alticinae). *Elytra, New Series* 11 (supplement): 217–229. DP: 25 October 2021 (paper title page).

Takizawa, H. & Basu, C. R. 1987: Notes on chrysomelid-beetles (Coleoptera, Chrysomelidae) of India and its neighboring areas. Part 4. *Kontyû* 55 (2): 266–283. DP: 25 June 1987 (paper title page).

Takizawa, H. & Daccordi, M. 1998: Description of a new species of the genus Gonioctena Chevrolat from Japan (Coleoptera: Chrysomelidae). *Entomological Science* 1 (1): 105–108. DP: 25 March 1998 (issue cover).

Tamanuki, K. 1927: Über sechs neue Käfer aus Sachalin mit der Beschreibung zwei neuer Aberrationen. *Dobutsugaku Zasshi* 39 (460): 76–82. DP: 15 February 1927 (issue cover).

Tamanuki, K. 1928: [Studies on Galerucella rubi Tamanuki n. sp.]. *Government of Karahuto, Department of Agriculture, Report* 1: 1–19 (in Japanese). DP: 6 August 1928 (stamp on the J. Bezděk's copy).

Tan, C.-C. [= J.-J.] & Kung, Y.-T. 1961: [new taxon]. In: Chen, S.-H., Tan, C.-C. [= J.-J.] & Yu, P.-Y.: Results of the zoologico-botanical expedition to Southwest China 1955–57 (Coleoptera, Hispinae I). *Acta Entomologica Sinica* 10 (4/6): 457–481. DP: [31] December 1961 (paper title page).

Tan, C.-C. [= J.-J.] & Sun, T.-H. 1962: [new taxa]. In: Chen, S.-H., Tan, C.-C. [= J.-J.], Yu, P.-Y. & Sun, T.-H.: Results of the zoologico-botanical expedition to southwest China 1955–57 (Coleoptera, Hispinae II). *Acta Entomologica Sinica* 11 (supplement): 120–138. DP: [31] December 1962 (paper title page).

Tan, J.-J. 1964: On the Lamprosomatinae of China. *Acta Zootaxonomica Sinica* 1 (1): 100–105. DP: [31] July 1964 (paper title page).

Tan, J.-J. 1981a: On a new genus and two new species of Eumolpidae (Coleoptera). *Sinozoologia* 1: 51–54. DP: [31] May 1981 (paper title page).

Tan, J.-J. 1981b: [new taxa]. In: Tan, J.-J. & Wang, S.-Y. 1981: Coleoptera: Eumolpidae. Pp. 433–456. In: Qinghai-Xizang Plateau Expedition Group, Chinese Academy of Sciences (eds.): *The series of the scientific expedition to the Qinghai-Xizang Plateau. Insects of Xizang. Vol. 1.* Peking: Science Press, xi + 600 pp. DP: [31] October 1981 (colophon).

Tan, J.-J. 1982a: New species of Eumolpidae (Coleoptera) from Yunnan, China. *Acta Zootaxonomica Sinica* 7 (1): 83–91. DP: [31] January 1982 (paper title page).

Tan, J.-J. 1982b: Two new species of Dactylispa Weise (Coleoptera: Hispidae) from China. *Sinozoologia* 2: 39–42. DP: [31] March 1982 (paper title page).

Tan, J.-J. 1982c: New species and subspecies of Chinese Platycorynus (Coleoptera: Eumolpidae). *Acta Zootaxonomica Sinica* 7 (3): 285–292. DP: [31] July 1982 (paper title page).

Tan, J.-J. 1982d: On the genus Platycorynus Chevr. of China (Coleoptera: Eumolpidae). *Acta Zootaxonomica Sinica* 7 (4): 391–395. DP: [31] October 1982 (paper title page).

Tan, J.-J. 1983a: New species and new record of Colasposoma from Yunnan, China (Coleoptera: Eumolpidae). *Acta Zootaxonomica Sinica* 8 (2): 173–176. DP: [30] April 1983 (paper title page).

Tan, J.-J. 1983b: [new taxon]. In: Tan, J.-J. & Wang, S.-Y.: Two new species of Parascela Baly from Yunnan, China (Coleoptera: Eumolpidae). *Acta Zootaxonomica Sinica* 8 (4): 392–394. DP: [31] October 1983 (paper title page).

Tan, J.-J. 1983c: On a newly recorded genus and two new species of Eumolpidae (Coleoptera) from Fujian province, China. *Wuyi Science Journal* 3: 129–132. DP: [31] December 1983 (paper title page).

Tan, J.-J. 1984: On a newly recorded genus and two new species of Eumolpidae from Yunnan, China (Coleoptera). *Entomotaxonomia* 6 (2/3): 93–96. DP: [30] September 1984 (paper title page).

Tan, J.-J. 1985: [new taxon]. In: Huang, D.-S., Han, Y. & Zhang, X.: The insect fauna of the Mt. Tuomuer area in Tianshan. Pp. 53–165. In: *Dengshan Kexue Kaochadui. [Biota of Tuomuer region, Tianshan]*. Beijing: Xinjiang People's Press, 353 pp. DP: [31] March 1985 (colophon).

Tan, J.-J. 1987: Coleoptera: Eumolpidae. Pp. 39–45. In: *Agricultural insects, spiders, plant diseases and weeds of Xizang. Vol. II*. Lasa: Xizang People Publishing House, 416 pp. DP: [31] August 1987 (paper title page).

Tan, J.-J. 1988a: Coleoptera: Eumolpidae. Pp. 309–333. In: Huang, F.-S., Wang, P.-Y., Yin, W.-Y., Yu, P.-Y., Lee, T.-S., Yang, C.-K. & Wang, X.-J. (eds.): *Insects of Mt. Namjagbarwa region of Xizang*. Beijing: Science Press, xii + 621 pp. DP: [30] June 1988 (colophon).

Tan, J.-J. 1988b: Studies on the genus Basilepta of China (Coleoptera: Eumolpidae). *Sinozoologia* 6: 149–175. DP: [31] December 1988 (paper title page).

Tan, J.-J. 1992a: Coleoptera: Eumolpidae. Pp. 590–607. In: Peng, J.-W. & Liu, Y.-Q. (eds.): *Iconography of forest insects in Hunan China*. Beijing: Academia Sinica & Hunan Forestry Institute, 4 + 60 + 1473 pp. DP: [30] April 1992 (colophon).

Tan, J.-J. 1992b: Coleoptera: Eumolpidae. Pp: 754–829. In: Chen, S.-H. (ed.): *Insects of Hengduan Mountain Region. Volume I*. Beijing: Science Press, xii + 865 pp. DP: [31] August 1992 (colophon).

Tan, J.-J. 1993a: Coleoptera: Eumolpidae. Pp. 357–367. In: Huang, F.-S. (ed.): *Insects of Wuling Mountains area, southwestern China*. Beijing: Science Press, x + 777 pp. DP: [30] April 1993 (colophon).

Tan, J.-J. 1993b: [new taxa]. In: Tan, J.-J. & Yu, P.-Y.: Coleoptera: Hispidae – Hispinae. Pp. 352–356. In: Huang, F.-S. (ed.): *Insects of Wuling Mountains area, southwestern China*. Beijing: Science Press, x + 777 pp. DP: [30] April 1993 (colophon).

Tan, J.-J. & Pu, F.-J. 1991: Two new species of the genus Melixanthus Suffrian from Hengduan Mountains, China (Coleoptera: Eumolpidae: Cryptocephalinae). *Sinozoologia* 8: 253–256. DP: [31] May 1991 (paper title page).

Tan, J.-J. & Wang, S.-Y. 1981: Coleoptera: Eumolpidae. Pp. 433–456. In: Qinghai-Xizang Plateau Expedition Group, Chinese Academy of Sciences (eds.): *The series of the scientific expedition to the Qinghai-Xizang Plateau. Insects of Xizang. Vol. 1*. Peking: Science Press, xi + 600 pp. DP: [31] October 1981 (colophon).

Tan, J.-J. & Wang, S.-Y. 1983: Two new species of Parascela Baly from Yunnan, China (Coleoptera: Eumolpidae). *Acta Zootaxonomica Sinica* 8 (4): 392–394. DP: [31] October 1983 (paper title page).

Tan, J.-J. & Wang, S.-Y. 1984: New species of Eumolpidae (Coleoptera) from the Hengduan Mountains, Yunnan (I). *Acta Zootaxonomica Sinica* 9 (1): 55–58. DP: [31] January 1984 (paper title page).

Tan, J.-J. & Wang, S.-Y. 2005: [new taxon]. In: Tan, J.-J., Wang, S.-Y. & Zhou, H.-Z.: *Coleoptera Eumolpidae Eumolpinae. Fauna Sinica. Insecta Vol. 40*. Beijing: Science Press, xvi + 415 pp., 8 pls. DP: [31] August 2005 (colophon).

Tan, J.-J., Wang, S.-Y. & Zhou, H.-Z. 2005: *Coleoptera Eumolpidae Eumolpinae. Fauna Sinica. Insecta Vol. 40*. Beijing: Science Press, xvi + 415 pp., 8 pls. DP: [31] August 2005 (colophon).

Tan, J.-J. & Yu, P.-Y. 1992: Coleoptera: Hispidae: Hispinae. Pp. 607–610. In: Peng, J.-W. & Liu, Y.-Q. (eds.): *Iconography of forest insects in Hunan China*. Beijing: Academia Sinica & Hunan Forestry Institute, 4 + 60 + 1473 pp. DP: [30] April 1992 (colophon).

Tan, J.-J. & Zhou, H.-Z. 1995: [new taxa]. In: Zhou, H.-Z. & Tan, J.-J.: Four new species of Chlamisus Rafinesque from Fujian Province, China. (Chlamisinae: Eumolpidae: Coleoptera). *Entomologia Sinica* 2 (3): 228–235. DP: 1 September 1995 (journal website).

Tan, J.-J. & Zhou, H.-Z. 1997: [new taxa]. In: Zhou, H.-Z. & Tan, J.-J. 1997: Faunistic notes and three new species of Eumolpidae from Fujian province, China (Coleoptera). *Acta Zootaxonomica Sinica* 22 (2): 189–193. DP: [30] April 1997 (paper title page).

Tappes, G. 1869: Extrait d'un travail sur les cryptocéphalides d'Europe et des pays limitrophes. *Annales de la Société Entomologique de France, Quatrième Série* 9 (1): 5–20, pl. 1. DP: 23 June 1869 (Lefèvre 1885).

Tappes, G. 1872: Cryptocéphalides d'Europe et des pays limitrophes. *Annales de la Société Entomologique de France, Cinquième Série* 1 [1871] (3): 253–270, pl. 3. DP: 14 February 1872 (Lefèvre 1885).

Tappes, G. 1874: [Description qui suit d'une nouvelle espèce de coléoptères]. *Bulletin de la Société Entomologique de France* 1874 (23): 284. DF: 15 December 1874 (Evenhuis 2002).

Tappes, G. 1884: Cryptocéphalides recueillis en Portugal et au Maroc. *Bulletin ou Comptes-Rendus des Scéances de la Société Entomologique de Belgique, Série III* 1884 (48): cclvi–cclviii. DP: 10 September 1884 (*Bull. Soc. Ent. Fr.* 1884: 176).

Tempère, G. 1954: Sur quelques coléoptères anthribides et bruchides français. *Revue Française d'Entomologie* 21 (1): 47–51. DP: 15 April 1954 (footers on pp. 33 and 49).

Tempère, G. & Rapilly, M. 1981: Pachybrachis aragonicus, nouvelle espèce de la faune Espagnole. *Nouvelle Revue d'Entomologie* 11 (3): 313–314. DP: 30 October 1981 (issue cover).

Tenenbaum, S. 1915: Nowe gatunki chrząszczy z Wysp Balearskich. Neue Käferarten von den Balearen. *Bulletin International de l'Académie des Sciences de Cracovie. Classe des Sciences Mathématiques et Naturelles. Série B: Sciences Naturelles* [1914] (8): 837–843, pls. 63–64. DP: 5 February 1915 (verso of issue title page).

Tenenbaum, S. 1927: Neue Aberrationen der polnischen Käfer. *Polskie Pismo Entomologiczne* 5 [1926] (3/4): 151–153. DP: 1 April 1927 (issue title page).

Ter-Minassian, M. E. 1950: Materialy k faune listoedov Armyanskoy SSR (podsemeystvo Chrysomelinae). [Materials on the leaf-beetles fauna of Armenian SSR (subfamily Chrysomelinae)]. *Zoologicheskiy Sbornik Instituta Fitopatologii i Zoologii Akademii Nauk Armyanskoy SSR* 7: 127–138. SFP: 7 February 1950, DP: 12 June 1950 (registered at ZIN library).

Ter-Minassian, M. E. 1960: Novyy vid roda Kytorrhinus Fisch. (Coleoptera, Bruchidae) iz Primorskogo Kraya. (New species of the genus Kytorrhinus Fisch. (Coleoptera, Bruchidae) from Ussuri land). *Entomologicheskoe Obozrenie* 39 (2): 417. SFP: 29 April 1960, DP: 28 May 1960 (registered at ZIN library).

Ter-Minassian, M. E. 1968: Novyy vid zhuka-zernovki roda Bruchus L. v faune SSSR (Coleoptera, Bruchidae). (A new species of genus Bruchus L. (Coleoptera, Bruchidae) from the fauna of

U.S.S.R.). *Entomologicheskoe Obozrenie* **47** (1): 181–183. SFP: 26 March 1968, DP: 19 April 1968 (registered at ZIN library).

Ter-Minassian, M. E. 1970: Novye vidy zhukov-zernovok (Coleoptera, Bruchidae) v faune Kavkaza. [New species of seed-beetles (Coleoptera, Bruchidae) in Caucasian fauna]. *Doklady Akademii Nauk Armyanskoy SSR* **49** [1969] (5): 273–276. SFP: 19 January 1970, DP: 9 March 1970 (registered at ZIN library).

Ter-Minassian, M. E. 1973: 212. Bruchidae. Ergebnisse der zoologischen Forschungen von Dr. Z. Kaszab in der Mongolei (Coleoptera). *Reichenbachia* **14**: 75–83. DP: 10 June 1973 (paper title page).

Ter-Minassian, M. E. 1975: Obzor zhukov-zernovok (Coleoptera, Bruchidae) Mongolii. (Review of Bruchidae (Coleoptera) of Mongolia). Pp. 243–248. In: Kerzhner, I. M. (ed.): *Nasekomye Mongolii 3*. (*Insects of Mongolia 3*). St. Petersburg: Academy of Sciences of the USSR, Zoological Institute, 675 pp. DP: 25 December 1975 (Kerzhner 1994).

Ter-Minassian, M. E. 1976: Novye vidy zhukov-zernovok (Coleoptera, Bruchidae). (New species of genus Bruchidius Tchils. (Coleoptera, Bruchidae)). *Doklady Akademii Nauk Armyanskoy SSR* **61** [1975] (5): 301–305. SFP: 26 January 1976, DP: [31 December] 1976 (paper title page).

Ter-Minassian, M. E. 1978: Novyy vid zernovok roda Caryedon Schoenh. (Coleoptera, Bruchidae) iz Uzbekistana. (A new species of seed-beetles of the genus Caryedon Schoenh. (Coleoptera, Bruchidae) from Uzbekistan). *Trudy Zoologicheskogo Instituta Akademii Nauk SSSR* **71**: 55–56. SFP: 25 November 1977, DP: 30 June 1978 (Asanovich et al. 2007).

Ter-Minassian, M. E. & Egorov, A. B. 1981: [new taxa]. In: Egorov, A. B. & Ter-Minassian, M. E.: Novye vidy zhukov-zernovok (Coleoptera, Bruchidae) iz Primorskogo Kraya i Yakutii. [New species of seed-beetles (Coleoptera, Bruchidae) from Primorskiy Kray and Yakutia]. *Trudy Zoologicheskogo Instituta Akademii Nauk SSSR* **92**: 20–23. SFP: 16 September 1981, DP: [31] December 1981 (Asanovich et al. 2007).

Théry, A. 1895: Descriptions de quelques buprestides nouveaux de Ho-Chan (Chine). *Bulletin de la Société Entomologique de France* **1895** (4): cxi–cxv. DP: 13 March 1895 (*Bull. Soc. Ent. Fr.* 1895 (5): cliii – errata mentioned in the issue 4).

Théry, A. 1930: Notes d'entomologie marocaine et nord-africaine (Septième note). *Bulletin de la Société des Sciences Naturelles du Maroc* **9** [1929] (1/6): 90–91. DP: 15 February 1930 (last page of the issue).

Thierriat, A. 1943: Donacia malinowskyi Ahr. en France. *Bulletin Scientifique de Bourgogne* 10 [1941–1945] (2): 17–22. DP: 21 October 1943 (volume table of contents).

Thomson, C. G. 1859: *Skandinaviens Coleoptera, synoptiskt bearbetade. I. tom.* Lund: Berlingska Boktryckeriet, [3] + 290 pp. DP: 14 December 1859 (Bousquet 2016).

Thomson, C. G. 1866: *Skandinaviens Coleoptera, synoptiskt bearbetade. Tom. VIII.* Lund: Lundbergska Boktryckeriet, 409 + lxxv pp. DP: 10 October 1866 (Bousquet 2016).

Thomson, C. G. 1868: *Skandinaviens Coleoptera, synoptiskt bearbetade. Tom. X.* Lund: Lundbergska Boktryckeriet, 420 pp. DP: [31 December] 1868 (Bousquet 2016).

Thomson, C. G. 1884: [Une notice sur la Donacia impressa et les espèces voisines]. *Bulletin de la Société Entomologique de France* **1884** (23): 221–222. DP: 20 December 1884 (Evenhuis 2002).

Thomson, J. 1856: Description de quelques coléoptères nouveaux ou peu connus. *Annales de la Société Entomologique de France, Troisième Série* 4 (2): 317–334, pls. 7–8. DP: 12 November 1856 (Evenhuis 2022).

Thomson, J. 1858: Deuxième partie. Insects. I. Ordre Coléoptères. Pp. 29–343. In: Voyage au Gabon. Histoire naturelle des insectes et des arachnides recueillis pendant un voyage fait au Gabon en 1856 et en 1857 par M. Henry C. Deyrolle sous les auspices de MM. Le Comte de Mniszech et James Thomson, précédée de l'histoire du voyage par J. Thomson; Arachnide par H. Lucas. *Archives Entomologiques ou recueil contenant des illustrations d'insectes nouveaux ou rares. Tome deuxième.* Paris: Bureau du Trésorier de la Société entomologique de France, 469 + [1] + [1] pp., XV pls. DP (pp. 9–256): [30] June 1858, (pp. 257–336): [30] November 1858, (pp. 337–472): [31 December] 1858 (Bousquet 2016).

Thunberg, C. P. 1784: Novae insectorum species descriptae. *Nova Acta Regiae Societatis Scientiarum Upsaliensis* 4: 1–28, 1 pl. DP: [31 December] 1784 (volume title page).

Thunberg, C. P. 1787a: *D. D. Museum Naturalium Academiae Upsalensis. Cujus partem tertiam.* Upsaliae: Joh. Edman, 33–42 pp. DP: 21 June 1787 (title page).

Thunberg, C. P. 1787b: *D. D. Museum Naturalium Academiae Upsalensis. Cujus partem quartam.* Upsaliae: Joh. Edman, 43–58 pp., 1 pl. DP: 19 December 1787 (title page).

Thunberg, C. P. 1789: *Dissertatio entomologica novas insectorum species sistens, cujus partem quintam.* Upsaliae: Joh. Edman, 85–106 pp., pl. 5. DP: 13 June 1789 (title page).

Thunberg, C. P. 1792: Descriptiones insectorum svecicorum. *Nova Acta Regiae Societatis Scientiarum Upsaliensis* 5: 85–119. DP: [31 December] 1792 (volume title page).

Thunberg, C. P. 1794: *D. D. Dissertatio entomologica sistens insecta Suecica. Quorum partem octavam.* Upsaliae, 99–104 pp. DP: 15 December 1794 (title page).

Thunberg, C. P. 1805: [Editorial abstract of manuscript "Illustrationes generum aliquot insectorum coleoptratorum"]. *Göttingische Gelehrte Anzeigen* 1805 (29): 281–282. DP: 21 February 1805 (issue title page).

Thunberg, C. P. 1814: Beskrifning på tvänne nya Insect-slågten, Gnatocerus och Taumacera från Goda-Hopps Udden. *Kongliga Vetenskaps Academiens Handlingar* 1814: 46–50, pl. IVa. DP: [31 December] 1814 (volume title page).

Thunberg, C. P. 1815: De coleopteris rostratis. *Nova Acta Regiae Societatis Scientiarum Upsaliensis* 7: 104–125. DP: [31 December] 1815 (volume title page).

Thunberg, C. P. 1816: Fyra nya arter af Bruchus slågtet. *Kongliga Vetenskaps Academiens Handlingar* 1816: 43–47. DP: [31 December] 1816 (volume title page).

Thunberg, C. P. 1827: Coleoptera Capensia, antennis filiformibus, enumerata et nova descripta. *Nova Acta Regiae Societatis Scientiarum Upsaliensis* 9: 1–52. DP: [31 December] 1827 (volume title page).

Tiberghien, G. 1969: Contribution à la connaissance des Chrysomelidae (Col.). Remarques sur quelques Clytra ibériques et description d'une nouvelle sous-espèce de Clytra (Clytrella) atraphaxidis Pall. *Bulletin de la Société Entomologique de France* 74 (5/6): 121–123. DP: 14 November 1969 (*dates de la publication* on p. 260).

Tiberghien, G. 1971: Notes biologiques et descriptions de quelques formes nouvelles de Cryptocephalus (Col. Chrysomelidae, Cryptocephalinae) (8e contribution a la connaisance des Chrysomeloidea). *Bulletin de la Société Linnéenne de Bordeaux* 1 (8): 189–192. DP: [31] October 1971 (issue cover).

Tiberghien, G. 2014: Timarchini (Timarcha). Pp. 601–602. In: Tronquet, M. (ed.): Catalogue des coléoptères de France. *Revue de l'Association Roussillonnaise d'Entomologie* 23 (supplément): 1–1052. DP: [31] May 2014 (colophon).

Tominaga, O. & Katsura, K. 1984: Studies on the Japanese Donaciinae (Coleoptera: Chrysomelidae). 2. Notes on the geographical diversity of Plateumaris constricticollis, with description of an allied new species. *Bulletin of the Osaka Museum of Natural History* 37: 25–40. DP: 31 March 1984 (verso of volume cover).

Tomov, V. 1975: Eine neue Luperus-Art aus Südjugoslawien (Coleoptera, Chrysomelidae, Galerucinae). *Entomologische Arbeiten aus dem Museum G. Frey* 26: 188–190. DP: 1 November 1975 (table of contents).

Tomov, V. 1976: Eine neue Cryptocephalus-Art (Coleoptera, Chrysomelidae) aus der Türkei. *Acta Zoologica Bulgarica* 5: 83–85. DP: [30] November 1976 (paper title page).

Tomov, V. 1978: Über die Nomenklatur zweier Cryptocephalus-Arten (Coleoptera, Chrysomelidae). *Acta Zoologica Bulgarica* 11: 60–64. DP: [31] October 1978 (paper title page).

Tomov, V. 1979: Lamellosus subg. n., eine neue Untergattung der Gattung Cryptocephalus Geoffr. (Coleoptera, Chrysomelidae). *Reichenbachia* 17: 43–47. DP: 22 March 1979 (paper title page).

Tomov, V. 1982a: Chrysomelidae (Coleoptera) of Korea preserved in the Hungarian Natural History Museum, Budapest, II. *Entomological Review of Japan* 37 (1): 37–48. DP: 30 June 1982 (General index 1981–1985).

Tomov, V. 1982b: Lachnaia (Barathraea) padillai sp. n., eine neue Art aus Tunesien, mit einigen Bemerkungen über die restlichen Arten der Untergattung (Coleoptera, Chrysomelidae). *Reichenbachia* 20 (1): 107–110. DP: 27 September 1982 (paper title page).

Tomov, V. 1983: Antipa urikkana sp. n. (Coleoptera, Chrysomelidae), eine neue Art aus Marokko. *Acta Zoologica Bulgarica* 22: 64–66. DP: [30] September 1983 (paper title page).

Tomov, V. 1984: Cryptocephalinae and Galerucinae from Turkey (Coleoptera, Chrysomelidae). *Fragmenta Entomologica* 17 (2): 373–378. DP: 15 October 1984 (table of contents on the cover).

Tomov, V. 1987: Eine neue Labidostomis-Art aus Jugoslawien (Insecta, Coleoptera, Chrysomelidae, Clytrinae). *Reichenbachia* 25 (1): 25–26. DP: 30 October 1987 (paper title page).

Tomov, V. 1990: Zwei neue Labidostomis-Arten aus Griechenland (Insecta, Coleoptera, Chrysomelidae: Clytrinae). *Reichenbachia* 27 (2): 135–136. DP: 10 September 1990 (paper title page).

Tottenham, C. E. 1941: Two Coleoptera new to science in Britain. *Entomologist's Monthly Magazine* 77 (920): 14–15. DP: 6 January 1941 (*dates of publication* on p. ii).

Tournier, H. 1875: Diagnoses de trois coléoptères nouveaux. *Petites Nouvelles Entomologique* 1 [1869–1875] (7) (119): 475. DP: 1 March 1875 (issue title page).

Tower, W. L. 1906: *An investigation of evolution in chrysomelid beetles of the genus Leptinotarsa*. Washington: Carnegie Institution of Washington, x + 320 pp. DP: [31 December] 1906 (book title page).

Townson, R. 1797: *Travels in Hungary, with a short account of Vienna in the year 1793*. London: G. G. and J. Robinson, xviii + 506 pp., 16 pls., 1 map. DP: [31] March 1797 (Bousquet 2016).

Trella, T. 1930: Wykaz chrząszczów okolic Przemyśla. Uzupełnienia do dotychczasowych wykazów oraz rodzina: Chrysomelidae. (Verzeichnis der Käfer der Umgebung von Przemyśl. Ergänzungen und die Familie: Chrysomelidae). *Polskie Pismo Entomologiczne* 9 (3/4): 186–200. DP: 15 December 1930 (first page of issue).

Trella, T. 1932: Wykaz chrząszczów okolic Przemyśla. Uzupełnienia do wykazów dotychczasowych. (Verzeichnis der Käfer der Umgebung von Przemyśl. Ergänzungen zu den bisher erschienenen Verzeichnissen). *Polskie Pismo Entomologiczne* 10 (3/4): 223–229. DP: 20 February 1932 (issue title page).

Trella, T. 1936: Wykaz chrząszczów okolic Przemyśla. Uzupełnienia dotychczasowych wykazów. (Verzeichnis der Käfer der Umgebung von Przemyśl. Ergänzungen zu den bisher erschienenen Verzeichnissen). *Polskie Pismo Entomologiczne* 13 [1934] (1/4): 85–97. DP: 22 April 1936 (table of contents).

Truqui, E. 1852: Novae Histerinorum et Cryptocephalorum species. *Annales de la Société Entomologique de France, Deuxième Série* 10 (1): 61–68, pl. 2. DP: 19 July 1852 (Evenhuis 2022).

Tuda, M. 2003: A new species of Callosobruchus (Coleoptera: Bruchidae) feeding on seeds of Dunbaria (Fabaceae), a closely related species to a stored-bean pest, C. chinensis. *Applied Entomology and Zoology* 38 (2): 197–201. DP: 25 February 2003 (colophon).

Tuda, M. & Morimoto, K. 2004: A new species Megabruchidius sophorae (Coleoptera, Bruchidae), feeding on seeds of Styphnolobium (Fabacae) new to Bruchidae. *Zoological Science* 21 (1): 105–110. DP: 25 January 2004 (last page of the cover).

Tunkl, F. von 1929: Vier neue Variationen von Lema (Crioceris) asparagi L. (Col.). *Entomologischer Anzeiger* 9 (8): 163–164. DP: 20 April 1929 (issue cover).

Uhagón, S. 1887: Coleópteros de Badajoz. Tercera parte. *Anales de la Sociedad Española de Historia Natural* 16 (2): 373–404. DP: 30 September 1887 (issue cover).

Uhmann, E. 1927a: Hispinen des Deutschen Entomologischen Institutes Berlin-Dahlem. 3. Beitrag zur Kenntnis der Hispinen. *Entomologische Blätter* 23 (2): 58–65. DP: 30 June 1927 (issue cover).

Uhmann, E. 1927b: H. Sauter's Formosa-Ausbeute: Hispinae (Col.). 5. Beitrag zur Kenntnis der Hispinen. *Supplementa Entomologica* 16: 108–116. DP: 10 November 1927 (volume cover).

Uhmann, E. 1928a: Eine neue Dactylispa-Art aus Ostindien. 7. Beitrag zur Kenntnis der Hispinen. *Entomologisk Tidskrift* 49 (2): 102–103. DP: 15 June 1928 (last page of the cover).

Uhmann, E. 1928b: Eine neue palaearktische Dactylispa-Art von Amur-Gebiet (Chrysomelid.-Hispinae). (9. Beitrag zur Kenntnis der Hispinen). *Coleopterologisches Centralblatt* 3 [1928–1929] (1/2): 35–37. DP: 20 August 1928 (table of contents).

Uhmann, E. 1928c: Hispinen aus Ostafrika. 13. Beitrag zur Kenntis der Hispinen (Col. Chrys.). *Archiv für Naturgeschichte, Abteilung A* 92 [1926] (12): 65–93. DP: [31] December 1928 (issue cover).

Uhmann, E. 1929: Hispinen von den Sunda-Inseln. 14. Beitrag zur Kenntnis der Hispinen (Col. Chrys.). *Zeitschrift für Wissenschaftliche Insektenbiologie* 24 (1/3): 143–149. DP: 25 January 1929 (footer on p. 1).

Uhmann, E. 1930a: Ueber Arten der Gattungen Anisodera Chevr. und Lasiochila Ws. 18. Beitrag zur Kenntnis der Hispinen (Col. Chrys.). *Zoologische Mededeelingen* 13 (1/2): 76–93. DP: 15 August 1930 (table of contents).

Uhmann, E. 1930b: Hispinen des Deutschen Entomologisches Institutes in Berlin-Dahlem. II. Teil. 23. Beitrag zur Kenntis der Hispinen. *Entomologische Blätter* 26 (3): 125–136. DP: 30 September 1930 (issue cover).

Uhmann, E. 1931: Nachtrag zum „Katalog der Lasiochila-Arten" und eine neue Callispa-Art. 34. Beitrag zur Kenntnis der Hispinae (Col. Chrys.). *Zoologische Mededeelingen* 14 (3): 193. DP: 31 December 1931 (table of contents).

Uhmann, E. 1934: Hispinen aus Tonkin. 39. Beitrag zur Kenntnis der Hispinen (Col. Chrys.). *Mitteilungen der Deutschen Entomologischen Gesellschaft* 5 (3/4): 28–30. DP: 25 May 1934 (footer on p. 48).

Uhmann, E. 1936: Hispinen von der Insel Ukerewe im Victoria-See (Ostafrika). 59. Beitrag zur Kenntnis der Hispinen (Coleoptera: Chrysomelidae). *Arbeiten über Morphologische und Taxonomische Entomologie aus Berlin-Dahlem* 3 (2): 123–125. DP: 8 June 1936 (volume title page).

Uhmann, E. 1938: Hispinen aus China: Yunnan. 75. Beitrag zur Kenntnis der Hispinen (Col. Chrys.). *Entomologisk Tidskrift* 59 (3/4): 224–227. DP: 5 December 1938 (verso of the volume title page).

Uhmann, E. 1939a: Entomological results from the Swedish Expedition 1934 to Burma and British India. Coleoptera: Hispinae gesammelt von René Malaise. 74. Beitrag zur Kenntnis der Hispinen (Col. Chrys.). *Arkiv för Zoologi* 30A (26): 1–9. DP: 26 January 1939 (last page of the paper).

Uhmann, E. 1939b: Hispinen von Java aus der Sammlung des Herrn F. C. Drescher. 78. Beitrag zur Kenntnis der Hispinen (Col. Chrys.). *Treubia* 17 (1): 37–43. DP: [31] March 1939 (issue cover).

Uhmann, E. 1940a: Die Genotypen der von mir aufgestellten Hispinen-Gattungen. 88. Beitrag zur Kenntnis der Hispinen (Col. Chrys.). *Entomologisk Tidskrift* 61 (3/4): 143–144. DP: 15 June 1940 (verso of the volume title page).

Uhmann, E. 1940b: Die Hispinen der Klapperich-Fukien (China)-Expedition. I. Teil. 91. Beitrag zur Kenntnis der Hispinen (Col. Chrys.). *Entomologische Blätter* 36 (4): 125–128. DP: 20 August 1940 (issue cover).

Uhmann, E. 1943: Neue amerikanische Hispinen. 100. Beitrag zur Kenntnis der Hispinen (Coleoptera: Chrysomelidae). *Folia Zoologica et Hydrobiologica* 12 (1): 115–121. DP: 1 September 1943 (issue cover).

Uhmann, E. 1951a: Hispinae aus dem Britishen Museum. – VI. Teil. 131. Beitrag zur Kenntnis der Hispinae (Coleopt., Chrysom.). *Annals and Magazine of Natural History, Twelfth Series* 4 (37): 66–76. DP: 1 January 1951 (Evenhuis 2003).

Uhmann, E. 1951b: [new taxon]. In: Barber, H. S.: Hispella a synonym of Hispa Linnaeus, and a new species of Dactylispa from China (Coleoptera: Chrysomelidae). *Pan-Pacific Entomologist* 27 (1): 17–19. DP: 31 March 1951 (*mailing dates* in table of contents).

Uhmann, E. 1954: Hispinae aus dem Britishen Museum. – VIII. Teil. 156. Beitrag zur Kenntnis der Hispinae (Coleopt. Chrysom.). *Annals and Magazine of Natural History, Twelfth Series* 7 (79): 497–518. DP: 22 July 1954 (Evenhuis 2003).

Uhmann, E. 1955a: Die Hispinae der Klapperich-Fukien (China)-Expedition April-Juni 1938. II. Teil. 128. Beitrag zur Kenntnis der Hispinae (Coleopt. Chrysom.). *Entomologische Blätter* 50 [1954] (2/3): 186–215. DP: 1 June 1955 (issue cover).

Uhmann, E. 1955b: Die Hispinae (Coleoptera) der Afghanistan-Expedition (1952–1953) J. Klapperichs. 162. Beitrag zur Kenntnis der Hispinae (Coleoptera, Chrysomelidae). *Acta Entomologica Musei Nationalis Pragae* 29 [1953–1954]: 171–175. DP: 20 June 1955 (colophon).

Uhmann, E. 1958: Chrysomelidae: Hispinae. Hispinae Africanae, Eurasiaticae, Australicae. In: Hincks, W. D. (ed.): *Coleopterorum Catalogus Supplementa. Pars 35, Fasc. 2. Editio secunda.* 's-Gravenhage: W. Junk, ix + 155–398 pp. DP: 30 December 1958 (book cover).

Ulanowski, A. R. S. von 1886: Mittheilung. *Societas Entomologica* 1 [1886–1887] (3): 22. DP: 15 May 1886 (issue title page).

Ulrich, W. 1924: Die Verbreitung und Rassenbildung des Labidostomis taxicornis F. (Col. Chrysomel.). *Zeitschrift für Wissenschaftliche Insektenbiologie* 19 (2): 153–156. DP: 15 June 1924 (footer on p. 147).

Uyttenboogaart, D. L. 1935: Report on Canarian Coleoptera collected by R. Frey and R. Storå in 1931 for Museum Zoologicum Universitatis Helsingfors (Contributions to the knowledge of the fauna of the Canary-Islands. XVI). *Commentationes Biologicae* 6 [1937] (2): 1–17. DP: [30] September 1935 (last page of the paper).

Van Dyke, E. C. 1938: A review of the genus Chrysolina Motschulsky in North America (Coleoptera-Chrysomelidae). *Bulletin of the Brooklyn Entomological Society, New Series* 33 (2): 45–58. DP: 11 June 1938 (issue title page).

Vauloger de Beaupré, M. 1895: Description de deux Clytra nouveaux du Nord de l'Afrique. *L'Abeille, Journal d'Entomologie* 28 [1892–1896] (11): 194–195. DP: 8 October 1895 (stamp on the issue cover of the BHL copy).

Vazirani, T. G. 1970: A new species of Merista Chapuis (1895) (Insecta: Coleoptera: Chrysomelidae) from India. *Records of the Zoological Survey of India* 64 [1966] (1/4): 111–112. DP: [31] December 1970 (issue title page).

Vazirani, T. G. 1972: Notes on a collection of Hispinae and Cassidinae (Coleoptera, Chrysomelidae) from N. E. India. *Records of the Zoological Survey of India* 66 [1968] (1/4): 143–152. DP: [30] June 1972 (volume title page).

Vázquez, A. 1904: Un nuevo coleóptero. *Boletín de la Real Sociedad Española de Historia Natural* 1904: 374. DP: [30] November 1904 (footer on p. 369).

Vela, J. M. 2018: On the identity of Calomicrus fallax (Joannis, 1865) with description of a new, widespread species of Calomicrus from Spain and Portugal and comments on the Iberian species (Coleoptera, Chrysomelidae, Galerucinae). *Graellsia* 74 (2): e074. DP: 28 August 2018 (paper title page).

Vela, J. M., Alonso-Zarazaga, M. A. & Daccordi, M. 2020: The species of Timarcha Samouelle, 1819 described by Linnaeus (Coleoptera, Chrysomelidae). *ZooKeys* **986**: 55–80. DP: 5 November 2020 (paper title page).

Vela, J. M. & Bastazo, G. 1993: El género Otiothraea Warchalowski 1990 (Coleoptera: Chrysomelidae, Clytrinae) en la Península Ibérica: descripción de una especie nueva y asignación de otra previamente conocida. *Zoologica Baetica* 4: 31–37. DP: [31 December] 1993 (paper title page).

Vela, J. M. & Bastazo, G. 2009: Two new species of the genus Cyrtonus from Andalusia, southern Spain (Coleoptera: Chrysomelidae). *Genus* 20 (1): 77–84. DP: 30 March 2009 (paper title page).

Vela, J. M. & Bastazo, G. 2012: A new Cryptocephalus of the C. curvilinea – group (Coleoptera, Chrysomelidae) from the Atlantic coastal marshes of Southern Spain and Portugal. *Boletín de la Asociación Española de Entomología* **36** (1/2): 31–41. DP: [30] June 2012 (issue cover).

Vela, J. M. & Bastazo, G. 2017: Luperus doeberli, new species from southeastern Turkey (Coleoptera: Chrysomelidae: Galerucinae). *Entomologische Blätter und Coleoptera* **113** (1): 241–244. DP: [30] November 2017 (issue cover).

Vela, J. M & Daccordi, M. 2021: Description of a new species of the genus Timarcha Samouelle, 1819 from Northeastern Spain (Coleoptera, Chrysomelidae). *Boletín de la Asociación Española de Entomología* 45 (3/4): 293–298. DP: 14 December 2021 (footers throughout the paper).

Vela, J. M & García Becerra, R. 1996: Calomicrus doramasensis n. sp., a new leaf-beetle from the Canary Islands (Coleoptera, Chrysomelidae, Galerucinae). *Vieraea* **25**: 147–152. DP: [31] December 1996 (paper title page).

Verma, K. K., Gómez-Zurita, J., Jolivet, P. & Vig, K. 2005: Biology of Eupales ulema (Germar, 1813) and its taxonomic placement among Eumolpinae (Coleoptera, Chrysomelidae). *Nouvelle Revue d'Entomologie, Nouvelle Série* **22** (2): 155–164. DP: 30 September 2005 (last page of the issue cover).

Vig, K. 2002: Beetle collection of the Savaria Museum (Szombathely, Hungary) II. Leaf beetle collection of Attila Podlussány (Coleoptera, Chrysomelidae). *Praenorica Folia Historico-Naturalia* 5: 5–171. DP: [31 December] 2002 (issue cover).

Vigors, N. A. 1825: Descriptions of some rare, interesting, or hitherto uncharacterised subjects of zoology. *Zoological Journal* 2 [1825–1826] (6): 234–241, pl. IX. DP: [31] July 1825 (issue title page).

Villa, A. & Villa, G. B. 1833: Coleopterorum species novae in catalogo dupletorum extantes diagnosibus, adumbrationibus atque observationibus illustratae. Pp. 32–36. In: Villa, A. & Villa, G. B.:

Coleoptera Europae dupleta in collectione Villa quae pro mutua commutatione offerri possunt. Mediolani, 36 pp. DP: [31 December] 1833 (Bousquet 2016).

Villa, A. & Villa, G. B. 1835: Coleopterorum species novae in supplemento salutatae diagnosibus atque observationibus illustratae. Pp. 47–50. In: Villa, A. & Villa, G. B.: *Supplementum Coleopterorum Europae dupletorum catalogo collectionis Villa idest species aliae, quae nunc pro mutua commutatione itidem offerri possunt; nec non emendationes aliquarum specierum in catalogo anni 1833 extantium*. Mediolani, 37–50 pp. DP: [31 December] 1835 (Bousquet 2016).

Villa, A. & Villa, G. B. 1839: Coleopterorum species novae in hoc altero supplemento commemoratae diagnosibus atque observationibus illustratae. Pp. 61–63. In: Villa, A. & Villa, G. B.: *Alterum supplementum Coleopterorum Europae sive additio ad Catalogum et Supplementum I. Dupletorum collectionis Villa continens species alias, nunc pro mutua commutatione adhuc offerrendas; nec non aliquarum specierum emendationes et synonima quae in Catalogo anni 1833 et supplemento anni 1835 oblatae fuerunt*. [1838], Mediolani, pp. 51–66. DP: [31] May 1839 (Bousquet 2016).

Villers, C. J. de 1789: *Caroli Linnaei entomologia, faunae suecicae descriptionibus aucta; DD. Scopoli, Geoffroy, de Geer, Fabricii, Schrank, &c. speciebus vel in systemate non enumeratis, vel nuperrime detectis, vel speciebus Galliae australis locupletata, generum specierumque rariorum iconibus ornata. Tomus primus*. Lugduni: Piestre & Delamolliere, xiv + 765 pp., 3 pls. DP: 3 May 1789 (Bousquet 2017).

Vitale, F. 1930: Nuove aberrazioni di coleotteri. Nota 1a. *Bolletino della Società Entomologica Italiana* **62** (7): 135–137. DP: 20 July 1930 (issue cover and Poggi 2008).

Vitale, F. 1933: I Chrysomelidae siciliani. Catalogo sinonimico (Parte 1ª). *Atti della Regia Accademia Peloritana* (Messina) **35**: 39–66, 1 pl. DP: [30] November 1933 (reprint cover).

Voet, J. E. 1806: *Catalogus systematicus coleopterorum. Catalogue systématique des coléoptères. Systematische naamlijst van dat geslacht van insecten dat men torren noemt. Tomus II*. Haag: Bakhuysen, 82 (Latin text) + 86 (French text) + 87 (Dutch text) + 20 pp. + 50 pls. DP: [31 December] 1806 (Bousquet 2016).

Vogel, E. 1861: [new taxon]. In: Schaufuss, L. W.: [Folgendes über seine neuesten entomologischen Forschungen]. *Sitzungs-Berichte der Naturwissenschaftlichen Gesellschaft Isis zu Dresden* **1861** (7/9): 47–51. DP: [31 December] 1861 (issue title page).

Vogel, E. 1875: [new taxon]. In: Schaufuss, L. W.: Beitrag zur Käfer-Fauna Spaniens. *Nunquam Otiosus* **2** [1872–1877]: 318–344. DP: 24 December 1875 (table of contents).

Voronova, N. V. 1982: Reviziya vidov Chrysolina gruppy Chrysolina rufilabris (Coleoptera, Chrysomelidae). [Revision of Chrysolina rufilabris species group (Coleoptera, Chrysomelidae)]. Pp. 119–129. In: Sokolov, V. E., Medvedev, L. N. & Roginskaya, E. Ya. (eds.): *Zoologicheskie issledovaniya v MNR* (*Biologicheskie Resursy i Prirodnye Usloviya Mongol'skoy Narodnoy Respubliki 18*). [*Zoological research in Mongolia* (*Biological resources and natural conditions in Mongolian National Republic 18*)]. Moskva: Nauka, 160 pp. SFP: 10 October 1982, DP: [31 December] 1982 (book title page).

Voronova, N. V. & Zaytsev, Yu. M. 1982: Zhuki Leptispa – vrediteli bambuka vo V'etname. [Leptispa beetles – the pests of bamboo in Vietnam]. Pp. 115–124. In: Medvedev, L. N. (ed.):

Zhivotnyy mir V'etnama. [*Animal world of Vietnam*]. Moskva: Nauka, 168 pp. SFP: 20 July 1982, DP: [31 December] 1982 (book title page).

Vuillefroy, F. de 1868: Insectes nouveaux de la faune européenne. *L'Abeille, Journal d'Entomologie* 5 [1868–1869]: 289–295. DP: [31 December] 1868 (footers on p. 289).

Wagener, B. 1877: Cassididae. *Mittheilungen des Münchener Entomologischen Vereins* 1: 49–79. DP: [31 December] 1877 (volume title page).

Wagener, B. 1881: Cassididae. *Mittheilungen des Münchener Entomologischen Vereins* 5: 17–85. DP: [31] May 1881 (Natur. Novit. 1881, 3 (10): 83).

Wagner, H. 1919: [new aberration]. In: Neresheimer, J. & Wagner, H.: Beiträge zur Coleopterenfauna der Mark Brandenburg. VIII. *Entomologische Mitteilungen* 8 (4/6): 65–75. DP: 14 Juni 1919 (issue cover).

Wagner, H. 1927: Eine Sammelreise nach Zentral-Spanien (Sommer 1925). *Coleopterologisches Centralblatt* 2 [1927–1928] (2): 101–112. DP: 10 September 1927 (table of contents).

Walckenaer, C. A. 1802: *Faune parisienne, insectes. Ou histoire abrégée des insectes des environs de Paris, classés d'après le système de Fabricius; précédée d'un discours sur les insectes en général, pour servir d'introduction à l'étude de l'entomologie; accompagnée de sept planches gravées. Tome premier*. Paris: Dentu, vii + cxxx + 303 pp. + 7 pls. DP: 22 September 1802 (Bousquet 2016).

Walker, F. 1859a: Characters of some apparently undescribed Ceylon insects. *Annals and Magazine of Natural History, Third Series* 3 (13): 50–56. DP: 1 January 1859 (Evenhuis 2003).

Walker, F. 1859b: Characters of some apparently undescribed Ceylon insects. *Annals and Magazine of Natural History, Third Series* 3 (16): 258–265. DP: 1 April 1859 (Evenhuis 2003).

Walker, F. 1871: *List of Coleoptera collected by J. K. Lord, Esq. in Egypt, Arabia and near the African shore of the Red Sea. With characters of the undescribed species*. London: E. W. Janson, 19 pp. DP: 1 July 1871 (Bousquet 2016).

Waltl, J. 1835: *Reise durch Tyrol, Oberitalien und Piemont nach dem südlichen Spanien. Nebst einem Anhange zoologischen Inhalts*. Passau: Pustet'sche Buchhandlung, [8] + 247 pp. (Erster Theil) + 120 pp. (Zweiter Theil: Ueber die Thiere Andalusien). DP: [28] February 1835 (Bousquet 2016).

Waltl, J. 1838: Beyträge zur Kenntniss der Coleopteren der Türkey. *Isis von Oken* 1838 (6): 449–472. DP: [30] June 1838 (published monthly).

Waltl, J. 1839: Käfer um Passau. *Isis von Oken* 1839 (3): 221–227. DP: [31] March 1839 (published monthly).

Wang, F.-Y. & Zhou, H.-Z. 2011: A synopsis on the Chinese species of Clytra Laicharting, with description of two new species (Coleoptera: Chrysomelidae: Cryptocephalinae: Clytrini). *Zootaxa* 3067: 1–25. DP: 25 October 2011 (footer on the paper title page).

Wang, F.-Y. & Zhou, H.-Z. 2012: Taxonomy of the genus Aetheomorpha Lacordaire (Coleoptera: Chrysomelidae: Cryptocephalinae: Clytrini) from China, with description of five new species. *Journal of Natural History* 46 (23/24): 1407–1440. DP: 11 May 2012 (paper title page).

Wang, F.-Y. & Zhou, H.-Z. 2013: Four new species of the genus Smaragdina Chevrolat, 1836 from China (Coleoptera: Chrysomelidae: Cryptocephalinae: Clytrini). *Zootaxa* 3737 (3): 251–260. DP: 20 November 2013 (footer on the paper title page).

Wang, F.-Y. & Zhou, H.-Z. 2020: Taxonomy of the leaf beetle genus Exomis Weise (Coleoptera: Chrysomelidae: Clytrini) with description of six new species from China. *Zootaxa* **4748** (2): 351–364 DP: 6 March 2020 (footer on the paper title page).

Wang, H.-J., Li, W.-Z. & Yang, X.-K. 2000: Study on the genus Sphenoraia Clark (Coleoptera: Chrysomelidae: Galerucinae) from China. Pp. 117–121. In: Zhang, Y.-L. (ed.): *Systematic and faunistic research on Chinese insects. Proceedings of the 5th National Congress of Insect Taxonomy.* [place of publication not given]: China Agriculture Press, xvi + 331 pp. DP: [31] October 2000 (colophon).

Wang, S.-Y. 1990a: Two new species of Podagricomela Heikertinger (Coleoptera: Chrysomelidae) from China. *Entomotaxonomia* **12** (2): 123–126 DP: [30] June 1990 (paper title page).

Wang, S.-Y. 1990b: The alticine genus Trachyaphthona from the Hengduan Mountains (Coleoptera: Chrysomelidae). *Sinozoologia* 7: 127–135 DP: [30] September 1990 (paper title page).

Wang, S.-Y. 1992a: Two new species of leaf beetles from Wuling Mountain of China (Coleoptera: Chrysomelidae). *Sinozoologia* **9**: 175–178. DP: [30] April 1992 (paper title page).

Wang, S.-Y. 1992b: Coleoptera: Chrysomelidae – Chrysomelinae. Pp. 628–645. In: Chen, S.-H. (ed.): *Insects of Hengduan Mountain Region. Volume 1.* Beijing: Science Press, xii + 865 pp. DP: [31] August 1992 (colophon).

Wang, S.-Y. 1992c: Coleoptera: Chrysomelidae – Alticinae. Pp. 675–753. In: Chen, S.-H. (ed.): *Insects of Hengduan Mountain Region. Volume I.* Beijing: Science Press, xii + 865 pp. DP: [31] August 1992 (colophon).

Wang, S.-Y. 1993: [new taxa]. In: Wang, S.-Y. & Yu, P.-Y.: Coleoptera: Chrysomelidae – Alticinae. Pp. 315–330. In: Huang, F.-S. (ed.): *Insects of Wuling Mountains area, southwestern China.* Beijing: Science Press, x + 777 pp. DP: [30] April 1993 (colophon).

Wang, S.-Y. 1997: [new taxa]. In: Wang, S.-Y. & Yu, P.-Y.: Coleoptera: Chrysomelidae: Alticinae. Pp. 905–931. In: Yang, X.-K. (ed.): *Insects of the Three Gorge Reservoir area of Yangtze river. Part 1.* Chongqing: Chongqing Publishing House, xx + 974 pp. DP: [31] January 1997 (colophon).

Wang, S.-Y. 1998: [new taxon]. In: Wang, J.-S. & Yang, X.-K. 1998 (eds.): *Fauna of Chrysomelidae of Wuyishan Nature Reserve in China.* Beijing: China Forestry Publishing House, 213 pp., 8 pls. DP: [31] March 1999 (colophon).

Wang, S.-Y. 2002a: [new taxon]. In: Wang, S.-Y. & Yang, X.-K.: Chrysomelidae: Alticinae. Pp. 310–317. In: Li, W.-Z. & Jin, D. (eds.): *Insects from Maolan Landscape.* Guiyang: Guizhou Science and Technology Publishing House, x + 615 pp. DP: [30] June 2002 (colophon).

Wang, S.-Y. 2002b: [new taxon]. In: Wang, S.-Y. & Yu, P.-Y.: Chrysomelidae: Alticinae. Pp. 663–693. In: Huang, P.-K. (ed.): *Fauna of insects in Fujian province of China. Vol. 6.* Fuzhou: Fujian Science & Technology Press, 894 pp. DP: [31] December 2002 (colophon).

Wang, S.-Y. 2004a: [new taxon]. In: Wang, S.-Y. & Li, W.-Z.: Coleoptera: Chrysomelidae – Chrysomelinae. Pp. 72–73. In: Yang, X.-K. (ed.): *Insects of the Great Yarlung Zangbo Canyon of Xizang, China.* Beijing: China Science and Technology Press, [12] + 339 pp., 12 pls. DP: [30] June 2004 (colophon).

Wang, S.-Y. 2004b: [new taxa]. In: Wang, S.-Y. & Li, W.-Z.: Coleoptera: Chrysomelidae – Alticinae. Pp. 77–81. In: Yang, X.-K. (ed.): *Insects of the Great Yarlung Zangbo Canyon of Xizang, China*. Beijing: China Science and Technology Press, [12] + 339 pp., 12 pls. DP: [30] June 2004 (colophon).

Wang, S.-Y. 2004c: [new taxa]. In: Wang, S.-Y., Li, W.-Z.& Cui, J.-Z. 2004: Coleoptera: Chrysomelidae: Alticinae. Pp. 378–384. In: Yang, X.-K. (ed.): *Insects from Mt. Shiwandashan area of Guangxi*. Beijing: China Forestry Publishing House, 668 + [2] pp. DP: [30] September 2004 (colophon).

Wang, S.-Y. 2006: [new taxon]. In: Wang, H.-J. & Yang, X.-K. (eds.): *Fauna of Chrysomelidae of Gansu*. Lanzhou: Gansu Science Technology Press, 296 pp., 8 pl. DP: [31] August 2006 (colophon).

Wang, S.-Y. 2007: [new taxon]. In: Wang, S.-Y. & Li, W.-Z.: Notes on the Chinese flea beetles of the genus Acrocrypta Baly (Coleoptera, Chrysomelidae, Alticinae). *Acta Zootaxonomica Sinica* **32** (2): 462–464. DP: [30] April 2007 (paper title page).

Wang, S.-Y. & Ge, S.-Q. 2002: [new taxon]. In: Wang, S., Ge, S.-Q., Yang, X.-K., Li, W.-Z. & Zhang, P.-F. 2002: Notes on Chrysomelinae (Coleoptera: Chrysomelidae) of Guangxi, China. *Entomotaxonomia* **24** (2): 116–124. DP: [30] June 2002 (paper title page).

Wang, S.-Y. & Ge, S.-Q. 2009a: [new taxon]. In: Wang, S.-Y., Ge, S.-Q., Cui, J.-Z., Li, W.-Z. & Yang, X.-K.: A new species of genus Lipromela Chen, from China (Chrysomelidae, Alticinae). *Acta Zootaxonomica Sinica* **34** (4): 766–768. DP: [31] October 2009 (paper title page).

Wang, S.-Y. & Ge, S.-Q. 2009b: [new taxa]. In: Wang, S.-Y., Li, W.-Z., Cui, J.-Z., Ge, S.-Q. & Yang, X.-K.: The species of the genus Argopus Fischer von Waldheim from China (Coleoptera, Chrysomelidae, Alticinae). *Acta Zootaxonomica Sinica* **34** (4): 898–904. DP: [31] October 2009 (paper title page).

Wang, S.-Y. & Ge, S.-Q. 2010: [new taxa]. In: Wang, S.-Y., Cui, J.-Z., Li, W.-Z. Ge, S.-Q. & Yang, X.-K.: The species of Luperomorpha Weise from China (Coleoptera, Chrysomelidae, Alticinae). *Acta Zootaxonomica Sinica* **35** (1): 190–201. DP: [31] January 2010 (paper title page).

Wang, S.-Y. & Ge, S.-Q. 2012a: [new taxa]. In: Wang, S.-Y., Ge, S.-Q., Li, W.-Z., Cui, J.-Z. & Yang, X.-K.: Study of the pepper pest genus, Lanka Maulik, and descriptions of two new species from China (Coleoptera, Chrysomelidae, Alticinae). *Acta Zootaxonomica Sinica* **37** (2): 331–336. DP: [30] April 2012 (paper title page).

Wang, S.-Y. & Ge, S.-Q. 2012b: [new taxon]. In: Wang, S.-Y., Ge, S.-Q., Li, W.-Z., Cui, J.-Z. & Yang, X.-K.: New record genus Microcrepis Chen from China, and description of a new species (Coleoptera, Chrysomelidae, Alticinae). *Acta Zootaxonomica Sinica* **37** (2): 337–340. DP: [30] April 2012 (paper title page).

Wang, S.-Y., Ge, S.-Q. & Cui, J.-Z. 2010: [new taxon]. In: Wang, S.-Y., Cui, J.-Z., Li, W.-Z., Ge, S-Q. & Yang, X.-K.: Three new species of Sphaeroderma apicale species group from China and Vietnam (Coleoptera, Chrysomelidae). *Acta Zootaxonomica Sinica* **35** (4): 905–910. DP: [31] October 2010 (paper title page).

Wang, S.-Y., Ge, S.-Q. & Li, W.-Z. 2010: [new taxon]. In: Wang, S.-Y., Cui, J.-Z., Li, W.-Z., Ge, S-Q. & Yang, X.-K.: Three new species of Sphaeroderma apicale species group from China and Vietnam (Coleoptera, Chrysomelidae). *Acta Zootaxonomica Sinica* 35 (4): 905–910. DP: [31] October 2010 (paper title page).

Wang, S.-Y. & Li, W.-Z. 2004: Coleoptera: Chrysomelidae – Alticinae. Pp. 77–81. In: Yang, X.-K. (ed.): *Insects of the Great Yarlung Zangbo Canyon of Xizang, China*. Beijing: China Science and Technology Press, [12] + 339 pp., 12 pls. DP: [30] June 2004 (colophon).

Wang, S.-Y. & Yang, X.-K. 2007: [new taxon]. In: Wang, S.-Y. Ge, D.-Y., Li, W.-Z. & Yang, X.-K.: A new record genus and a new species of flea beetles from China (Coleoptera, Chrysomelidae, Alticinae). *Acta Zootaxonomica Sinica* 32 (4): 952–954. DP: [31] October 2007 (paper title page).

Wang, S.-Y. & Yang, X.-K. 2008a: [new taxon]. In: Wang, S.-Y., Ge, D.-Y., Li, W.-Z. & Yang, X.-K.: A new species of the genus Euphitrea Baly from China and description of male of E. rufomarginata Wang (Coleoptera, Chrysomelidae, Alticinae). *Acta Zootaxonomica Sinica* 33 (1): 46–48. DP: [31] January 2008 (paper title page).

Wang, S.-Y. & Yang, X.-K. 2008b: [new taxa]. In: Ge, D.-Y., Wang, S.-Y., Li, W.-Z. & Yang, X.-K.: Study of Phygasia (Coleoptera: Chrysomelidae) from China, with descriptions of eight new species. *Biologia* (Bratislava) 63 (4): 553–565. DP: [31] August 2008 (issue cover).

Wang, S.-Y. & Zhang, Y. 2006: [new taxon]. In: Zhang, Y. & Yang, X.-K.: A study on the flea beetle genus Euphitrea Baly from China, with the description of five new species (Coleoptera: Chrysomelidae: Alticinae). *Proceedings of the Entomological Society of Washington* 108 (4): 842–859. DP: 12 October 2006 (verso of the issue cover).

Wang, X.-L. & Ge, S.-Q. 2020: [new name]. In: Wang, X.-L., Ge, S.-Q. & Li, S.-Q.: A replacement name for Chrysolina mikhailovi Lopatin, 2011 (Coleoptera: Chrysomelidae: Chrysomelinae), primary junior homonym of Ch. mikhailovi Ge & Daccordi, 2011. *Zoological Systematics* 45 (1): 69–70. DP: 24 January 2020 (journal website).

Wanka, T. von 1917: Ein neuer Longitarsus aus Schlesien. *Entomologische Blätter* 13 (4/6): 74–75. DP: 16 June 1917 (issue cover).

Wankowicz, J. 1880: [Une nouvelle espèce européenne de coléoptère du genre Mniophila]. *Bulletin de la Société Entomologique de France* 1880 (20): 170–171. DP: 6 November 1880 (Evenhuis 2002).

Warchałowski, A. 1966: Les espèces noires du genre Longitarsus Latr. du sud-est asiatique (Col., Chrysomelidae). *Polskie Pismo Entomologiczne* 36 (1): 93–110. DP: [31] August 1966) (issue colophon).

Warchałowski, A. 1967a: Beitrag zur Kenntnis der Halticinen Nordpersiens (Col., Chrysomelidae). *Polskie Pismo Entomologiczne* 37 (1): 53–64. DP: 15 August 1967 (paper title page).

Warchałowski, A. 1967b: Die geographischen Rassen von Longitarsus fuscoaeneus Redt. und Longitarsus corynthius (Reiche) (Coleoptera, Chrysomelidae). *Polskie Pismo Entomologiczne* 37 (4): 625–632. DP: 30 December 1967) (paper title page).

Warchałowski, A. 1968: Ein neuer Longitarsus aus Italien (Coleoptera, Chrysomelidae). *Memorie del Museo Civico di Storia Naturale di Verona* 16: 209–211. DP: 31 December 1968 (reprint cover).

Warchałowski, A. 1969: Beitrag zur Kenntnis der koreanischen Halticinen (Coleoptera, Chrysomelidae). *Annales Zoologici* (Warszawa) 27 [1969–1970]: 225–236. DP: 30 September 1969 (paper title page).

Warchałowski, A. 1972: Neue asiatische Arten aus der Gattung Longitarsus Latr. (Coleoptera, Chrysomelidae). *Polskie Pismo Entomologiczne* 42 (2): 313–318. DP: 30 June 1972 (paper title page).

Warchałowski, A. 1973a: Zur Kenntnis der orientalischen Arten aus der Verwandtschaft von Chaetocnema (Tlanoma) basalis (Baly) (Coleoptera, Chrysomelidae). *Polskie Pismo Entomologiczne* 43 (1): 41–53. DP: 30 March 1973 (paper title page).

Warchałowski, A. 1973b: Zweiter Beitrag zur Kenntnis der Halticinen Nordpersiens (Coleoptera, Chrysomelidae). *Polskie Pismo Entomologiczne* 43 (4): 659–687. DP: 30 November 1973 (paper title page).

Warchałowski, A. 1974: Übersicht der Blattkäfer Bulgariens (Coleoptera, Chrysomelidae). *Polskie Pismo Entomologiczne* 44 (3): 473–542. DP: 30 September 1974 (paper title page).

Warchałowski, A. 1978: *Klucze do oznaczania owadow Polski. Cześć XIX. Chrzaszcze – Coleoptera. Zeszyt 94c. Stonkowate – Chrysomelidae. Podrodziny: Halticinae, Hispinae i Cassidinae.* Warszawa: Państwowe Wydawnictwo Naukove, 157 pp. DP: [30] September 1978 (colophon).

Warchałowski, A. 1985: Revision der Gattung Labidostomis Germar, 1824 (Coleoptera, Chrysomelidae, Clytrinae). *Polskie Pismo Entomologiczne* 55 (4): 621–765. DP: 30 December 1985 (paper title page).

Warchałowski, A. 1990: Otiothraea ghardaiensis, eine neue Gattung und Art aus der nördlichen Sahara (Coleoptera, Chrysomelidae, Clytrinae). *Polskie Pismo Entomologiczne* 59 [1989] (4): 793–797. DP: 30 March 1990 (paper title page).

Warchałowski, A. 1991a: Kurze Übersicht der oberseits hellen Calomicrus Steph.-Arten (Coleoptera: Chrysomelidae: Galerucinae). *Genus* 2 (1): 41–55. DP: 31 May 1991 (paper title page).

Warchałowski, A. 1991b: L'établissement d'un nouveau sous-genre Homalopus Chevrolat, 1837 (Coleoptera, Chrysomelidae, Cryptocephalinae). *Polskie Pismo Entomologiczne* 61 (1): 75–78. DP: 30 June 1991 (paper title page).

Warchałowski, A. 1991c: Über die rot und schwarz gefleckten Arten der Untergattung Coptocephala s. str. (Coleoptera: Chrysomelidae: Clytrinae). *Genus* 2 (3): 229–279. DP: 27 December 1991 (paper title page).

Warchałowski, A. 1991d: Eine nomenklatorische Änderung innerhalb der Gattung Chrysolina Motschulsky (Coleoptera: Chrysomelidae: Chrysomelinae). *Genus* 2 (3): 281–282. DP: 27 December 1991 (paper title page).

Warchałowski, A. 1991e: *Chrysomelidae – Stonkowate (Insecta: Coleoptera). Część II (podrodziny: Clythrinae i Cryptocephalinae). Fauna Polski 13.* Warszawa: PWN, 347 pp. DP: [31 December] 1991 (book cover).

Warchałowski, A. 1993a: Smaragdina uyguni n. sp., eine neue Art aus Südanatolien (Coleoptera: Chrysomelidae). *Genus* 4 (2): 129–132. DP: 30 September 1993 (paper title page).

Warchałowski, A. 1993b: Psylliodes dogueti sp. n., eine neue Erdflohart aus der Türkei (Coleoptera: Chrysomelidae: Halticinae). *Genus* 4 (4): 359–361. DP: 30 December 1993 (paper title page).

Warchałowski, A. 1993c: *Chrysomelidae – Stonkowate (Insecta: Coleoptera). Część III (podrodziny: Lamprosomatinae, Eumolpinae; z podrodziny Chrysomelinae plemię Timarchini oraz część plemenia Chrysomelini: podplemiona Doryphorina i Chrysolinina). Fauna Polski 15*. Warszawa: Polska Akademia Nauk, Muzeum i Institut Zoologii, 279 pp. DP: [31 December] 1993 (book cover).

Warchałowski, A. 1995a: Bemerkungen über die Gattung Phratora Chevrolat 1837 (Coleoptera: Chrysomelidae). *Genus* 6 (3/4): 455–461. DP: 15 December 1995 (paper title page).

Warchałowski, A. 1995b: Bemerkungen zur Systematik und Nomenklatur der Erdflöhe (Coleoptera: Chrysomelidae: Halticinae). *Genus* 6 (3/4): 463–468. DP: 15 December 1995 (paper title page).

Warchałowski, A. 1996: Übersicht der westpaläarktischen Arten der Gattung Longitarsus Berthold, 1827 (Coleoptera: Chrysomelidae: Halticinae). *Genus* **Supplement** [6]: 1–266. DP: 31 August 1996 (colophon).

Warchałowski, A. 1997: Der erste afrikanische Vertreter der Untergattung Asionus Lopatin, 1988, Cryptocephalus mroczkowskii sp. nov. (Coleoptera: Chrysomelidae: Cryptocephalinae). *Annales Zoologici* (Warszawa) 47 (1/2): 273–274. DP: 20 September 1997 (verso of the issue cover).

Warchałowski, A. 1998a: Über einige kleinasiatische Arten der Gattung Pachybrachis Chevrolat, 1837 (Coleoptera: Chrysomelidae: Cryptocephalinae). *Annales Zoologici* (Warszawa) 48 (1/2): 85–90. DP: 30 August 1998 (verso of the issue cover).

Warchałowski, A. 1998b: Die westpaläarktischen Arten der Gattung Phyllobrotica Chevrolat, 1837 (Coleoptera: Chrysomelidae: Galerucinae). *Annales Zoologici* (Warszawa) 48 (1/2): 91–98. DP: 30 August 1998 (verso of the issue cover).

Warchałowski, A. 1998c: Longitarsus kippenbergi n. sp., eine neue Art aus Marokko (Coleoptera: Chrysomelidae: Halticinae). *Genus* 9 (3): 351–360. DP: 30 September 1998 (paper title page).

Warchałowski, A. 1998d: Beschreibung der neuen Art Derocrepis (Aeschrocnemis) whiteheadi sp. n. und ihres Standortes in Kleinasien (Coleoptera: Chrysomelidae: Halticinae). *Genus* 9 (3): 361–365. DP: 30 September 1998 (paper title page).

Warchałowski, A. 1999a: Description d'une espèce nouvelle de Longitarsus de Turquie (Coleoptera, Chrysomelidae). *Nouvelle Revue d'Entomologie, Nouvelle Série* 15 [1998] (3): 291–293. DP: 12 January 1999 (last page of the cover).

Warchałowski, A. 1999b: Übersicht der westpaläarktischen Arten der Untergattung Burlinius Lopatin, 1965 (Coleoptera, Chrysomelidae: Cryptocephalinae). *Genus* 10 (4): 529–627. DP: 20 December 1999 (paper title page).

Warchałowski, A. 2000a: *Chrysomelidae. Stonkowate Cześć VII (Insecta: Coleoptera). Fauna Polski 22*. Warszawa: Polskie Towarzystwo Entomologiczne, 357 pp. DP: 20 December 2000 (colophon).

Warchałowski, A. 2000b: A short review of the genus Chilotomina Reitter, 1912 (Coleoptera: Chrysomelidae: Clytrinae). *Genus* 11 (4): 573–583. DP: 26 December 2000 (paper title page).

Warchałowski, A. 2000c: A few nomenclatorial changes in Chrysomelidae (Coleoptera: Chrysomelidae). *Genus* 11 (4): 587–588. DP: 26 December 2000 (paper title page).

Warchałowski, A. 2001a: Cryptocephalus siedei sp. nov., a new Asiatic representative of subgenus Homalopus Chevrolat, 1837 (Coleoptera: Chrysomelidae: Cryptocephalinae). *Annales Zoologici* (Warszawa) 51 (1): 81–83. DP: 26 March 2001 (verso of the issue cover).

Warchałowski, A. 2001b: Geinella trapezicollis sp. nov., a new species from Tibet (Coleoptera: Chrysomelidae: Galerucinae). *Annales Zoologici* (Warszawa) 51 (2): 185–187. DP: 28 June 2001 (verso of the issue cover).

Warchałowski, A. 2001c: Macrocoma djurdjurensis sp. nov., a new North-African species of subfamily Eumolpinae (Coleoptera: Chrysomelidae). *Annales Zoologici* (Warszawa) 51 (2): 189–190. DP: 28 June 2001 (verso of the issue cover).

Warchałowski, A. 2001d: Macrocoma hormuziaca sp. nov., a new Eumolpinae-species from Iran (Coleoptera: Chrysomelidae). *Annales Zoologici* (Warszawa) 51 (3): 345–346. DP: 28 September 2001 (verso of the issue cover).

Warchałowski, A. 2001e: A preliminary review of Western Palaearctic Macrocoma Chevrolat, 1837 (Coleoptera: Chrysomelidae: Eumolpinae). *Genus* 12 (4): 449–477. DP: 28 December 2001 (paper title page).

Warchałowski, A. 2003: *Chrysomelidae: the leaf-beetles of Europe and the Mediterranean area.* Warszawa: Natura Optima Dux Foundation, 600 pp. DP: [31 December] 2003 (book title page).

Warchałowski, A. 2004a: Nomenclatorial notes in Chrysomelidae (Coleoptera: Chrysomeloidea). *Genus* 15 (1): 113–114. DP: 30 March 2004 (paper title page).

Warchałowski, A. 2004b: Labidostomis kantneri sp. nov. from Iran (Coleoptera: Chrysomelidae: Clytrinae). *Annales Zoologici* (Warszawa) 54 (3): 557–559. DP: 30 September 2004 (verso of the issue cover).

Warchałowski, A. 2005: Bezdekia tenebrosa, new Palaearctic genus and species of the subfamily Eumolpinae (Insecta: Coleoptera: Chrysomelidae). *Annales Zoologici* (Warszawa) 55 (2): 303–304. DP: 30 June 2005 (verso of the issue cover).

Warchałowski, A. 2006: Stylosomus arnoldi, a new North-African species of the subfamily Cryptocephalinae (Insecta: Coleoptera: Chrysomelidae). *Annales Zoologici* (Warszawa) 56 (3): 497–498. DP: 30 September 2006 (verso of the issue cover).

Warchałowski, A. 2007: Doeberlia subopaca, a new genus and species of leaf-beetles from China (Coleoptera: Chrysomelidae: Chrysomelinae). *Annales Zoologici* (Warszawa) 57 (3): 481–482. DP: 30 September 2007 (verso of the issue cover).

Warchałowski, A. 2008: A review of the subgenus Pachybrachis Chevrolat, 1837 s. str. of Palaearctic region (Coleoptera: Chrysomelidae: Cryptocephalinae). *Genus* 19 (2): 213–283. DP: 31 July 2008 (paper title page).

Warchałowski, A. 2010a: Remarks on the genus Pseudaoria Jacoby, 1908 with description of a new species from China (Chrysomelidae: Eumolpinae). *Annales Zoologici* (Warszawa) 60 (3): 337–340. DP: 30 September 2010 (verso of the issue cover).

Warchałowski, A. 2010b: *The Palaearctic Chrysomelidae. Identification keys. Volume 1 & 2.* Warszawa: Natura Optima Dux Foundation, 1212 pp., 102 pls. DP: 1 November 2010 (publisher website).

Warchałowski, A. 2010c: Two nomenclatorial changes in the genus Cryptocephalus from Asia (Chrysomelidae: Cryptocephalinae). *Genus* 21 (4): 583–584. DP: 27 December 2010 (paper title page).

Waterhouse, G. R. 1838: Descriptions of some of the insects brought to this country by C. Darwin, Esq. *Transactions of the Entomological Society of London* 2 [1837–1840] (2): 131–135. DP: 5 September 1838 (Wheeler 1912).

Waterhouse, G. R. 1861: *Catalogue of British Coleoptera.* [1858–1862], London: Taylor and Francis, iv + 117 pp. DP (pp. 81–96): 7 January 1861 (Bousquet 2016).

Weber, F. 1801: *Observationes entomologicae, continentes novorum quae condidit generum characteres, et nuper detectarum specierum descriptiones.* Kiliae: Bibliopoli Academici Novi, xii + 116 + [1] pp. DP: 29 August 1801 (Bousquet 2016).

Weidenbach, C. von 1859: Sechs neue Käfer aus der Umgegend von Augsburg. *Bericht des Naturhistorischen Vereins in Augsburg* 12: 84–86. DP: 5 October 1859 (*Verh. Zool.-Bot. Ges. Wien* 1859, 9: 105).

Weill, J. 1951: Un nouveau Chrysochloa de France (Col. Chrysomelidae). *Revue Française d'Entomologie* 18 (3): 154–156. DP: 15 November 1951 (footer on p. 141).

Weise, J. 1875: Phaedon carniolicus Germ. und seine Varietäten. *Deutsche Entomologische Zeitschrift* 19 (2): 365–366. DP: [31] December 1875 (table of contents).

Weise, J. 1878: [new taxa]. In: Schneider, O. & Leder, H.: *Beiträge zur Kenntniss der kaukasischen Käferfauna.* Brünn: W. Burkart, 359 + [1] pp., 6 pls. DP: [31 December] 1878 (Döberl 2009, see also the comments in the Dating of publications chapter).

Weise, J. 1880a: [new taxa]. In: Leder, H.: Beitrag zur Kaukasischen Käfer-Fauna. Unter Mitwirkung hervorragender Fachgenossen. *Verhandlungen der Kaiserlich-Königlichen Zoologisch-Botanischen Gesellschaft in Wien* 29 [1879] (2): 451–488. DP: 15 April 1880 (Evenhuis 2023).

Weise, J. 1880b: Acanthodon nov. gen. eur. Chrysomelidarum. *Entomologische Monatsblätter* 2 (15): 76. DP: [31] May 1880 (issue title page).

Weise, J. 1880c: Die Verschiedenheiten des Forceps der Orina-Arten. *Entomologische Monatsblätter* 2 (17): 97–102. DP: [31] July 1880 (issue title page).

Weise, J. 1880d: Synonymische Bemerkungen über Chrysomelinen. *Entomologische Monatsblätter* 2 (20): 158. DP: [31] October 1880 (issue title page).

Weise, J. 1881: Chrysomelidae. Pp. 1–192 [Lieferung 1]. In: *Naturgeschichte der Insekten Deutschlands. Erste Abtheilung. Coleoptera. Sechster Band.* [1881–1893], Berlin: Nicolaische Verlags-Buchhandlung R. Stricker, xiv + 1161 pp., 1 pl. DP: 1 October 1881 (*Vorrede* p. vi).

Weise, J. 1882a: Ueber Phyllodecta viennensis Schrank. *Deutsche Entomologische Zeitschrift* 26 (1): 157–158. DP: 8 March 1882 (*Bull. Soc. Ent. Fr.* 1882: 92).

Weise, J. 1882b: Chrysomelidae. Pp. 193–368 [Lieferung 2]. In: *Naturgeschichte der Insekten Deutschlands. Erste Abtheilung. Coleoptera. Sechster Band.* [1881–1893], Berlin: Nicolaische Verlags-Buchhandlung R. Stricker, xiv + 1161 pp., 1 pl. DP: 1 December 1882 (*Vorrede* p. vi).

Weise, J. 1883a: Bemerkungen über die im Glatzer Gebirge lebenden Orina-Arten und ihre Larven. *Deutsche Entomologische Zeitschrift* 27 (2): 210–218. DP: [31] July 1883 (table of contents).

Weise, J. 1883b: [Psylliodes napi var. flavicornis]. *Deutsche Entomologische Zeitschrift* 27 (2): 218–219. DP: [31] July 1883 (table of contents).

Weise, J. 1883c: Die Orina-Arten der Schweiz. *Deutsche Entomologische Zeitschrift* 27 (2): 243–247. DP: [31] July 1883 (table of contents).

Weise, J. 1883d: Bemerkungen über Chrysomeliden. *Deutsche Entomologische Zeitschrift* 27 (2): 251–254. DP: [31] July 1883 (table of contents).

Weise, J. 1883e: Ueber die mit Galeruca Geoffr. verwandten Gattungen. *Deutsche Entomologische Zeitschrift* 27 (2): 315–316. DP: [31] July 1883 (table of contents).

Weise, J. 1884a: Beitrag zur Chrysomeliden-Fauna von Amasia. *Deutsche Entomologische Zeitschrift* 28 (1): 157–160. DP: [31] January 1884 (table of contents).

Weise, J. 1884b: Einige neue Chrysomeliden und Coccinelliden. *Deutsche Entomologische Zeitschrift* 28 (1): 161–166. DP: [31] January 1884 (table of contents).

Weise, J. 1884c: [new taxa]. In: Heyden, L. F. J. D. von & Kraatz, G.: Neue Käfer-Arten aus Osch (Turkestan). *Deutsche Entomologische Zeitschrift* 28 (1): 216–228. DP: [31] January 1884 (table of contents).

Weise, J. 1884d: Chrysomelidae. Pp. 369–568 [Lieferung 3]. In: *Naturgeschichte der Insekten Deutschlands. Erste Abtheilung. Coleoptera. Sechster Band.* [1881–1893], Berlin: Nicolaische Verlags-Buchhandlung R. Stricker, xiv + 1161 pp., 1 pl. DP: 15 October 1884 (*Vorrede* p. vi).

Weise, J. 1885a: Zwei neue Labidostomis-Arten. *Deutsche Entomologische Zeitschrift* 29 (2): 313–315. DP: [31] December 1885 (table of contents).

Weise, J. 1885b: Leptosonyx hirtus Weise n. sp. *Deutsche Entomologische Zeitschrift* 29 (2): 315–316. DP: [31] December 1885 (table of contents).

Weise, J. 1885c: Zwei neue Xenomela (Chrysomelinen) aus Turkestan und Samarkand. *Deutsche Entomologische Zeitschrift* 29 (2): 406–408. DP: [31] December 1885 (table of contents).

Weise, J. 1886a: Chrysomelidae. Pp. 569–768 [Lieferung 4]. In: *Naturgeschichte der Insekten Deutschlands. Erste Abtheilung. Coleoptera. Sechster Band.* [1881–1893], Berlin: Nicolaische Verlags-Buchhandlung R. Stricker, xiv + 1161 pp., 1 pl. DP: 15 June 1886 (*Vorrede* p. vi).

Weise, J. 1886b: [new taxa]. In: Heyden, L. F. J. D. von & Kraatz, G.: Beiträge zur Coleopteren-Fauna von Turkestan, namentlich des Alai-Gebirges. *Deutsche Entomologische Zeitschrift* 30 (1): 177–194. DP: [30] June 1886 (table of contents).

Weise, J. 1886c: Cryptocephalus abdominalis. *Deutsche Entomologische Zeitschrift* 30 (1): 208. DP: [30] June 1886 (table of contents).

Weise, J. 1886d: Synonymische und andere kleine Mittheilungen über Chrysomeliden. *Deutsche Entomologische Zeitschrift* 30 (1): 249–250. DP: [30] June 1886 (table of contents).

Weise, J. 1886e: Vier neue Pachybrachys-Arten. *Deutsche Entomologische Zeitschrift* 30 (2): 21–25. DP: [31] December 1886 (table of contents).

Weise, J. 1887a: Einige Chrysomelen-Formen aus Galizien. *Deutsche Entomologische Zeitschrift* **31** (1): 185–186. DP: [30] June 1887 (table of contents).

Weise, J. 1887b: Hydrothassa planiuscula Weise n. sp. *Deutsche Entomologische Zeitschrift* **31** (1): 188. DP: [30] June 1887 (table of contents).

Weise, J. 1887c: Mittheilungen über das Sammeln von Käfern und über die Fangstellen im Glatzer Gebirge. *Zeitschrift für Entomologie, Neue Folge* (Breslau) **12**: 47–60. DP: [31] October 1887 (*Natur. Novit.* 1887, 9 (19): 248).

Weise, J. 1887d: [new taxa]. In: Heyden, L. F. J. D. von: XI. Beitrag zur Coleopteren-Fauna von Turkestan. *Deutsche Entomologische Zeitschrift* **31** (2): 305–336. DP: [30] November 1887 (table of contents).

Weise, J. 1887e: Orina virgulata Germ. var. nov. Papei Weise. *Deutsche Entomologische Zeitschrift* **31** (2): 364. DP: [30] November 1887 (table of contents).

Weise, J. 1887f: Anthobiodes n. gen. Halticarum. *Deutsche Entomologische Zeitschrift* **31** (2): 402. DP: [30] November 1887 (table of contents).

Weise, J. 1887g: [new taxon]. In: Heyden, L. F. J. D. von: Verzeichniss der von Herrn Otto Herz auf der chinesischen Halbinsel Korea gesammelten Coleopteren. *Horae Societatis Entomologicae Rossicae* **21** (3/4): 243–273. DP: 27 December 1887 (Kerzhner 1984, Standfuss & Kerzhner 2004).

Weise, J. 1887h: Neue sibirische Chrysomeliden und Coccinelliden nebst Bemerkungen über früher beschriebene Arten. *Archiv für Naturgeschichte* **53** [1] (2): 164–214. DP: [31 December] 1887 (issue cover).

Weise, J. 1888a: Timarcha strophium. *Deutsche Entomologische Zeitschrift* **1888** (1): 124. DP: [30] April 1888 (table of contents).

Weise, J. 1888b: Chrysomelidae. Pp. 769–960 [Lieferung 5]. In: *Naturgeschichte der Insecten Deutschland. Erste Abtheilung. Coleoptera. Sechster Band.* [1881–1893]. Berlin: Nicolaische Verlags-Buchhandlung R. Stricker, xiv + 1161 pp., 1 pl. DP: 10 May 1888 (*Vorrede* p. vi).

Weise, J. 1888c: Cryptocephalus tardus Weise n. sp. from Monte Rosa. *Deutsche Entomologische Zeitschrift* **1888** (2): 368. DP: 28 November 1888 (*Bull. Soc. Ent. Fr.* 1888: clxxxv).

Weise, J. 1889a: Neue Chrysomeliden aus Circassien. Gesammelt von Hans Leder im Jahre 1887. *Wiener Entomologische Zeitung* 8 (1): 1–4. DP: 28 January 1889 (footers throughout the paper).

Weise, J. 1889b: Griechische Chrysomelidae und Cocinellidae. *Deutsche Entomologische Zeitschrift* **1889** (1): 58–65. DP: [31] May 1889 (table of contents).

Weise, J. 1889c: Zwei neue deutsche Chrysomelinen-Varietäten. *Deutsche Entomologische Zeitschrift* **1889** (1): 127. DP: [31] May 1889 (table of contents).

Weise, J. 1889d: [Monolepta angustula Weise n. sp. vom Alai-Gebirge]. *Deutsche Entomologische Zeitschrift* **1889** (1): 128. DP: [31] May 1889 (table of contents).

Weise, J. 1889e: [Gynandrophthalma flavicollis Charp. var. picticollis]. *Deutsche Entomologische Zeitschrift* **1889** (1): 128. DP: [31] May 1889 (table of contents).

Weise, J. 1889f: Kleine Mittheilungen zur thüringer Käferfauna (Weimar). *Deutsche Entomologische Zeitschrift* **1889** (1): 217–219. DP: [31] May 1889 (table of contents).

Weise, J. 1889g: Insecta, a Cl. G. N. Potanin in China et in Mongolia novissime lecta. IX. Chrysomelidae et Coccinellidae. *Horae Societatis Entomologicae Rossicae* **23** (3/4): 560–653. DP: 24 August 1889 (Kerzhner 1984, Standfuss & Kerzhner 2004).

Weise, J. 1889h: Neue Chrysomeliden und Coccinelliden aus dem Kaukasus. *Wiener Entomologische Zeitung* 8 (7): 259–262. DP: 25 August 1889 (footers throughout the paper).

Weise, J. 1889i: [Coptocephala punctata]. *Deutsche Entomologische Zeitschrift* **1889** (2): 336. DP: [30] November 1889 (table of contents).

Weise, J. 1889j: Aphthona metallica. *Deutsche Entomologische Zeitschrift* **1889** (2): 415. DP: [30] November 1889 (table of contents).

Weise, J. 1889k: Synonymische Bemerkungen. *Deutsche Entomologische Zeitschrift* **1889** (2): 416. DP: [30] November 1889 (table of contents).

Weise, J. 1889l: Synonymische Bemerkungen über Chrysomelini. *Deutsche Entomologische Zeitschrift* **1889** (2): 419–420. DP: [30] November 1889 (table of contents).

Weise, J. 1889m: [new taxa]. In: Reitter, E.: Coleopterologische Ergebnisse der im Jahre 1886 und 1887 in Transcaspien von Dr. G. Radde, Dr. A. Walter und A. Konchin ausgeführten Expedition. *Verhandlungen des Naturforschenden Vereines in Brünn* **27** [1888]: 95–133. DP: [31 December] 1889 (volume title page).

Weise, J. 1890a: Insecta, a Cl. G. N. Potanin in China et in Mongolia novissime lecta. XVI. Chrysomelidae et Coccinellidae (Appendix). *Horae Societatis Entomologicae Rossicae* **24** [1889–1890] (3/4): 477–492. DP: 12 May 1890 (Kerzhner 1984, Standfuss & Kerzhner 2004).

Weise, J. 1890b: Bekannte und neue Halticiden. *Wiener Entomologische Zeitung* **9** (4): 111–116. DP: 15 May 1890 (footers throughout the paper).

Weise, J. 1890c: Ueber einige Chrysomelinen aus Siebenbürgen. *Deutsche Entomologische Zeitschrift* **1890** (1): 30. DP: 28 May 1890 (*Bull. Soc. Ent. Fr.* 1890: c).

Weise, J. 1890d: [new taxon]. In: Schönfeldt, H. von.: Ein Beitrag zur Coleopterenfauna der Liu-Kiu-Inseln. *Entomologische Nachrichten* **16** (11): 168–175. DP: [31] June 1890 (issue title page).

Weise, J. 1890e: Xenomela Fausti Weise nov. spec. *Deutsche Entomologische Zeitschrift* **1890** (2): 23. DP: [31] October 1890 (table of contents).

Weise, J. 1890f: [Crepidodera norica]. *Deutsche Entomologische Zeitschrift* **1890** (2): 24. DP: [31] October 1890 (table of contents).

Weise, J. 1890g: [new taxa]. In: Heyden, L. F. J. D. von 1890: XIII. Beitrag zur Coleopteren-Fauna von Turkestan. *Deutsche Entomologische Zeitschrift* **1890** (2): 353–367. DP: [31] October 1890 (table of contents).

Weise, J. 1891a: [new name]. In: Reitter, E. (ed.): *Catalogus Coleopterorum Europae, Caucasi et Armeniae rossicae*. Berlin: Friedländer & Sohn, Mödling: Edmund Reitter, Caen: Revue d'Entomologie, viii + 420 pp. DP: 13 May 1891 (Bousquet 2016).

Weise, J. 1891b: Bekannte und neue Chrysomeliden aus Spanien I. *Deutsche Entomologische Zeitschrift* **1891** (1): 145–150. DP: [30] June 1891 (table of contents).

Weise, J. 1891c: Wieviel Arten enthält die Gattung Sclerophaedon? *Deutsche Entomologische Zeitschrift* **1891** (1): 158–159. DP: [30] June 1891 (table of contents).

Weise, J. 1891d: Ueber Varietäten von Phytodecta. *Deutsche Entomologische Zeitschrift* **1891** (1): 160. DP: [30] June 1891 (table of contents).

Weise, J. 1891e: Bemerkungen zur Gattung Cassida. *Wiener Entomologische Zeitung* 10 (6): 203–205. DP: 31 July 1891 (footers throughout the paper).

Weise, J. 1892a: Bekannte und neue Chrysomeliden aus Spanien. II. *Deutsche Entomologische Zeitschrift* [**1891**] (2): 373. DP: [31] March 1892 (table of contents).

Weise, J. 1892b: Beiträge zur Kenntniss der Gattung Orina. *Deutsche Entomologische Zeitschrift* [**1891**] (2): 374–376. DP: [31] March 1892 (table of contents).

Weise, J. 1892c: Ueber einige Haltica-Arten etc. von Weimar. *Deutsche Entomologische Zeitschrift* [**1891**] (2): 379. DP: [31] March 1892 (table of contents).

Weise, J. 1892d: Beitrag zur Coleopteren-Fauna von Turkestan. *Deutsche Entomologische Zeitschrift* **1892** (1): 131–140. DP: [31] August 1892 (table of contents).

Weise, J. 1892e: [new taxa]. In: Schilsky, J.: VII. Beitrag zur Kenntniss der deutschen Käferfauna. *Deutsche Entomologische Zeitschrift* **1892** (1): 193–208. DP: [31] August 1892 (table of contents).

Weise, J. 1892f: [Beitrag zur deutschen Käferfauna]. *Deutsche Entomologische Zeitschrift* **1892** (1): 208. DP: [31] August 1892 (table of contents).

Weise, J. 1892g: Zwei neue Cassida-Arten. *Wiener Entomologische Zeitung* **11** (8): 238–239. DP: 31 October 1892 (footer on p. 239).

Weise, J. 1892h: Pachygnatha, eine neue Chrysomelen-Gattung. *Deutsche Entomologische Zeitschrift* **1892** (2): 352. DP: [31] December 1892 (table of contents).

Weise, J. 1892i: Chrysomeliden und Coccinelliden von der Insel Nias, nebst Bemerkungen über andere, meistens südost-asiatischen Arten. *Deutsche Entomologische Zeitschrift* **1892** (2): 385–400. DP: [31] December 1892 (table of contents).

Weise, J. 1892j: Chrysomelen aus dem hohen Caucasus. *Deutsche Entomologische Zeitschrift* **1892** (2): 405–406. DP: [31] December 1892 (table of contents).

Weise, J. 1892k: Synonymische Bemerkungen. *Deutsche Entomologische Zeitschrift* **1892** (2): 413–414. DP: [31] December 1892 (table of contents).

Weise, J. 1893a: Chrysomelidae. Pp. 961–1161 [Lieferung 6]. In: *Naturgeschichte der Insecten Deutschland. Erste Abtheilung. Coleoptera. Sechster Band.* [1881–1893], Berlin: Nicolaische Verlags-Buchhandlung R. Stricker, xiv + 1161 pp., 1 pl. DP: [31] March 1893 (Bousquet 2016).

Weise, J. 1893b: [Pachybrachys karamani von Spalato]. *Deutsche Entomologische Zeitschrift* **1893** (2): 348. DP: 26 July 1893 (*Bull. Soc. Ent. Fr.* 1893: cclxix).

Weise, J. 1894a: [new taxa]. In: Hauser, F.: Beitrag zur Coleopteren-Fauna von Transcaspien und Turkestan. Zusammengestellt unter mitwirkung der Herren Dr. Eppelsheim, Escherich, Faust, Dr. von Heyden, Kuwert, Reitter und Weise. *Deutsche Entomologische Zeitschrift* **1894** (1): 17–74, pl. 1. DP: 10 January 1894 (*Bull. Soc. Ent. Fr.* 1894: xiv).

Weise, J. 1894b: Gynandrophthalma algirica. *Deutsche Entomologische Zeitschrift* **1894** (1): 84. DP: 10 January 1894 (*Bull. Soc. Ent. Fr.* 1894: xiv).

Weise, J. 1894c: Psylliodes Sicardi Weise n. sp. von Tunis. *Deutsche Entomologische Zeitschrift* **1894** (1): 90. DP: 10 January 1894 (*Bull. Soc. Ent. Fr.* 1894: xiv).

Weise, J. 1894d: Chrysomeliden aus Amasia. *Deutsche Entomologische Zeitschrift* **1894** (1): 91–92. DP: 10 January 1894 (*Bull. Soc. Ent. Fr.* 1894: xiv).

Weise, J. 1894e: Cryptocephalus castilianus Weise n. sp. *Deutsche Entomologische Zeitschrift* **1894** (1): 93–94. DP: 10 January 1894 (*Bull. Soc. Ent. Fr.* 1894: xiv).

Weise, J. 1894f: Chiridula 11-notata. *Deutsche Entomologische Zeitschrift* **1894** (1): 96. DP: 10 January 1894 (*Bull. Soc. Ent. Fr.* 1894: xiv).

Weise, J. 1894g: Pachybrachys tesselatus Ol. *Deutsche Entomologische Zeitschrift* **1894** (1): 144. DP: 10 January 1894 (*Bull. Soc. Ent. Fr.* 1894: xiv).

Weise, J. 1894h: Ueber Halticinen. *Deutsche Entomologische Zeitschrift* **1894** (1): 152. DP: 10 January 1894 (*Bull. Soc. Ent. Fr.* 1894: xiv).

Weise, J. 1894i: Chrysomelidae sibiricae I. *Wiener Entomologische Zeitung* **13** (5): 153–155. DP: 20 May 1894 (footers throughout the paper).

Weise, J. 1894j: Ueber zwei japanische Galeruca-Arten. *Deutsche Entomologische Zeitschrift* **1894** (2): 168. DP: [31] August 1894 (table of contents).

Weise, J. 1894k: Zur Gattung Orina. *Deutsche Entomologische Zeitschrift* **1894** (2): 250–256, pl. 1. DP: [31] August 1894 (table of contents).

Weise, J. 1895a: Zwei neue Cryptocephalinen-Gattungen. *Deutsche Entomologische Zeitschrift* **1895** (1): 57–58. DP: [28] February 1895 (table of contents).

Weise, J. 1895b: Luperus fiorii n. spec. *Deutsche Entomologische Zeitschrift* **1895** (1): 212. DP: [28] February 1895 (table of contents).

Weise, J. 1895c: Gynandrophthalma Korbi n. spec. *Deutsche Entomologische Zeitschrift* **1895** (1): 216. DP: [28] February 1895 (table of contents).

Weise, J. 1895d: Ueber Gynandrophtalma tibialis Brullé. *Deutsche Entomologische Zeitschrift* **1895** (1): 224. DP: [28] February 1895 (table of contents).

Weise, J. 1895e: Neue Chrysomeliden nebst synonymischen Bemerkungen. *Deutsche Entomologische Zeitschrift* **1895** (2): 327–352. DP: [31] October 1895 (table of contents).

Weise, J. 1896a: Chrysomelidae sibiricae. II. *Wiener Entomologische Zeitung* **15** (2): 79–82. DP: 29 February 1896 (footers throughout the paper).

Weise, J. 1896b: Beschreibung neuer Cassida-Arten und synonymische Bemerkungen. *Deutsche Entomologische Zeitschrift* **1896** (1): 15–32. DP: [31] May 1896 (table of contents).

Weise, J. 1896c: Orina cacaliae var. lissahorensis. *Deutsche Entomologische Zeitschrift* **1896** (1): 110. DP: [31] May 1896 (table of contents).

Weise, J. 1897a: Synonymische Bemerkungen über europäische Chrysomelinen. *Deutsche Entomologische Zeitschrift* **[1896]** (2): 293–296. DP: [31] January 1897 (table of contents).

Weise, J. 1897b: Eine neue Podontia aus Jericho. *Wiener Entomologische Zeitung* **16** (3): 81–82. DP: 31 March 1897 (footer on p. 81).

Weise, J. 1897c: Nachtrag. Neue Chrysomeliden aus Angora. Pp. 63–68. In: Escherich, K.: Zoologische Ergebnisse einer von Dr. K. Escherich und Dr. L. Kathariner nach Central-Kleinasien unternommenen Reise. IV. Teil. Coleopteren. *Entomologische Zeitung* (Stettin) 58 (1/3): 3–69. DP: 20 May 1897 (last page of the issue).

Weise, J. 1897d: Kritisches Verzeichniss der von Mr. Andrewes eingesandten Cassidinen und Hispinen aus Indien. *Deutsche Entomologische Zeitschrift* 1897 (1): 97–150. DP: [31] July 1897 (table of contents).

Weise, J. 1898: Ueber neue und bekannte Chrysomeliden. *Archiv für Naturgeschichte* 64 [1] (2): 177–224. DP: [31] August 1898 (issue cover).

Weise, J. 1899a: Chrysomeliden aus Creta. *Deutsche Entomologische Zeitschrift* 1899 (1): 137–140. DP: [31] August 1899 (table of contents).

Weise, J. 1899b: Einige neue Cassidinen-Gattungen und Arten. *Archiv für Naturgeschichte* 65 [1] (3): 268–273. DP: [31] October 1899 (issue cover).

Weise, J. 1900a: Synonymische Bemerkungen. *Deutsche Entomologische Zeitschrift* [1899] (2): 379–380. DP: 28 February 1900 (*Bull. Soc. Ent. Fr.* 1900: 113).

Weise, J. 1900b: Beschreibungen von Chrysomeliden u. synonymische Bemerkungen. *Archiv für Naturgeschichte* 66 [1] (2): 267–296. DP: [30] June 1900 (issue cover).

Weise, J. 1900c: Neue Coleopteren aus Kleinasien. *Deutsche Entomologische Zeitschrift* 1900 (1): 132–140. DP: 25 July 1900 (*Bull. Soc. Ent. Fr.* 1900: 289).

Weise, J. 1900d: [new taxa]. In: Bodemeyer, E. von.: *Quer durch Klein-Asien in den Bulghar-Dagh. Eine naturwissenschatliche Studien-Reise*. Emmendingen: Verlangs-Aktiengesellschaft vormals Dölter, v + 169 pp. DP: 5 November 1900 (Bousquet 2016).

Weise, J. 1901a: Beschreibungen africanischer Chrysomelidae nebst synonymischen Bemerkungen. *Deutsche Entomologische Zeitschrift* [1900] (2): 446–459. DP: [28] February 1901 (table of contents).

Weise, J. 1901b: Cassidinen aus Ceylon gesammelt von Dr. Horn. *Deutsche Entomologische Zeitschrift* 1901 (1): 49–56. DP: 10 July 1901 (*Bull. Soc. Ent. Fr.* 1901: 241).

Weise, J. 1901c: Ostafrikanische Criocerinen. *Archiv für Naturgeschichte* 67 [1] (2): 145–163. DP: [31] July 1901 (issue cover).

Weise, J. 1901d: Ein Beitrag zur Kenntniss von Paropsis Oliv. *Archiv für Naturgeschichte* 67 [1] (2): 164–174. DP: [31] July 1901 (issue cover).

Weise, J. 1902a: Nachtrag zum Verzeichnisse kleinasiatischer Coleopteren von Hrn. E. v. Bodemeyer, Freiburg i. Br. 1900. *Deutsche Entomologische Zeitschrift* 1901 (2): 203–204. DP: 22 January 1902 (*Bull. Soc. Ent. Fr.* 1902: 30).

Weise, J. 1902b: Afrikanische Hispinen. *Deutsche Entomologische Zeitschrift* 1901 (2): 225–240. DP: 22 January 1902 (*Bull. Soc. Ent. Fr.* 1902: 30).

Weise, J. 1902c: Chrysochloa alpestris var. vinariensis. *Deutsche Entomologische Zeitschrift* 1901 (2): 240. DP: 22 January 1902 (*Bull. Soc. Ent. Fr.* 1902: 30).

Weise, J. 1902d: Einige neue afrikanische Chrysomeliden. *Deutsche Entomologische Zeitschrift* 1901 (2): 301–310. DP: 22 January 1902 (*Bull. Soc. Ent. Fr.* 1902: 30).

Weise, J. 1902e: Afrikanische Chrysomeliden. *Archiv für Naturgeschichte* 68 [1] (2): 119–174. DP: [30] June 1902 (issue cover).

Weise, J. 1902f: Biologische Mittheilungen. *Deutsche Entomologische Zeitschrift* **1902** (1): 103–109. DP: 1 July 1902 (table of contents).

Weise, J. 1902g: Uebersicht der Eumela-Arten. *Deutsche Entomologische Zeitschrift* **1902** (1): 109–110. DP: 1 July 1902 (table of contents).

Weise, J. 1902h: Zwei neue Chrysomeliden aus Tonkin. *Deutsche Entomologische Zeitschrift* **1902** (2): 367–368. DP: [31] October 1902 (table of contents).

Weise, J. 1903a: Verzeichniss der von Dr. Horn auf Ceylon gesammelten Chrysomeliden I. Eupoda et Camptosomata. *Deutsche Entomologische Zeitschrift* **1903** (1): 17–34. DP: 25 February 1903 (*Bull. Soc. Ent. Fr.* 1903: 103).

Weise, J. 1903b: Afrikanische Galerucinen. *Deutsche Entomologische Zeitschrift* **1903** (1): 35–56, pl. 1. DP: 25 February 1903 (*Bull. Soc. Ent. Fr.* 1903: 103).

Weise, J. 1903c: Ueber die mit lichenis Richter verwandten Chrysomelen. *Deutsche Entomologische Zeitschrift* **1903** (1): 161–163. DP: 25 February 1903 (*Bull. Soc. Ent. Fr.* 1903: 103).

Weise, J. 1903d: Afrikanische Chrysomeliden. *Archiv für Naturgeschichte* **69** [1] (2): 197–226. DP: [31] May 1903 (issue cover).

Weise, J. 1904a: Pseudocolaspis substriata sp. n. *Deutsche Entomologische Zeitschrift* **1904** (1): 100. DP: [31] January 1904 (table of contents).

Weise, J. 1904b: Über bekannte und neue Chrysomeliden. *Archiv für Naturgeschichte* **70** [1] (2): 157–178. DP: [31] July 1904 (issue cover).

Weise, J. 1904c: [Haptoscelis melanocephala Panz. var. Baltica Ws.]. *Deutsche Entomologische Zeitschrift* **1904** (2): 368. DP: [31] July 1904 (table of contents).

Weise, J. 1904d: Einige neue Cassidinen und Hispinen. *Deutsche Entomologische Zeitschrift* **1904** (2): 433–452. DP: [31] July 1904 (table of contents).

Weise, J. 1904e: Beschreibung einiger Hispinen. *Archiv für Naturgeschichte* **71** [1905] [1] (1): 49–104. DP: [31] December 1904 (issue cover).

Weise, J. 1905a: Neue afrikanische Chrysomeliden und Coccinelliden. *Deutsche Entomologische Zeitschrift* **1905** (1): 33–54. DP: 22 February 1905 (*Bull. Soc. Ent. Fr.* 1905: 73).

Weise, J. 1905b: Protorina plagiata var. croatica m. *Deutsche Entomologische Zeitschrift* **1905** (1): 80. DP: 22 February 1905 (*Bull. Soc. Ent. Fr.* 1905: 73).

Weise, J. 1905c: Zweites Verzeichnis der Hispinen und Cassidinen aus Vorder-Indien. *Deutsche Entomologische Zeitschrift* **1905** (1): 113–129. DP: 22 February 1905 (*Bull. Soc. Ent. Fr.* 1905: 73).

Weise, J. 1905d: Chirida mimica n. sp. *Deutsche Entomologische Zeitschrift* **1905** (2): 210. DP: [30] September 1905 (table of contents).

Weise, J. 1905e: Ueber die Gattung Potaninia Ws. *Deutsche Entomologische Zeitschrift* **1905** (2): 216. DP: [30] September 1905 (table of contents).

Weise, J. 1905f: Bemerkungen über Hispinen. *Deutsche Entomologische Zeitschrift* **1905** (2): 317–320. DP: [30] September 1905 (table of contents).

Weise, J. 1905g: Ueber Chrysomeliden und Coccinelliden aus Rhodesia. *Deutsche Entomologische Zeitschrift* **1905** (2): 321–335. DP: [30] September 1905 (table of contents).

Weise, J. 1906a: Ostafrikanische Chrysomeliden und Coccinelliden. *Deutsche Entomologische Zeitschrift* **1906** (1): 35–64. DP: [28] February 1906 (table of contents).

Weise, J. 1906b: Zwei neue Hispiden aus Ostindien. *Deutsche Entomologische Zeitschrift* **1906** (2): 404. DP: 10 September 1906 (table of contents).

Weise, J. 1906c: Ergänzung der Beiträge zur Käferfauna von Kleinasien. *Deutsche Entomologische Zeitschrift* **1906** (2): 434. DP: 10 September 1906 (table of contents).

Weise, J. 1906d: Pachybrachis bodemeyeri. *Deutsche Entomologische Zeitschrift* **1906** (2): 472. DP: 10 September 1906 (table of contents).

Weise, J. 1906e: [new names]. In: Heyden, L. F. J. D. von, Reitter, E. & Weise, J.: *Catalogus Coleopterorum Europae, Caucasi et Armeniae Rossicae.* Berlin: Friedländer & Sohn; Paskau: Edmund Reitter; Caen: Revue d'Entomologie, [4] pp. + 750 columns + pp. 750–774. DP: [31 December] 1906 (book title page).

Weise, J. 1907: Kleine Mitteilungen und synonymische Bemerkungen. *Wiener Entomologische Zeitung* **26** (1): 13–15. DP: 1 January 1907 (table of contents).

Weise, J. 1908a: [new taxa]. In: Kolbe, H. J., Obst, P. & Weise, J.: Coleoptera. Pp. 82–96, pl. III. In: Filchner, W. (ed.): *Wissenschaftliche Ergebnisse der Expedition Filchner nach China und Tibet 1903–1905. X. Band. 1. Teil. Zoologische und botanische Sammlungen.* Berlin: Ernst Siegfried Mittler und Sohn, xii + 288 pp., 26 pls. DP: [31] January 1908 (*Natur. Novit.* 1908, 30 (1): 11).

Weise, J. 1908b: Coleopteren aus Ostindien. *Entomologische Zeitung* (Stettin) **69** (2): 213–230. DP: [31] May 1908 (last page of the issue).

Weise, J. 1908c: Description of new Cassididae of the Philippine Islands. *Philippine Journal of Science (A – General Science)* **3** (4): 259–260. DP: 16 November 1908 (*Contents and index of Philipp. J. Sci. 1906–1915*).

Weise, J. 1908d: Chrysomelidae und Coccinellidae. Pp. 1–13. In: Michaelsen, W. & Hartmeyer, R. (eds.): *Die Fauna Südwest-Australiens. Ergebnisse der Hamburger südwest-australischen Forschungsreise 1905. Band II. Lieferung 1–4.* [1907–1908]. Jena: Gustav Fischer, 68 pp. + 10 pls. DP: [31] December 1908 (*Natur. Novit.* 1908, 30 (24): 622).

Weise, J. 1909: 7. Coleoptera. 12. Chrysomelidae und Coccinellidae. Pp. 153–265, pl. 4. In: Sjöstedt, Y. (ed.): *Wissenschaftliche Ergebnisse der Schwedischen Zoologischen Expedition nach dem Kilimandjaro, dem Meru und den umgebenden Massaisteppen Deutsch-Ostafrikas 1905–1906 unter Leitung von Prof. Dr. Yngve Sjöstedt. 1. Band.* Stockholm: P. Palmquists Aktiebolag, 442 pp. DP: [31 December] 1909 (preprint cover).

Weise, J. 1910a: Chrysomelidae von Madagascar, den Comoren und den Inseln Ostafrikas. Pp. 419–506. In: Voeltzkow, A. (ed.): *Reise in Ostafrika in den Jahren 1903–1905 mit Mitteln der Hermann und Elise geb. Heckmann Wentzel-Stiftung ausgeführt. Wissenschaftliche Ergebnisse. Band II. Systematische Arbeiten. Heft V.* Stuttgart: E. Schweizerbartsche Verlagsbuchhandlung, Nägele & Dr. Sproesser, pp. [9] + 419–644, pls. 25–27. DP: [30] June 1910 (*Natur. Novit.* 1910, 32 (12): 317).

Weise, J. 1910b: Verzeichniss von Coleopteren aus den Philippinen, nebst zwei neuen Arten aus Niederländisch Ostindien. *Philippine Journal of Science (D – Ethnology, Anthropology, and General Biology*) 5 (2): 139–148. DP: 7 October 1910 (*Contents and index of Philipp. J. Sci. 1906–1915*).

Weise, J. 1910c: Chrysomeliden und Coccinelliden. *Verhandlungen des Naturforschenden Vereines in Brünn* 48 [1909]: 25–53. DP: [31 December] 1910 (volume title page).

Weise, J. 1911a: Neue Chrysomeliden. *Annales de la Société Entomologique de Belgique* **55** (6): 166–175. DP: 1 July 1911 (footer on p. 165).

Weise, J. 1911b: Chrysomelidae: Hispinae. In: Schenkling, S. (ed.): *Coleopterorum Catalogus. Pars 35*. Berlin: W. Junk, 94 pp. DP: 5 August 1911 (book cover).

Weise, J. 1912: Beitrag zur Kenntnis der Chrysomeliden. *Archiv für Naturgeschichte, Abteilung A* 78 (2): 76–98. DP: [30] June 1912 (issue cover).

Weise, J. 1913a: Synonymische Bemerkungen. *Wiener Entomologische Zeitung* **32** (1): 17–18. DP: 1 January 1913 (footer on p. 17).

Weise, J. 1913b: Chrysomeliden und Coccinelliden. Pp. 423–446. In: *Nova Guinea. Résultats de l'Expédition Scientifique Néerlandaise à la Nouvelle-Guinée en 1907 et 1909 sous les auspices de Dr. H. A. Lorentz. Vol. IX Zoologie*. [1909–1914], Leide: E. J. Brill, 666 pp., 16 pls. DP (Livraison III): [31] May 1913 (*Natur. Novit.* 1913, 35 (9): 273).

Weise, J. 1913c: Über Chrysomeliden und Coccinelliden der Philippinen: II. Teil (Coleoptera). *Philippine Journal of Science (D – General Biology, Ethnology, and Anthropology)* 8 (3): 215–242. DP: 8 July 1913 (*Contents and index of Philipp. J. Sci. 1906–1915*).

Weise, J. 1913d: Synonymische Mitteilungen. *Wiener Entomologische Zeitung* **32** (7/9): 218–220. DP: 15 July 1913 (footer on p. 219).

Weise, J. 1914: Coleoptera, I: Chrysomelidae und Coccinellidae. Pp. 251–275. In: Michaelsen, W. (ed.): *Beiträge zur Kenntnis der Land- und Süsswasserfauna Deutsch-Südwestafrikas. Ergebnisse der hamburger deutsch-südwestafrikanischen Studienreise 1911. Band I.* [1914–1916], Hamburg: L. Friederichsen & Co., iv + [map] + 452 pp. DP (Lieferung 2): [31 December] 1914 (*Inhaltverzeichnis* and the title page of each Lieferung).

Weise, J. 1915a: Lieferung 7: Chrysomelidae und Coccinellidae. Pp. 155–184. In: Schubotz, H. (ed.): *Ergebnisse der Zweiten Deutschen Zentral-Afrika-Expedition 1910–1911 unter Führung Adolf Friedrichs, Herzogs zu Mecklenburg. Band I. Zoologie, Teil I.* [1913–1917], Leipzig: Klinkhardt & Biermann, 597 pp, 27 pls. DP: 26 February 1915 (cover of Lieferung 7).

Weise, J. 1915b: Übersicht der Chrysomelini. *Deutsche Entomologische Zeitschrift* **1915** (4): 434–436. DP: 15 August 1915 (table of contents).

Weise, J. 1916a: Synonymische Mitteilungen. *Deutsche Entomologische Zeitschrift* **1916** (1): 37–41. DP: 1 April 1916 (table of contents).

Weise, J. 1916b: Chrysomelidae: 12. Chrysomelinae. In: Schenkling, S. (ed.): *Coleopterorum Catalogus. Pars 68.* Berlin, W. Junk, 255 pp. DP: 8 September 1916 (book cover).

Weise, J. 1919: Afrikanische Chrysomeliden und Coccinelliden. *Archiv für Naturgeschichte, Abteilung A* **83** [1917] (4): 174–207. DP: [30] June 1919 (issue cover).

Weise, J. 1921: XXVIII. Chrysomeliden und Coccinelliden. Pp. 2–5. In: Strand, E.: Neue Beiträge zur Arthropoden-Fauna Norwegens nebst gelegentlichen Bemerkungen über deutsche Arten.

XXVIII–XXX. *Archiv für Naturgeschichte, Abteilung A* 86 [1920] (12): 1–15. DP: [31] March 1921 (issue cover).

Weise, J. 1922a: Hispinen der alten Welt. *Philippine Journal of Science* 21 (1): 57–85. DP: 12 September 1922 (table of contents).

Weise, J. 1922b: Chrysomeliden der indo-malayischen Region. *Tijdschrift voor Entomologie* 65 (1/3): 39–130. DP: 1 October 1922 (issue cover).

Weise, J. 1922c: Chrysomeliden der Philippinen, III. *Philippine Journal of Science* 21 (5): 423–490. DP: 15 November 1922 (table of contents).

Weise, J. 1922d: Chrysomelidae. Pp. 154–157. In: Friedrichs, K.: Die von Dr. K. Friedrichs in Samoa und Indochina gesammelten Käfer. *Archiv für Naturgeschichte, Abteilung A* 88 (10): 147–159. DP: [31] December 1922 (issue cover).

Weise, J. 1923: Results of Dr. E. Mjöberg's Swedish Scientific Expedition to Australia 1910–1913. 31. Chrysomeliden und Coccinelliden aus Queensland. *Arkiv för Zoologi* 15 (12): 1–150. DP: 16 May 1923 (last page of the paper).

Weise, J. 1924a: Chrysomelidae: 13. Galerucinae. In: Schenkling, S. (ed.): *Coleopterorum Catalogus. Pars 78*. Berlin: W. Junk, 225 pp. DP: 30 May 1924 (book cover).

Weise, J. 1924b: Zoological results of the Swedish expedition to Central Africa 1921. Insecta. 7. Chrysomelidae und Coccinellidae. *Arkiv för Zoologi* 16 (22): 1–30. DP: 19 September 1924 (last page of the paper).

Wencker, J. A. 1866: [new taxon]. In: Wencker, J. A. & Silbermann, G. H. R.: *Catalogue des coléoptères d'Alsace et des Vosges. Suivi de descriptions de plusieurs espèces nouvelles par Ch. Brisout de Barneville et Wencker*. Strasbourg: G. Silbermann, vi + 142 pp. DP: [31 December] 1866 (book title page).

Westhoff, F. 1882: Die Käfer Westfalens. II. Abtheilung. *Verhandlungen des Naturhistorischen Vereins der Preussischen Rheinlande und Westfalens und des Regierungebezirks Osnabrück* (Bonn) 38 [1881–1882] (Supplement): xxviii + 141–323 + [1] pp. DP: [28] February 1882 (Bousquet 2016).

Westwood, J. O. 1838a: [Characters of new insects from Manilla collected by Mr. Cuming]. *Proceedings of the Zoological Society of London* 5 [1837] (59): 127–130. DP: 14 June 1838 (Duncan 1937).

Westwood, J. O. 1838b: Lupère. Luperus Geoffroy. L. a nez, L. Nasutus. Westwood. *Magasin de Zoologie* 7 [1837]: class 9, pl. 177: 1–2. DP: [31 December] 1838 (Dickinson et al. 2011).

Westwood, J. O. 1864: Descriptions of some new species of coleopterous insects belonging to the eupodous Phytophaga, natives of the Old World and Australia. *Transactions of the Entomological Society of London, Third Series* 2 [1864–1866] (3): 271–280. DP: 28 November 1864 (Wheeler 1912).

White, B. 1942: A new genus and species of Coleoptera (Chrysomelidae) from southwestern United States. *Entomological News* 53 (1): 16–21. DP: 23 January 1942 (verso of the volume title page).

White, R. E. 1981: Homonymy in world species-group names of Criocerinae (Coleoptera: Chrysomelidae). *United States Department of Agriculture, Technical Bulletin* **1629**: ii + 1–69. DP: 12 March 1981 (stamp on the BHL copy).

White, R. E. 1996: A revision of the genus Chaetocnema of America north of Mexico (Coleoptera: Chrysomelidae). *Contribution of the American Entomological Institute* **29** (1): 1–158. DP: [31 December] 1996 (issue cover).

Wiedemann, C. R. W. 1819: Neue Käfer aus Bengalen und Java. *Zoologisches Magazin* 1 (3): 157–183. DP: 2 May 1819 (*Allg. Verz. Bücher, Ostermesse* 1819: 142 and Evenhuis 2014).

Wiedemann, C. R. W. 1821: [new taxon]. In: Wiedemann, C. R. W. & Germar, E. F.: Neue exotische Käfer. *Magazin der Entomologie* **4**: 107–183. DP: [31 December] 1821 (see the comments in Dating of publications).

Wiedemann, C. R. W. 1823: Zweihundert neue Käfer von Java, Bengalen und dem Vorgebirge der Guten Hoffnung. *Zoologisches Magazin* **2** (1): 3–133. DP: 20 April 1823 (*Allg. Verz. Bücher, Ostermesse* 1823: 109 and Evenhuis 2014).

Wiel, P. van der 1956: Bijdrage tot de kennis der nederlandse kevers, IV (16e vervolg op het aanhangsel in Coleoptera Neerlandica III). *Tijdschrift voor Entomologie* **99** (1/2): 1–21. DP: 12 May 1956 (volume title page).

Wilcox, J. A. 1965: A synopsis of the North American Galerucinae (Coleoptera: Chrysomelidae). *Bulletin of the New York State Museum and Science Service* 400: iv + 1–226. DP: [31] August 1965 (volume cover).

Wilcox, J. A. 1971: Chrysomelidae: Galerucinae (Oidini, Galerucini, Metacyclini, Sermylini). In: Wilcox, J. A. (ed.): *Coleopterorum Catalogus Supplementa. Pars 78 (1), Editio seconda.* 's-Gravenhage: W. Junk, 1–220. DP: 15 June 1971 (issue title page).

Wilcox, J. A. 1972: A review of the North American chrysomeline leaf beetles (Coleoptera: Chrysomelidae). *Bulletin of the New York State Museum and Science Service* **421**: iv + 1–37. DP: [31] December 1972 (volume cover).

Wilcox, J. A. 1973: Chrysomelidae: Galerucinae (Luperini: Luperina). In: Wilcox, J. A. (ed.): *Coleopterorum Catalogus Supplementa. Pars 78 (3). Editio seconda.* 's-Gravenhage: W. Junk, 433–664 pp. DP: 20 August 1973 (issue title page).

Wilcox, J. A. 1975: Chrysomelidae: Galerucinae. Addenda et index. In: Wilcox, J. A. (ed.): *Coleopterorum Catalogus Supplementa. Pars 78 (4). Editio seconda.* 's-Gravenhage: W. Junk, 667–770 pp. DP: 20 June 1975 (issue title page).

Wingelmüller, A. 1937: Monographie der paläarktischen Arten der Tribus Cionini. *Koleopterologische Rundschau* **23** (4/5): 143–221. DP: 31 October 1937 (issue cover).

Winkelman, J. & Debreuil, M. 2008: Les Chrysomelinae de France (Coleoptera, Chrysomelidae). *Rutilans* **Supplément**: 1–188. DP: [30] September 2008 (colophon).

Winkler, A. 1929: Pars 10. Pp. 1137–1264. In: *Catalogus Coleopterorum regionis palaearcticae.* [1924–1932], Wien: Albert Winkler, vi + [2] + 1698 pp. DP: 19 October 1929 (*Ausgabe-Termine* on the verso of the book title page).

Winkler, A. 1930: Pars 11. Pp. 1265–1392. In: *Catalogus Coleopterorum regionis palaearcticae*. [1924–1932], Wien: Albert Winkler, vi + [2] + 1698 pp. DP: 11 May 1930 (*Ausgabe-Termine* on the verso of the book title page).

Wittmer, W. 1936: Neue Koleopteren aus Aegypten und dem Sinaï aus der Sammlung des Museums "Pietro Rossi", Duino. *Bulletin de la Société Royale Entomologique d'Egypte* 20: 16–18. DP: 28 December 1936 (verso of the volume title page).

Wollaston, T. V. 1854: *Insecta Maderensia; being an account of the insects of the islands of the Madeiran group*. London: J. Van Voorst, xliii + 634 pp., 13 pls. DP: 3 September 1854 (Bousquet 2016).

Wollaston, T. V. 1857: *Catalogue of the coleopterous insects of Madeira in the collection of the British Museum*. London: Taylor & Francis, xvi + 234 pp., 1 pl. DP: 7 September 1857 (Bousquet 2016).

Wollaston, T. V. 1860a: On the Halticidae of the Canary Islands. *Journal of Entomology* 1 [1860–1862] (1): 1–12. DP: [30] April 1860 (issue title page).

Wollaston, T. V. 1860b: On additions to the Madeiran Coleoptera. *Annals and Magazine of Natural History, Third Series* 5 (30): 448–459. DP: 1 June 1860 (Evenhuis 2003).

Wollaston, T. V. 1862a: On the Euphorbia-infesting Coleoptera of the Canary Islands. *Transactions of the Entomological Society of London, Third Series* 1 [1862–1864] (2): 136–189. DP: 29 May 1862 (Wheeler 1912).

Wollaston, T. V. 1862b: Brief diagnostic characters of new Canarian Coleoptera. *Annals and Magazine of Natural History, Third Series* 9 (54): 437–442. DP: 1 June 1862 (Evenhuis 2003).

Wollaston, T. V. 1864: *Catalogue of the coleopterous insects of the Canaries in the collection of the British Museum*. London: Trustees of the British Museum of Natural History, xiii + 684 pp. DP: 25 June 1864 (Sherborn 1934, Bousquet 2016).

Wollaston, T. V. 1865: *Coleoptera Atlantidum, being an enumeration of the coleopterous insects of the Madeiras, Salvages and Canaries*. London: John van Voorst, xlvii + 526 + 140 pp. [Appendix]. DP: 30 December 1865 (Bousquet 2016).

Wollaston, T. V. 1868: *Coleoptera Hesperidum, being an enumeration of the coleopterous insects of the Cape Verde Archipelago*. [1867], London: John van Voorst, xxxix + 285 pp. DP: 28 January 1868 (Williams 2017).

Wollaston, T. V. 1870: On the Coleoptera of St. Helena. *Annals and Magazine of Natural History, Fourth Series* 5 (25): 18–37. DP: 1 January 1870 (Evenhuis 2003).

Würmli, M. 1975: Ergebnisse der Bhutan-Expedition 1972 des Naturhistorisches Museum in Basel. Coleoptera: fam. Chrysomelidae subf. Hispinae. *Entomologica Basiliensia* 1: 371–381. DP: 1 October 1975 (table of contents).

Würmli, M. 1976a: Revision der Tribus Botryonopini (Coleoptera: Chrysomelidae: Hispinae). *Entomologische Arbeiten aus dem Museum G. Frey* 27: 70–99. DP: 1 December 1976 (table of contents).

Würmli, M. 1976b: Revision der Gattung Chaeridiona. *Entomologische Arbeiten aus dem Museum G. Frey* 27: 402–405. DP: 1 December 1976 (table of contents).

Xu, S.-Y. & Nie, R.-E. 2022: [new taxa]. In: Xu, S.-Y., Nie, R.-E. & Yang, X.-K.: Notes on spotted-elytron species of Gallerucida Motschulsky with the description of six new species from

China (Coleoptera, Chrysomelidae, Galerucinae). *ZooKeys* **1116**: 33–55. DP: 4 August 2022 (paper title page).

Xu, S.-Y. & Yang, X.-K. 2022: [new taxa]. In: Xu, S.-Y., Nie, R.-E. & Yang, X.-K.: Notes on spotted-elytron species of Gallerucida Motschulsky with the description of six new species from China (Coleoptera, Chrysomelidae, Galerucinae). *ZooKeys* **1116**: 33–55. DP: 4 August 2022 (paper title page).

Xu, Y., Bi, W.-X. & Liang, H.-B. 2021: New record of the genus Manipuria Jacoby (Chrysomelidae, Criocerinae) from China, with description of a new species. *ZooKeys* **1009**: 29–43 DP: 4 January 2021 (paper title page).

Xu, Y. & Liang, H.-B. 2024: Three new species and five new records within the genus Lilioceris (Coleoptera, Chrysomelidae, Criocerinae) from China. *ZooKeys* **1189**: 55–81. DP: 12 January 2024 (paper title page).

Yang, C.-K., Liao, C.-Q., Xu, J.-S., Liu, P., Staines, C. L. & Dai, X.-H. 2023: Field survey of Cassidinae beetles (Coleoptera, Chrysomelidae) and their host plants in southern Guangxi, China. *Biodiversity Data Journal* **11**: e107523. DP: 31 July 2023 (paper title page).

Yang, X.-K. 1991a: Study on the genus Paridea Baly from China (Chrysomelidae: Galerucinae). *Sinozoologia* **8**: 267–295. DP: [31] May 1991 (paper title page).

Yang, X.-K. 1991b: Notes on Chinese species of the genus Sermyloides Jacoby (Coleoptera, Chrysomelidae: Galerucinae). *Sinozoologia* **8**: 297–305. DP: [31] May 1991 (paper title page).

Yang, X.-K. 1991c: Study on the genus Pseudocophora Jacoby from China (Coleoptera, Chrysomelidae: Galerucinae). *Sinozoologia* **8**: 307–312. DP: [31] May 1991 (paper title page).

Yang, X.-K. 1991d: Sermyloides pilifera, a new species from China (Coleoptera: Chrysomelidae, Galerucinae). *Acta Entomologica Sinica* **34** (3): 349–351. DP: [31] August 1991 (paper title page).

Yang, X.-K. 1991e: Review on the genus Cneoranidea (Chrysomelidae: Galerucinae). *Scientific Treatise on Systematic and Evolutionary Zoology* **1**: 199–206 DP: [31] August 1991 (colophon).

Yang, X.-K. 1992a: The Chinese species of the genus Strobiderus (Coleoptera: Chrysomelidae, Galerucinae). *Sinozoologia* **9**: 187–189 DP: [30] April 1991 (paper title page).

Yang, X.-K. 1992b: Chrysomelidae Galerucinae. Pp. 552–589. In: Peng, J.-W. & Liu, Y.-Q. (eds.): *Iconography of forest insects in Hunan China*. Beijing: Academia Sinica & Hunan Forestry Institute, 4 + 60 + 1473 pp. DP: [30] April 1992 (colophon).

Yang, X.-K. 1992c: The Chinese known species of the genus Macrima and description of one new species (Coleoptera: Chrysomelidae, Galerucinae). *Zoological Research* **13** (3): 257–261 DP: [30] May 1992 (journal website).

Yang, X.-K. 1992d: Galerucid beetles of Mount Mogan and descriptions of two new species (Coleoptera: Chrysomelidae, Galerucinae). *Journal of Zhejiang Forestry College* **9** (4): 409–413. DP: [31 December] 1992 (paper title page).

Yang, X.-K. 1993a: Coleoptera: Chrysomelidae – Galerucinae. Pp. 331–346. In: Huang, F.-S. (ed.): *Insects of Wuling Mountains area, southwestern China*. Beijing: Science Press, x + 777 pp. DP: [30] April 1993 (colophon).

Yang, X.-K. 1993b: Three new species of the genus Paridea from China (Coleoptera: Chrysomelidae: Galerucinae). *Acta Zootaxonomica Sinica* 18 (2): 196–200. DP: [30] April 1993 (paper title page).

Yang, X.-K. 1993c: Study on the genus Laphris and its phylogenetic relationship with related genera (Coleoptera: Chrysomelidae: Galerucinae). *Acta Zootaxonomica Sinica* 18 (3): 362–369. DP: [31] July 1993 (paper title page).

Yang, X.-K. 1993d: Notes on the genus Fleutiauxia Laboissiere (Coleoptera: Chrysomelidae: Galerucinae). *Entomotaxonomia* 15 (3): 219–227. DP: [30] September 1993 (paper title page).

Yang, X.-K. 1993e: A new species of the genus Sermyloides (Coleoptera: Chrysomelidae: Galerucinae). *Acta Zootaxonomica Sinica* 18 (4): 456–458. DP: [31] October 1993 (paper title page).

Yang, X.-K. 1994a: Study on the genus Gallerucida (I). Species of which the elytra with black punctures (Coleoptera: Chrysomelidae: Galerucinae). *Acta Zootaxonomica Sinica* 19 (2): 202–205. DP: [30] April 1994 (paper title page).

Yang, X.-K. 1994b: Study on the genus Gallerucida (II). Descriptions of five new species (Coleoptera: Chrysomelidae: Galerucinae). *Acta Zootaxonomica Sinica* 19 (3): 343–350. DP: [31] July 1994 (paper title page).

Yang, X.-K. 1995a: Studies on the subfamily Galerucinae I. Complementary description of the genus Aplosonyx and the descriptions of two new species (Coleoptera: Chrysomelidae). *Acta Zootaxonomica Sinica* 20 (1): 90–94. DP: [31] January 1995 (paper title page).

Yang, X.-K. 1995b: [new taxon]. In: Yang, X.-K., Li, W.-Z., Yang, J., Yu, Y.-X. & Cao, Z.-P.: Sastracella Jacoby, the first record from China, and description of one new species (Coleoptera: Chrysomelidae). *Sinozoologia* 12: 210–214. DP: [31] July 1995 (paper title page).

Yang, X.-K. 1995c: Coleoptera: Chrysomelidae – Galerucinae. Pp. 259–263. In: Wu, H. (ed.): *Insects of Baishanzu Mountain, Eastern China*. Beijing: China Forestry Publishing House, xiii + 586 pp. DP: [31] December 1995 (colophon).

Yang, X.-K. 1996: Revision of the genus Arthrotidea Chen (Coleoptera: Chrysomelidae: Galerucinae). *Bulletin of the National Museum of Natural Science* (Taichung) 8: 29–34. DP: [31] December 1996 (volume cover).

Yang, X.-K. 1997: [new taxa]. In: Yang, X.-K., Li, W.-Z., Zhang, B.-Q. & Xiang, Z.-Q.: Coleoptera: Chrysomelidae: Galerucinae. Pp. 863–904. In: Yang, X.-K. (ed.): *Insects of the Three Gorge Reservoir area of Yangtze river. Part 1*. Chongqing: Chongqing Publishing House, xx + 974 pp. DP: [31] January 1997 (colophon).

Yang, X.-K. 2001: [new taxon]. In: Yang, X.-K., Ge, S.-Q. & Li, W.-Z.: Revision of the genus Agetocera Hope (Coleoptera: Chrysomelidae: Galerucinae). *Oriental Insects* 35: 105–154. DP: 31 July 2001 (volume cover).

Yang, X.-K. 2004: [new taxa]. In: Yang, X.-K. & Li, W.-Z.: Coleoptera: Chrysomelidae-Galerucinae. Pp. 73–77. In: Yang, X.-K. (ed.): *Insects of the Great Yarlung Zangbo Canyon of Xizang, China*. Beijing: China Science and Technology Press, [12] + 339 pp., 12 pls. DP: [30] June 2004 (colophon).

Yang, X.-K. 2010: [new name]. In: Xue, H.-J. & Yang, X.-K.: Species catalogue of Pyrrhalta Joannis (Coleoptera: Chrysomelidae: Galerucinae) of the world. *Entomotaxonomia* 32 (supplement): 119–136. DP: 10 August 2010 (paper title page).

Yang, X.-K. 2021: [new taxon]. In: Fan, C.-X., Lin, Z.-B., Yu, Y.-L., Li, Z.-Q., Xu, S.-Y., Yang, X.-K. & Nie, R.-E.: One new species in the leaf-beetle genus Sphenoraia Clark (Coleoptera: Chrysomelidae: Galerucinae) from the Haizhu Wetland of Guangzhou, China. *Entomotaxonomia* 43 (4): 245–249. DP: 16 October 2021 (footer on the paper title page).

Yang, X.-K., Ge, S.-Q., Nie, R.-E., Ruan, Y.-Y. & Li, W.-Z. (eds.) 2015: *Chinese leaf beetles*. Beijing: Science Press, [14] + 507 pp., 83 pls. DP: [31] December 2015 (Rui-E Nie 2022, pers. comm.).

Yang, X.-K., Ge, S.-Q., Wang, S.-Y., Li, W.-Z. & Cui, J.-Z. 2014: *Coleoptera Chrysomelidae Chrysomelinae. Fauna Sinica. Insecta Vol. 61*. Beijing: Science Press, xvi + 641 pp., 8 pls. DP: [30] June 2014 (colophon).

Yang, X.-K. & Li, W.-Z. 1997a: [new taxon]. In: Yang, X.-K., Li, W.-Z. & Yao, J.: The galerucine beetles of Xishuangbanna, Yunnan province (Coleoptera: Chrysomelidae). *Acta Zootaxonomica Sinica* 22 (4): 384–391. DP: [31] October 1997 (paper title page).

Yang, X.-K. & Li, W.-Z. 1997b: [new taxa]. In: Yang, X.-K., Li, W.-Z., Zhang, B.-Q. & Xiang, Z.-Q.: Coleoptera: Chrysomelidae: Galerucinae. Pp. 863–904. In: Yang, X.-K. (ed.): *Insects of the Three Gorge Reservoir area of Yangtze river. Part 1*. Chongqing: Chongqing Publishing House, xx + 974 pp. DP: [31] January 1997 (colophon).

Yang, X.-K. & Li, W.-Z. 1998: Coleoptera: Chrysomelidae: Galerucinae. Pp. 128–135. In: Wu, H. (ed.): *Insects of Longwangshan Nature Reserve*. Beijing: China Forestry Publishing House, [8] + 404 pp. DP: [31] October 1998 (colophon).

Yang, X.-K. & Li, W.-Z. 2002: [new taxon]. In: Li, W.-Z. & Yang, X.-K.: Chrysomelidae: Galerucinae. Pp. 299–309. In: Li, W.-Z. & Jin, D. (eds.): *Insects from Maolan Landscape*. Guiyang: Guizhou Science and Technology Publishing house, x + 615 pp. DP: [30] June 2002 (colophon).

Yang, X.-K. & Nie, R.-E. 2015: [new taxon]. In: Yang, X.-K., Ge, S.-Q., Nie, R.-E., Ruan, Y.-Y. & Li, W.-Z. (eds.): *Chinese leaf beetles*. Beijing: Science Press, [14] + 507 pp., 83 pls. DP: [31] December 2015 (Rui-E Nie 2022, pers. comm.).

Yang, X.-K. & Wang, J.-S. 1998: [new taxa]. In: Yang, X.-K., Wang, J.-S. & Wu, Y.-Y.: Three new species of Galerucinae (Coleoptera: Chrysomelidae) from Wuyi Mountain. *Entomotaxonomia* 20 (4): 261–264. DP: [31] December 1998 (paper title page).

Yang, X.-K. & Wu, Y.-Y. 1998a: [new taxon]. In: Wu, Y.-Y., Yang, X.-K. & Li, W.-Z.: A new species of Galerucinae (Coleoptera: Chrysomelidae) from Wuyi Mountains. *Entomotaxonomia* 20 (3): 182–184. DP: [30] September 1998 (paper title page).

Yang, X.-K. & Wu, Y.-Y. 1998b: [new taxon]. In: Yang, X.-K., Wang, J.-S. & Wu, Y.-Y.: Three new species of Galerucinae (Coleoptera: Chrysomelidae) from Wuyi Mountain. *Entomotaxonomia* 20 (4): 261–264. DP: [31] December 1998 (paper title page).

Yang, X.-K. & Yao, J. 1997: [new taxon]. In: Yang, X.-K., Li, W.-Z. & Yao, J.: The galerucine beetles of Xishuangbanna, Yunnan province (Coleoptera: Chrysomelidae). *Acta Zootaxonomica Sinica* 22 (4): 384–391. DP: [31] October 1997 (paper title page).

Yasutomi, K. 1952: Contributions to the knowledge of Japanese Cassidinae. *Kontyû* **19** [1951–1952] (3/4): 93–99. DP: 30 March 1952 (issue cover).

Yu, P.-Y. 1985: Four new species of Hispidae. *Entomotaxonomia* 7 (1): 1–5. DP: [31] March 1985 (paper title page).

Yu, P.-Y. 1992a: Crioceridae. Pp. 526–532. In: Peng, J.-W. & Liu, Y.-Q. (eds.): *Iconography of forest insects in Hunan China*. Beijing: Academia Sinica & Hunan Forestry Institute, 4 + 60 + 1473 pp. DP: [30] April 1992 (colophon).

Yu, P.-Y. 1992b: [new taxon]. In: Tan, J.-J. & Yu, P.-Y.: Coleoptera: Hispidae: Hispinae. Pp. 607–610. In: Peng, J.-W. & Liu, Y.-Q. (eds.): *Iconography of forest insects in Hunan China*. Beijing: Academia Sinica & Hunan Forestry Institute, 4 + 60 + 1473 pp. DP: [30] April 1992 (colophon).

Yu, P.-Y. 1993: [new taxon]. In: Wang, S.-Y. & Yu, P.-Y.: Coleoptera: Chrysomelidae – Alticinae. Pp. 315–330. In: Huang, F.-S. (ed.): *Insects of Wuling Mountains area, southwestern China*. Beijing: Science Press, x + 777 pp. DP: [30] April 1993 (colophon).

Yu, P.-Y. 1997a: Coleoptera: Crioceridae. Pp. 837–845. In: Yang, X.-K. (ed.): *Insects of the Three Gorge Reservoir area of Yangtze river. Part 1*. Chongqing: Chongqing Publishing House, xx + 974 pp. DP: [31] January 1997 (colophon).

Yu, P.-Y. 1997b: [new taxon]. In: Wang, S.-Y. & Yu, P.-Y.: Coleoptera: Chrysomelidae: Alticinae. Pp. 905–931. In: Yang, X.-K. (ed.): *Insects of the Three Gorge Reservoir area of Yangtze river. Part 1*. Chongqing: Chongqing Publishing House, xx + 974 pp. DP: [31] January 1997 (colophon).

Yu, P.-Y. 2002: Coleoptera: Hispidae: Cassidinae. Pp. 463–467. In: Huang, F.-S. (ed.): *Forest insects of Hainan*. Beijing, National Science Foundation of China, xv + 1064 pp. DP: [31] July 2002 (colophon).

Yu, P.-Y. & Huang, H. 2002: Coleoptera: Hispidae: Anisoderinae. Pp. 455–457. In: Huang, F.-S. (ed.): *Forest insects of Hainan*. Beijing: National Science Foundation of China, xv + 1064 pp. DP: [31] July 2002 (colophon).

Yu, P.-Y. & Kung, Y.-T. 1961: [new taxon]. In: Chen, S.-H., Tan, C.-C. [= J.-J.] & Yu, P.-Y.: Results of the zoologico-botanical expedition to Southwest China 1955–57 (Coleoptera, Hispinae I). *Acta Entomologica Sinica* 10 (4/6): 457–481. DP: [31] December 1961 (paper title page).

Yu, P.-Y. & Li, W.-Z. 2002: Coleoptera: Hispidae: Callispinae. Pp. 461–462. In: Huang, F.-S. (ed.): *Forest insects of Hainan*. Beijing: National Science Foundation of China, xv + 1064 pp. DP: [31] July 2002 (colophon).

Yuasa, H. 1930: Two new species of eumolpid-beetles noxious to the mulberry-tree in the Liu-kiu Islands. *Proceedings of the Imperial Academy* (Tokyo) **6** (7): 293–295. DP: [31] July 1930 (issue title page).

Yuasa, H. 1938: The asparagus beetles in Nippon. *Kontyû* **12** (2): 37–42. DP: [31] March 1938 (issue title page).

Yuasa, H. 1939: Notes on the Chrysomelidae in Nippon (IV). On Lema diversa and L. lewisii. *Kontyû* **13** (4): 161–164. DP: [30] September 1939 (issue cover).

Yus Ramos, R. 1978: Contribución al conocimiento de los brúquidos (Col. Bruchidae) del Mediterráneo occidental: I. Notas taxonómicas sobre el género Bruchus L. *Nouvelle Revue d'Entomologie* 8 (3): 315–320. DP: 15 November 1978 (issue cover).

Yus Ramos, R. 2007a: Las especies de Bruchidius Schilsky del grupo serraticornis: revisión de la fauna Ibero-Balear (Coleoptera: Bruchidae). *Boletín de la Sociedad Entomológica Aragonesa* **41**: 321–333. DP: 31 October 2007 (colophon).

Yus Ramos, R. 2007b: Bruchidius pardoi, una nueva especie de brúquido en el sur de la Península Ibérica (Coleoptera: Bruchidae). *Boletín de la Asociación Española de Entomología* **31** (3/4): 37–50. DP: 15 December 2007 (issue title page).

Yus Ramos, R. 2008: Catálogo comentado de los brúquidos de las islas Canarias (Coleoptera: Bruchidae). *Vieraea* **36**: 29–64. DP: [31] October 2008 (paper title page).

Yus Ramos, R. 2009: Revisión del género Megabruchidius Borowiec, 1984 (Coleoptera: Bruchidae) y nuevas citas para la fauna Europea. *Boletín de la Sociedad Entomológica Aragonesa* **45**: 371–382. DP: 30 November 2009 (colophon).

Yus Ramos, R. 2010a: Correcciones al Catálogo de Coleópteros Bruchinae Paleárticos de Löbl & Smetana (2010) (Coleoptera: Bruchidae). *Boletín de la Asociación Española de Entomología* **34** (1/2): 219–234. DP: 21 June 2010 (paper title page).

Yus Ramos, R. 2010b: Caryedon omanensis nov. sp., una nueva especie para la fauna de brúquidos del Sultanato de Omán, Península Arábiga (Coleoptera: Bruchidae). *Boletín de la Asociación Española de Entomología* **34** (3/4): 257–265. DP: [31] December 2010 (issue cover).

Yus Ramos, R. 2010c: Estado actual de conocimiento sobre los brúquidos (Coleoptera: Bruchidae) de Marruecos. Nuevas citas. *Boletín de la Sociedad Entomológica Aragonesa* **47**: 273–392. DP: 31 December 2010 (colophon).

Yus Ramos, R. 2012a: Novedades de la fauna ibero-balear de brúquidos (Coleoptera, Bruchidae) a incluir en el Catálogo de Coleópteros Paleárticos de Löbl & Smetana (2010). *Boletín de la Asociación Española de Entomología* **36** (3/4): 335–350. DP: 22 December 2012 (paper title page).

Yus Ramos, R. 2012b: *Los brúquidos (Coleoptera: Bruchidae) de la Macaronesia*. La Laguna (Tenerife), 245 pp. DP: 30 June 2012 (*Bol. Soc. Ent. Arag.* 2012, 50 (1): 582).

Yus Ramos, R. 2013a: Serratobruchidius, un género nuevo de Bruchidiini para los Bruchidius Schilsky (s. l.) del grupo serraticornis (Coleoptera: Bruchidae). *Boletín de la Sociedad Entomológica Aragonesa* **52**: 17–21. DP: 30 June 2013 (paper title page).

Yus Ramos, R. 2013b: Bruchidius gavirai nov. sp., un brúquido depredador de semillas de Cistáceas en el sur de la Península Ibérica (Coleoptera: Bruchidae). *Boletín de la Sociedad Entomológica Aragonesa* **53**: 57–62. DP: 31 December 2013 (paper title page).

Yus Ramos, R. 2018: Sobre la validez del género Butiobruchus Prevett, 1966 y propuesta de nueva caracterización (Coleoptera: Bruchidae). *Boletín de la Sociedad Entomológica Aragonesa* **62**: 249–252. DP: 30 June 2018 (paper title page).

Yus Ramos, R. 2019: Lista actualizada de especies de brúquidos (Coleoptera: Bruchidae) de la Península Ibérica, Islas Baleares e Islas Canarias. *Boletín de la Sociedad Entomológica Aragonesa* **65**: 1–19. DP: 31 December 2019 (paper title page).

Yus Ramos, R. 2021: Revisión de los Bruchini Bridwell, 1946 (Coleoptera: Bruchidae) de la Península Ibérica, Islas Baleares e Islas Canarias. *Boletín de la Sociedad Entomológica Aragonesa* **68**: 86–142. DP: 30 June 2021 (paper title page).

Yus Ramos, R. & Angelini, F. 2017: Contribucion al conocimiento de los brúquidos (Coleoptera: Bruchidae) de Turquía. *Boletín de la Sociedad Entomológica Aragonesa* **61**: 67–74. DP: 31 December 2017 (paper title page).

Yus Ramos, R. & Angelini, F. 2018: Contribución al conocimiento de los brúquidos (Coleoptera, Bruchidae) de Grecia. *Boletín de la Asociación Española de Entomología* **42** (3/4): 351–389. DP: 26.xii.2018 (footers throughout the paper).

Yus Ramos, R. & François, A. 2015: Confirmación de la validez taxonómica de Bruchidius leprieuri (Jacquet, 1886) (Coleoptera: Bruchidae). *Boletín de la Asociación Española de Entomología* **39** (1/2): 221–224. DP: 3 June 2015 (last page of the paper).

Yus Ramos, R. & García Becerra, R. 2019: Primer registro de un brúquido (Coleoptera: Bruchidae) para Europa continentale e Islas Canarias: Acanthoscelides macrophthalmus (Schaeffer, 1907). *Boletín de la Sociedad Entomológica Aragonesa* **65**: 185–189. DP: 31 December 2019 (paper title page).

Yus Ramos R. & Hadulla, K. 2019: Algunos registros nuevos de brúquidos para tres países de la región paleártica occidental (Coleoptera: Bruchidae). *Boletín de la Asociación Española de Entomología* **43** (1/2): 143–146. DP: 12 June 2019 (last page of the paper).

Yus Ramos, R. & Labrique, H. 2015: Catalogue des coléoptères du Maroc. Volume 1 (Coleoptera: Bruchidae). *Mémoires de la Société Linnéenne de Lyon* **6**: 1–142. DP: [31] October 2015 (last page of the cover).

Yus Ramos, R., Ventura, D., Bensusan, K., Coello-García, P., György, Z. & Stojanova, A. 2014: Alien seed beetles (Coleoptera: Chrysomelidae: Bruchinae) in Europe. *Zootaxa* **3826** (3): 401–448. DP: 1 July 2014 (footer on the paper title page).

Zabarinskiy, P. A. 1891: O nekotorykh vrednykh nasekomykh 1891 g. [On some insect pests from the year 1891]. *Zapiski Imperatorskago Obshchestva Sel'skago Khozyaystva Yuzhnoy Rossii* (Odessa) **1891** (12): 54–63. DP: [31] December 1891 (issue title page).

Zacher, F. 1930: Untersuchungen zur Morphologie und Biologie der Samenkäfer (Bruchidae-Larridae). Beiträge zur Kenntnis der Vorratsschädlinge. 6. Beitrag. *Arbeiten aus der Biologischen Reichsanstalt für Land- und Forstwirtschaft* (Berlin-Dahlem) **18** (3): 233–384. DP: [31] October 1930 (issue title page).

Zacher, F. 1949: Ein südamerikanischer Samenkäfer, Megacerus eulophus Er., in Hamburg. *Entomon, Internationale Zeitschrift für die gesamte Insektenkunde* **1** (6): 138–139. DP: [30] June 1949 (issue table of contents).

Zacher, F. 1952: Die Nährpflanzen der Samenkäfer. *Zeitschrift für Angewandte Entomologie* **33** (3): 460–480. DP: [30] April 1952 (issue cover).

Zampetti, M. F. 1979a: Bruchidius osellai n. sp. del Kashmir (Coleoptera Bruchidae). *Bollettino del Museo Civico di Storia Naturale Verona* 5 [1978]: 391–394. DP: 31 March 1979 (table of contents).

Zampetti, M. F. 1979b: Contributo alla conoscenza dei Bruchidae del Libano (Coleoptera). *Fragmenta Entomologica* **15** [1979–1980] (1): 201–207. DP: 30 September 1979 (table of contents in the cover).

Zampetti, M. F. 1982: Posizione sistematica di alcune specie appartenenti al genere Bruchidius Schilsky (Coleoptera, Bruchidae). *Bollettino del Museo Civico di Storia Naturale Verona* 8 [1981]: 383–410. DP: 31 December 1982 (table of contents).

Zampetti, M. F. 1984: Contributo alla conoscenza dei Bruchidi di Turchia. II. (Coleoptera, Bruchidae). *Fragmenta Entomologica* 17 (2): 395–406. DP: 15 October 1984 (table of contents in the cover).

Zampetti, M. F. 1992: Bruchidius kieneri – nuova specie di bruchide di Israele (Coleoptera, Bruchidae). *Lambillionea* **92** (3): 248–250. DP: 30 September 1992 (paper title page).

Zampetti, M. F. 1993: Una nova specie di Bruchus del Libano (Coleoptera, Bruchidae). *Fragmenta Entomologica* **24** [1992] (2): 215–218. DP: 30 April 1993 (table of contents in the cover).

Zampetti, M. F. 1998: Un nuovo Bruchidius del Parco Nazionale del Circeo (Coleoptera, Bruchidae). *Bollettino dell'Associazione Romana di Entomologia* **52** [1997] (1/4): 55–58. DP: 15 April 1998 (table of contents).

Zampetti, M. F. & Ricci, M. S. 2012: *Guida ai coleotteri bruchidi della fauna Italiana. Sistematica e biologia. Gestione e controllo*. Roma: Pandion – Darwin Edizioni, 430 pp. DP: [31 December] 2012 (publisher website).

Zampetti, M. F. & Toma, L. 2020: Description of seven new species of seed beetles (Coleoptera Bruchidae) from Mediterranean basin and Africa. *Biodiversity Journal* **11** (2): 553–564. DP: 30 June 2020 (paper title page).

Zaytsev, F. A. 1930: K rasprostraneniyu na Kavkaze vidov triby Donaciini (Coleoptera, Chrysomelidae). (Zur Verbreitung der Donaciini in den Kaukasus-Ländern). *Bulletin du Musée de Georgie* 5 [1928]: 105–114. DP: [31 December] 1930 (volume title page).

Zaytsev, F. A. 1938: Rasprostranenie na Kavkaze predstaviteley podsem. Hispini i Cassidini (Coleoptera, Chrysomelidae). (The distribution of the species of the subfamilies Hispini and Cassidini (Coleoptera, Chrysomelidae) in Caucasus). *Trudy Zoologicheskogo Sektora Gruzinskogo Filiala Akademii Nauk SSSR* **2**: 95–107. SFP: 15 April 1938, DP: 17 October 1938 (registered at ZIN library).

Zaytsev, Yu. M., Medvedev, L. N. & Muravitsky, O. S. 2013: K poznaniyu lichinok evropeyskikh vidov zhukov-listoedov roda Cryptocephalus (Coleoptera, Chrysomelidae). (Notes on leaf-beetle larvae of European species of the genus Cryptocephalus (Coleoptera, Chrysomelidae)). *Euroasian Entomological Journal* **12** (4): 361–372. DP: 9 September 2013 (journal website).

Zebe, G. 1855: Cryptocephalus saliceti Zebe. nov. spec. *Entomologische Zeitung* (Stettin) **16** (1): 28–29. DP: [31] January 1855 (issue title page).

Zerche, L. 2003: Neue und alte Deliphrosoma-Arten aus der Kaukasus-Region und neue Funde anderer Deliphrosoma-Arten (Col., Staphylinidae, Omaliinae). *Entomologische Nachrichten und Berichte* 47 (2): 81–90. DP: 20 August 2003 (verso of the issue cover).

Zetterstedt, J. W. 1828: *Fauna insectorum Lapponica. Pars I.* Hammone: Schulz, Hammone, xx + [1] + 563 pp. DP: [31 December] 1828 (Bousquet 2016).

Zetterstedt, J. W. 1837: [Hefte 2. Columns 193–348]. In: *Insecta Lapponica.* [1837–1840], Lipsiae: Leopoldi Voss, vi + [2] pp. + 1140 columns. DP: 22 December 1837 (Bousquet 2016).

Zhang, H., Song, N. & Yin, X., 2022: Higher-level phylogeny of Chrysomelidae based on expanded sampling of mitogenomes. *PLoS ONE* 17 (1): e0258587. DP: 21 January 2022 (paper title page).

Zhang, L.-J., Li, W.-Z. & Yang, X.-K. 2008: Two new species of the genus Liroetis Weise, 1889, from China (Coleoptera: Chrysomelidae: Galerucinae). *Pan-Pacific Entomologist* 84 (1): 22–25. DP: 12 June 2008 (last page of the paper).

Zhang, L.-J. & Yang, X. K. 2004: A review of the genus Paragetocera Laboissiere (Coleoptera: Chrysomelidae: Galerucinae). *Oriental Insects* 38: 289–302. DP: 1 May 2004 (volume cover).

Zhang, L.-J. & Yang, X.-K. 2005a: A new species of the genus Agetocera Hope (Coleoptera: Chrysomelidae: Galerucinae) from China. *Proceedings of the Entomological Society of Washington* 107 (1): 119–122. DP: 10 January 2005 (verso of the issue cover).

Zhang, L.-J. & Yang, X.-K. 2005b: A new species of the genus Cneoranidea Chen, 1942 from China with a key to the known species (Coleoptera: Chrysomelidae: Galerucinae). *Pan-Pacific Entomologist* 81 (1/2): 47–53. DP: 29 September 2005 (verso of the issue cover).

Zhang, L.-J. & Yang, X.-K. 2006: [new taxon]. In: Zhang, L.-J., Yang, X.-K., Cui, J. Z. & Li, W.-Z. 2006: A key to the genus Mimastra Baly (Coleoptera: Chrysomelidae: Galerucinae) from China, with the description of a new species. *Entomological News* 117 (2): 203–210. DP: 17 May 2006 (footer on the paper title page).

Zhang, L.-J. & Yang, X.-K. 2008: [new taxon]. In: Zhang, L.-J., Li, W.-Z. & Yang, X.-K.: A new species of Siemssenius Weise (Coleoptera: Chrysomelidae: Galerucinae) from China, and a key to the known species. *Proceedings of the Entomological Society of Washington* 110 (1): 126–129. DP: 7 January 2008 (verso of the issue cover).

Zhang, M.-N., Ruan, Y.-Y., Bai, M., Chen, X.-Q., Li, L.-X., Yang, X.-K., Meng, Z.-Y., Liu, Y. & Du, X.-Y. 2023: Geometric morphometric analysis of genus Chaetocnema (Coleoptera: Chrysomelidae: Alticini) with insights on its subgenera classification and morphological diversity. *Diversity* 15 (8): 1–17. DP: 8 August 2023 (paper title page).

Zhang, S.-F. & Liu, Y.-P. 1987: Notes on a new species of the genus Sulcobruchus from China (Coleoptera: Bruchidae). *Entomotaxonomia* 9 (3): 191–193. DP: [30] September 1987 (paper title page).

Zhang, Y., Wang, S.-Y. & Yang, X.-K. 2006: A new species of the genus Altica Geoffroy (Coleoptera, Chrysomelidae, Alticinae) from China. *Acta Zootaxonomica Sinica* 31 (4): 855–858. DP: [31] October 2006 (paper title page).

Zhang, Y. & Yang, X.-K. 2006: A study on the flea beetle genus Euphitrea Baly from China, with the description of five new species (Coleoptera: Chrysomelidae: Alticinae). *Proceedings of the Entomological Society of Washington* **108** (4): 842–859. DP: 12 October 2006 (verso of the issue cover).

Zhang, Y. & Yang, X.-K. 2007: A key to species of Nisotra Baly (Coleoptera: Chrysomelidae: Alticinae) from China, with description of a new species and two new records for China. *Proceedings of the Entomological Society of Washington* **109** (4): 841–846. DP: 5 October 2007 (verso of the issue cover).

Zhou, H.-Z. & Tan, J.-J. 1995: Four new species of Chlamisus Rafinesque from Fujian Province, China. (Chlamisinae: Eumolpidae: Coleoptera). *Entomologia Sinica* **2** (3): 228–235. DP: 1 September 1995 (journal website).

Zhou, H.-Z. & Tan, J.-J. 1997: Faunistic notes and three new species of Eumolpidae from Fujian province, China (Coleoptera). *Acta Zootaxonomica Sinica* **22** (2): 189–193. DP: [30] April 1997 (paper title page).

Zimsen, E. 1964: *The type material of I. C. Fabricius*. Copenhagen: Munksgaard, 656 pp. DP: [31 December] 1964 (colophon).

Zoia, S. 2007: A revision of the Pachnephorus from the Afrotropical Region (Coleoptera, Chrysomelidae). *Fragmenta Entomologica* **39** (1): 1–156. DP: 30 October 2007 (last page of the cover).

Zoia, S. 2012: Eumolpinae (Coleoptera: Chrysomelidae) of Socotra Island. Pp. 449–501. In: Hájek, J. & Bezděk, J. (eds.): Insect biodiversity of the Socotra Archipelago. *Acta Entomologica Musei Nationalis Pragae* **52** (supplementum 2): i–vi + 1–557. DP: 17 December 2012 (colophon and paper title page).

Zoia, S. 2014: The Mediterranean Colaspidea (Coleoptera Chrysomelidae Eumolpinae). *Entomologia* **2** (1): 1–34. DP: 11 April 2014 (paper title page).

Zoia, S. 2017: The Macrocoma Chapuis from the Canary Islands, with description of a new species (Coleoptera Chrysomelidae Eumolpinae). *Bollettino della Società Entomologica Italiana* **149** (2): 75–92. DP: 31 August 2017 (paper title page).

Zschach J. J. 1788: *Museum N. G. Leskeanum. Pars entomologica ad systema entomologiae Cl. Fabricii ordinata*. Lipsiae: J. G. Müller, 136 pp., 3 pls. DP: 13 April 1788 (Bousquet 2017).

Zubkov, V. P. [= Zoubkoff, B.], 1833: *Nouveaux coléoptères recueillis en Turcménie. Bulletin de la Société Impériale des Naturalistes de Moscou* **6**: 310–340. DP: [31 December] 1833 (Bousquet & Davies 2021).

Zoufal, V. 1926: Orsodacne cerasi a. nova Paláseki. *Věstník Klubu Přírodovědeckého v Prostějově* **19** [1922–1925]: 49. DP: [31 December] 1926 (volume title page).

Zulueta, A. de 1925: La herencia ligada al sexyo en el coleóptero Phytodecta variabilis (Ol.). *Eos, Revista Española de Entomología* **1** (2): 203–229, pl. 5. DP: 25 June 1925 (issue cover).

www.ingramcontent.com/pod-product-compliance
Lightning Source LLC
LaVergne TN
LVHW010639300926
849161LV00005B/992